ESSENTIALS OF MICROBIOLOGY

by

Dr. Anuradha De

M. B. B. S., M. D.

Professor

Department of Microbiology

T. N. Medical College & BYL Nair Ch. Hospital, Mumbai - 400 008.
Ex-Professor (Addl.) in Microbiology,
L. T. M. Medical College & Hospital, Sion, Mumbai 400 022.

SECOND EDITION

www.nationalbookdepct.com

CBS

CBS Publishers & Distributors Pvt Ltd.

Essential of Microbiology

First Edition   : 2008
Second Edition : 2018

Published by

Raju Shah

The National Book Depot

Opp. Wadia Children's Hospital, Parel, Mumbai - 400 012.

Tel : (+91-22) 2416 5274 / 2413 1362 / 2413 2411

Fax : 2413 0877

E-mail: nationalbook55@gmail.com

www.nationalbookdepot.com

and

Satish Kumar Jain

CBS Publishers & Distributors Pvt Ltd

4819/XI Prahlad Street, 24 Ansari Road, Daryaganj, New Delhi - 110 002.

Ph: 23289259 / 23266861 / 23266867 | Fax: 011-23243014

E-mail: delhi@cbspd.com; cbspubs@airtelmail.in

www.cbspd.com

ISBN : 978-93-80206-88-2

Printed by : Neel Graphics

Dedicated to
The Budding Microbiologists
of Today

****

"Life is a journey in the darkness of the night. Wake up to the inner light.

Open your heart and the light will enter and dwell within it".

✍ Swami Vivekananda

PREFACE TO THE FIRST EDITION

Microbiology has stepped into a world of modern amenities and can aptly be called as 'Microbiology in the millenium'. Medical Microbiology has tremendously expanded during the last two decades. With advancement in the fields of bacteriology, immunology, virology and parasitology and with development of molecular biology as a full-fledged discipline, Microbiology has got a new lease of life. Again, due to HIV epidemic, many fungal diseases have also come into the limelight as opportunistic invaders.

Diarrhoeal diseases, tuberculosis, upper respiratory tract infections, malaria, dengue, HIV and other diseases are a major threat in densely populated cities in India. The infectious diseases have crossed all limits and many newly emerging and re-emerging diseases have come in the forefront. The development of multidrug resistant bacteria has further complicated the situation. Medical microbiologists are struggling hard to face the various challenges posed by such problems.

Swami Vivekananda says, 'Knowledge should spread like the glow of a lamp, so as to brighten and enlighten. A candle loses nothing of its light by lighting another candle'. This book is named as 'Essentials of Microbiology', as it essentially deals with the important aspects of Microbiology catering to the needs of the present era. I dedicate this book to all medical students, postgraduates and teachers, whose thirst for Microbiology is never-ending and never-satisfying. We all are educated. In fact, 'Education is the manifestation of the perfection already in man'. We all know a lot about Microbiology. But, 'Every soul is a young eagle soaring higher and higher, gathering more and more strength till it reaches the Glorious Sun'.

In this book, I have tried to cover some topics in details, like bacterial and viral vaccines, HIV infection and AIDS, enteric fever, viral hepatitis, leptospirosis, malaria, tuberculosis and dengue – the last three being among the ten major tropical diseases.

I would like to thank Mr. Raju Shah of National Book Depot for his keen interest for publishing this book. I also thank Mr. Viren Shah and Mr. Manoj of Neel Graphics for their constant help in computer typing and printing of this book. I wish to thank Dr. M. E. Yeolekar, Dean of L.T.M. Medical College & Hospital for permitting me to relase this book. I would be failing in my duties if I do not thank my husband Mr. Swapan De, who always encourages me in such endeavours. Lastly, I thank my one and only daughter Debolina De who spared me for writing this book.

January 2008   **Anuradha De**

PREFACE TO THE SECOND EDITION

It is indeed a great pleasure to bring out the Second Edition of **"Essentials of Microbiology"**. In this edition, I have included many tables in different colour, so that it is highlighted from the other contents of the page and also incorporated coloured photographs in most of the chapters, which will help the postgraduate students to have a better understanding of the subject.

Keeping in pace with the advancing technologies and the different programs implemented by the Govt. of India, I have revised some chapters accordingly, e.g. Automation, Microbial virulence factors, QA/QC, Biosafety measures, BMW management, RNTCP, DOTS and NACP.

I have also added some new chapters like Specimen collection, transport & processing; Clinical Pathology; IAEC; Nanotechnology; Biofilms and Quorum sensing, which is the need of the hour. On some existing chapters, I have added some relevant topics like Antimicrobial drugs, Antimicrobial stewardship, Autoimmunity, Chikungunya and Protozoal infections in AIDS.

I sincerely thank my postgraduate students Dr. Shaoli, Dr. Rajni, Dr. Deepti, Dr. Swati, Dr. Sapna and Dr. Pradnya for helping me out in some of the chapters. I thank Dr. Shanta Shubhra Das for designing the coverpage of this book. I thank Mr. Raju Shah of The National Book Depot for playing a pivotal role in bringing out this edition and Mr. Viren Shah & Mr. Sanjay Shah of Neel Graphics for printing this edition. I also thank my husband Mr. Swapan De & my daughter Debolina De, who constantly supported me in this venture.

2018

Anuradha De

CONTENT

SCIENTISTS AND THEIR CONTRIBUTIONS TO MICROBIOLOGY

1
Chapter

1.	Varo	1st century B.C.	'Animalia minuta'
2.	Hippocrates	5th century B.C.	Oral thrush
3.	Jensen	1590 A.D.	Hand lens
4.	Galileo	1608 A.D.	Telescope
5.	Antony van Leeuwenhoek	1683 A.D.	Microscope
6.	Linnaeus	1758 A.D.	i) Ascaris
			ii) Enterobius
			iii) Dracanculus
		1771 A.D.	iv) Trichuris
7.	Edward Jenner	1798 A.D.	Small pox vaccine
8.	Augustino Bassi	1835 A.D.	Pathogenic role of microorganisms
9.	Gill	1842 A.D.	Mycetoma
10.	Dubini	1843 A.D.	*Ancylostoma duodenale*
11.	Louis Pasteur (French chemist)	1822-1895 A.D.	Father of modern Microbiology
			i) Microbial fermentation
			ii) Sterilization
			iii) Attenuation
			iv) Vaccine for anthrax and hydrophobia (1885)
12.	Lord Joseph Lister	1867 A.D.	Father of antiseptic surgery
13.	Robert Koch (German physicist)	1843-1910 A.D.	Father of medical Microbiology
			i) Staining techniques
			ii) Use of agar for preparation of solid media
		1876 A.D.	iii) Anthrax bacilli
		1882 A.D.	iv) Tubercle bacilli
		1883 A.D.	v) Cholera vibrio
			vi) Koch's postulates
		1890 A.D.	vii) Koch's phenomenon
14.	Bilharz	1852 A.D.	*Schistosoma hematobium*
15.	Lambl	1859 A.D.	*Giardia lamblia*
16.	Hansen	1874 A.D.	Leprosy bacillus
17.	Cobbold	1877 A.D.	*Wuchereria bancrofti*
18.	Neisser	1881 A.D.	Gonococcus
19.	Kleb	1883 A.D.	
20.	Loeffler	1884 A.D.	Diphtheria bacillus
21.	Christian Gram	1884 A.D.	Gram stain
22.	Ehrlich	1886 A.D.	Acid fast stain (modified by Ziehl and Neelsen)
23.	Fraenkel	1886 A.D.	Pneumococcus
24.	Adolf Weil	1886 A.D.	*Leptospira*
25.	Weichselbaum	1887 A.D.	Meningococcus

26.	Kitasato	1889 A.D.	Tetanus bacilli
27.	Yersin	1890 A.D.	Plague bacilli
28.	Wilch & Nuttal	1892 A.D.	*Clostridium perfringens*
29.	Sir Ronald Ross	1897 A.D.	i) Mosquito cycle of malaria parasites
		1903 A.D.	ii) Leishmania
30.	Emil von Behring	1901 A.D.	Antiserum therapy
31.	Stiles	1902 A.D.	*Necator americanus*
32.	Walter Reed	1902 A.D.	Yellow fever virus
33.	Schaudinn	1903 A.D.	*Entamoeba histolytica*
34.	Schaudinn & Hoffmann	1905 A.D.	*Treponema*
35.	Tyzzer	1907 A.D.	*Cryptosporidium*
36.	Paul Ehrlich & Elie Metchnikoff	1908 A.D.	i) Theories of immunity ii) Phagocytosis
37.	Nicollé & Mancéaux	1908 A.D.	*Toxoplasma*
38.	Landsteiner & Popper	1909 A.D.	Polio virus
39.	Chagas	1909 A.D.	*Trypanosoma cruzi*
40.	Richet	1913 A.D.	Anaphylaxis
41.	Twort & d'Herelle	1915 and 1917 A.D.	Bacteriophage
42.	Bordet	1919 A.D.	Complement
43.	Alexander Fleming	1929 A.D.	Fungus Penicillium
44.	Landsteiner	1930 A.D.	Human blood groups
45.	Goodpasture	1930 A.D.	Cultivation of viruses in chick embryos
46.	Ruska	1934 A.D.	Electron microscope
47.	Enders, Weller & Robbins	1949 A.D.	Cultivation of polio virus in tissue culture
48.	Theiler	1951 A.D.	Yellow fever vaccine
49.	Salk	1952 A.D.	Killed polio vaccine
50.	Sabin	1953 A.D.	Oral polio vaccine
51.	Beadle, Lederberg & Tatum	1958 A.D.	Conjugation in bacteria with E. coli K 12
52.	Runyon	1959 A.D.	Classification of atypical mycobactaria
53.	Burnet and Medawar	1960 A.D.	Immunological tolerance
54.	Crick, Watson & Wilkins	1962 A.D.	Nobel Prize in Physiology or Medicine for discoveries concerning the molecular structure of nucleic acids and its significance for information transfer in living material
55.	Maurice Hilleman	1963 A.D.	Father of vaccines
56.	Blumberg	1965 A.D.	Australia antigen
57.	Delbrück, Hershey & Luria	1969 A.D.	Replica plating technique
58.	Porter and Edelman	1972 A.D.	Structure of immunoglobulins
59.	Bishop	1973 A.D.	Rotavirus
60.	Benacerraf, Dausset & Snell	1980 A.D.	Histocompatibility
61.	Luc Montagnaier	1983 A.D.	Lymphadenopathy associated virus(LAV)
62.	Robert Gallo	1984 A.D.	Human T cell lympho-tropic virus (HTLV)-III
63.	Kohler, Milstein & Jerne	1984 A.D.	Hybridoma technology
64.	Pierre Tiollais & colleagues	1985 A.D.	Hepatitis B recombinant vaccine

For entries 61 and 62: } Human Immuno-deficiency Virus (HIV)

65.	Susumu Tonegawa	1987 A.D.	Discovered the genetic basis of antibody diversity
66.	Bishop & Varmus	1989 A.D.	Discovered the cellular origin of retroviral oncogenes, the proto-oncogenes
67.	Kary Mullis & Michael Smith	1993 A.D.	Nobel Prize in Chemistry for inventing the Polymerase Chain Reaction (PCR) technique
68.	Stanley Prusiner	1997 A.D.	Prions (Proteinaceous infectious materials)
69.	Marshall & Warren	2005 A.D.	Identification of Helicobacter pylori and its role in gastritis and peptic ulcer disease
70.	Hausen, Barré-Sinoussi & Montagnier	2008 A.D.	Nobel Prize for discovering that human papillomaviruses can cause cervical cancer and for their discovery of HIV
71.	Beutler, Jules, Hoffmann & Steinman	2011 A.D.	Discoveries concerning the activation of innate immunity and discovered the role of dendritic cell in adaptive immunity

REFERENCES

1. Ananthanarayan & Paniker's Textbook of Microbiology. 9th Ed. Kapil A. Universities Press, Hyderabad 2013.

2. Medals and awards in Microbiology. microbeonline.com

STERILIZATION & DISINFECTION AND HEALTHCARE ASSOCIATED INFECTION (HCAI)

2
Chapter

STERILIZATION

Heat sterilization is the most reliable method.

Vegetative organisms are killed at 65°C by coagulation of proteins. *E. fecalis* (72°C in 3 minutes) and certain viruses are heat resistant. Spores of *C. tetani* require temperature of more than 100°C.

A. Dry Heat

1. **Hot air oven**. Temperature 160°C-180°C. Holding time 1½ - 2 hours. Items are wrapped in kraft paper or aluminium foil. Dry heat is used for solids, non-aqueous liquids, fine instruments (optical instruments), glass, metal, hollow needles, heat-stable powder, waxes and petroleum jelly.

    It is unsuitable for rubber, plastics, combustible substances and glycerol.

2. **Radiant heat** (infrared and microwaves) may be used for glass syringes, but is not common.

B. Moist heat. Uses steam which can penetrate and kill bacteria at temperatures lower than that required with dry heat,

1. **Autoclave**. Saturated steam is best as it has 100% relative humidity, has high latent heat and therfore penetrates better. Packs remain dry after autoclaving.

    Temperature   - 121°C for 15 minutes

                            - 126°C for 10 minutes

                            - 134°C for 3 minutes

    It is used for linen, bedding, gown, all media except serum / egg-containg or sugar / gelatin - containing media.

2. **Low temperature steam formaldehyde (LTSF) sterilizer**.

C. Ethylene oxide gas sterilization (ETO)
Temperature 55°C, relative humidity 30-60%, time 24 hours, Gas concentration 450-800 mg/l. Sterilises heat-sensitive products and those which are non immersible in liquid, e.g. plastic syringe, tubing, telescopic instruments, bronchoscopes, etc. It is carcinogenic.

Tests for sterilization efficiency

1. Thermocouples.

2. Bowie - Dick autoclave tapes.

3. Browne's tube used for steam sterilization and hot air sterilizer (yellow spot and green spot respectively)

4. Chemical indicators in the form of strips or tapes that change colour when a particular temperature is reached. Available for testing colour when a perticular temperature is reached. Available for testing ethylene oxide, dry heat and steam sterilizer.

5. Biological indicators – Spores of *C. tetani* (10^6) for dry heat and spores of *B. stearothermophilus* (10^6) for moist heat.

After sterilization, they are placed in a broth (RCM) incubated for 7 days at 37°C for *C. tetani* and 56°C for *B. stearotheromophilus*. The end result should be total kill.

Sterilization, Disinfection and Waste disposal

The process by which an article, surface or medium in completaty free of microorganisms whether in the vegetative or spore state is called **sterilization**.

The reduction of number of microbes on an inanimate surface, achieved by action of their structure or metabolism, to a level judged to be appropriate for a defined purpose, that is to a level which is not harmful to health and/or to the quality of perishable good, is called **chemical disinfection**, A product which is capable of disinfecting is called a **disinfectant**.

A. Hand and skin disinfection. Aim is to reduce transient microbial flora. This is done by :

    a) Hygienic handrub

    b) Hygienic handwash

Types:

1) Post-contamination treatment of head (PCTH) – first disinfected, then cleaned.

2) Surface hand disinfection. Aim is to reduce transient and residual microbial flora.

3) Pre-operation skin disinfection. Aim is reduction of resident microbial flora of patient's skin in the area of and before a "medical injury".

B. **Instrument disinfection**. Disinfection by immersing the instruments.
   Aims :
   a) Protection of staff preparing the instruments for sterilisation.
   b) Substitution sterilisation if it is not applicable (e.g. flexible endoscopes).

C. **Surface disinfection**. Aim is to reduce the microorganisms present on the surface to prevent spreading.
   1) Wiping of large surface
   2) Wiping of small surfaces
   3) Spray disinfection for small surfaces only
   4) Flooding only in food industry.

D. **Washing disinfection** - linen and clothes going to the laundry.

E. **Excretion disinfection** - disinfection of sputum of a case of tuberculosis.

F. **Water disinfection** - largest, for producing potable water.

"The procedure of disinfection in more important than the disinfectant itself."

Common Disinfectants:

1. **Alcohols**. Ethanol, 1-propanol, 2-propanol - Microbicidal activity against bacteria including tubercle bacilli, viruses (partly) and fungi. Not bacterial spores. Inflamnable, easily evaporating (than explosive).

   Applications - Hands, skin, small surfaces, thermometers, lights or electrical cards.

2. **Aldehydes**. Formaldehyde, glutaraldehyde, glyoxal. Microbicidal activity against bacteria and fungi. Not viruses and bacterial spores.

   Applications - instrument, surfaces, excretion, laundry (fabrics, linen, etc.).
   Formaldelyde - hemodialysis water systems.
   Glutaraldehyde – respirators, anaesthesia equipments, cystoscopes, bronchoscopes, etc.

3. **Quaternary ammonium compounds**.
   Benzalkonium chloride, didecyl dimethyl ammonium chloride. Microbicidal activity against bacteria but not tubercle bacilli, fungi, viruses (only limited), not bacterial spores. It has pleasant smell, activity is restricted in presence of organic materials, has limited biodegradability.

   Applications - instruments, surfaces, additive to surgical and skin disinfection, walls, floors and furnitures.

4. **Amphotesides, Biguanides and Tertiary ammonium compounds**.
   Microbicidal activity - against bacteria excluding tubercle bacilli and some problem with staphylococci, fungi, viruses (only limited), not bacterial spores. Smelly.

   Applications – instruments, surfaces.

5. **Phenols** – Orthopheyl phenol, benzylchlor phenol. Microbicidal activity – bacteria, fungi, viruses (very limited), not bacterial spores.

   Applications – floors, walls, furnitures, especially for fecal contamination.

6. **Chlorine** – Dichlor isocyanurate, chloramine T, sodium hypochlorite. Microbicidal activity against all organisms. Fast acting and corrosive. More faster the action, it is more corrosive. Easily inactivated in presence of organic material. Smelly.

   Applications – Water, surface, hands, mucus membrane, instruments, excretion, cleaning in place (CIP), renal dialysis equipments, toilets, lavatories and bathtubs.

7. **Oxidising agents, Iodine, Iodophores**. Peracetic acid, hydrogen peroxide, ozone, polyvinyl pyrrolidone iodine (PPI).

   Applications – same as chlorides, thermometres, kitchen or nursery items.

Good Disinfection Practice (GDP)

1. Formation of infection control team.
2. Guidelines and rules for staff especially development of "hygiene plans".

Every hospital, medical practice and institution needs a special way of GDP.

Chemicals give Level I (inactivation of vegetative bacteria, fungi, some viruses) or

Level II (inactivation of vegetative bacteria, fungi, all viruses and mycobacteria) disinfection.

In-use test for disinfectants – Used in the hospitals, e.g. thermometer fluid, cheatle forceps fluid, laboratory discard jar fluid, mop fluid, etc. Thermometer and cheatle forceps are best to immerse in disinfectant solution and then wiped dry.

Procedure:

One ml of disinfectant fluid is mixed with 9 ml of diluent.
a) Nutrient broth for alcohols, aldehydes, hypochlorites and phenolics
b) Nutrient broth with Tween 80 (3%) for diguanides, hypochlorite - detergent mix, iodophores, phenolic mix, detergent mix and quaternary ammonium compounds.

A pipette delivering 50 drops per ml is taken. Ten drops are placed separately on the surface of nutrient agar plates in duplicate. Both the plates are incubated for 72 hours – one at room temperature and other at 37°C. Growth in more than 5 drops, out of 10 drops on either plate, indicates failure of disinfection.

HEALTHCARE ASSOCIATED INFECTION (HCAI)

HCAI is also known as nosocomial or hospital infection. "An infection occurring in a patient during the process of care in a hospital or other health-care facility which was not present or incubating at the time of admission. This includes infections acquired in the health-care facility but appearing after discharge, and also occupational infections among health-care workers of the facility".

- At any time, over 1.4 million people worldwide are suffering from infections acquired in health-care facilities
- In modern health-care facilities in the developed world: 5–10% of patients acquire one or more infections
- In developing countries the risk of HCAI is 2-20 times higher than in developed countries and the proportion of patients affected by HCAI can exceed 25%
- In intensive care units, HCAI affects about 30% of patients and the attributable mortality may reach 44%.

Common healthcare associated infections

Infections	Organisms
Urinary tract infections	*E. coli, Klebsiella, Serratia, Proteus, Pseudomonas aeruginosa*
Lower respiratory tract infections (nosocomial pneumonia)	*H. influenzae, S. pneumoniae, S. aureus,* Enterobacteriaceae, Respiratory viruses
Wound and skin infections	*S. aureus* (MRSA), *E. coli, Proteus,* Enterococcus (VRE), Anaerobes, *S. epidermidis*
Burns	*S. aureus* (MRSA), *P. aeruginosa, Acinetobacter, S. pyogenes.*
Gastrointestinal infections	*Salmonella, Shigella sonnei,* Viruses causing diarrhoea.

Each hospital should have an effective **Infection Control Policy**

1. Hospital Acquired Infection Control Committee **(HAICC)** should meet at least once a month to formulate and update policies for pevention of hospital infection.

   They also formulate new antibiotic policies according to drug resistance pattern at a particular time and monitoring sterilisation and disinfection practices.

Members:
   a) Chair person (Medical Superintendent)
   b) Infection control officer (Microbiologist)
   c) Members (Chiefs of all clinical units/departments, Chief of blood bank service, Microbiologist, Medical record officer, Chief of nursing services, Infection control sister
   d) Invited Members – Chiefs of all supportive services (O.T., C.S.S.D, Dietetics, Laundry, House keeping, etc.)

2. Hospital Acquired Infection Control Team **(HAICT)** takes day to day responsibilities and should meet at least once a week.

Members:
a) Infection Control Officer (Microbiologist)
b) Infection Control Sister (Hospital sister)
c) Clinician (one clinician from hospital)

They will look after surveillance and role of laboratory, control of infection, monitoring hygienic practices, educating all staff in the hospital regarding universal safety precautions and precautions to be taken for high risk procedures.

Common Infectious Diseases

Disease	Mode of transmission
Tuberculosis	Airborne, droplet contact
Hepatitis B	Direct or indirect contact
AIDS	Direct contact (blood or body fluids)
Malaria	Insect vectors
Chicken pox	Direct, indirect, or droplet contact
Influenza	Airborne, droplet contact

Infections in Intensive Care Unit (ICU)

The incidence of infection in the ICU is one of the highest in a hospital. Incidence of nosocomial pneumonia in Mumbai is 1.8% in general wards, whereas it is 16.7% in the ICU. Usually antibiotic - resistant bacteria like methicillin resistant *Staphylococcus aureus* (MRSA), *Pseudomonas aeruginosa, Serratia marcescens* and vancomycin - resistant enterococci (VRE) survive and persist in the environment leading to recurrent outbreaks.

One of the major reasons for increased risk of nosocomial infections in the ICU is inappropriate use of antibiotics. Ventilator associated pneumonia and blood stream infections are particularly dangerous in critically ill patients. Selective decontamination of digestive tract (SDD) is done by giving non-absorbable antibiotics in the intestinal canal and oropharyngeal cavity, which kills the potentially pathogenic microorganisms. Limiting unnecessary use of antibiotics, increasing compliance with infection control practices, maintaining proper ventilation and environmental hygiene and modifying architectural design can go a long way in prevention of infection in the ICUs. It is a multidisciplinary approach. To reduce antimicrobial resistance, antibiotic cycling can be done by withdrawing an antibiotic class from use and subsequently reintroducing it at a later point in time. This allows resistance rates for specific antibiotics to decrease or at least remain stable, when their use is periodically eliminated from the ICU.

Protocol for Intensive Areas

- Doors and windows should be well sealed. Single room divided into separate cubicles with toilet facilities.
- Use of protective clothing by staff and visitors.
- Handwashing facilities at the entrance or chlorhexidine hand rub.
- Coarse filters (7μ) and HEPA filters (0.3μ) provided in A.C. system. Cleaned every 3-4 months. Ideally 30 air changes per hour. Environmental moitoring should be done after changing / cleaning the filters.
- Patients should be given sterile water. In burns unit – waterproof cover for mattresses, cleaned with hypochlorite when necessary and dressing materials sterilised separately for each patient.
- All areas should be wiped daily with clean damp cloth, separate cloth shuold be used for each patient, Mop should be cleaned thoroughly and dried.
- All drugs and I.V. therapy must be prepared in the laminar flow cabinet by a person wearing sterile protective clothing.
- Patient should be screened weekly for bacterial pathogens – nose, throat and perineal swabs, hair, urine and feces are the samples to be screened. When infection is seen samples should be sent for microbiological investigations.
- Staff with cough, cold, fever, skin infection or any commmunicable disease should not be allowed inside.

Outbreak of multiresistant gram negative organisms

Reservoir:

a) Hands of staff
b) Stool of patients on broad spectrum antibiotics
c) Drains, sinks
d) Poorly disinfected non-clinical equipments
e) Open containers of contaminated disinfectants and other fluids
f) Bars of soap lying in pool of water

Routes of spread:

a) Via hands and non-compliance with hand disinfection procedure
b) Bed pans and urinals
c) Bed clothes that become contaminated with urine and feces
d) Staff sitting on bed of colonized patients
e) Use of antibiotics which are resistant

Control of multiresistant gram negative organisms:

1. Meticulous hand disinfection
2. Cleaning of urinals / bed pans with hot water (about 80°C)
3. Provision of adequate gloves and plastic aprons to health care workers (HCWs).
4. Aseptic procedure for insertion of catheters, plastic catheters changed every 48 hours and silicon-coated catheters changed every 2 weeks.
5. Urinary drainage bag should be emptied fully, using separate container / jug for each patient.
6. Antibiotics should be restricted to minimal use, only those antibiotics should be used to which the organism is sensitive.
7. Freshly prepared disinfectant should be used, always prepared in a small quantity and container should not be kept open.
8. Non-clinical equipments should be disinfected.
9. All equipments should be stored dry.
10. Patient should not be transferred from one ward to another.
11. Stool culture – 3 specimens per patient per week.

Risk factors for Surgical Site Infections (SSI)

1. **Type of operation :**

   Clean – Operation in which respiratory, alimentary or genitourinary tracts are not entered and there is no break in aseptic technnique.

   Clean contaminated – Operation in which respiratory, alimentary or genitourinary tracts are entered but without any significant spillage.

   Contaminated – Operation in which 'pus' is encountered or where there is visible contamination of a wound, e.g. gross spillage from a hollow viscus or compound or open injuries operated on within 4 hours.

   Directory – Operation in the presence of pus, where there is a previous perforated hollow viscus or compound or open injuries more than four hours old.

2. **Operative factors :**
   - Insertion of prosthetic implants
   - Duration of surgery exceeding 3 hours
   - Poor surgical technique involving excessive cauterization and tissue necrosis.

3. **Co-morbidities in the patient :** ASA score > 2 is associated with increased risk. Comorbid diseases are diabetes mellitus, ischemic heart disease, hypertension, anemia, impaired renal function.

4. **Additional patient risk factors:** Extremes of age, malnutrition, obesity and prolonged stay in hospital prior to surgery.

Factors favouring antibiotic resistance in ICU

- Area of maximum antibiotic use in the hospital
- Excessive use leads to development of antibiotic resistance
- Because of sick patients antibiotic use cannot be curtailed in ICU
- Prolonged length of stay (> 7 days)
- Endotracheal tubes, vascular lines, urinary catheters
- Mechanical ventilation > 7 days
- Previous use of broad spectrum antibiotics.

Ten ways to reduce antibiotic resistance

1. Try to establish an etiological diagnosis
2. Once culture and sensitivity pattern is known, antibiotic should be changed from broad spectrum to narrow spectrum
3. Positive cultures obtained from non sterile sties should be interpreted with care
4. Quantitative cultures should be done to distinguish between colonisation and infection
5. Leucocytosis and fever should be reviewed after 3 days, continued treatment should not be done
6. Undrained abscesses or foreign body infections should not be continuously treated
7. Outbreaks (> 3 cases) of resistant infections e.g. MRSA should be detected very fast and preventive measures taken accordingly
8. Compliance with infection control practices:
   - Hand rub solution
   - Barrier precautions (gown, gloves, mask)
   - Environmental disinfection
9. Regular patient surveillance should be done which helps to:
   - Define high risk patients
   - Identify their underlying diseases
   - Time of onset of infection
10. Guidelines followed for antibiotic selection, dosage and duration:
    - Pharmacokinetic (PK) and pharmacodynamic (PD) principles
    - Time above MIC (Minimum Inhibitory Concentration for β-lactams and glycopeptides
    - Ratio of area under curve (AUC)/MIC for fluoroquinolones
    - Peak MIC ratios for aminoglycosides.

Occupational health risks for hospital staff

Relative risk of infection for a healthy person and their respective route of infection is as follows :

High risk	Route of infection	Prevention
Chicken pox	Inhalation	Restrict non-immune staff
Measles	Inhalation	Immunization
Rubella	Inhalation	Immunization
Hemorrhagic fevers	Blood borne	Avoid needle stick, sharps injuries and use of gloves.
Hepatitis B	Blood borne	Immunization
RSV, Viral conjunctivitis	Contact	Handwashing
Impetigo	Contact	No-touch technique, handwashing.
Scabies	Contact	Handwashing, use of gloves.
Herpes zoster	Contact	No-touch technique, handwashing.
Viral diarrhoea, Dysentery	Feco-oral	Handwashing
Moderate to low risk		
Pulmonary tuberculosis	Inhalation	Immunization. For close contacts, use of special masks.
Salmonella	Feco-oral	Handwashing
Cholera	Feco-oral	Handwashing
Hepatitis A	Feco-oral	Handwashing and immunization
Poliomyelitis	Feco-oral	Handwashing and immunization
Herpes simplex, CMV	Contact	Handwashing
Meningococcal meningitis	Contact/droplets	Immunization
Diphtheria	Contact/droplets	Immunization
Hepatits C	Blood borne	Avoid needle stick, sharps injuries and use of gloves.

Protocol for Operating Room and Sterile Preparation Room

- Entry restricted to working team only.
- Staff should wear sterile mask, gloves, gowns and protective eye wear.
- Ideal temperature of OT is 20-22°C, shoyld be cooler than outside area to aid the outward movement of air. Humidity not less than 55%. Air supply by **Plenum ventilation**, 20-24 air changes per hour. Exhaust fan should be in the corvidor.

- Unnecessary opening and closing of outer door is avoided.
- All apparatus should be clean and sterilised.
- Minimum equipments shold be kept.
- Bacterial load in air should be monitored periodically.

Daily cleaning of OT

- Before start of first case, all surfaces should be carbolized with 7% lysol.
- Walls wiped upto 3m every day.
- Floor scrubbed with warm water and detergent and kept dry.
- OT table and non-clinical equipments should be wiped and kept dry.

Weekly cleaning of OT

- Complete OT complex is cleaned with detergent and warm water.
- All the shelves emptied, cleaned and restacked.
- Working of all equipments checked and cleaned.
- Ventilation filters checked and charged when required.

At the end of surgery

- Linen, gloves, etc. should be preferably discarded in colour coded plastic bags.
- All instruments immersed in 7% lysol for 30 mins. and then autoclaved.
- OT should be cleaned thoroughly at the end of all operations. If clostridia is grown from a patient operated in main OT, fumigation should be carried out before starting any other operation.

REFERENCES

1. Ananthanarayan & Paniker's Textbook of Microbiology. 9th Ed. Kapil A. Universities Press, Hyderabad 2013.

2. Chakraborty PA. Textbook of Microbiology. 4th Ed. New Central Book Agency (P) Ltd., Kolkata 2009.

3. Block S. S. Disinfection, Sterilization and Preservation. 4th Ed. Lea & Febiger, Philadelphia 1991.

4. Education Program for Infection Control: Basic concepts and Training - International Federation of Infection Control. Eds. Ayliff GA, Hambraeus A, Mehtar S. 1995.

5. Bennett JV, Brachman PS. Hospital Infections. 2nd Ed. Little, Brown & Co. Boston/Toronto. 1986.

UNIVERSAL SAFETY PRECAUTIONS (USP) AND BIO-MEDICAL WASTE (BMW) MANAGEMENT

UNIVERSAL SAFETY PRECAUTIONS (USP)

An approach to infection control is taking precautions against potentially infective sources, e.g. all blood and body fluids. All human blood and certain body fluids are treated as if known to be infectious. A set of precautions to protect health care workers from occupational exposure to blood borne pathogens.

Five components of USP:

1. Hand washing before and after examination of each patient, before and after surgery, etc.

2. Barrier protection – wearing sterile gowns, plastic aprons, gloves, masks, protective eye wear and footwear.

3. Needles never to recap. Used needles should be cut with a needle - cutter and other sharps should be immersed in a puncture - proof container containing 2% sodium hypochlorite, for at least 30 minutes.

4. Spillage of blood and body fluids should be first treated with 2% sodium hypochlorite for 20 minutes and then that area should be washed wtih hot, water and soap. Blood soiled linen should be covered with 5% sodium hypochlorite solution for 30 minutes and then sent to laundry.

5. All instruments should be sterilised by autoclaving or boiling for at least 20 minutes. Heat sensitive instruments should be immersed in 2% glutaraldehyde for 30 minutes. Equipments and work surfaces should be disinfected properly.

Standard Precautions

- Hand hygiene
- Barrier precautions
- Safe handling and disposal of sharps
- Management of spills
- Management of exposure

Three new elements have been added:

- Respiratory hygiene/cough etiquette
- Safe injection practices
- Use of masks for insertion of catheters or injection into spinal or epidural areas.

According to WHO 600,000 to 800,000 cut and puncture injuries/ per year, of which approximately 50% are not registered.

Risk of transmission

Infection	Percutaneous	Mucocutaneous
HIV	0.3%	0.05%
HBV	9-30%	-
HCV	3-10%	-

Handwashing and Hand hygiene

- One of the most important ways to protect against transmission of microbes and disease is hand hygiene.

- Cleaning of hands protect the patient against harmful germs carried on our hands or present on our own skin and protect ourselves and the health-care environment from harmful germs.

- It is important because, minor cuts on hands are invisible, the organisms present on hands beneath the gloves multiply in the warm and humid condition provided by the glove cover and when we remove the glove, hands actually have more microbes. So, it is a standard precaution.

- Handrubbing with alcohol-based handrub is the preferred routine method of hand hygiene if hands are not visibly soiled.

- Handwashing with soap and water is essential when hands are visibly dirty or visibly soiled (following visible exposure to body fluids).

- If exposure to spore forming organisms e.g. *Clostridium difficile* is strongly suspected or proven, including during outbreaks – hands should be cleaned using soap and water.

- When to wash hands?
  - Before touching a patient
  - Before clean/aseptic procedure
  - After body fluid exposure risk
  - After touching a patient
  - After touching patient surroundings

- Compliance with hand hygiene differs across facilities and countries, but is globally < 40%

- Main reasons for non-compliance reported by health-care workers are:
  - Too busy
  - Skin irritation
  - Glove use
  - Don't think it is necessary
- Use alcoholic hand rub before and after any procedure.
- Always wash hands before and after removing gloves.
- Gloves are not a substitute for hand washing.
- Gloves need to be removed between patients.

Barrier Precautions

Entry via:

- Percutaneous – Practices, Gloves
- Mucous membrane – Masks, Goggles
- Non intact skin – Dressings, Gloves, Gown, Leg cover

Personal Protective Equipment (PPE)

Consist of:

- Gloves
- Gowns, usually impermeable
- Aprons, usually impermeable
- Face shields
- Eye wear, such as goggles to protect eyes
- Masks such as N 95, which should be appropriately fitted
- Boots or shoe coverings
- Leggings
- Head covering

Mask

- Worn by healthcare providers and visitors to protect against microbes transmitted by airborne or droplet means
- May also be worn by patient with airborne or droplet transmissible diseases, especially under certain circumstances such as during direct care or transport

Airborne Precautions

- Used to prevent or reduce the transmission of microorganisms that are airborne in small droplet nuclei (5 µ or smaller in size) or dust particles containing the infectious agent.
- These can remain suspended in the air or be dispersed widely by air currents and can be inhaled by or deposited on a host in the same room or further away.
- Includes diseases like pulmonary tuberculosis, rubeola and varicella

- A single room under negative pressure ventilation: 6 air change/hour and discharge air to outside or filter it using HEPA filter.
- A high efficiency mask or powered air-purifying respirator (PAPR) when entering the room for all persons entering the room (N 95 respiratory mask).
- The door must be kept closed at all times.
- Only personnel that have immunity against varicella, TB and measles should care for these patients
- If private room absolutely not available, cohort patients with similar disease.
- Limit patient movement or transport only if necessary
- Use surgical or N 95 mask on patient if transport is needed
- Known susceptible health care workers should not enter room of patients with varicella or rubella if other workers are available

Droplet Precautions

- Used to reduce the risk of transmission of microorganisms transmitted by large particle droplets (larger than 5 µ in size).
- This type of transmission usually requires close contact between the source person and the recipient because droplets do not remain suspended in the air. They usually travel 3 feet or less within the air and thus special air handling is not required, however newer recommendations suggest a distance of 6 feet be used for safety.
- Droplet transmission involves contact of the conjunctiva of the eyes or the mucous membranes of the nose or mouth of a person with the microorganism generated from the infected source person during coughing, sneezing or talking, or during the performance of procedures such as suctioning and bronchoscopy.
- Includes such diseases as influenza, rubella, parvovirus B19 and mumps.
- Place the patient in a private room (Isolate)
- If not available, cohort with patient with active infection with same microorganism
- Use of respiratory protection such as a mask when entering the room recommended and definitely if within 3 feet of patient
- Limit movement and transport of the patient. Use a mask on the patient if they need to be moved and follow respiratory hygiene/cough etiquette
- Keep patient at least 3 feet apart between infected patient and visitors
- Room door may remain open

Respiratory Hygiene/Cough Etiquette/Patient Teaching

- Cover mouth/nose with tissue
- Throw tissues properly when coughing or sneezing
- Sneeze or cough into the sleeve
- Use surgical masks while cough
- Use alcoholic hand rub in between

What type of PPE would you wear?

- While suctioning oral secretions, gloves and mask/goggles or a face shield, sometimes gown should be worn.
- While responding to an emergency where blood is spurting, gloves, fluid-resistant gown, mask/goggles or a face shield should be worn.
- While drawing blood from a vein, gloves should be worn.

Safe Handling and Disposal of Sharps

Sharps

- A sharp is any device having corners, edges, or projections capable of cutting or piercing the skin, e.g. needles, needles with syringes, needles with attached tubing, blades / lancets (razors, scalpels, blades), etc.
- Additionally, all broken glass items or glass items with sharp edges or points contaminated with biohazardous waste must be disposed into a sharps container including:
- Contaminated Pasteur pipettes
- Contaminated glass slides
- Contaminated broken and unbroken glassware

Needlestick injuries occur 40% before use, 40% after use and before disposal, 15% during disposal and 5% others.

- Avoid sharp injuries
- No recapping of needles
- Never pass sharps directly, put in a container to pass

'Sharps' Disposal – Puncture Proof Containers

- Closable
- Puncture resistant and resistant to disinfectants
- Leak proof from sides and bottom
- Appropriately labeled and color-coded
- Opening that is large enough to accommodate disposal of the known sharps likely to be generated
- Easily accessible to the immediate area where sharps are used
- At all places of generation
- Should contain sodium hypochlorite 1% solution
- Dispose needle as an assembly
- All needles should be burnt or mutilated or cut in a needle cutter
- Other sharps should be disposed in 'sharps' container.

Management of Spills

- Wear PPE
- Place absorbent over spillage
- Allow to soak
- Discard as infectious waste
- Again place absorbent over spillage area
- Pour 1-5% sodium hypochlorite
- Leave in contact for 10 minutes
- Remove
- Discard as infectious waste
- Use clean mop
- Wipe with disinfectant again
- Mop to dry
- Clean the mop
- Allow mop to dry.

BIO-MEDICAL WASTE (BMW) MANAGEMENT

Any waste generated during diagnosis, treatment, immunization, research activities and health camps is known as Bio-Medical Waste (BMW).

Ministry of environment & forests notification, New Delhi, 20th July, 1998 on the Bio-Medical Waste (Management and Handling) Rules, 1998 conferred by section 6, 8 and 25 of the Environment Protection Act, 1986 (29 of 1986), and in supersession of the Bio-Medical Waste (Management and Handling) Rules, 1998, the Central Government hereby makes the following rules, namely, 'Bio-Medical Waste Management Rules, 2016', published in the Official Gazette on 28th March 2016.

These rules provide a system for regulating handling BMW which includes collection, segregation at source, norms for packaging labeling and options for treatment and disposal along with the standard for treatment technologies and for proper management & handling of Bio-Medical Waste. This is applicable to all persons who generate, collect, receive, store, transport, treat, dispose or handle Bio-Medical Waste in any form. This is applicable to hospitals, nursing homes, clinics, dispensaries, veterinary institutions, animal houses, pathological laboratories, blood banks, ayush hospitals, health camps, blood donation camps, vaccination camps, first aid rooms of schools, forensic and research laboratories.

Contents of BMW Rules 2016:

- SCHEDULES
    - I: BMW categories and their segregation, collection, treatment, processing and disposal options
    - II: Standards for treatment and disposal of BMW
    - III: List and duties of prescribed authorities
    - IV: Labels for BMW containers/bags and transportation

- FORMS
  - I: Accident reporting
  - II: Application for authorisation or renewal of authorization
  - III: Authorization
  - IV: Annual report
  - V: Application for appeal.

SCHEDULE I

BMW Categories and their segregation, collection, treatment, processing and disposal options

Category	Type of Waste	Type of Bag/Container	Treatment/ Disposal options
Yellow	a) Human anatomical waste – Human tissues, Organs, body parts, Foetus below viability period (as per MTP Act 1971) b) Animal anatomical waste – Experimental animal carcasses, Organs, body parts, tissues c) Soiled Waste – Items contaminated with blood, body fluids (Dressings, plaster casts, cotton swabs, bags containing residual/ discarded blood and/or components)	Yellow coloured non-chlorinated plastic bags	Incineration or Plasma Pyrolysis or deep burial
	d) Expired or Discarded Medicines – Antibiotics, Cytotoxic drugs, Glass or plastic ampoules, vials, etc. contaminated with cytotoxic drugs	Yellow coloured non-chlorinated plastic bags	Cytotoxic drugs returned back to manufacturer. All other discarded medicines either sent back to manufacturer or disposed by incineration.
	e) Chemical waste – Production of biological and used/ discarded disinfectants	Yellow coloured containers or non-chlorinated plastic bags	Incineration or Plasma Pyrolysis or encapsulation in hazardous waste treatment, storage and disposal facility
	f) Chemical liquid waste – Chemicals in production of biologicals, Used or discarded disinfectants, X-ray film developers, Formalin, Infected secretions/body fluids, House keeping liquids	Separate collection system leading to effluent treatment system	Liquid waste shall be pre-treated before mixing with other waste water
	g) Discarded linen, mattresses, beddings contaminated with blood or body fluids	Yellow coloured non-chlorinated plastic bags or suitable packing material	
	h) Microbiology, Biotechnology and Other clinical laboratory waste	Autoclave in safe plastic bags or containers	On-site pre-treatment to sterilze with non-chlorinated chemicals as per NACO or WHO guidelines and thereafter incineration
Red	Contaminated Waste (Recyclable) – Disposable items like tubings, bottles, I/V sets, catheters, urine bags, syringes (without needles), vacutainers (with needles cut), gloves	Red coloured non-chlorinated plastic bags or containers	Autoclaving or micro-waving/ hydroclaving followed by shredding or mutilation or combination of sterilization and shredding
White (Trans-lucent)	Waste sharps including Metals, needles, syringes with fixed needles, needles from needle tip cutter or burner, scalpels, blades	Puncture proof, leak proof, tamper proof containers	Autoclaving followed by shredding or mutilation or encapsulation Combination of shredding cum autoclaving Final disposal to iron foundries/ sanitary landfill / concrete waste sharp pit
Blue	a) Glassware Broken/discarded and contaminated glass (medicine vials, ampoules) except cytotoxic wastes b) Metallic body implants	Cardboard boxes with blue coloured marking	Disinfection (by soaking the washed glass waste after cleaning with detergent and Sodium Hypochlorite treatment) or through autoclaving or microwaving or hydroclaving and then sent for recycling

Segregation, Packaging, Transportation and Storage

- Segregation at point of generation
- Containers or Bags shall be labelled as specified in Schedule IV
- Bar coding and GPS shall be added by the occupier and operator of CBMWTF within one year time

- Transport vehicle details
- Untreated human anatomical waste, animal anatomical waste, soiled waste and biotechnology waste shall not be stored beyond 48 hours
- Clinical laboratory waste shall be pre-treated by sterilization to Log 6 or disinfection to Log 4 as per WHO guidelines.

Label for transport of Bio-medical Waste containers/bags

Day   Month   Year   Date of generation

Waste category no.

Waste quantity

Sender's Name & Address   **Receiver's Name & Address**

Phone no: Phone no:

Telex no: Telex no:

Fax no: Fax no:

Contact Person: **Contact Person:**

In case of emergency please contact

Name & Address:

Phone no:

Note: Label shall be non-washable and prominently visible.

Biohazard symbol

Cytotoxic Hazard Symbol

Annual report

- To be submitted by occupier/operator on or before 30th June every year
- Prescribed authority shall compile, review and analyse information and send to Central Pollution Control Board on or before 31st July every year
- CPCB shall send information to Ministry of Environment, Forest and Climate Change on or before 31st August every year

- Online availability on websites of Occupiers, State Pollution Control Boards and CPCB.

Legal liability

Occupier/operator of CBMWTF liable for damages to environment/public due to improper handling of BMW Liable for action under section 5 and section 15 of the Act, in case of any violation.

REFERENCES

1. Ministry of Environment and Forests. Government of India. Draft Biomedical Waste (Management and Handling) Rules, 2011. Available from: http://moef.nic.in/downloads/publicinformation/salient-features-draft-bmwmh.pdf.

2. Townend WK. Healthcare waste management: policies, legislations, principles and technical guidelines, Waste Management World Magazine 2009; 10(4). Available from: http://www.waste-management-world.com.

3. De A. Practical and Applied Microbiology. 5th Ed. The National Book Depot, Mumbai 2014.

4. http://mpcb.gov.in/biomedical/pdf

SPECIMEN COLLECTION, TRANSPORT & PROCESSING

Chapter 4

Key to the diagnosis of infection

Culture = gold standard

- Representative from lesion
- Avoid contamination with normal flora
- Speedy delivery to laboratory
- Normally sterile sites – culture media
- Sites with normal flora – culture media + selective culture media
- Special techniques – Mycobacteria, Leptospira, Fungi

General rules for collection of specimens

- Label – Name/Time/Date/Ward details/Nature of specimen/Site/Test required
- Before starting antibiotics
- Proper disinfection
- Bypass normal flora
- Quantitative culture results
- Correlate with smear findings

Collect specimens for any Microbiological investigation in sterile containers, e.g. autoclaved glassware – Sterile swab, sterile test tube, sterile petri dish.

Criteria for specimen rejection

- Improper transport time, temperature, medium
- Inadequate volume/specimen
- Specimen in formalin
- Unlabeled/Mislabeled specimen
- Leaking specimen
- Broken container
- Contamination of specimen
- Dry Swabs

Specimens

- Abscess pus
- Blood
- Bone marrow
- Body fluids (pleural, peritoneal, pericardial, synovial)
- I/V Catheter tip
- Cerebrospinal fluid (CSF)
- Ear pus/swab
- Eye swab
- Genital tract (Cervix, Cul de sac, Uterine products, Placenta, etc.)
- Intrauterine device
- Urethra
- Vagina
- Vulva
- Prostratic fluid
- Tissue (surgical/biopsy)
- Urine
- Intestinal tract (feces, rectal swab, swabs from operative sites, gastric aspirate)
- Respiratory tract (throat, pharynx, nasal sinuses, nasopharynx, nose, sputum, endotracheal secretion, bronchial and transtracheal aspirates, bronchoalveolar lavage)
- Oral swab
- Superficial – skin and soft tissue

Transport media

- Stuart's transport media – for anaerobes and gonococci
- Cary Blair – for Salmonella, Shigella, Y. pestis, V.cholerae, Campylobacter
- Venkatraman-Ramakrishnan medium – for V. cholerae

Shipping should be done in Double mailing containers.

Air transport of infectious substances

- International Air Transportation Association (IATA)
- Infectious Substances Shipping Guidelines

Transport regulations

Transport of infectious substances is subject to strict national and international regulations:

- Proper use of packaging materials
- Proper labelling, notification

Triple packaging

- Main goals
  - Protects the environment, the carrier
  - Protects the sample
  - Arrival in good condition for analysis

- If triple packaging not available
  - Prepare according to international dangerous goods transportation rules (see IATA guidelines)

The basic triple packaging system

Three layers of protection are needed:

- Primary receptacle – Leak-proof specimen container; Packaged with sufficient absorbent material to absorb the entire content of the primary receptacle in case of breakage.

- Secondary packaging – Leak-proof secondary container; Encloses and protects the primary receptacle(s); Several cushioned primary receptacles may be placed; Additional absorbent material to absorb all fluid in case of breakage; Labelled.

- Outer packaging – Outer packaging protects contents from outside influences, physical damage, while in transit.

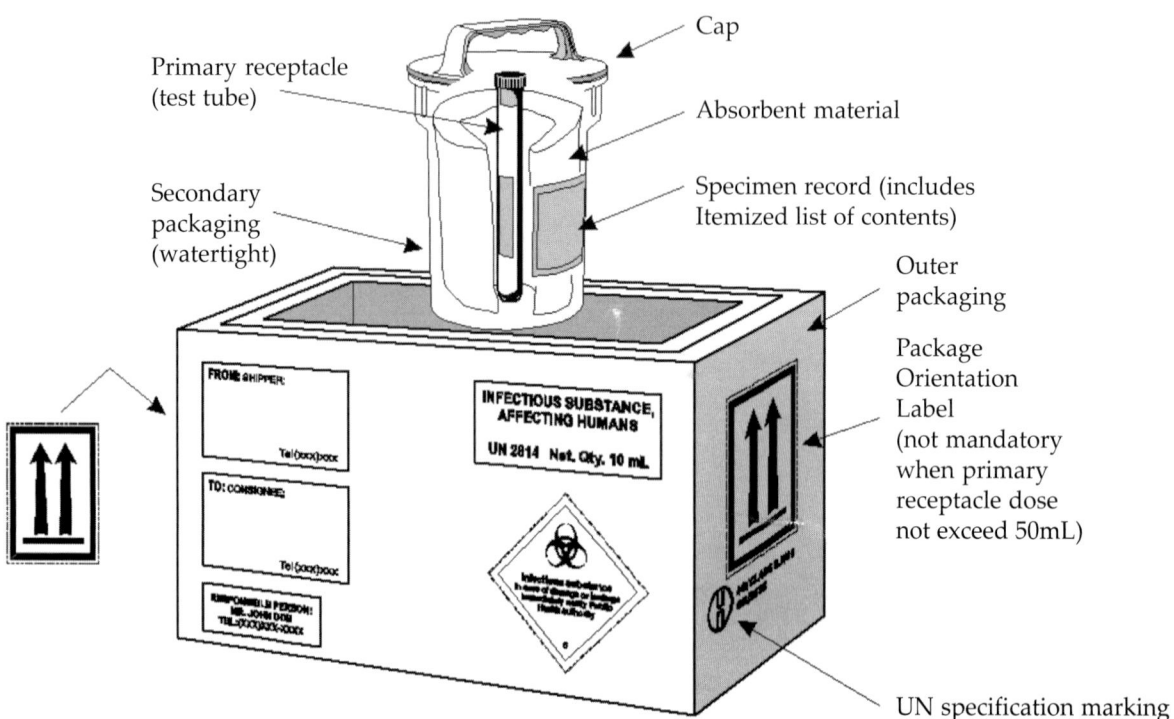

Laboratory Organization

- Collection and transport of specimens-
- Avoid contamination of material and self by spillage, aerosol or needle-stick injury
- Label containers as Hepatitis risk or HIV risk when known
- Reception of specimens-
- Not in public corridor
- Staff must wear gown and gloves
- In case of leakage, discard the sample. But when sample is precious like CSF, transfer it to another sterile container.

Specimens of Lower Respiratory Tract

Sputum, Bronchoalveolar lavage (BAL), Gastric lavage (GL), Endotrachial aspirate (ETA)

- Prompt delivery, within 2 hours of collection. If delayed, should be refrigerated
- > 10 squamous epithelial cells in gram stained smear in 10X objective – unacceptable

- BAL – centrifuge
- ETA > 10 squamous epithelial cells – reject
- Induced sputum – 15% NaCl + 10% glycerine aerosol
- GL – in children

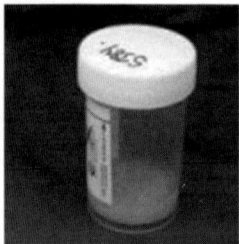

Sputum container

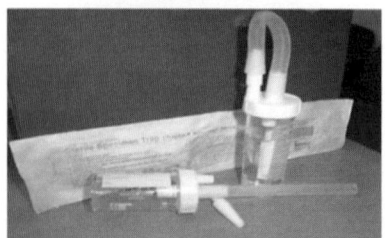

Suction container for selection of BAL

Upper Respiratory tract
- Cotton tipped swabs/dacron, alginate.
- Processed within 4 hours

Blood Culture
- No. of collections:
  - Endocarditis – Three in 24 hours
  - Pyrexia of Unknown Origin (PUO) – First day 2 and second day 2
- Additives:
  - Sodium Polyanethol Sulphonate (SPS) 0.03% – anticoagulant, antiphagocytic
  - Resins – inhibit antibiotics
- Incubation at 37°C
- Three subcultures - 2nd, 4th and 7th days
- Modification : Biphasic media (Castaneda's method), Lysis centrifugation, Instrument based (BACTEC, BacT/Alert)

Conventional blood culture   **Castaneda's method**

Cerebrospinal fluid (CSF)
- Three tubes are used
- Mycobacteria & fungi – 5ml
- To process rapidly, if not – kept in incubator (37°C)/room temperature
- **CSF should never be refrigerated** (except for viruses)
- Gross Examination : Color, turbidity, blood stained
- Centrifuged deposit processed for direct examination and culture

Urine
- To process within 2 hours of collection
- If delay, refrigerate (not > 18 hrs.)
- Midstream clean catch – after proper anogenital wash
- Catheterised patient – disinfect site & collect catheterised aspirate with sterile needle & syringe. Seal catheter puncture site
- Suprapubic aspiration in infants & children – gentle suprapubic tapping

- Direct examination :
  - Wet mount: centrifuged deposit
  - Gram stain: uncentrifuged urine
  - Screening tests: quantitative culture
  - Colony count: standard loop, pour plate, pipette, etc.

Stool
Purged stool/Rectal swab
- Collection in sterile wide mouth container
- To be processed within 2 hours
- If delay, swab swirled in sample and inoculated in alkaline peptone water, Cary Blair medium, Gram Negative broth, Selenite F broth
- Gross examination: colour, consistency, presence of blood and mucous and parasites (adult worm, segments of worm)
- Microscopic examination: saline/iodine mount, hanging drop.
- Culture: direct plating, plating from enrichment media.

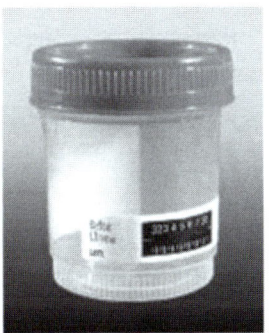

Sterile wide mouth container

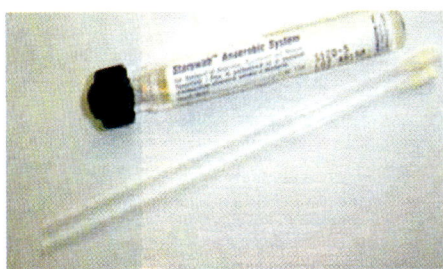

Sterile swabs and container

STD Specimens
Urethral discharge: If purulent discharge, collect on swab

If no obvious discharge, sterile loop inserted 3 cm in urethra and sample is collected
- To be inoculated immediately on routine and selective media. If delay, to be collected in transport media like Stuart's and Amie's media
- To be subcultured within 12 hours.

Vaginal swab

- Speculum moistened with warm water (creams/antiseptics not to be used)
- Swabs collected from posterior fornix.

Cervical swabs – Sterile swab inserted in cervical canal, rotated for 10 seconds and withdrawn.

Genital ulcer – Clean with saline and collect from base of ulcer or aspirate from bubo.

Pus/Wounds

Aspirates preferred over swabs in sterile tubes
Biopsy specimen

- From active part of lesion in sterile container
- To collect in sterile saline and not in formalin
- Direct examination
- Staining
- Culture

Anerobic specimens

Specimens to be collected:

- Frank pus is collected in sterile syringe and needle. Needle end is sealed with a sterile rubber cork and transported to the laboratory and processed within two hours.

- Fluids in sterile small containers filled upto the brim without leaving any air space.
- Films from the muscles at the edge of the affected area.
- Exudates from the active part of capillary pipette or a swab
- Necrotic tissue and muscle fragments
  - Gram stain
  - Aerobic and anaerobic culture (RCM broth, thioglycollate broth).

Culture Methods

- Streak culture – For isolation of bacteria in pure culture (Hexagonal and T-method)
- Lawn or carpet culture – For antibiotic susceptibility testing
- Stab culture – For maintenance of stock cultures and in motility medium
- Stroke culture – On agar slope for slide agglutination and other tests
- Pour plate method – For viable bacterial count and colony count in urine
- Sweep plate method – For wards and OT surveillance in hospitals.

REFERENCES

1. Koneman EW, Allen SD, Janda WM, Schreckenberger P, Winn Jr. WC. In Color Atlas & Textbook of Diagnostic Microbiology. 6th Ed. Lippincott, Philadelphia 2006.
2. Collee JG, Marion BP, Fraser AG, Simmons A. Mackie and McCartney's Practical Medical Microbiology. 14th Ed. Churchill Livigstone, U.K. 1996.
3. Forbes B. A., Sahm D. F. & Weissfeld A. S. In Bailey & Scott's Diagnostic Microbiology. 12th Ed. Mosby, Inc. An Affiliate of Elsevier Science 2007.

DISEASES CAUSED BY IMPORTANT BACTERIA

Chapter

A)	Gram-positive cocci	Important diseases caused
1.	S. aureus	Boils, carbuncles, abscesses, bullous impetigo, pemphigus neonatorum, osteomyelitis, tonsillitis, pharyngitis, pneumonia, meningitis, endocarditis. Food poisoning. Toxic shock syndrome. Stitch abscesses.
2.	S. epidermidis	Cystitis in persons with abnormalities of the urinary tract. Subacute endocarditis and septicemia.
3.	S. saprophyticus	Acute urinary tract infection in women of child-bearing age.
4.	S. pyogenes	Cellulitis, erysipelas, impetigo. Sore throat, tonsillitis, (Group A) pharyngitis, otitis media, mastoiditis, Ludwig's angina, pneumonia, meningitis. Puerperal sepsis. Nonsuppurative complications – acute rheumatic fever and acute glomerulonephritis.
5.	S. agalactiae (Group B)	Neonatal meningitis Puerperal sepsis Mastitis in cattle.
6.	Enterococci	Urinary tract infections Wound infections Septicemia, endocarditis.
7.	Viridans streptococci	Subacute endocarditis Caries tooth.
8.	S. pneumoniae	Lobar pneumonia and bronchopneumonia Otitis media, meningitis.
9.	Peptostreptococcus	Puerperal sepsis Brain abscess Pleuropulmonary infections.
B)	**Gram-negative cocci**	**Important diseases caused**
10.	N. meningitidis	Meningitis Septicemia
11.	N. gonorrhoeae	Gonorrhoea Ophthalmia neonatorum, otitis media, meningitis, endocarditis.
12.	B. catarrhalis	Respiratory tract infection in adults with chronic obstructive airway disease.
13.	Veilonella	Usually nonpathogenic. Occasionally septicemia, endocarditis and abscesses in internal organs.
C)	**Gram Positive Bacilli**	**Important diseases caused**
14.	C. diphtheriae	Diphtheria – faucial commonest, followed by laryngeal, nasal, otitic, conjunctival, genital and cutaneous.
15.	B. anthracis	Anthrax - cutaneous (malignant pustule) - pulmonary (Woolsorters' disease) - septicemic.
16.	B. cereus	Food poisoning - emetic type - diarrhoeal type.
17	C. perfringens	Gas gangrene Food poisoning (Types A and C) Septicemia

18.	*C. tetani*	Tetanus
19.	*C. botulinum*	Botulism - Foodborne (Types A, B and E) - Infant - Wound
20.	*C. difficile*	Pseudomembranous enterocolitis (due to prolonged therapy with oral clindamycin, lincomycin, ampicillin and cephalosporins).
21.	*Propionibacterium acnes*	Acne, pustules of skin.

D)	**Gram Negative Bacilli**	**Important diseases caused**
22.	*E. coli*	Urinary tract infections (O groups 1, 2, 4, 6 and 7) Diarrhoea (by EPEC, ETEC, EIEC, EHEC, EAEC) Septicemia, neonatal meningitis, wound infections.
23.	*K. pneumoniae*	Pneumonia, UTI, wound infections, septicemia, meningitis, diarrhoea rarely.
24.	*Proteus spp.*	UTI, wound and burn infections, meningitis. Summer diarrhoea in children (*M. morganii*).
25.	*Salmonella spp.*	Enteric fever (typhoid and paratyphoid fevers) Food poisoning Septicemia, meningitis Osteomyelitis.
26.	*Shigella spp.*	Bacillary dysentery Severe type by *S. dysenteriae*. Common in India : *S. flexneri*, then *S. boydii*.
27.	*Virbio cholerae*	Cholera - Classical more severe form, no chronic carrier state. - El Tor less severe but chronic carrier state & survive for longer duration in the environment.
28.	*Vibrio parahemolyticus*	Food poisoning (Kanagawa positive strains).
29.	*Aeromonas hydrophila*	Red leg disease in frogs. Diarrhoea and pyogenic infections in man.
30.	*Campylobacter jejuni*	Diarrhoea
31.	*Helicobacter pylori*	Associated with duodenal and gastric ulcer diseases.
32.	*Pseudomonas*	UTI after instrumentation or catheterization, wound and *aeruginosa* burn infections, otitis media, septicemia, eye infections, necrotizing pneumonia, acute purulent meningitis after lumbar puncture or cranial injury, ecthyma gangrenosum (haemorrhagic infarction of skin).
33.	*Burkholderia cepacia*	Cystic fibrosis
34.	*B. pseudomallei*	Melioidosis (Pneumoenteritis) in rats, guinea pigs, rabbits and man.
35.	*B. mallei*	Glanders and farcy in horses and mules.
36.	*Bacteroides spp.*	Glanders in man. Intraabdominal infections, brain and lung abscess, deep abscesses.
37.	*Fusobacterium spp.*	Pleuropulmonary and oropharyngeal infections, chronic suppurative otitis media, genital infections.
38.	*Y. pestis*	Plague - Bubonic, pneumonic, septicemic.
39.	*Y. enterocolitica*	Diarrhoea
40.	*H. influenzae*	Chronic bronchitis, acute epiglottitis, otitis media, sinusitis, pneumonia, meningitis (predominantly type b).
41.	*H. ducreyi*	Soft chancre
42.	*Bordetella pertussis*	Whooping cough
43.	*Brucella melitensis*	Malta fever
44.	*B. abortus*	Abortion in cattle, undulant fever in man.

E)	Acid Fast Bacilli	Important diseases caused
45.	M. tuberculosis	Tuberculosis - Pulmonary - Extrapulmonary: meningeal, renal, bone, intestinal, etc.
46.	M. leprae	Leprosy - Lepromatous (LL), Borderline Lepromatous (BL), Borderline (BB), Borderline Tuberculoid (BT), Tuberculoid (TL).
47.	M. kansasii	Pulmonary infection like T.B.
48.	M. scrofulaceum	Cervical lymphadenitis in children.
49.	M. avium intracellulare	Pulmonary disease in immunocompromised patients.
50.	M. ulcerans	Buruli ulcer
51.	M. marinum	Swimming pool granuloma.
52.	M. fortuitum and M. chelonei	Post injection abscesses, post traumatic abscesses.
F)	Spirochaetes	Important diseases caused
53.	Treponema pallidum	Syphilis (hard chancre) - Acquired syphilis - primary, secondary, latent, tertiary or late stages. - Congenital syphilis - Endemic or nonvenereal syphilis (bejel).
54.	T. pertenue	Yaws
55.	T. carateum	Pinta
56.	B. recurrentis	Relapsing fever
57.	B. vincentii	Vincent's angina in association with anaerobic fusiform bacilli
58.	B. burgdorferi	Lyme disease
59.	L. icterohaemorrhagiae	Weil's disease
60.	L. canicola	Meningitis
G)	Other bacteria	Important diseases caused
61.	Spirillum minus	Rat bite fever
62.	Streptobacillus moniliformis	Rat bite fever
63.	Rickettsia prowazeki	Epidemic typhus
64.	R. mooseri	Endemic typhus
65.	R. rickettsii	Rocky mountain spotted fever.
66.	R. tsutsugamushi	Scrub typhus
67.	Rochalimaea quintana	Trench fever
68.	Coxiella burnetii	Q fever
69.	Chlamydia trachomatis	Trachoma (A, B, Ba, C); Inclusion conjunctivitis (D-K) Non gonococcal urethritis (D-K); Cervicitis, proctitis, salpingitis (D-K) Pneumonia in neonates (D-K); Lymphogranuloma venereum (L1-L3)
70.	C. psittaci	Pneumonia, pharyngitis in man Psittacosis in birds.
71.	C. pneumoniae	Acute respiratory disease.
72.	Mycoplasma pneumoniae	Primary atypical pneumonia.
73.	M. hominis & U. urealyticum	Nongonococcal urethritis.

REFERENCES

1. Ananthanarayan & Paniker's Textbook of Microbiology. 9th Ed. Kapil A. Universities Press, Hyderabad 2013.

2. Chakraborty PA. Textbook of Microbiology. 4th Ed. New Central Book Agency (P) Ltd., Kolkata 2009.

FLOW CHARTS FOR IDENTIFICATION OF COMMON AEROBIC AND ANAEROBIC BACTERIA

6

Chapter

Flow Chart - I

Important genera of bacteria
Gram-positive

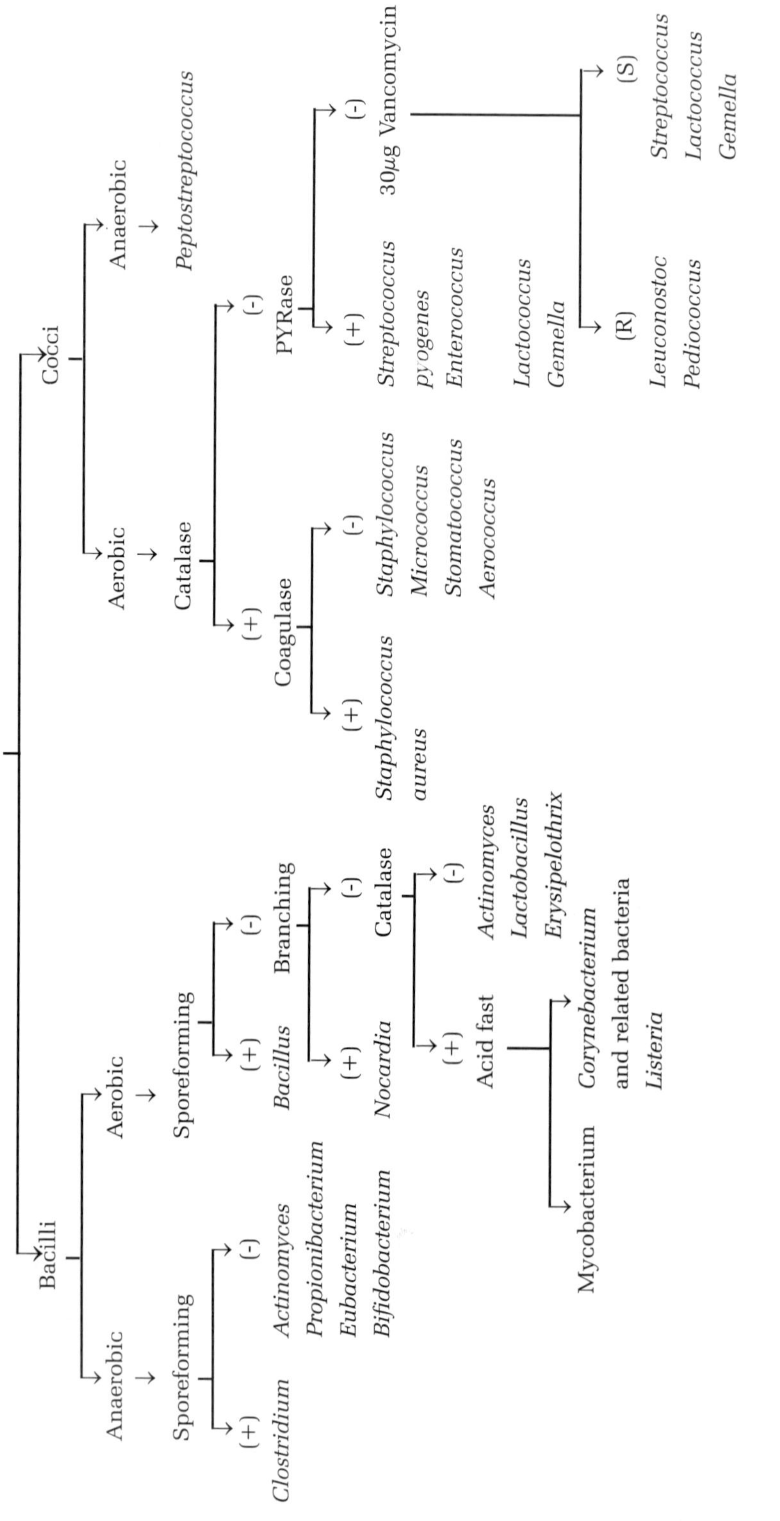

Bacilli

Anaerobic
Sporeforming
- (+) *Clostridium*
- (-) Sporeforming
  - (-) *Actinomyces*, *Propionibacterium*, *Eubacterium*, *Bifidobacterium*

Aerobic
Sporeforming
- (+) *Bacillus*
- (-) Branching
  - (+) *Nocardia*
  - (-) Catalase
    - (+) Acid fast
      - *Mycobacterium*
      - *Corynebacterium* and related bacteria, *Listeria*
    - (-) *Actinomyces*, *Lactobacillus*, *Erysipelothrix*

Cocci

Aerobic
Catalase
- (+) Coagulase
  - (+) *Staphylococcus aureus*
  - (-) *Staphylococcus*, *Micrococcus*, *Stomatococcus*, *Aerococcus*
- (-) PYRase
  - (+) *Streptococcus pyogenes*, *Enterococcus*, *Lactococcus*, *Gemella*
  - (-) 30µg Vancomycin
    - (R) *Leuconostoc*, *Pediococcus*
    - (S) *Streptococcus*, *Lactococcus*, *Gemella*

Anaerobic
Peptostreptococcus

Contd.......

Flow Chart - I

Important genera of bacteria

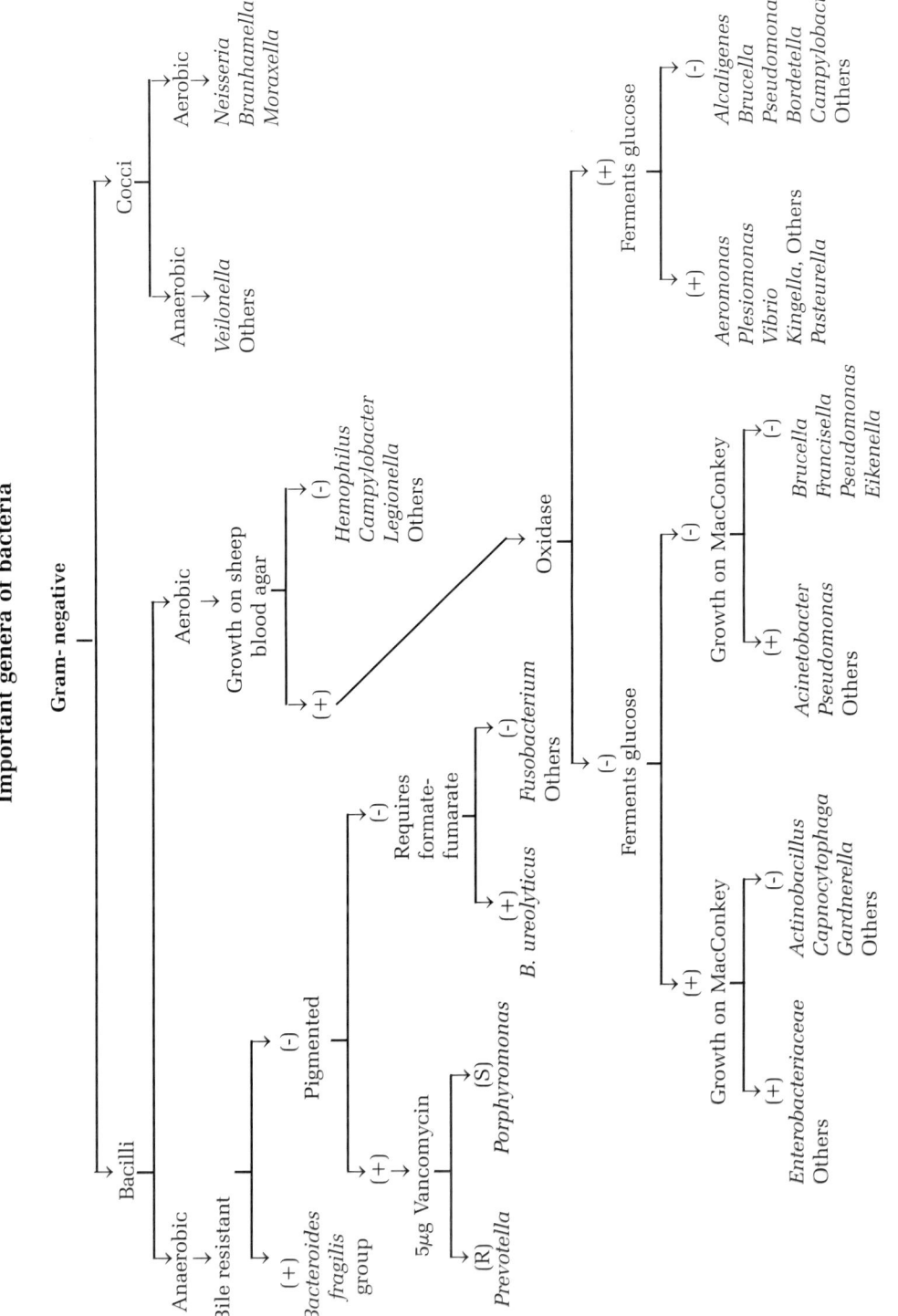

Flow Chart - II

Identification of common Gram – positive cocci and Gram – negative bacilli

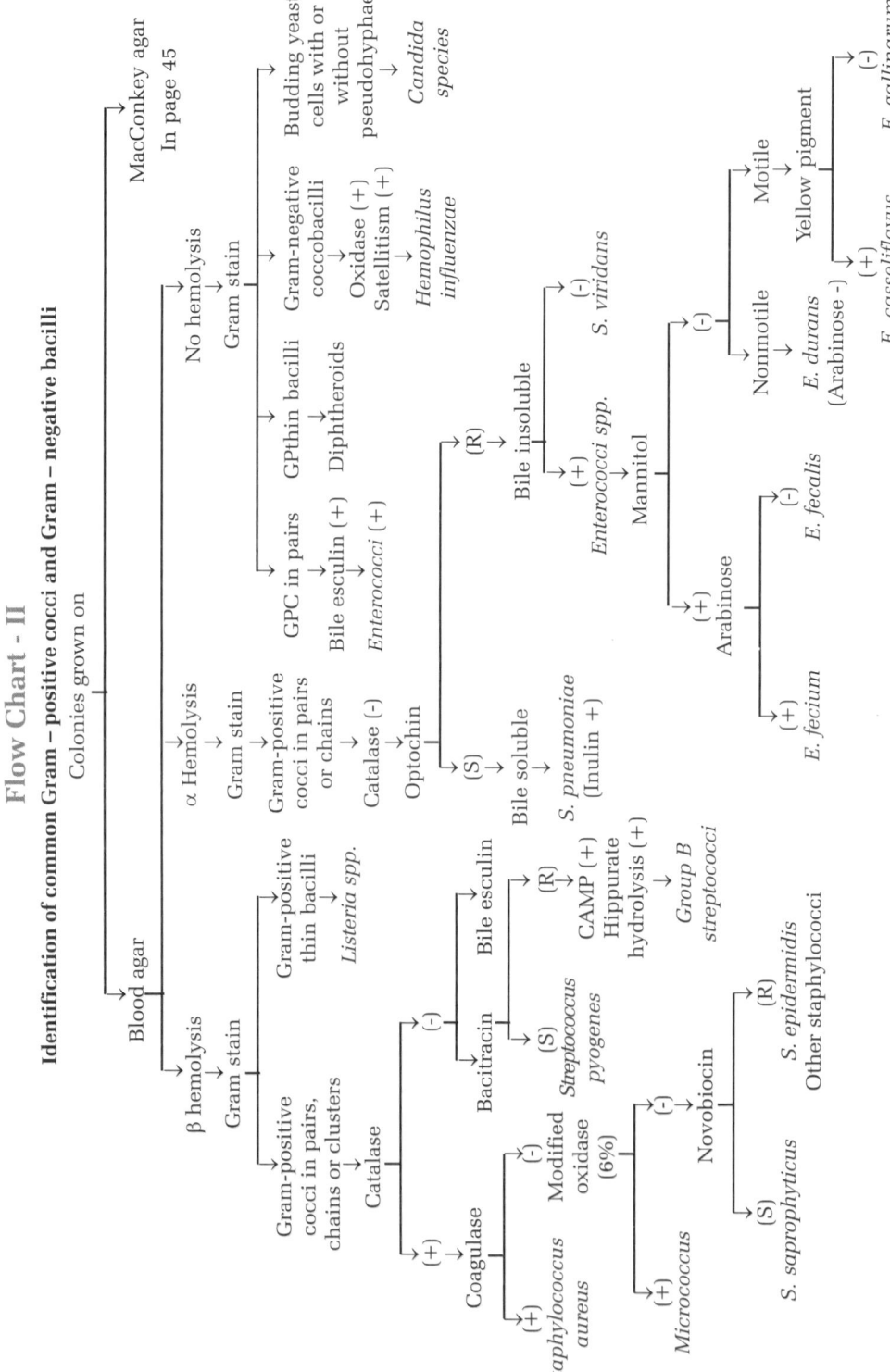

Contd.......

Flow Chart - II

Identification of common Gram – positive cocci and Gram – negative bacilli

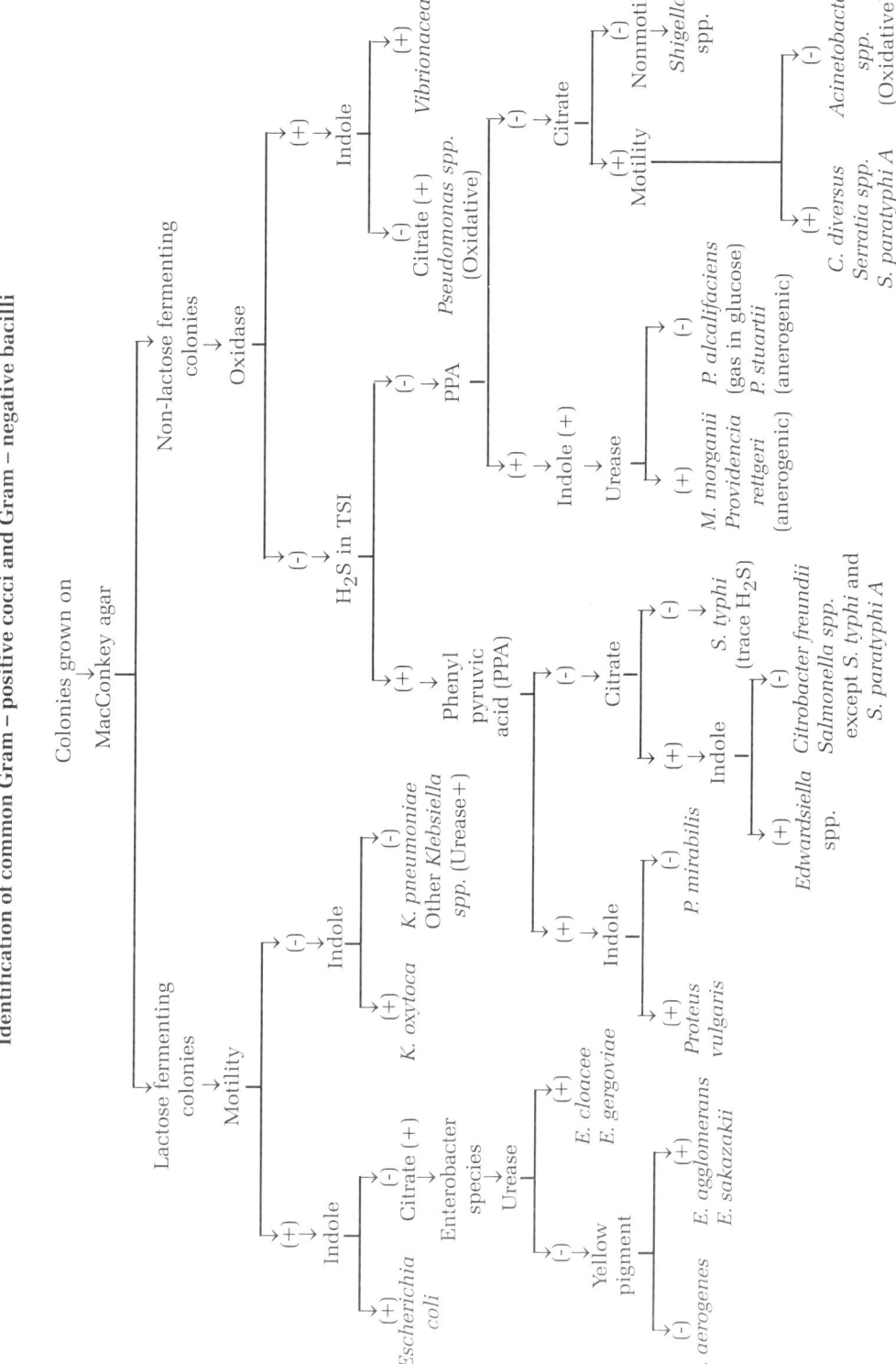

Flow Chart - III

Identification of Shigella species and Yersinia enterocolitica

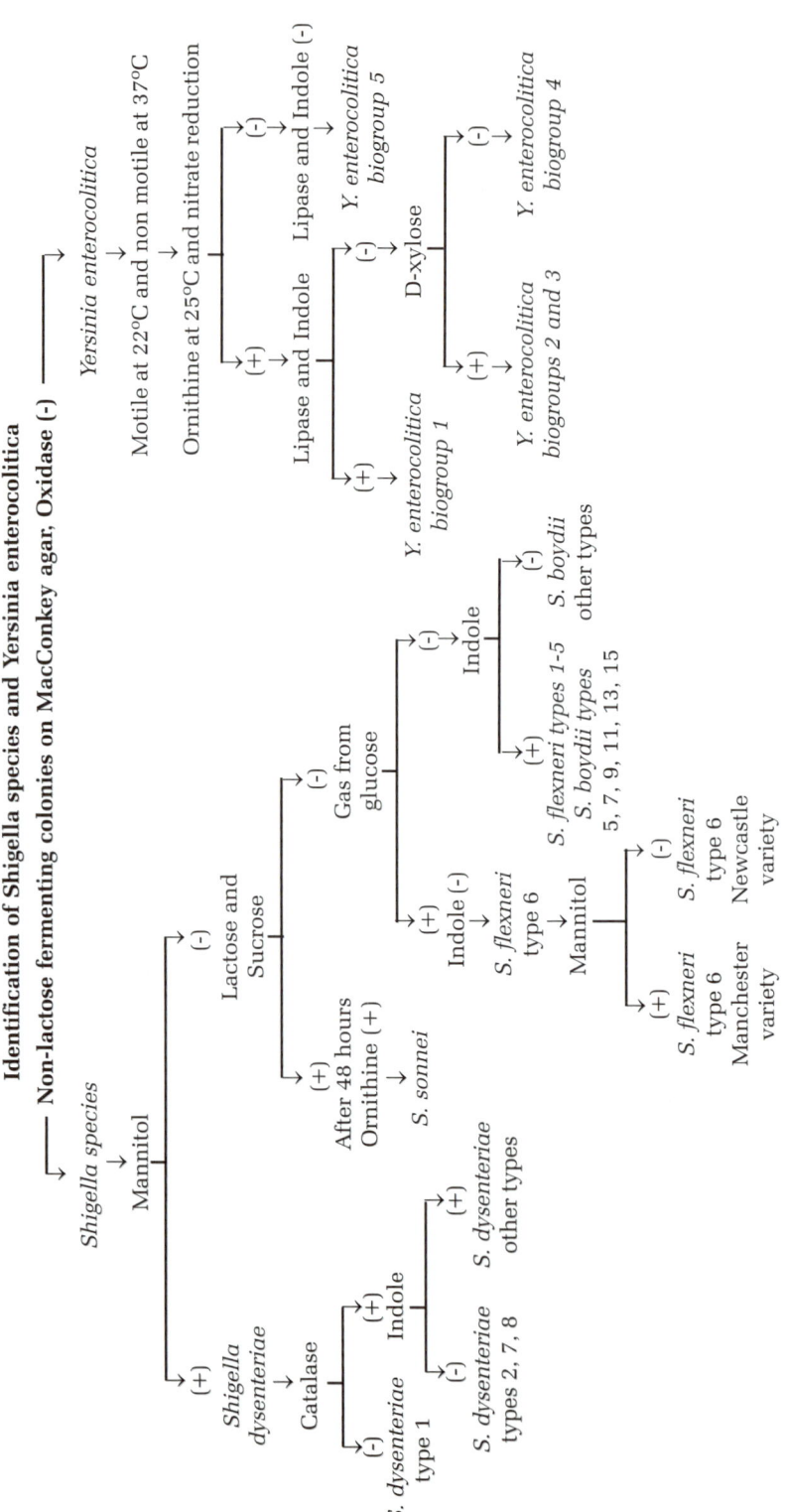

Exceptions :

S. *flexneri* type 4a – catalase (-)

S. *dysenteriae* type 3 – mannitol (+)

Flow Chart - IV

Identification of Vibrionaceae
Non-lactose fermenting colonies on MacConkey agar, Oxidase (+)
Mannitol and Sucrose

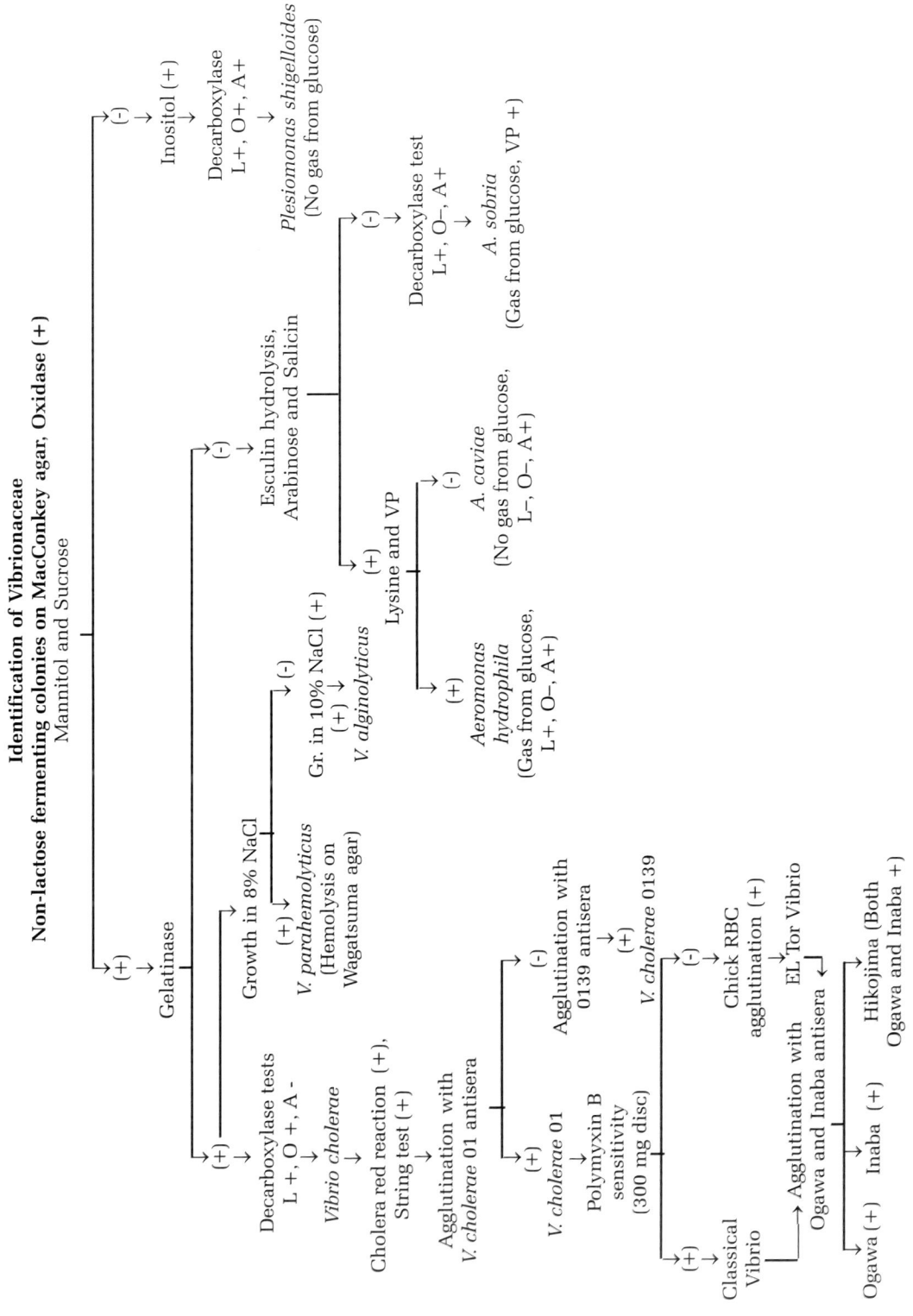

Flow Chart - V

Identification of Campylobacter species

Gram stain – Delicate gram - negative spiral bacilli

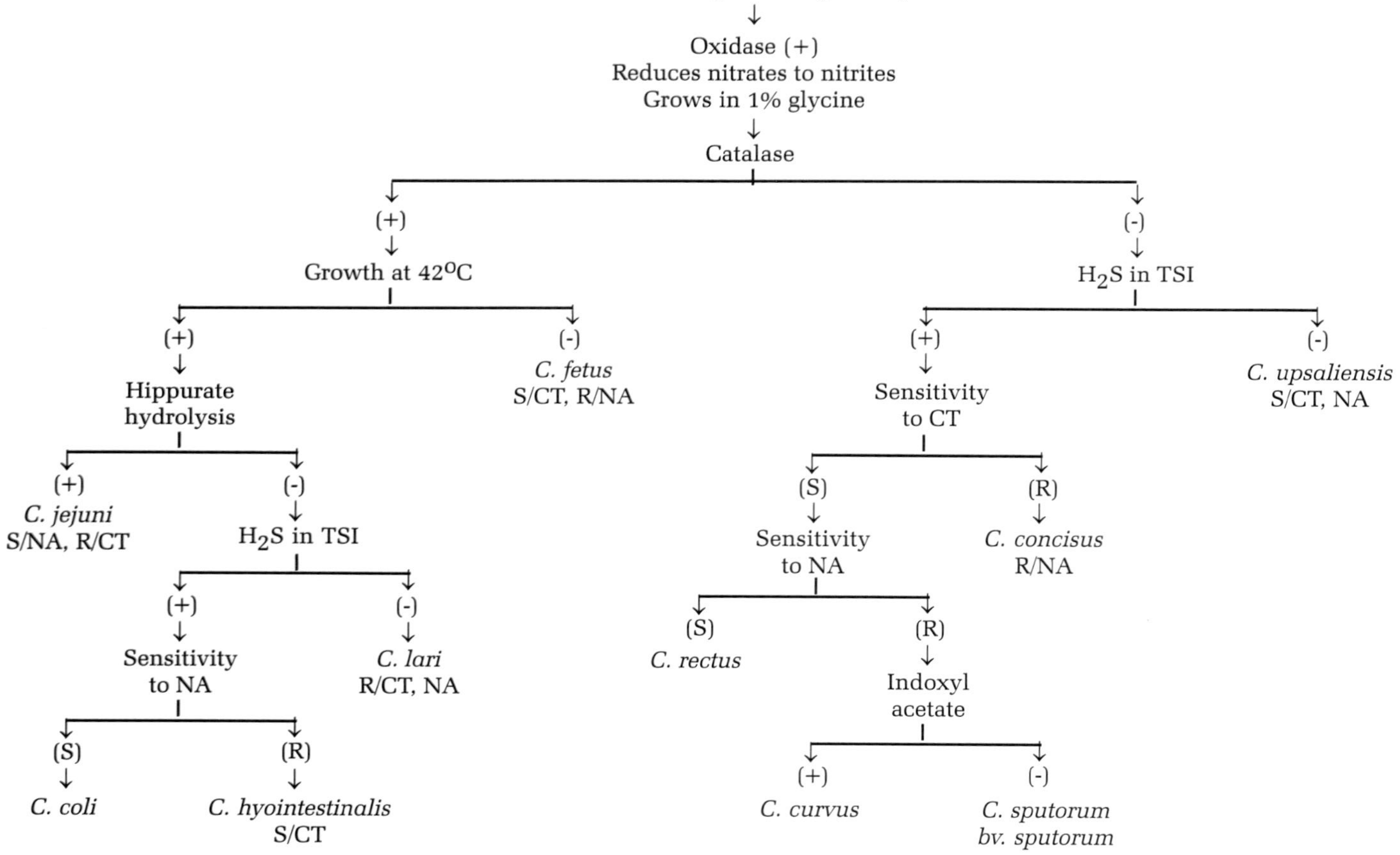

→

Oxidase (+)
Reduces nitrates to nitrites
Grows in 1% glycine
→
Catalase

Growth at 42°C

(+) →

Hippurate hydrolysis

(+) →
C. jejuni
S/NA, R/CT

(−) →
H₂S in TSI

(+) →
Sensitivity to NA

(S) →
C. coli

(R) →
C. hyointestinalis
S/CT

(−) →
C. lari
R/CT, NA

(−) →
C. fetus
S/CT, R/NA

H₂S in TSI

(+) →
Sensitivity to CT

(S) →
Sensitivity to NA

(S) →
C. rectus

(R) →
Indoxyl acetate

(+) →
C. curvus

(−) →
C. sputorum
bv. *sputorum*

(R) →
C. concisus
R/NA

(−) →
C. upsaliensis
S/CT, NA

CT = Cephalothin
NA = Nalidixic acid

Flow Chart - VI A

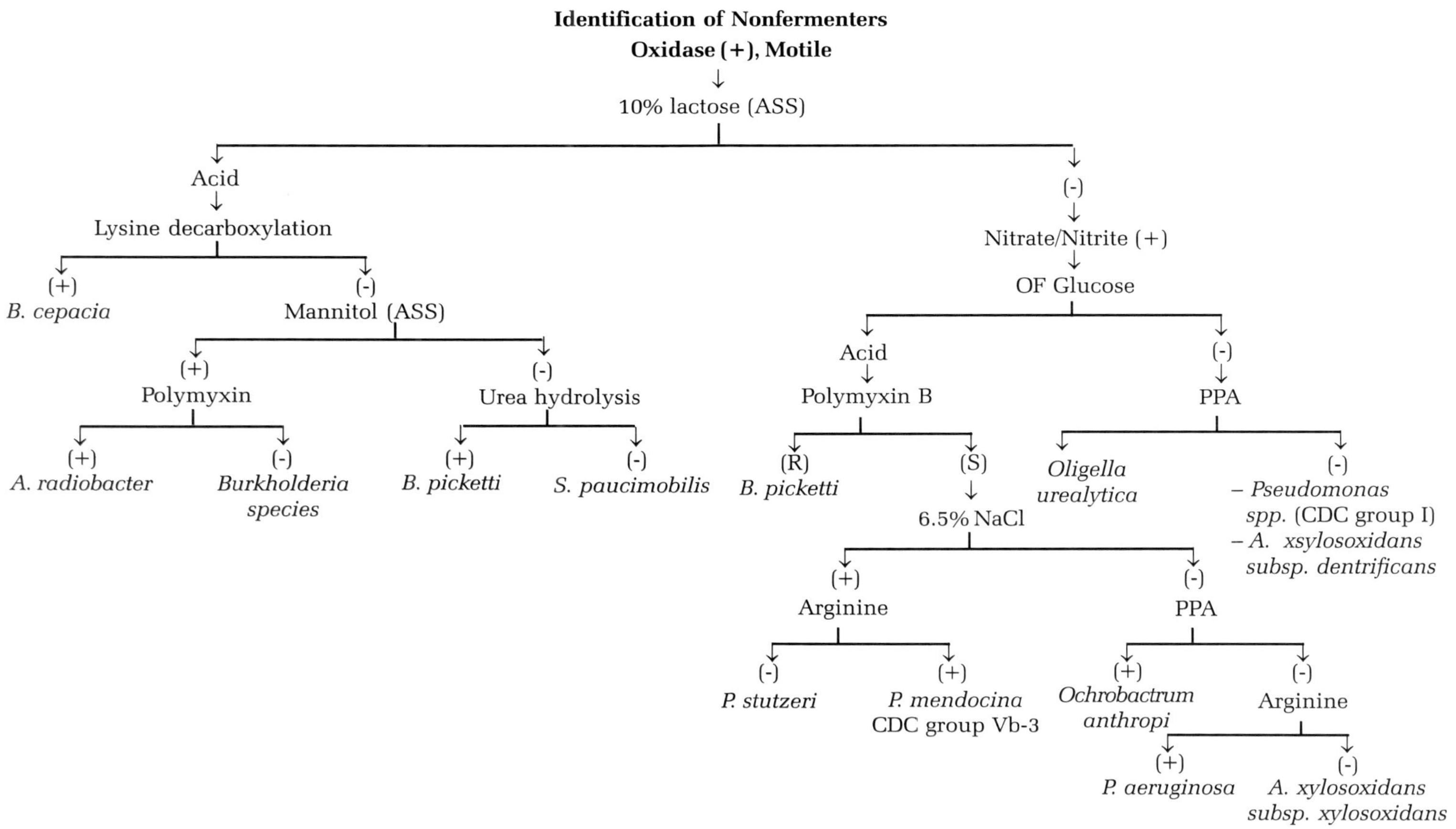

Identification of Nonfermenters
Oxidase (+), Motile

10% lactose (ASS)

Acid (+) → B. cepacia

Lysine decarboxylation (-) → Mannitol (ASS)

Mannitol (+) → Polymyxin
- Polymyxin (+) → A. radiobacter
- Polymyxin (-) → Burkholderia species

Mannitol (-) → Urea hydrolysis
- Urea hydrolysis (+) → B. picketti
- Urea hydrolysis (-) → S. paucimobilis

Nitrate/Nitrite (-) → OF Glucose

OF Glucose Acid → Polymyxin B
- Polymyxin B (R) → B. picketti
- Polymyxin B (S) → 6.5% NaCl

6.5% NaCl (+) → Arginine
- Arginine (-) → P. stutzeri
- Arginine (+) → P. mendocina CDC group Vb-3

6.5% NaCl (-) → PPA
- PPA (+) → Ochrobactrum anthropi
- PPA (-) → Arginine
  - Arginine (+) → P. aeruginosa
  - Arginine (-) → A. xylosoxidans subsp. xylosoxidans

OF Glucose (-) → PPA
- Oligella urealytica
- PPA (-) → – Pseudomonas spp. (CDC group I)
  – A. xsylosoxidans subsp. dentrificans

ASS = Ammonium salt sugars

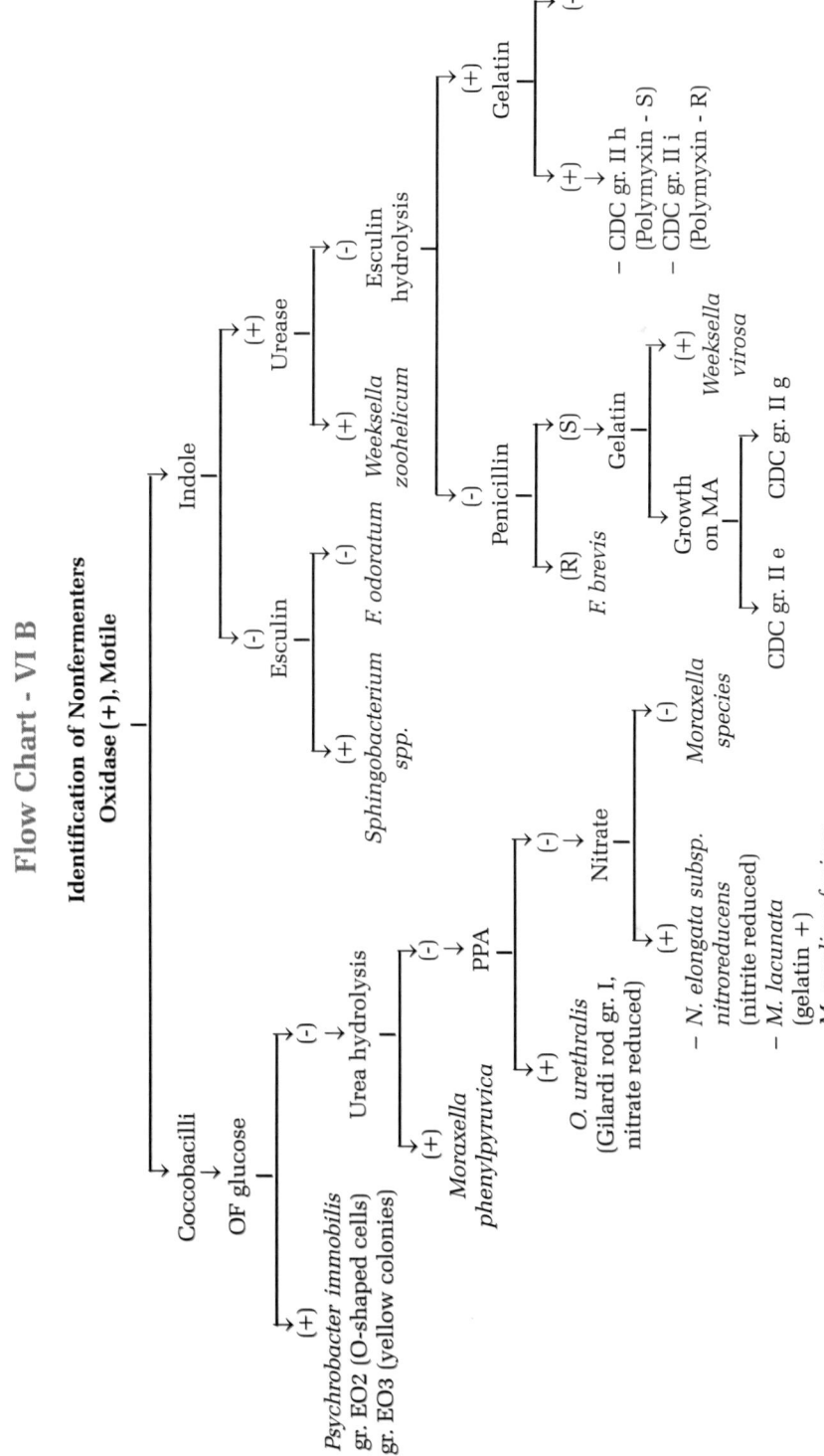

Flow Chart - VI B

Identification of Nonfermenters
Oxidase (+), Motile

gr. = Group
S   = Sensitive
R   = Resistant
MA = MacConkey agar

Flow Chart - VI C

Identification of Nonfermenters

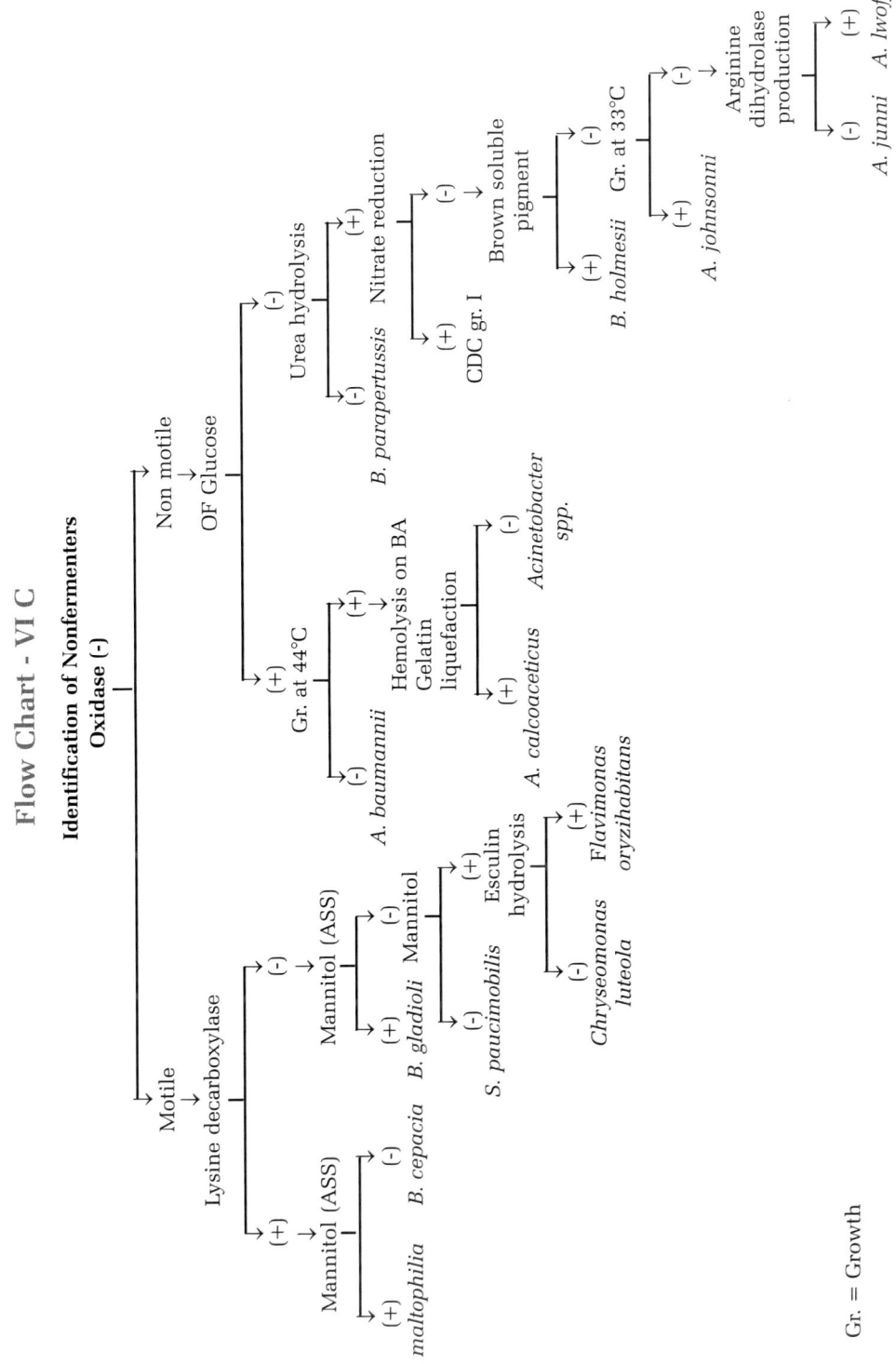

Gr. = Growth

Flow Chart - VII A

Identification of common Gram-positive anerobic bacteria

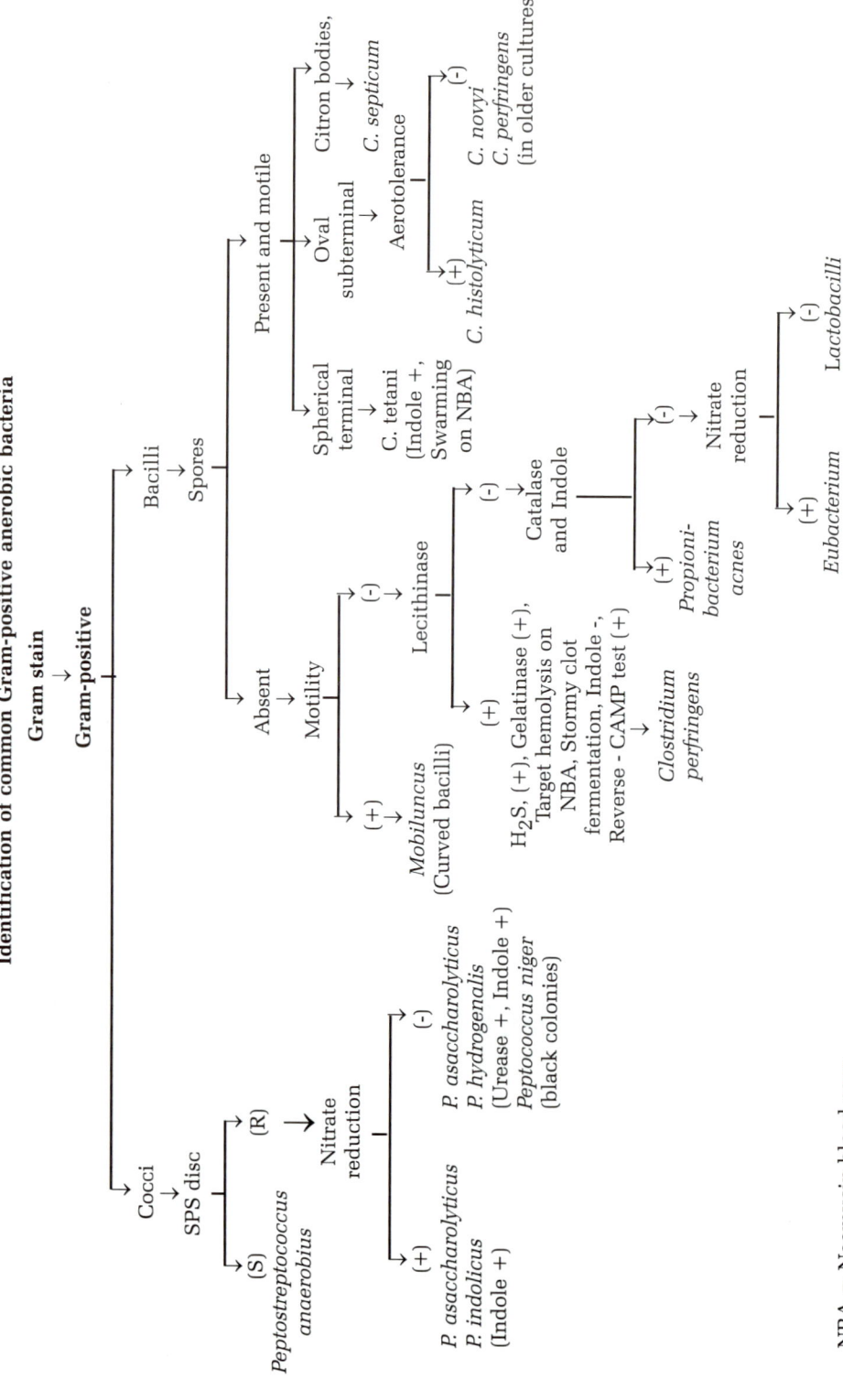

NBA = Neomycin blood agar
SPS = Sodium polyanethol suplhonate (1mg disc)

Flow Chart - VII B

Identification of common gram-negative anerobic bacteria

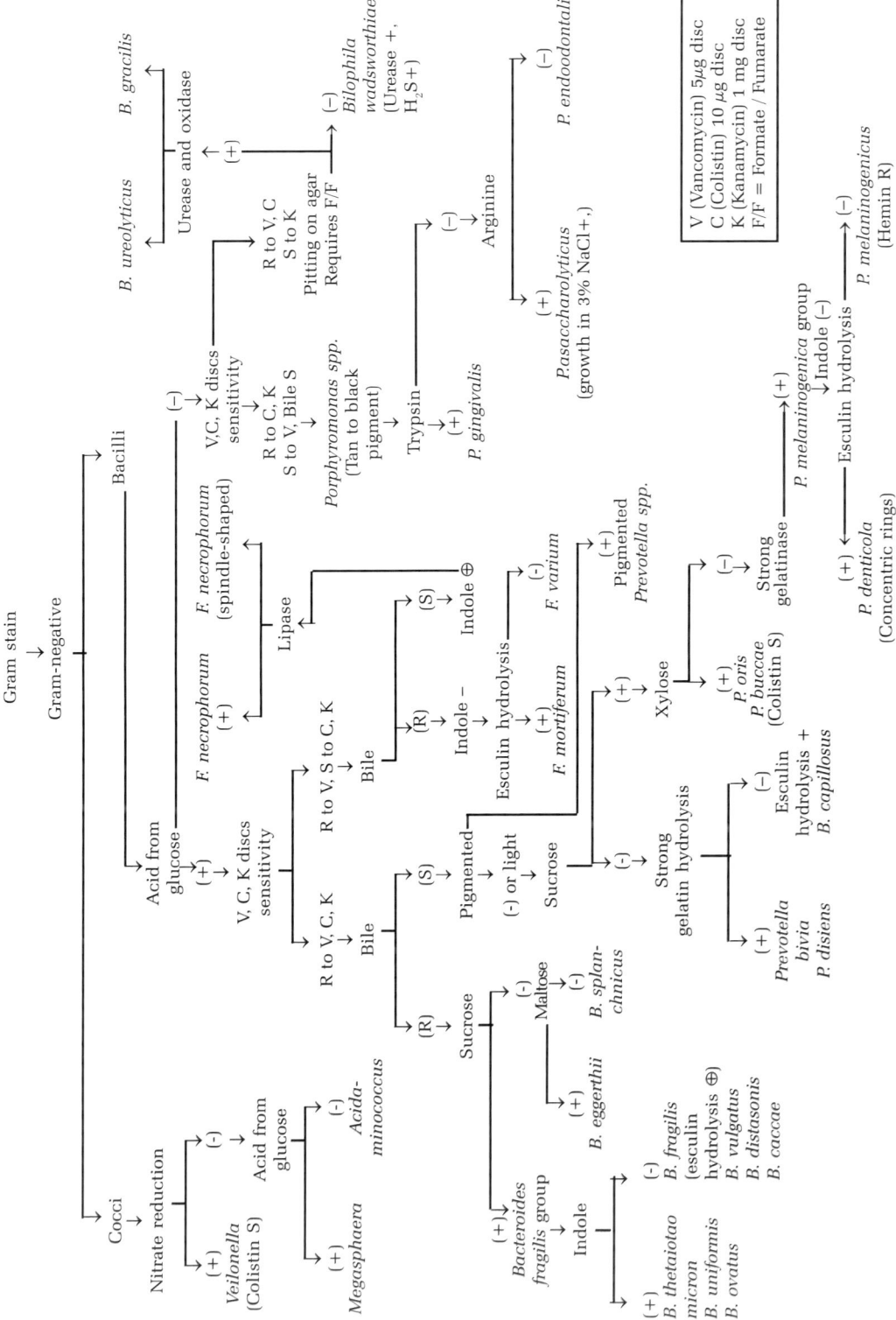

Flow Chart - VII C

In Robertson's Cooked Meat (RCM) broth

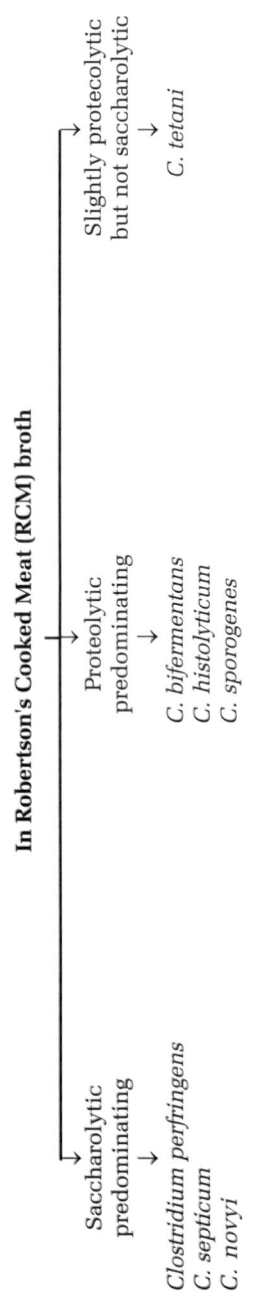

Saccharolytic predominating

Clostridium perfringens
C. septicum
C. novyi

Proteolytic predominating

C. bifermentans
C. histolyticum
C. sporogenes

Slightly protecolytic but not saccharolytic

C. tetani

Flow Chart VII D
On Egg Yolk Agar (EYA)

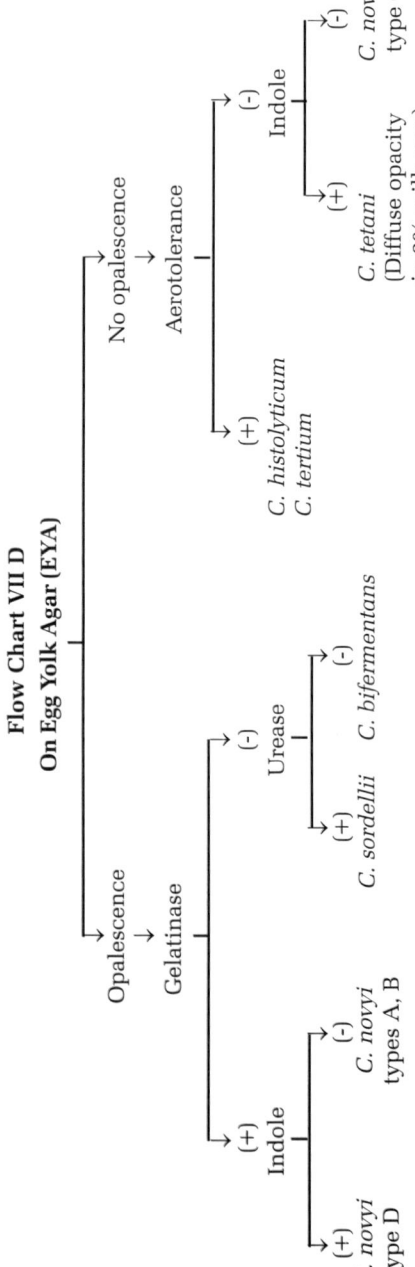

Opalescence

Gelatinase

(+) Indole

C. novyi type D

(−) *C. novyi* types A, B

Urease

(+) *C. sordellii*

(−) *C. bifermentans*

No opalescence

Aerotolerance

(+) *C. histolyticum* *C. tertium*

(−)

Indole

(+) *C. tetani* (Diffuse opacity in 3% milk agar)

(−) *C. novyi* type C

Flow Chart - VIII

Identification of fastidious Gram-negative bacteria

Growth in MacConkey agar

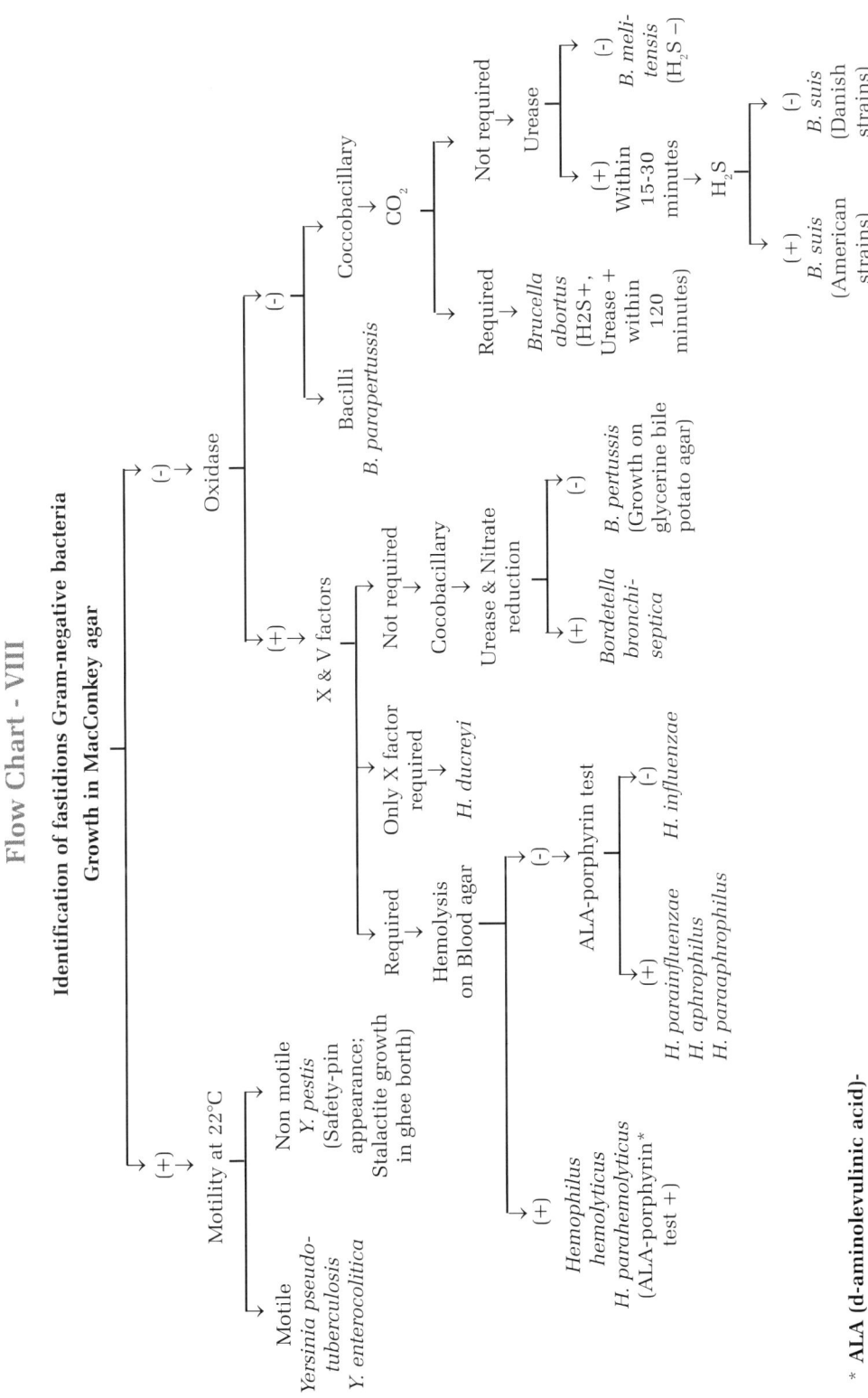

* **ALA (d-aminolevulinic acid)-porphyrin** impregnated filter paper discs – Brick-red fluorescence under Wood's lamp – positive test. No fluorescence – negative test.

Flow Chart - IX

Identification of Mycobacteria

Primary cultures on L-J medium incubated in dark at 37°C

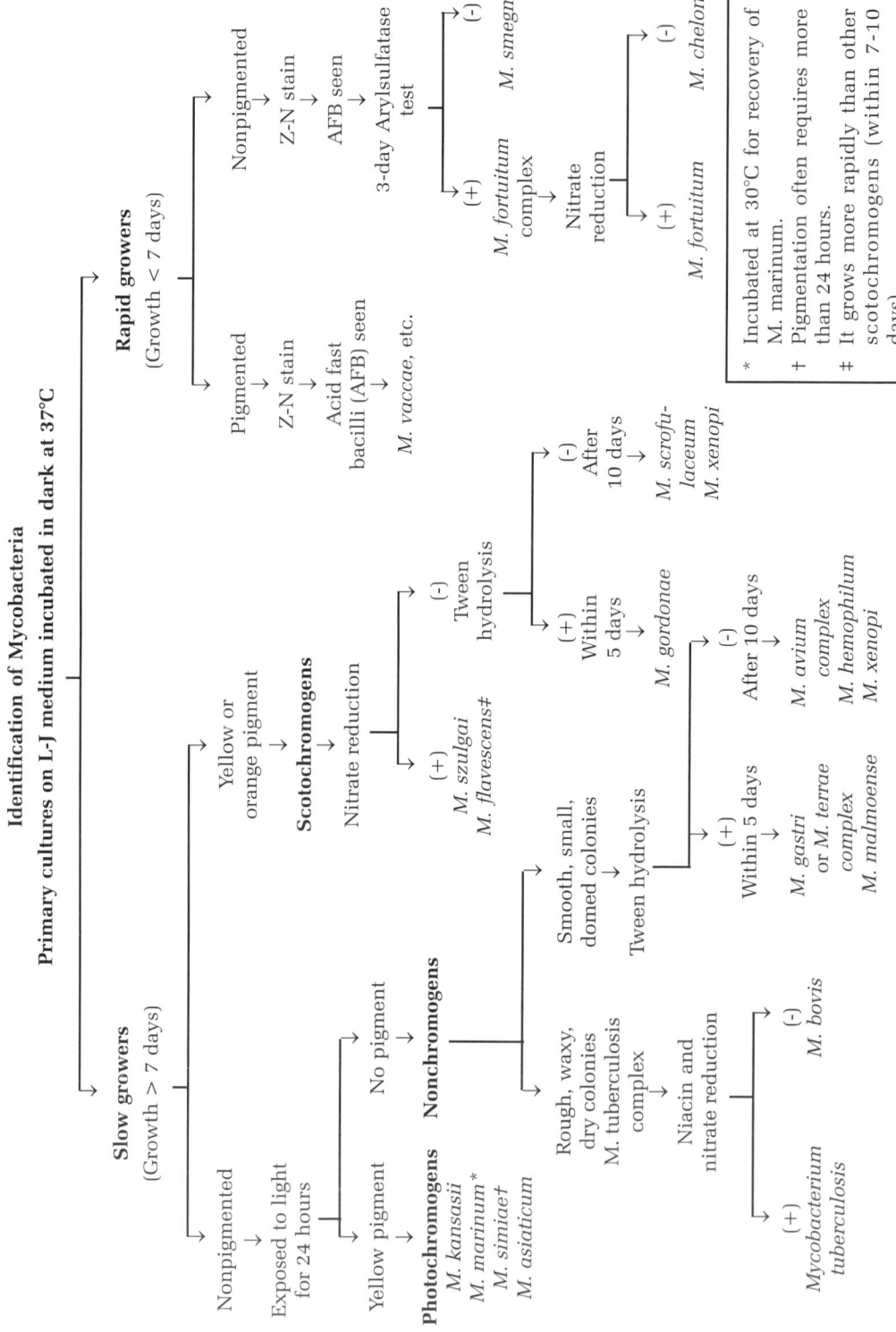

REFERENCES

1. Koneman EW, Allen SD, Janda WM, Schreckenberger P, Winn Jr. WC. In Color Atlas & Textbook of Diagnostic Microbiology. 6th Ed. Lippincott, Philadelphia. 2006.
2. Chakraborty PA. Textbook of Microbiology. 4th Ed. New Central Book Agency (P) Ltd., Kolkata. 2009.
3. Wadsworth Anaerobic Bacteriology Manual Paperback. 5th Ed. Eds. Paula Summanen & Ellen Jo Baron. Star Publishing Co., California. 1993.

MICROBIAL VIRULENCE FACTORS AND PATHOGENESIS OF IMPORTANT BACTERIAL DISEASES

Chapter

MICROBIAL VIRULENCE FACTORS

A Virulence Factor is any moiety produced by a pathogen that is essential for causing disease in a host.

- 'Pathogenicity' is generally employed to refer to the ability of a microbial species to produce disease.
- 'Virulence' refers to the ability of a strain of microorganism to produce disease.
- Enhancement of virulence is known as **exaltation** and can be demonstrated experimentally by serial passage in susceptible hosts.
- Reduction of virulence is known as **attenuation**.

Adherence and colonization factors

- The first host barrier for many invading pathogens is usually a mucosal surface, such as the gut or respiratory tract.
- Epithelial cell turnover is around 48 hours in these environments. Therefore, the bacterium must attach and replicate sufficiently to avoid being swept away.
- Simple attachment is mediated through a receptor on the host cell surface, and an adhesin on the bacterial one.
- Binding of the receptor on the host cell surface and the bacterial adhesin is often highly specific and accomplished by complimentary molecules
- Some may be species or even strain specific, while others exhibit tissue trophism e.g. *S.mutans* will colonize teeth but not tongue epithelium

Fimbriae/Pili

- Common pili (almost always called fimbriae) are usually involved in specific adherence (attachment).
- They are major determinants of bacterial virulence because they allow pathogens to attach to (colonize) tissues and/or to resist attack by phagocytic white blood cells.
- For example, pathogenic *Neisseria gonorrhoeae* adheres specifically to the human cervical or urethral epithelium by means of its fimbriae.
- Enterotoxigenic strains of *E. coli* adhere to the mucosal epithelium of the intestine by means of specific fimbriae.
- The M-protein and associated fimbriae of *Streptococcus pyogenes* are involved in adherence and to resistance to engulfment by phagocytes.

- Others e.g. fimbrial protein subunit of *Vibrio* and *Pseudomonas spp*. that binds D-Mannose on host cell surfaces.
- Colonization factors may act in conjunction with host derived polymers and those from other bacteria to produce a community or biofilm, e.g. in dental plaque.

Invasins

- Invasins differ from adherence factors in that they act extracellularly, breaking down host defenses at the local level, thus allowing infection to spread.
- Most invasins are enzymes that affect physical barriers such as tissue matrices and cell membranes. The bacterium can then spread through intracellular spaces.
- Highly invasive pathogens typically produce spreading or generalised lesions e.g. streptococcal septicemia following wound infection.
- Less invasive pathogens cause more localised lesions e.g. staphylococcal abscess.
- Some pathogens lack invasiveness altogether; yet, cause serious, even fatal diseases e.g. in case of tetanus bacillus.

Once, organism is located inside the host……..subsequent factors act (to promote virulence). These include:

- Evasion of the host defence system. Bacterial capsule plays an important role in this.
- Ability of the pathogen to compete with the commensal flora for nutrients and space.

Some bacteria are classically intracellular and evade the host defense mechanisms by their intracellular location.

Aggresins

- In order to survive and multiply within the host, many organisms produce a variety of substances that allow them to avoid or circumvent host defense mechanisms.
- These substances are termed aggresins.
- They include capsules, extracellular slime substances, surface proteins and carbohydrates, enzymes, toxins and other small molecules.

Polysaccharide Virulence Factors

Exopolysaccharides

- include discrete capsules of many pathogenic bacteria.
- loosely associated slime produced by many mucoid bacteria.

Capsules

First known bacterial virulence factor.

The capsular structures of some bacteria enable the organisms to avoid phagocytosis by preventing interaction between the bacterial cell surface and phagocytic cells or by concealing bacterial cell-surface components that would otherwise interact with phagocytic cells or complement and lead to their ingestion.

Capsules prevent bacteria from being engulfed by phagocytes.

Capsulate or slime producing bacteria → usually grow in microcolonies in vivo

Complement and opsonic antibodies → cannot get access to the cell envelope

Bacterial surface is more hydrophilic → less readily can be phagocytosed, e.g. alginate slime of mucoid *Pseudomonas aeruginosa* in lungs of children with cystic fibrosis.

Capsule allows the bacterium to evade the immune system by making it resistant to the lytic action of complement.

LPS → longer the chain more resistant to complement à complement cascade is activated at a distance from the bacterial membrane → no lysis occurs.

Capsular polysaccharide confers the property of **serum resistance**, e.g. K1 antigen of *E. coli.*

- Lipopolysaccharides of some *E.coli* serotypes are incapable of protecting against serum lysis.
- For these strains the capsule (K1) confers protection
- Other serotypes of *E. coli* (e.g. O6) the capsule is less important.
- Lipopolysaccharides of some *E. coli* serotypes are incapable of protecting against serum lysis.
- For these strains the capsule (K1) confers protection
- Other serotypes of *E. coli* (e.g. O6) the capsule is less important.
- Capsule of *B. fragilis* and other *Bacteroids* species are said to have a role in inhibiting phagocytosis & intracellular killing of facultative anaerobes (predominantly *E. coli*) in mixed anaerobic and aerobic infections.

Mixed infections → synergistic action between anaerobes and aerobes is involved and a multifactorial mechanism with LPS, protein, metabolites, chemotaxis of neutrophils and redox potential of all systems all possibly having an influence.

Surface Proteins

Some bacteria possess surface proteins that play a role in adherence and other virulence-associated functions, e.g. M protein of group A streptococci is involved in the pathogenesis of infection with this organism.

Interference with phagosome-lysosome fusion

- The mycobacteria and brucella are able to adapt to an intracellular existence within host cells by producing substances that prevent intracellular destruction of organisms.
- In the mycobacteria, this may be due to presence of cell-wall-associated mycosides and sulpholipids that become incorporated into the inner aspect of the phagosome and prevent **lysosome/phagosome fusion.**
- *Listeria monocytogenes* expresses a surface protein called **internalin** that binds to glycoprotein receptors on epithelial cells and allows the organism to become internalized in a membrane-bound vacuole.
- The organism then produces a hemolysin called **listeriolysin O** which intercalates into the membrane of the vacuole and causes the formation of pores.
- *L. monocytogenes* then enters the cytoplasm of the cell, where it continues to grow, thereby escaping the toxic environment within the phagolysosome.
- *S. aureus* secretes **catalase and superoxide dismutase**, which inhibit organism destruction by the myeloperoxidase system of phagocytic cells.
- These mechanisms also contribute to virulence by protecting the organisms from destruction by specific antibodies and complement.

Intracellular existence and effect on therapy

- Intracellular existence has a significant influence on therapy.
- Infections with organisms such as *Brucella* species and *Francisella tularensis* must be treated with antibiotics that are able to act intracellularly in order to have an effect on these organisms.
- Many bacteria produce **enzymes or toxins or possess cellular constituents** that have direct or necrotizing effects on host inflammatory cells and other components of the immune system, e.g. Leucocidin produced by *S. aureus.*

Cellular components/Enzymes

- The lipopolysaccharide of gram-negative bacteria may delay or blunt the acute inflammatory response, allowing the organism to establish itself in the host with relative ease.
- The lipopolysaccharide is an amphipathic molecule anchored in the outer membrane of the gram negative bacteria & forms an essential component of the cell envelope.
- The lipid A part is embedded in the outer membrane and is the toxic part of the molecule.
- The **Lipid A** portion of endotoxin, in particular, can activate complement and stimulate the release of various cytokines i.e. Il-2, Il-6, Il-8 that lead to clinical manifestations of endotoxic shock, which includes hypotension, DIC and death.
- Gram-positive bacteria possess cell wall **peptidoglycan polymers** and membrane **teichoic acids** that cause a similar release of cytokines and can precipitate similar septic shock symptoms.
- Lipotechoic acids are implicated in the adherence of group A streptococci to fibronectin.

Invasive properties of some bacteria — enzymes that act intracellularly

- *S. aureus, S. pneumoniae*, group B streptococci, and *Propionibacterium acnes* produce an enzyme called **hyaluronate lyase** (hyaluronidase) during the log phase.
- Hyaluronic acid is the ground substance responsible for cell-to-cell adhesion.
- Hyaluronate lyase promotes the spread of the organism through connective tissues by depolymerizing hyaluronic acid.
- Group A streptococci and staphylococci also elaborate enzymes that hydrolyse fibrin clots (**streptokinase** and **staphylokinase**) which also facilitate the spread of organisms in the tissues.

Siderophores

- The host environment is ideal for bacteria, except that, iron though plentiful, is tightly bound in haem, ferritin, transferrin, or lactoferrin.
- Successful pathogenic bacteria have evolved ways to obtain iron from the environment or from the iron-restricted milieu of host tissues.
- Microorganisms scavenge this iron by the production of siderophores, which are small molecules that function as high-affinity iron chelators.
- Production of siderophores is, therefore, considered a virulence factor, e.g. enterochelin from *Escherichia* and *Salmonella* spp.

Plasmids

- Plasmids themselves are not virulence factors.
- However, R factors (plasmids that contain genes coding for resistance to antimicrobial agents) may be considered virulence factors because when resistance to antimicrobial factors is acquired growth and spread of the microorganism continues despite therapeutic interventions.

Toxins

Toxins of microbial origin fall into two groups:

- Exotoxins
- Endotoxins

Exotoxins

- Bacterial exotoxins are the most potent biologic toxins known and are mostly produced by gram-positive bacteria.
- Usually protein in nature
- Heat-labile
- Many can be inactivated or destroyed by proteolytic enzymes.
- Toxic activity can be destroyed by formaldehyde treatment (toxoid development) and neutralized by specific antibodies.
- In general bacterial exotoxins fall into two groups :
  - **Cytolytic toxins** act on cell membranes to cause pore formation and subsequent lysis of the cell, e.g. seven enterotoxins of *S. aureus*.
  - Two subunits A & B or **bipartite toxins.** These toxins contain a B (binding) subunit that attaches to a specific host-cell receptor, and an A (active) subunit, that passes into the cell and interacts with the target, e.g. cholera toxin of *V. cholerae*.
- Certain diseases, such as tetanus, botulism, diphtheria, and cholera, are due almost entirely to the effects of the toxins on their target organs and tissues.
- Tetanus is caused by the systemic effects of tetanus neurotoxin, the toxin produced by *Clostridium tetani*.
- Tetanus neurotoxin is released upon cell lysis after bacterial growth under anaerobic conditions (i.e., in deep puncture wounds).
- Diphtheria is another example of an illness that is due primarily to the action of a toxin.
- Only strains of *C.diphtheriae* that contain a lysogenic bacteriophage (ß-phage) are able to produce diphtheria toxin.

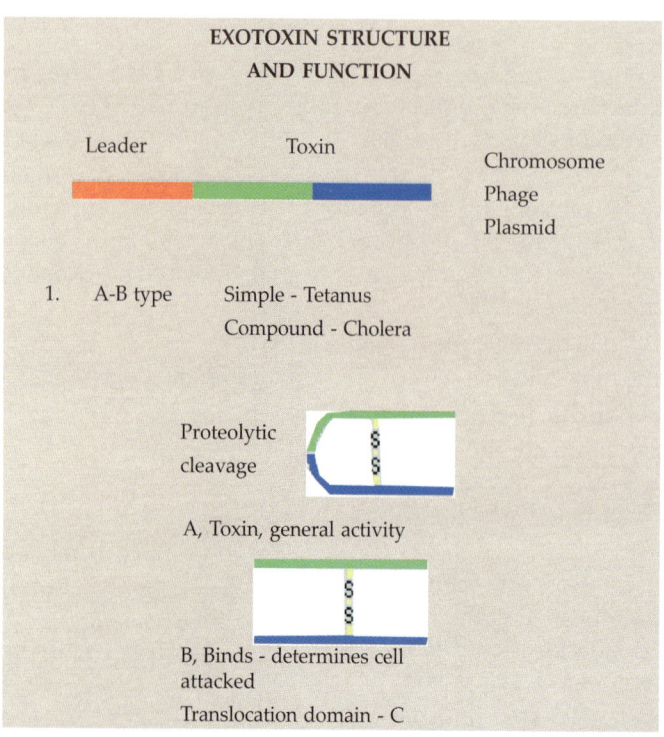

EXOTOXIN STRUCTURE AND FUNCTION

Leader   Toxin   Chromosome
                                       Phage
                                       Plasmid

1.   A-B type   Simple - Tetanus
                   Compound - Cholera

Proteolytic
cleavage

A, Toxin, general activity

B, Binds - determines cell
attacked

Translocation domain - C

Exotoxins	Endotoxins
• Proteins	• Lipopolysaccharides
• Heat labile	• Heat stable
• Actively secreted by cells; diffuse into the surrounding medium	• Form part of cell wall; do not diffuse into the surrounding medium
• Readily separated from cultures by physical means such as filteration	• Obtained only by cell lysis
• Action is often enzymic	• No enzymic action
• Specific pharmacological effect for each exotoxin	• Effect non-specific; action common to all endotoxins
• Specific tissue affinities	• No specific tissue affinity
• Active in very minute doses	• Active only in very large doses
• Highly antigenic	• Weakly antigenic
• Action specifically neutralised by antibody	• Neutralisation by antibody ineffective

Enterotoxins and diseases caused

Pathogen	Disease	Toxin
Vibrio cholerae	Cholera	Cholera toxin (CT)
Escherichia coli (LT)	Traveller's diarrhea (ETEC)	E. coli labile toxin (LT)
E. coli (EPEC)	Infant diarrhoea	Moderate level of Shiga-like toxin (SLTs)
E. coli (EIEC)	Dysentery-like	Multiple SLTs
E. coli (EHEC)	Haemorrhagic colitis	High level production of SLTs
Shigella dysenteriae	Dysentery	Shiga toxin
Clostridium perfringens enterotoxin	Food poisoning with diarrhea	Cytotoxic
Clostridium difficile	Pseudomembranous colitis	Toxin A, Toxin B
S. aureus enterotoxin	Food poisoning with vomiting & diarrhea	Seven
Bacillus cereus	Food poisoning with vomiting & diarrhea	Diarrhoeagenic toxin
Bacillus anthracis	Anthrax	Anthrax toxin
Bordetella pertussis	Whooping cough	Pertussis toxin
Clostridium botulinum	Botulism	Series of neurotoxins (A-G)
Clostridium tetani	Tetanus	Tetanospasmin
Corynebacterium diphtheriae	Diphtheria	Diphtheria toxin

Endotoxins

- Produced only by gram-negative bacteria.
- Consist primarily of lipopolysaccharide.
- Heat-stable
- Not detoxified by formaldehyde treatment.
- Only partially neutralized by specific antibodies.
- Relatively low toxicity.
- Biologic and toxic activities of endotoxins are broad.

"O" side chains (oligosaccharides)	Core polysaccharide	Lipid A
Species or serotype antigens	Genus-specific antigens	Toxic moiety

General Structure of Endotoxin

Colonization factors involved in adherence include:

- Fimbriae/pili
- Lipoteichoic acids
- Exopolysaccharides (capsules and glycocalyxes)
- Outer membrane proteins

Lipotechoic acids

- Lipotechoic acids are components of the cell wall in gram positive bacteria which serve as virulence factors.
- Lipotechoic acids play an important role in attachment of Gram positive organisms to host tissues thus acting as an adherence factor, e.g. in infection with Group B streptococci.

Outer Membrane Proteins

- Virulence-related outer membrane proteins are expressed in Gram-negative bacteria and are essential to bacterial survival within macrophages and for eukaryotic cell invasion.
- This family consists of several bacterial and phage Ail/Lom-like proteins.
- The *Yersinia enterocolitica* Ail protein is a known virulence factor.
- Members of this group include:
  - PagC, required by *Salmonella typhimurium* for survival in macrophages and for virulence in mice
  - Rck outer membrane protein of the *S. typhimurium* virulence plasmid
  - Ail, a product of the *Yersinia enterocolitica* chromosome capable of mediating bacterial adherence to and invasion of epithelial cell lines
  - OmpX from *Escherichia coli* that promotes adhesion to and entry into mammalian cells. It also has a role in the resistance against attack by the human complement system
  - A bacteriophage lambda outer membrane protein, Lom.

Virulence Factors of *Staphylococcus aureus*

A) Cell associated polymers

1. Cell wall polysaccharide peptidoglycan confers rigidity and structural integrity to the bacterial cell, activates complement and induces release of inflammatory cytokines.
2. Cell wall teichoic acid facilitates adhesion of the cocci to the host cell surface & protects them from complement - mediated opsonisation.
3. Capsular polysaccharide inhibits opsonisation.

B) Cell surface proteins

1. Protein A has chemotactic, antiphagocytic and anticomplementary effects. It binds to Fc portion of IgG, leaving Fab region free to combine with specific antigen.
2. Clumping factor, another surface protein is the 'bound coagulase'.

C) Extracellular enzymes

1. Coagulase (eight types, commonest type A) which clots human or rabbit plasma. It acts along with coagulase reacting factor present in plasma, binds to prothrombin and converts fibrinogen to fibrin. It is also known as 'free coagulase'.
2. Lipases help to inject skin and subcutaneous tissues.
3. Hyaluronidase breaks down connective tissue.
4. Staphylokinase (fibrinolysin) and proteases help in initiation and spread of infection.
5. Heat stable nuclease
6. Receptors for proteins like fibronectin, fibrinogen, IgG & C1q, which facilitate adhesion to host cell and tissues.

D) Toxins

1. Hemolysins - α, β, γ and δ.
   - α hemolysin lyses rabbit red blood cells. It is also leucocidal, cytotoxic, dermonecrotic, neurotoxic and it is toxic to macrophages, lysosomes, muscle tissues, renal cortex and circulatory system.
   - δ hemolysin acts on cell membranes of erythrocytes, leucocytes, macrophages and platelets.
2. Leucocidin (Panton - Valentine toxin) acts on leucocytes.
3. Enterotoxins - 8 types (A, B, C1, C2, C3, D, E & H). Act directly on antonomic nervous system to cause illness.

   It also has pyrogenic, mitogenic, hypotensive, thrombocytopenic and cytotoxic effects.
4. Toxic shock syndrome toxin (TSST type 1) - formerly known as enterotoxin F or pyrogenic exotoxin C. TSST -1 and enterotoxins are potent

activators of T lymphocytes. Releases interleukins 1 and 2, tumor necrosis factor and interferon gamma.

5. Exfoliative or epidermolytic toxin - outer layer of epidermis gets separated from underlying tissues.

Infections caused by *Staphylococcus aureus*

1. Cutaneous lesions - boils, carbuncles, pustules, abscesses, styes, wounds and burn infections.

2. Deep lesions - tonsillitis, pharyngitis, acute osteomyelitis, septic arthritis, pneumonia, lung abscess, empyema, meningitis, brain abscess, acute bacterial endocarditis, enterocolitis, septicemia and pyemia.

3. Toxic shock syndrome due to toxin TSST-1 (enterotoxin F), sometimes leading to acute renal failure, disseminated intravascular coagulation (DIC) and peripheral gangrene.

4. Food poisiong due to preformed toxin which in ingested with contaminated food caused by enterotoxins A-E.

5. Skin exfoliation (stripping of superficial layers of skin from underlying tissue e.g. bullous impetigo, pemphigus neonatorum and toxic epidermal necrosis.

Coagulse Negative Staphylococcus (CONS).

S.epidermidis, S. saprophyticus, S. hemolyticus, S. warneri, S. hominis, S.simulans, S.lugdunensis, S. saccharolyticus. S. capitis, S. cohnii, S. xylosus, S.ureolyticus.

Infections caused by CONS.

- Nosocomial & community acquired UTI - anatomical, structural & functional changes, nephropathies, in sexually active women, diabetes mellitus.
- Infection with indwelling devices - catheters, etc.
- Bacteremia in compromised host - fever, bacteria should be present in two or more samples.
- Native & prosthetic valve endocarditis (NVE and PVE).
- Osteomyelitis, cellulitis, breast abscess, vertebral-osteitis.
- Post surgical endophthalmitis.
- Peritonitis – 40% of CAPD, *S. epidermidis* followed by *S. hemolyticus*.
- V. P. shunt meningitis.
- Prosthetic joint infection, mediastinitis.
- Brain empyema.

Predisposing factors

Prosthesis, indwelling catheters, extremes of age, immunocompromised, case of chronic ambulatory peritoneal dialysis (CAPD), prolonged hospital stay, blood related malignancy, woman of reproductive age group, shunts, renal disorders (calculus, neurogenic bladder, etc.) systemic disorders (diabetes mellitus, etc.)

Sources of bacteremia

- I. V. devices - Central & peripheral I. V. catheters, Pacemaker, Swan Ganz catheter, angiography
- A. V. fistula
- Gastrointestinal and genitourinary surgeries
- V. P. shunt
- Wound and skin infections.

Native Valve Endocarditis (NVE)

Infective endocarditis occuring 48 hours or more after admission or endocarditis related to an intervention performed in hospital within 4-8 weeks before admission. High risk factors - age, bacteremia, early period of prosthesis.

Virulence Factors of *Streptococcus pyogenes*

A. Peptidoglycan (mucoprotein) confers rigidity to cell wall. It is also pyrogenic and has thrombolytic activity.

B. M protein in outer part of cell wall inhibits phagocytosis.

C. Fimbria projecting through capsule of group A streptococci helps in the attachment of the cocci to epithelial cells.

D. Antigenic cross reaction between capsular hyaluronic acid and human synovial fluid; cell wall protein and myocardium; group A carbohydrate and cardiac valves; cytoplasmic membrane antigens and vascular intima; and peptidoglycans and skin antigens. These accounts for some of the manifestations of rheumatic fever and acute glomerulonephritis, tissue damage being of an immunological nature.

E. **Toxins**

1. Hemolysins - Streptolysins 'O' (oxygen labile), responsible for virulence, has cardiotoxic and leucotoxic acitivity.

   Streptolysin 'S' (oxygen stable), responsible for hemolysis seen around colonies on the surface of blood agar plates. It is soluble in serum.

2. Pyrogenic exotoxin (erythrogenic, Dick, scarletinal toxin) - induces fever. Three types (A, B & C).

   They are **'superantigens'** - T cell mitogens which induce massive release of inflammatory cytokines causing fever, shock and tissue damage.

3. Streptokinase (fibrinolysin) - promotes lysis of human fibrin clots by activating a plasma precursor (plasminogen).

   It breaks down the fibrin barrier around the lesions, facilitating the spread of infection.

4. Deoxyribonucleases (Streptodornase or DNAase) cause depolymerisation of DNA. It helps to liquefy thick pus. 4 types (A, B, C & D). B & D also possess ribonuclease activity.

5. Nicotinamide adenine dinucleotidase (NADase) acts as coenzyme NAD and liberates nicotinamide from the molecule. It is believed to be leucotoxic.

6. Hyaluronidase breaks hyaluronic acid of the tissues and favour spread of infection along intercellular spaces.

7. Serum opacity factor - a lipoproteinase produced by some M types of *Streptococcus pyogenes*.

Infections caused by *Streptococcus pyogenes*

1. Suppurative infections - sore throat and tonsillitis, impetigo, erysipelas, wound and burn infections, chronic skin lesion (eczema, psoriasis), scarlet fever due to erythrogenic toxin, bone and joint infections, lymphadenitis, septicemia, acute endocarditis, abscesses in internal organs.

2. Non-suppurative complications - rheumatic fever, acute glomerulonephritis.

Infections caused by :

Group B streptococci - Neonatal meningitis, pneumonia in the neonates.

Streptococcus viridans - Subacute bacterial endocarditis.

Streptococcus fecalis - Endocarditis, urinary tract infection.

Virulence factors for *Streptococcus pneumoniae*

A) Type specific capsular polysaccharide, also known as 'specific soluble substance' (SSS). 90 different serotypes are present. Protects the cocci from phagocytosis. Type 3 pneumococcus has abundant capsular material and therefore is highly virulent.

B) Abnormal protein (b-globulin) that precipitates with somatic 'C' antigen of pneumococci, appears in acute phase sera of cases of pneumonia & are produced by hepatocytes - known as 'C-reactive protein' (CRP). It is stimulated in bacterial infections, inflammation, malignancies and tissue destruction.

C) Toxins
   1. Oxygen labile hemolysin and leucocidin are not responsible for virulence.
   2. Pneumolysin, a membrane damaging toxin has cytotoxic and complement activating properties.

3. Pneumococcal autolysins, by releasing bacterial components in infected tissues contribute to virulence.

Infections caused by *S. pneumoniae*

1. Commonest respiratory infections are otitis media and sinusitis (6, 14, 19F & 23F in the West).

2. Acute tracheobronchitis and acute exacerbations in chronic bronchitis.

3. Pneumonia (types 1-8 for 75% cases) & > 50% of all fatalities due to pneumococcal bacteremia. Types 1, 3, 7, 8 and 12 are invasive types and type 3 is most virulent.

4. Bronchopneumonia, usually a terminal event in aged and debilitated patients.

5. Meningitis – secondary to pneumonia, sinusitis, or otitis media. Untreated cases almost always fatal. Case fatality rate 25%, even with antibiotic therapy.

6. Other suppurative lesions – empyema, pericarditis, conjunctivitis, suppurative arthritis, peritonitis (usually as complications of pneumonia).

Virulence factors of *Bacteroides spp.*

1. Capsular polysaccharides – inhibits opsono-phgocytosis, promotes abscess formation and adherence to epithelial cells.

2. Pili and fimbriae – promotes adherence to epithelial cells and mucous membranes.

3. Endotoxin – potent inducers of DNA replication and polyclonal immunoglobulin production in murine B lymphocytes.

4. Succinic acid – inhibits phagocytosis and intracellular killing of the producer and others in its milieu.

5. Extracellular enzymes (neuraminidase, peroxidase, collagenase, phosphatase, proteinase, hyaluronidase, phospholipase A, DNAase, hemolysin, fibrinolysin, heparinase, chondroitin sulphatase) – contribute to tissue damage and promote spread.

6. Superoxide dismutase – defends against toxic oxygen radicals and enhances aerotolerance.

7. Enterotoxin – alters the cytoskeleton and the barrier function of the intestinal epithelium.

8. LPS cell wall – weak endotoxic activity.

Virulence factors of *Fusobacterium spp.*

1. Exotoxins and Exoenzymes – oxygen stable hemolysin, lipase, leucocidin, DNAase.

2. Endotoxin.

Virulence factor of *L. buccalis*

LPS – strong endotoxic activity – causes pyrexia and Schwartzmann reaction.

Virulence factors of *Helicobacter pylori*

Main factors of virulence are urease, vacuolating cytotoxin VacA and pathogenicity isoland (cag PAI) gene products.

Virulence factors of *Campylobacter species*

1. Initial colonization – diffuse adherence to epithelial cells and internalisation.

2. Tissue invasion – less than that of *Shigella spp.* and gives negative Sereny test.

3. Cholera – like enterotoxin (heat labile) by *C. jejuni* – still not proven. Causes fluid accumulation in rat ileal loop. Main pathological lesion is inflammatory enteritis. Nontoxigenic strains are also capable of causing disease.

4. Cytotoxin
    -   Heat labile, 50-70 kDa, hemolytic for rabbit red blood cells.
    -   Heat stable, 70 kDa, hemagglutinate rabbit reticulocytes.

PATHOGENESIS OF SOME IMPORTANT BACTERIAL DISEASES

Pathogenesis of Meningococcal Meningitis

Droplet spread via infected respiratory secretions
from cases and carriers (in nasopharynx)

Bacterial endotoxin (10-20%)

Through blood stream   Along perineural sheath or olfactory nerve

Meningococcaemia
(petechial rash in skin and mucosa)

Meninges

Complications – Waterhouse-Frederichsen Syndrome during fulminating meningococcal septicemia

Pathogenesis of Acne

Normal follicle (colonized by *P. acnes*, staphalococci,
micrococci and pityrosporum yeasts)

↓

Microcomedone
Abnormal shedding of keratin by epithelial cells in follicle
causes formation of keratin plug

↓

Change in physiology of *P. acnes*

↓

P. acnes breaks down sebum to form fatty acids which together with
other bacterial and polymorph products cause inflammation

↓

Inflamed lesion

Regression and healing of inflammation   Amplification

P. acnes is also an immunostimulator.

Pathogenesis of Tetanus

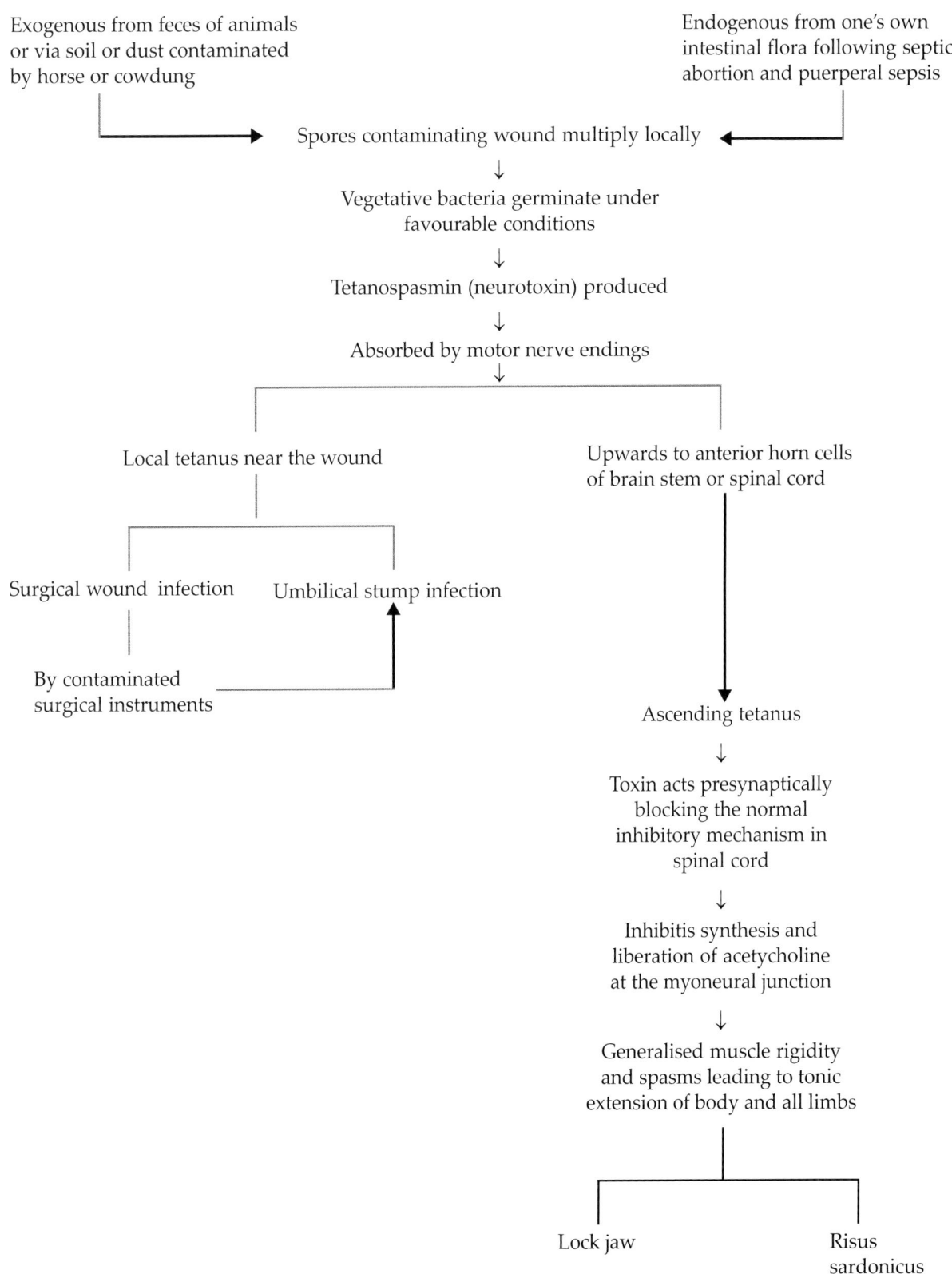

Exogenous from feces of animals
or via soil or dust contaminated
by horse or cowdung

Endogenous from one's own
intestinal flora following septic
abortion and puerperal sepsis

Spores contaminating wound multiply locally

↓

Vegetative bacteria germinate under
favourable conditions

↓

Tetanospasmin (neurotoxin) produced

↓

Absorbed by motor nerve endings
↓

Local tetanus near the wound

Upwards to anterior horn cells
of brain stem or spinal cord

Surgical wound infection   Umbilical stump infection

By contaminated
surgical instruments

Ascending tetanus

↓

Toxin acts presynaptically
blocking the normal
inhibitory mechanism in
spinal cord

↓

Inhibitis synthesis and
liberation of acetycholine
at the myoneural junction

↓

Generalised muscle rigidity
and spasms leading to tonic
extension of body and all limbs

Lock jaw   Risus
                                                      sardonicus

Complications: Tetanus neonatorum, post abortal and puerperal tetanus – 15-50% mortality rate.

Pathogenesis of Gas Gangrene

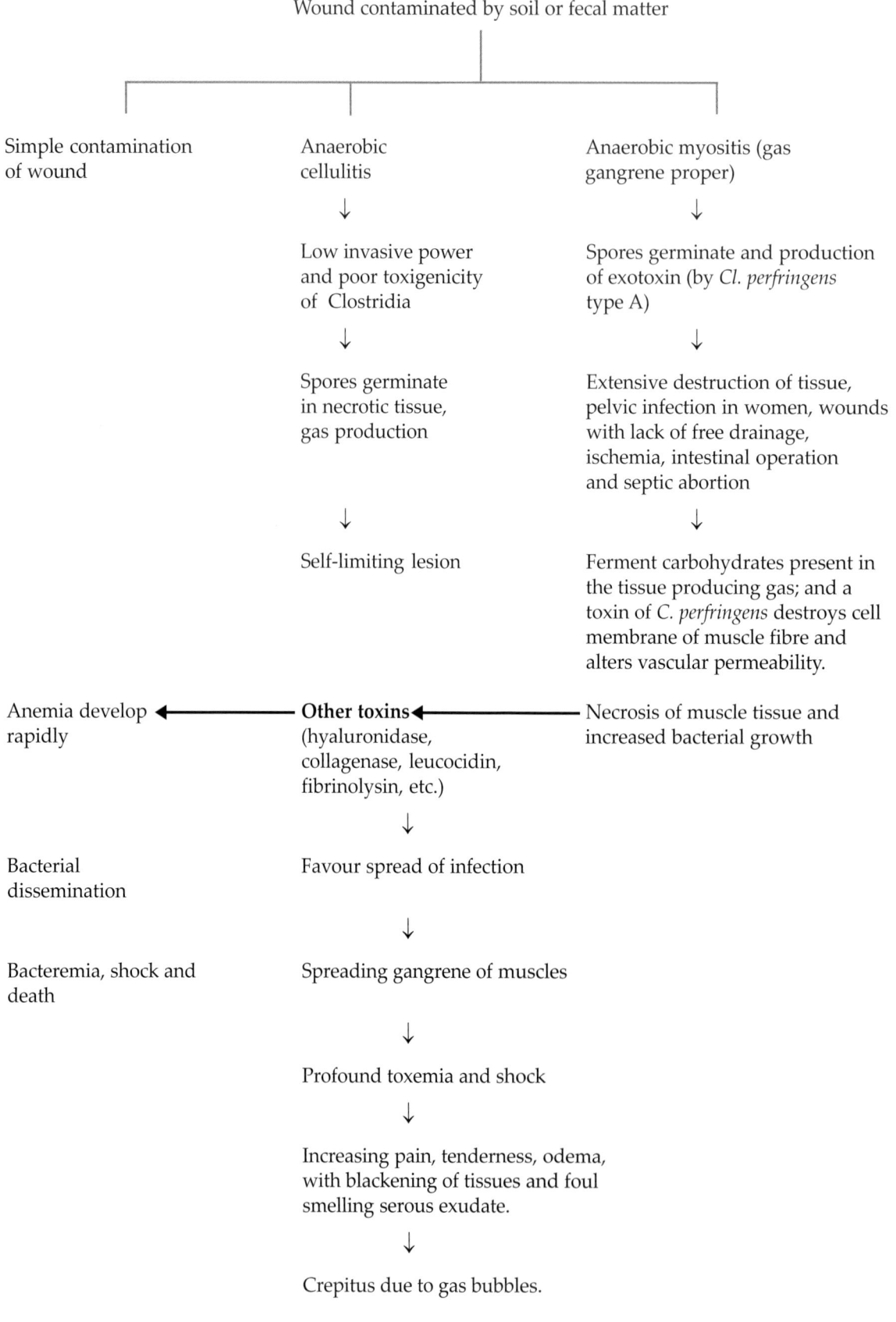

Wound contaminated by soil or fecal matter

Simple contamination
of wound

Anaerobic
cellulitis

↓

Low invasive power
and poor toxigenicity
of Clostridia

↓

Spores germinate
in necrotic tissue,
gas production

↓

Self-limiting lesion

Anaerobic myositis (gas
gangrene proper)

↓

Spores germinate and production
of exotoxin (by *Cl. perfringens*
type A)

↓

Extensive destruction of tissue,
pelvic infection in women, wounds
with lack of free drainage,
ischemia, intestinal operation
and septic abortion

↓

Ferment carbohydrates present in
the tissue producing gas; and a
toxin of *C. perfringens* destroys cell
membrane of muscle fibre and
alters vascular permeability.

Anemia develop ← **Other toxins** ← Necrosis of muscle tissue and
rapidly (hyaluronidase, increased bacterial growth
collagenase, leucocidin,
fibrinolysin, etc.)

↓

Bacterial
dissemination

Favour spread of infection

↓

Bacteremia, shock and
death

Spreading gangrene of muscles

↓

Profound toxemia and shock

↓

Increasing pain, tenderness, odema,
with blackening of tissues and foul
smelling serous exudate.

↓

Crepitus due to gas bubbles.

Pathogenesis of *Pseudomonas aeruginosa*

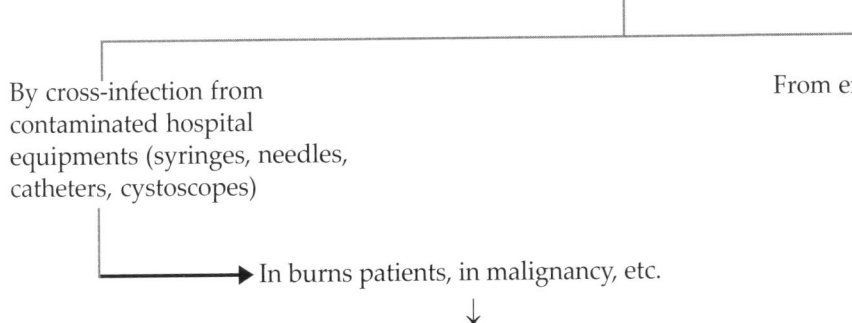

An opportunistic pathogen when body resistance is lowered and tissue is damaged

By cross-infection from contaminated hospital equipments (syringes, needles, catheters, cystoscopes)

From environment

→ In burns patients, in malignancy, etc.

↓

Exotoxin A acts as NADase and inhibits protein synthesis; Proteases and elastases; Hemolysins

↓

- UTI after catheterisation or instrumentation
- Wound and burns infections
- CSOM, otitis externa, eye infections
- Meningitis following lumbar puncture or after cranial injury
- Necrotising pneumonia from contaminated respirators
- Septicemia in immunocompromised patients
- Acute necrotising vasculitis leading to hemorrhagic infarction of skin (ecthyma gangrenosa) and of internal organs (liver and kidney).

Pathogenesis of Gonorrhoea

Due to various virulence factors :

1. **Pilus** – helps to attach to human mucosal epithelium; contains constant and hypervariable regions – analogous to Igs that contribute to antigenic diversity in gonococci.

2. **Protein I** – forms pores through outer membrane; antigenic; specific serotypes associated with virulence, present in T1 and T2 colonies.

3. **Protein II** – present in avirulent colonies (T3 and T4).

4. **Lipopolysaccharide (LPS)** – endotoxin activity.

5. **IgA protease** – core contains enzyme, released by cell to destroy IgA 1.

6. **Capsule** – resists phagocytosis, unless antibody is present.

Complications

Local spread	Systemic spread
In females - damage to fallopian tubes, pelvic inflammatory disease, anorectal infection) (watercan perineum In males - occasional epididymitis.	Skin lesions Endocarditis Arthritis Ophthalmia neonatorum in newborns, infection is acquired during delivery through mother's birth passage.

Chlamydia trachomatis infections and complications

	Infections	Complications
In men	Urethritis Epididymitis Proctitis Conjunctivitis	Systemic spread Reiter's Syndrome (urethritis, conjunctivitis, polyarthritis)
In women	Urethritis Cervicitis Salpingitis Conjunctivitis	Ectopic pregnancy Infertility Systemic spread - perihepatitis, arthritis, dermatitis
In neonates	Conjunctivitis	Interstitial pneumonitis.

Pathogenesis of Typhoid Fever

Via oral route with water or food contaminated by sewage or
via hands of a carrier
↓
Bypass gastric acid barrier in small intestine
↓
Multiply in small intestine
↓
Attach on surface of epithelial cells of villi.
↓
Pass to submucous coat and multiply
↓
Enter mesenteric lymph nodes
↓
Through thoracic duct
↓
Bacilli invade blood stream (**primary bacteremia**) 7-10 days
↓
Go to liver, lung, spleen, bone marrow, lymph node
and multiply in cells
↓
On 10th day, parasitised cells undergo necrosis and bacilli
again pass into blood **(secondary bacteremia)** on 14th day
↓
Organisms undergo lysis and liberate endotoxin
↓
Toxemia leads to step-ladder pyrexia, headache, anorexia,
hepatosplenomegaly, bradycardia, leucopenia, etc.
↓
Localise in organs e.g. gall bladder, liver, spleen, bone
↓
Caused arthritis, cholecystitis, meningitis,
endocarditis, nephritis, osteomyelitis.

From gall bladder
↓
Into intestine causing inflammation of Peyer's patches of
intestine and lymphoid follicles
↓
Necrosis and sloughing of follicles
↓
Typhoid ulcers
↓
Hemorrhage and perforation
(complications)

Bacilli are detected during 2nd to 3rd week in stool and during 3rd to 4th week in urine.

Paratyphoid fevers

Milder forms of febrile illness, shorter incubation period, diarrhoea more often seen.

Pathogenesis of Bacillary Dysentery

Ingestion of contaminated food or drink or from carriers through
contaminated objects or through flies (only 10 to 100 bacteria can caused disease)

↓

Organisms to terminal ileum and colon

↓

Attach to surface of epithelial cells and then
go inside the cells and multiply

↓

To lamina propria, replication and bacterial colonization takes place

↓

Acute inflammatory reaction with formation of micro-abscessess on
mucosal surface and capillary thrombosis

↓

Soft and friable necrotic epithelia slough out
in patches forming serpiginous ulcers

↓

Bleeding from mucosa

↓

Endotoxin irritates bowel causing diarrhoea.

S. dysenteriae type 1 produces
enterotoxin and neurotoxin

Granulation tissue is formed and the
ulcers heal by regeneration of the mucosal
epithelium

Enterotoxin causes transudation of fluid into lumen with blood	Neurotoxin damages endothelial cells of central nerous system	Entire mucous membrane of large gut slough out
↓	↓	↓
Complications	**Complications**	**Death.**
- Hemolytic uremic syndrome	- Polyneuritis	
- Arthritis	- Coma	
- Myocarditis	- Meningism	

S. dysenteriae - severe and fulminating disease.

S. flexneri and *S. boydii* – less severe disease, prevalent in tropical and sub-tropical countries.

S. sonnei - frequently seen in children.

Pathogenesis of Cholera

Organisms ingested in large numbers

↓

Sensitive to stomach acid and large dose is needed (105-106 organisms)
to cause disease unless patient is achlorhydric or taking antacids

↓

Colonization of small intestine depending on motility
(polar flagella). Production of mucinase (cause desquamation
of surface epithelium of tips of villi)
Attachment to specific receptors

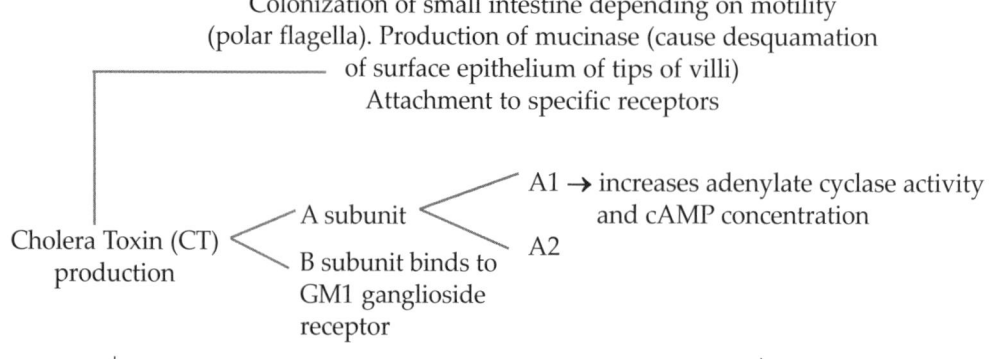

Cholera Toxin (CT)
production

A subunit

A1 → increases adenylate cyclase activity
and cAMP concentration

A2

B subunit binds to
GM1 ganglioside
receptor

↓

Increases capillary permeability

↓

Oedema and neutrophilic
infiltration of submucosa.

↓

Marked hypersecretion from
glands in intestinal lumen

↓

Massive loss of fluid and
electrolytes (high concentration
of sodium and bicarbonate and
lower conc. of chloride)

Hypokalemia   Base-deficit acidosis

↓

Muscle cramps

Colourless rice water stools
containing mucus flakes,
epithelial cells, large number
of vibrios (no damage to
enterocytes, so no blood
or WBC in stool)

Dehydration

↓

Hemoconcentration,
anuria, acute
tubular necrosis and
hypovolemic shock.

↓

Diarrhoea ++++

Vomiting +

Pathogenesis of Tuberculosis

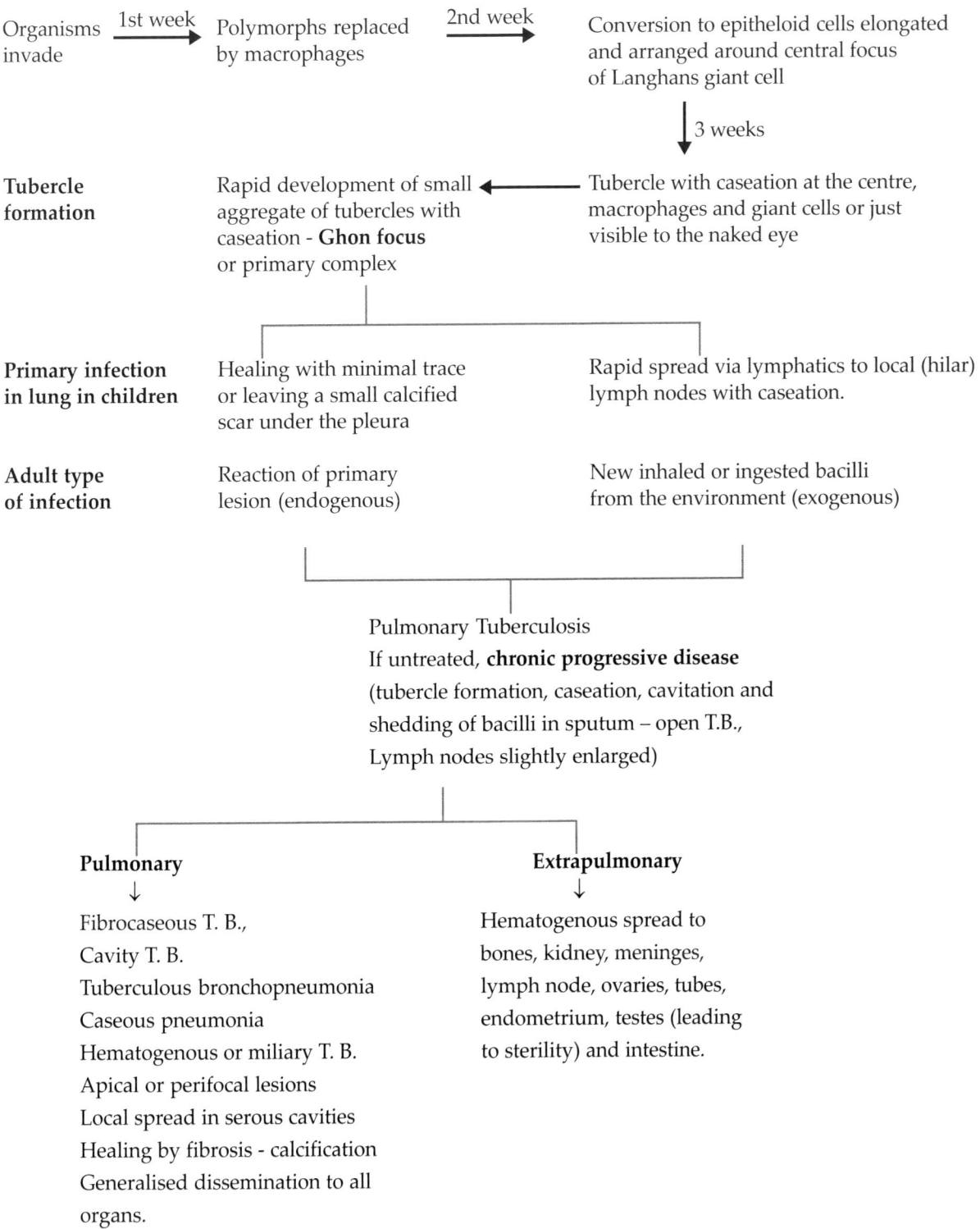

Pathogenesis of Leprosy

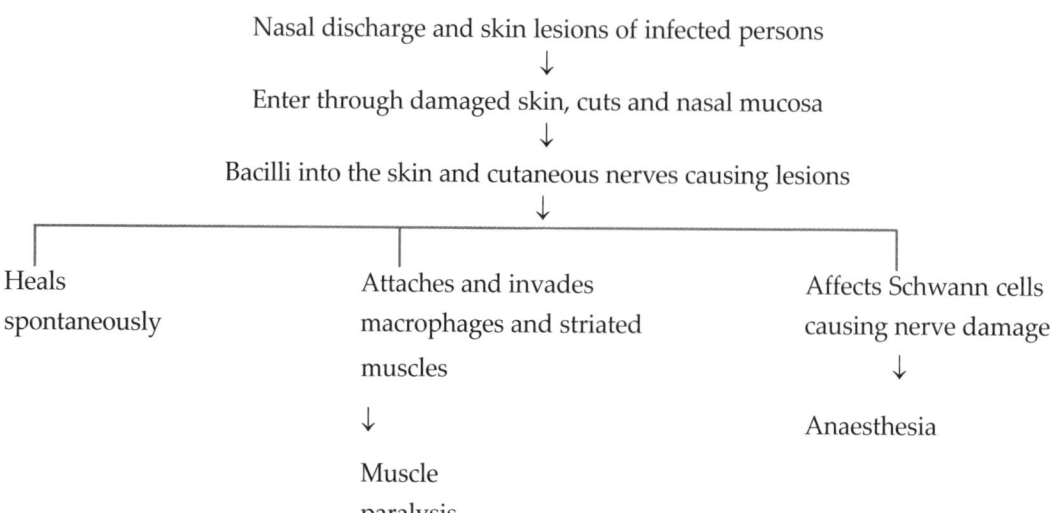

Nasal discharge and skin lesions of infected persons
↓
Enter through damaged skin, cuts and nasal mucosa
↓
Bacilli into the skin and cutaneous nerves causing lesions
↓

| Heals spontaneously | Attaches and invades macrophages and striated muscles ↓ Muscle paralysis | Affects Schwann cells causing nerve damage ↓ Anaesthesia |

Pathogenesis of Acquired Syphilis

Stage of disease	Signs and symptoms	Pathogenesis
Initial contact (2-10 weeks) ↓	Primary chancre	Multiplication of treponemes at site of infection; associated host response
Primary syphilis (1-3 months) ↓	Enlarged inguinal nodes, spontaneous healing	Proliferation of treponems in regional lymph nodes
Secondary syphilis (2-6 weeks) ↓	Myalgia, headache, fever, mucocutaneous rash. Spontaneous resolution	Multiplication and production of lesions in lymph nodes, liver, joints, muscles, skin and mucus membranes
Latent syphilis (3-30 years)	–	Treponemes dormant in liver and spleen
Tertiary syphilis	Neurosyphilis General paralysis of the insane Tabes dorsalis invasion and host response	Reawakening and multiplication of treponemes Further dissemination and (cell mediated hypersensitivity)
	Cardiovascular syphilis, Aortic lesions, Heart failure	
	Progressive destructive disease	Gummas in skin, bone, testes.

Pathogenesis of Diphtheria

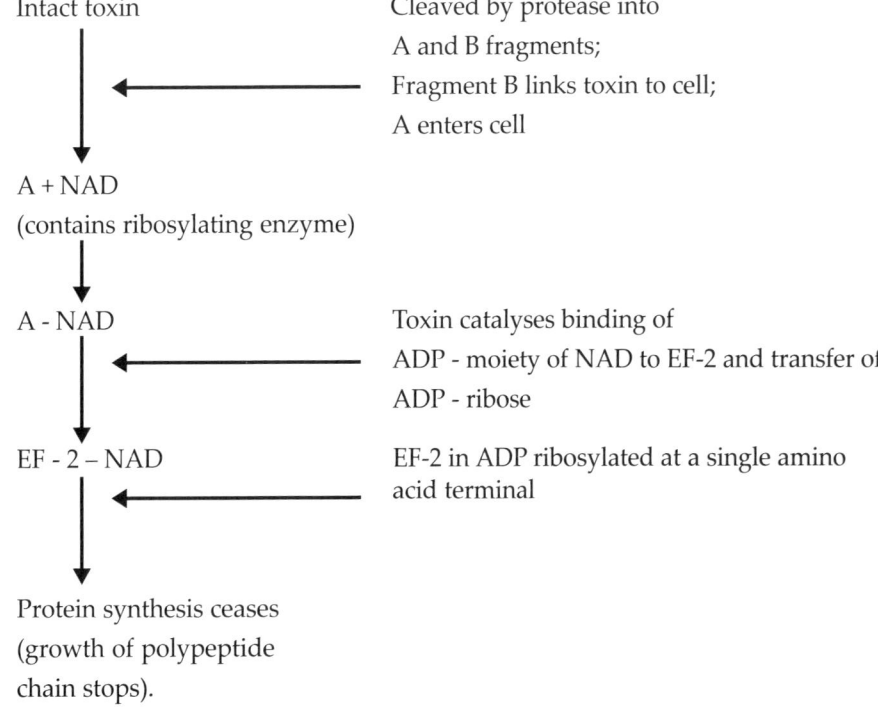

Mechanism of action
of diphtheria toxin

Intact toxin

Cleaved by protease into
A and B fragments;
Fragment B links toxin to cell;
A enters cell

A + NAD
(contains ribosylating enzyme)

A - NAD

Toxin catalyses binding of
ADP - moiety of NAD to EF-2 and transfer of
ADP - ribose

EF - 2 – NAD

EF-2 in ADP ribosylated at a single amino
acid terminal

Protein synthesis ceases
(growth of polypeptide
chain stops).

REFERENCES

1. Topley & Wilson's Microbiology and Microbial Infections. Bacteriology, Vol. 2. 10th Ed. Eds. Borriello SP, Murray PR, Funkay G. John Wiley & Sons Ltd., West Sussex, UK. 2009.

2. Koneman EW, Allen SD, Janda WM, Schreckenberger P, Winn Jr. WC. In Color Atlas & Textbook of Diagnostic Microbiology. 6th Ed. Lippincott, Philadelphia. 2006.

3. Chakraborty PA. Textbook of Microbiology. 4th Ed. New Central Book Agency (P) Ltd., Kolkata. 2009.

4. Cunningham, MW 2000. Pathogenesis of group A streptococcal infection. Clinical Microbiology Rev 13:470-511.

5. www.wikipedia.org.

ANTIMICROBIAL DRUGS AND ANTIMICROBIAL SUSCEPTIBILITY TESTING

8
Chapter

ANTIMICROBIAL DRUGS

Mechanisms of action of antibacterial drugs include:

- Inhibition of cell wall synthesis
- Inhibition of protein synthesis
- Inhibition of nucleic acid synthesis
- Inhibition of metabolic pathways
- Interference with cell membrane integrity

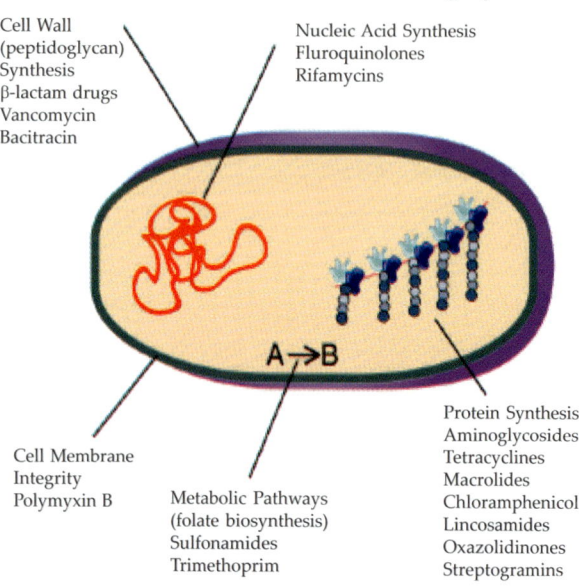

Cell Wall (peptidoglycan) Synthesis
β-lactam drugs
Vancomycin
Bacitracin

Nucleic Acid Synthesis
Fluroquinolones
Rifamycins

A→B

Cell Membrane Integrity
Polymyxin B

Metabolic Pathways (folate biosynthesis)
Sulfonamides
Trimethoprim

Protein Synthesis
Aminoglycosides
Tetracyclines
Macrolides
Chloramphenicol
Lincosamides
Oxazolidinones
Streptogramins

Inhibition of Cell wall synthesis

- Bacteria cell wall unique in construction - contains peptidoglycan
- Antimicrobials that interfere with the synthesis of cell wall do not interfere with eukaryotic cell
- Due to the lack of cell wall in animal cells and differences in cell wall in plant cells
- These drugs have very high therapeutic index, low toxicity with high effectiveness
- Antimicrobials of this class include:
  - β-lactam drugs
  - Vancomycin
  - Bacitracin

Penicillins and cephalosporins

- Part of group of drugs called β-lactams; Have shared chemical structure called β-lactam ring
- Competitively inhibits function of penicillin-binding proteins

- Inhibits peptide bridge formation between glycan molecules
- This causes the cell wall to develop weak points at the growth sites and become fragile.
- The weakness in the cell wall causes the cell to lyse.
- Penicillins and cephalosporins are considered bactericidal.
- Penicillins are more effective against Gram positive bacteria. This is because Gram positive bacteria have penicillin binding proteins on their walls.

The cephalosporins

- Chemical structures make them resistant to inactivation by certain â-lactamases
- Tend to have low affinity to penicillin-binding proteins of Gram positive bacteria, therefore, are most effective against Gram negative bacteria.
- Chemically modified to produce family of related compounds – First, second, third and fourth generation cephalosporins.

Vancomycin

- Inhibits formation of glycan chains
- Inhibits formation of peptidoglycans and cell wall construction
- Does not cross lipid membrane of Gram negative - Gram negative organisms innately resistant
- Important in treating infections caused by penicillin resistant Gram positive organisms
- Must be given intravenously due to poor absorption from intestinal tract
- Acquired resistance is most often due to alterations in side chain of NAM molecule and prevents binding of vancomycin to NAM component of glycan.

Bacitracin

- Interferes with transport of peptidoglycan precursors across cytoplasmic membrane
- Toxicity limits used for topical applications
- Common ingredient in non-prescription first-aid ointments.

Inhibition of protein synthesis

- Structure of prokaryotic ribosome acts as target for many antimicrobials of this class; differences in prokaryotic and eukaryotic ribosomes is responsible for selective toxicity.

- Drugs of this class include:
  - Aminoglycosides
  - Tetracyclines
  - Macrolides
  - Chloramphenicol

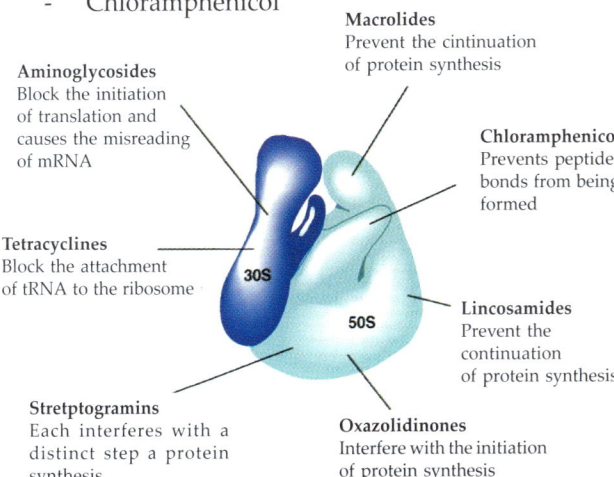

Aminoglycosides
Block the initiation of translation and causes the misreading of mRNA

Macrolides
Prevent the cintinuation of protein synthesis

Chloramphenicol
Prevents peptide bonds from being formed

Tetracyclines
Block the attachment of tRNA to the ribosome

Lincosamides
Prevent the continuation of protein synthesis

Stretptogramins
Each interferes with a distinct step a protein synthesis

Oxazolidinones
Interfere with the initiation of protein synthesis

30S
50S

Aminoglycosides

- Irreversibly binds to 30S ribosomal subunit
  - Causes distortion and malfunction of ribosome
  - Blocks initiation translation and causes misreading of mRNA
- Not effective against anaerobes, enterococci and streptococci
- Often used in synergistic combination with β-lactam drugs
- Allows aminoglycosides to enter cells that are often resistant
- Examples of aminoglycosides include gentamicin, streptomycin and tobramycin
- Side effects with extended use include ototoxicity and nephrotoxicity.

Tetracyclines

- Reversibly bind 30S ribosomal subunit and blocks attachment of tRNA to ribosome
- Prevents continuation of protein synthesis
- Effective against certain Gram positive and Gram negative
- Newer tetracyclines such as doxycycline have longer half-life, thus allowing for less frequent dosing
- Resistance due to decreased accumulation by bacterial cells
- Can cause discoloration of teeth if taken as young child.

Macrolides

- Reversibly binds to 50S ribosome and prevents continuation of protein synthesis
- Effective against variety of Gram positive organisms and those responsible for atypical pneumonia
- Often drug of choice for patients allergic to penicillin
- Macrolides include erythromycin, clarithromycin and azithromycin
- Resistance can occur via modification of RNA target
- Other mechanisms of resistance include production of enzyme that chemically modifies drug as well as alterations that result in decreased uptake of drug.

Chloramphenicol

- Binds to 50S ribosomal subunit and prevents peptide bonds from forming and blocking proteins synthesis
- Effective against a wide variety of organisms
- Generally used as drug of last resort for life-threatening infections
- Rare but lethal side effect is aplastic anemia.

Inhibition of nucleic acid synthesis

These include fluoroquinolones and rifamycins

Fluoroquinolones

- Inhibit action of topoisomerase DNA gyrase
- Topoisomerase maintains supercoiling of DNA
- Effective against Gram positive and Gram negative
- Examples include ciprofloxacin and ofloxacin
- Resistance due to alteration of DNA gyrase.

Rifamycins

- Block prokaryotic RNA polymerase and block initiation of transcription
- Rifampin is most widely used rifamycins
- Effective against many Gram positive and some Gram negative as well as members of genus *Mycobacterium*
- Primarily used to treat tuberculosis and Hansen's disease as well as preventing meningitis after exposure to *N. meningitidis*
- Resistance due to mutation coding RNA polymerase; Resistance develops rapidly.

Inhibition of metabolic pathways

- Relatively few
- Most useful are folate inhibitors
- Mode of action is to inhibit the production of folic acid
- Antimicrobials in this class include sulfonamides and trimethoprim

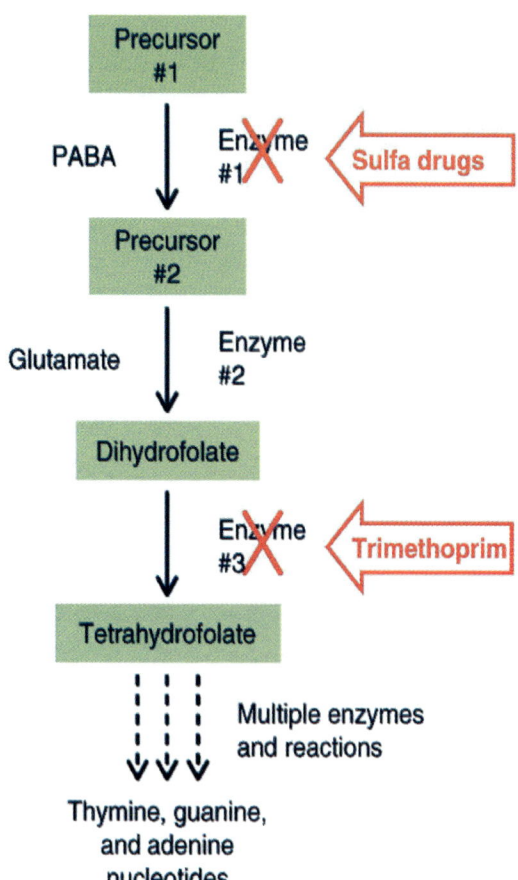

Sulfonamides

- Group of related compounds, collectively called sulfa drugs
- Inhibit growth of Gram positive and Gram negative organisms
- Through competitive inhibition of enzyme that aids in production of folic acid
- Structurally similar to para-aminobenzoic acid substrate in folic acid pathway
- Human cells lack specific enzyme in folic acid pathway - Basis for selective toxicity
- Resistance is due to plasmid, which codes for enzyme that has lower affinity to drug.

Trimethoprim

- Inhibits folic acid production and interferes with activity of enzyme following enzyme inhibited by sulfonamides
- Often used synergistically with sulfonamide
- Most common mechanism of resistance is plasmid encoded alternative enzyme
- Genes encoding resistant to sulfonamide and trimethoprim are often carried on the same plasmid.

Interference with cell membrane integrity

- Few damage cell membrane
- Polymyxin B is most common

- Common ingredient in first-aid skin ointments
- Binds membrane of Gram negative cells
- Alters permeability and leads to leakage of cell and cell death
- Also bind eukaryotic cells but to a lesser extent
- Limits use to topical application.

Mechanisms of drug resistance in different antibiotics

Antibiotic	Mechanism of drug resistance
β-lactams	1. β-lactamase
	2. Altered penicillin binding protein
	3. Altered gram negative outer membrane protein
	4. Active efflux
Macrolides	1. Alteration of target
	2. Active efflux
Chlaramphenicol	1. Chloramphenicol acetyltransferase
	2. Active efflux
Tetracycline	1. Active efflux
	2. Insensitivity of 30S ribosomal subunit
Aminoglycoside	1. Aminoglycoside modifying enzymes
	2. Decreased permeability through Gram negative outer membrane
	3. Active efflux
Sulfonamides & Trimethoprim	Production of insensitive targets (Dihydropterosis synthetase and Dihydrofolate reductase)
Quinolones	1. Insensitivity of the DNA gyrase gene
	2. Decreased intracellular drug accumulation (active efflux)
Polymyxins	Not defined

Mechanisms of resistance

- Drug inactivating enzymes
- Some organisms produce enzymes that chemically modify drug
- Penicillinase breaks β-lactam ring of penicillin antibiotics
- Alteration of target molecule
- Minor structural changes in antibiotic target can prevent binding
- Changes in ribosomal RNA prevent macrolides from binding to ribosomal subunits.

Decreased uptake of the drug → Alterations in porin proteins decrease permeability of cells → Prevents certain drugs from entering.

Increased elimination of the drug

- Some organisms produce efflux pumps
- Increases overall capacity of organism to eliminate drug

- Enables organism to resist higher concentrations of drug
- Tetracycline resistance.

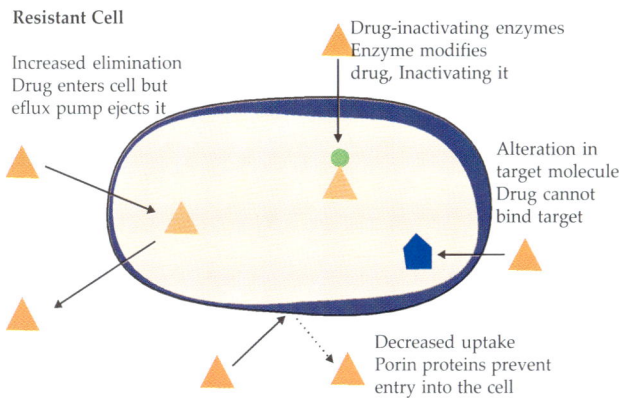

Non-Resistant Cell

Drug

Target

Drug binds target

Resistant Cell

Increased elimination
Drug enters cell but eflux pump ejects it

Drug-inactivating enzymes
Enzyme modifies drug, Inactivating it

Alteration in target molecule
Drug cannot bind target

Decreased uptake
Porin proteins prevent entry into the cell

ANTIMICROBIAL SUSCEPTIBILITY TESTING

- Probably the most widely used testing method is the disk-diffusion method, also known as the Kirby-Bauer test.
- Conventional disc diffusion method
  - Kirby-Bauer disc diffusion routinely used to qualitatively determine susceptibility
  - Standard concentration of strain uniformly spread of standard media
  - Discs impregnated with specific concentration of antibiotic placed on plate and incubated
  - Clear zone of inhibition around disc reflects susceptibility
  - Based on the size of zone, organism can be described as susceptible or resistant.

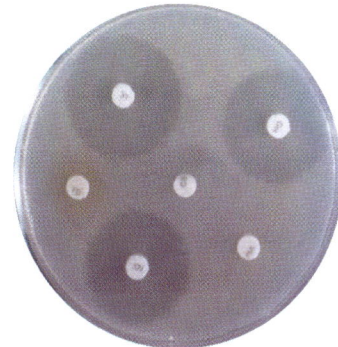

Mueller Hinton Agar (MHA) plate showing antibiotic susceptibility by Kirby Bauer Disc Diffusion method

The Disc Diffusion test

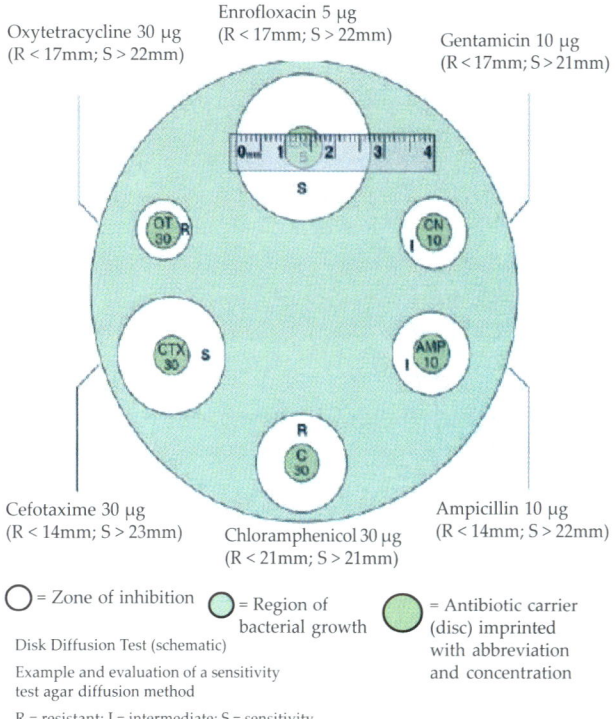

Oxytetracycline 30 µg
(R < 17mm; S > 22mm)

Enrofloxacin 5 µg
(R < 17mm; S > 22mm)

Gentamicin 10 µg
(R < 17mm; S > 21mm)

Cefotaxime 30 µg
(R < 14mm; S > 23mm)

Chloramphenicol 30 µg
(R < 21mm; S > 21mm)

Ampicillin 10 µg
(R < 14mm; S > 22mm)

⬭ = Zone of inhibition   ⬤ = Region of bacterial growth   ⬤ = Antibiotic carrier (disc) imprinted with abbreviation and concentration

Disk Diffusion Test (schematic)
Example and evaluation of a sensitivity test agar diffusion method
R = resistant; I = intermediate; S = sensitivity

Effects of combinations of drugs

- Sometimes the chemotherapeutic effects of two drugs given simultaneously is greater than the effect of either given alone.
- This is called synergism. For example, penicillin and streptomycin in the treatment of bacterial endocarditis. Damage to bacterial cell walls by penicillin makes it easier for streptomycin to enter.
- Other combinations of drugs can be antagonistic.
- For example, the simultaneous use of penicillin and tetracycline is often less effective than when wither drugs is used alone. By stopping the growth of the bacteria, the bacteriostatic drug tetracycline interferes with the action of penicillin, which requires bacterial growth.
- Combinations of antimicrobial drugs should be used only for:
  - To prevent or minimize the emergence of resistant strains.
  - To take advantage of the synergistic effect.
  - To lessen the toxicity of individual drugs.

Conventional Methods of Antimicrobial Susceptibility Testing

Main purpose of susceptibility tests :

- As a guide for treatment
  - Sensitivity of a given m.o. to known concentration of drugs
  - Its concentration in body fluids or tissues

- As an epidemiological tool
  - The emergence of resistant strains of major pathogens (e. g. Shigellae, *Salmonella typhi*)
  - Continued surveillance of the susceptibility pattern of the prevalent strains (e.g. Staphylococci, Gram-negative bacilli).

Antimicrobial susceptibility tests

- Solid media (diffusion)
  - Disk diffusion (Kirby-Bauer)
  - E-tests
  - MIC estimation (Agar dilution)
- Liquid media (broth dilution) allows MIC estimation
  - Minimum inhibitory concentration (MIC) – the smallest concentration of antibiotic that inhibits the growth of organism.
  - Minimum Bactericidal Concentration (MBC) or Minimum Lethal Concentration (MLC) – the lowest concentration of antimicrobial agent that allows less than 0.1% of the original inoculum to survive.
- Beta lactamase production: Clover leaf method, Nitrocefin method, etc.

Kirby Bauer Disc Diffusion Method (KBDDM)

Discovered by Kirby, Bauer, Sherris and Tuck in 1966.

- Used to determine the effectiveness of various antibiotics by comparing results to interpretive standard.
- The comparison allows the determination of whether a test organism is resistant or susceptible to a particular antibiotic.
- A Mueller Hinton Agar plate is uniformly and aseptically inoculated with the standardized bacterial isolate in a broth culture (matched with 0.5 McFarland standard) using a swab.
- Filter paper discs impregnated with specific concentrations of antibiotics are placed on the agar surface and incubated at 37°C overnight.
- The plates are then inspected for zones of inhibition around disks.
- After incubation, the diameter of the zone of growth inhibition is measured to the nearest millimetre and scored according to the size of the zone and the particular antibiotic, as sensitive, intermediate or resistant.
- If the zones overlap, measure the radius (from the center) and multiply by 2.
- The size of the zone of inhibition is directly proportional to the sensitivity of the organism to the antibiotic. If the organism is susceptible to a specific antibiotic, there will be no growth around the disc containing the antibiotic. Thus, a "zone of inhibition" can be observed and measured to determine the

susceptibility to an antibiotic for that particular organism. The measurement is compared to the criteria set by the Clinical Laboratory Standards Institute (CLSI). Based on the criteria, the organism can be classified as being Resistant (R), Intermediate (I) or Susceptible (S).

- **Standard strains include:**
  - *Staphylococcus aureus* ATCC 25923
  - *Escherichia coli* ATCC 25922
  - *Pseudomonas aeruginosa* ATCC 27853.

Turbidity standard for inoculum preparation

- To standardize the inoculum density for a susceptibility test, a $BaSO_4$ turbidity standard, equivalent to a 0.5 McFarland standard or its optical equivalent (e.g., latex particle suspension), should be used. A $BaSO_4$ 0.5 McFarland standard may be prepared as follows: A 0.5 ml aliquot of 0.048 mol/L $BaCl_2$ (1.175% w/v $BaCl.2H_2O$) is added to 99.5 ml of 0.18 mol/L H_2SO_4 (1% v/v) with constant stirring to maintain a suspension.
- The correct density of the turbidity standard should be verified by using a spectrophotometer with a 1 cm light path and matched accurate to determine the absorbance. The absorbance at 625 nm should be 0.008 to 0.10 for the **0.5 McFarland standard.**
- The Barium sulphate suspension should be transferred in 4 to 6 ml aliquots into screw-cap tubes of the same size as those used in growing or diluting the bacterial inoculum.
- These tubes should be tightly sealed and stored in the dark at room temperature.
- The barium sulphate turbidity standard should be vigorously agitated on a mechanical vortex mixer before each use and inspected for a uniformly turbid appearance. If large particles appear, the standard should be replaced. Latex particle suspensions should be mixed by inverting gently, not on a vortex mixer.
- The barium sulphate standards should be replaced or their densities verified monthly.

Factors Affecting Size of Zone of Inhibition

- **Inoculum density** - Larger zones with light inoculum and vice versa
- **Timing of disc application** - If after application of disc, the plate is kept for longer time at room temperature, small zones may form
- **Temperature of incubation** - Larger zones are seen with temperatures < 35°C
- **Incubation time** - Ideal 16-18 hours; less time does not give reliable results
- **Size of the plate** – Smaller plates accommodate less number of discs

- **Depth of the agar medium** (4 mm) – Thin media yield excessively large inhibition zones and vice versa
- **Proper spacing of the discs** (2.5 cm) – Avoids overlapping of zones
- **Potency of antibiotic discs** – Deterioration in contents leads to reduced size
- **Composition of medium** – Affects rate of growth, diffusion of antibiotics and activity of antibiotics
- **Acidic pH of medium** – Tetracycline, novobiocin, methicillin zones are larger
- **Alkaline pH of medium** – Aminoglycosides, erythromycin zones are larger
- **Reading of zones** – Subjective errors in determining the clear edge

Common interpretation problems

- Lack of standardization of the inoculums
- Thickness and quality of the culture media
- Quality and conservation of the disks
- Quality control with standardized strains
- Condition and duration of incubation.

Stoke's method of disc diffusion

- This method relies on the inhibition of bacterial growth measured under standard conditions. For this test also Mueller-Hinton agar is used.
- In this method, a control sensitive organism is inoculated on the same plate as the test organism. The zone of inhibition produced by test organism is then compared with that produced by control organism.
- The contol organism is inoculated in 2 bands on either side of the plate leaving a central band uninoculated.
- The test organism is seeded evenly in the band across the centre of the plate.
- Antibiotic discs are applied on the line between the test and the control organism. Four discs can be accommodated on a 9 cm petri dish. The petri dish is incubated at 37°C.
- If the test zones are obviously larger or give no zone of inhibition at all, then there is no need to measure the zone sizes. The zones are measured from the edge of the disc to the edge of the zones and interpreted as follows:

Interpretation:

Sensitive – Zone size ≥ or not more than 3 mm smaller than control.

Intermediate – Zone size > 3 mm, but smaller than the control by more than 3 mm

Resistant – Zone size 3 mm or less.

Quantitative Methods

In these tests, the minimum amount of antibiotic that inhibits the visible growth of an isolate or minimum inhibitory concentration (MIC) is determined. Bacterial isolate is subjected to various dilutions of antibiotics. The highest dilution of antibiotic that has inhibited the growth of bacteria is considered as MIC. These tests can be performed on broth or agar.

- Broth dilution methods
  - Macrobroth dilution MIC tests
  - Microbroth dilution MIC tests
- Agar dilution methods.

Macrobroth dilution tests

- A serial two-fold dilution of antibiotic are made in test tubes from 0 to maximum concentration that is achieved in vivo without toxic effect on patient.
- The inoculum density of bacterial isolate to be tested is standardized with 0.5 McFarland turbidity standard. The suspension should have a final inoculum of 5×10^5 cfu/ml.
- 1 ml of bacterial suspension is added to rows of antibiotic solution and incubated at 37°C overnight.
- The lowest concentration of antibiotic that completely inhibits visual growth of bacteria (no turbidity) is recorded as MIC.
- Antibiotic stock solution can be prepared using the formula:

$$\frac{1000}{P} \times V \times C = W,$$

where P = Potency given by the manufacturer in relation to the base; V = Volume in ml required; C = Final concentration of solution (multiples of 1000); W = Weight of the antimicrobial to be dissolved in the volume V.

- Example: For making 10 ml solution of the strength 10,000mg/l from powder base whose potency is 980 mg per gram, the quantities of the antimicrobials required is

$$W = \frac{1000}{980} \times 10 \times 10 = 102.04 \text{ mg}$$

- The stock solutions are made in higher concentrations to maintain their keeping qualities and stored in suitable aliquots at -20°C .Once taken out, they should not be refrozen or reused.

Microbroth dilution method

- A polystyrene tray containing 96 wells is filled with small volumes of serial two-fold dilutions of different antibiotics.
- The inoculum suspension and standardization is done according to McFarland standard.

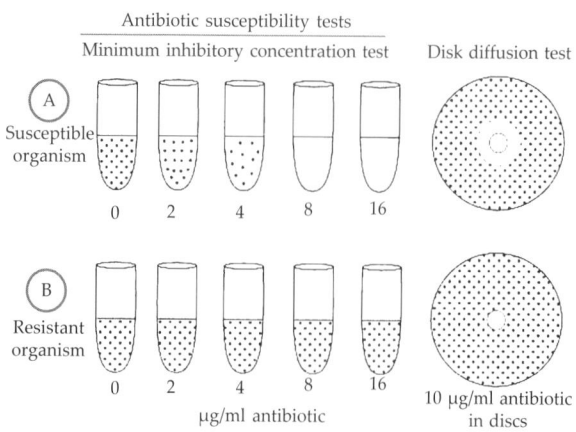

Antibiotic susceptibility tests

Minimum inhibitory concentration test — Disk diffusion test

A Susceptible organism
0 2 4 8 16

B Resistant organism
0 2 4 8 16

μg/ml antibiotic

10 μg/ml antibiotic in discs

- The bacterial inoculum is then inoculated into the wells and incubated at 37°C overnight.
- The lowest concentration of antibiotic that completely inhibits visual growth of bacteria (no turbidity) is recorded as MIC.
- Can be simultaneously performed with many tests organisms/specimens, less reagent required.

Agar dilution method

- A serial two-fold dilution of the antibiotic is prepared in melted Mueller-Hinton agar and poured in plates.
- The bacterial inoculum is standardized according to McFarland standard.
- Using calibrated loops a volume of 0.001-0.002 ml is inoculated on the surface of agar and incubated at 37°C overnight.
- The lowest concentration of antibiotic that inhibits visible growth on surface of agar is taken as MIC.
- One ml of desired drug dilutions to 19 ml of broth. The factor of agar dilution must be allowed for in the first calculation as follows:

Final volume of medium in plate   = 20 ml

Top antibiotic concentrations   = 64 mg/l

Total amount of drug = 1280μg to be added to 1 ml of water

2ml of 1280 μg/ml will be required to start the dilution = 2560μg in 2 ml = 1.28ml of 2000μg/ml ± 0.72 ml of water.

- 1 ml of this will be added to 19 ml agar.
- Stock dilution of 2000μg/ml is required for this range of MIC.
- One concentration of antibiotic/plate.
- Possible for several different strains/plate.

β-lactamase production

Rapid β-lactamase tests can yield clinically relevant information earlier than an MIC or disk diffusion test.

- Tests to detect beta-lactamases : Iodometric method, Acidometric method, Clover leaf method, chromogenic cephalosporin method.

  Iodometric methods are suitable for testing *N. gonorrhoeae*.

- Acidometric methods produce acceptable results with *Haemophilus* spp., *N. gonorrhoeae* and staphylococci.

- An agar plate is spread with standard strain of *Staphylococcus aureus* ATCC 25923 strain and an Ampicillin disc (30μg) is placed in the centre. Heavy inoculum of test strain is streaked radially from the disc and the plate is incubated at 37°C overnight. Deep indentation of growth of *S. aureus* in an otherwise circular zone of inhibition indicates β-lactamase production.

- Nitrocefin, a chromogenic cephalosporin, can be used to test *Neisseria gonorrhoeae*, *Moraxella Branhamella catarrhalis*, *Staphylococcus spp.*, *Haemophilus influenzae*, *Enterococcus spp.* and some anaerobic bacteria, and has been found effective in detecting all known β-lactamases. It is the only reliable test for detecting β-lactamase producing enterococci.

Newer Methods of Antimicrobial Susceptibility Testing

The first plasmid-mediated - lactamase in gram-negatives, TEM-1, was described in the early 1960s. The TEM-1 enzyme was originally found in a single strain of *E. coli* isolated from a blood culture from a patient named Temoniera in Greece, hence the designation TEM. Being plasmid and transposon mediated has facilitated the spread of TEM-1 to other species of bacteria. Within a few years after its first isolation, the TEM-1 - lactamase spread worldwide and is now found in many different species of members of the family *Enterobacteriaceae*, *Pseudomonas aeruginosa*, *Haemophilus influenzae*, and *Neisseria gonorrhoeae*. Another common plasmid - mediated - lactamase found in *Klebsiella pneumoniae* and *E. coli* is SHV-1 (for sulphydryl variable). The SHV-1 - lactamase is chromosomally encoded in the majority of isolates of *K. pneumoniae* but is usually plasmid mediated in *E. coli*.

Classification Schemes for Bacterial β-lactamases

Bush-Jaco-by Medei-ros group	1989 Bush group	Richmond-Sykes class	Mitsuhashi-Inoue type	Molecular class	Preferred substrates	Inhibited by CA EDTA		Representative enzymes
1	1	Ia, Ib, Id	CCase	C	Cephalosporins	–	–	AmpC enzymes from gram negative bacteria, MIR-1
2a	2a	Not included	PCase V	A	Penicillins	+	–	Penicillinases from gram positive bacteria
2b	2b	III	PCase I	A	Penicillins, cephalosporins	+	–	TEM-1, TEM-2, SHV-1
2be	2b	Not included except K1 in class IV	CXase	A	Penicillins, narrow-spectrum and extended - spectrum cephalosporins, monobactums	+	–	TEM-3 to TEM-26, SHV-2 to SHV-6, *Klebsiella oxytoca* K1
2br	Not included	Not included	Not included	A	Penicillins	+	–	TEM-30 to TEM-36, TRC-1
2c	2c	II, V	PCase IV	A	Penicillins, carbenicillin	+	–	PSE-1, PSE-3, PSE-4
2d	2d	V	PCase II, PCase III	D	Penicillins, cloxacillin	±	–	OXA-1 TO OXA-11, PSE-2 (OXA-10)
2e	2e	Ic	CXase	A	Cephalosporins	+	–	Inducible cephalosporinases from *Proteus vulgaris*
2f	Not included	Not included	Not included	A	Penicillins, cephalosporins, carbapenems	+	–	NMC-A from *Enterobacter cloacae*, Sme-1 from *Serratia marcescens*
3	3	Not included	Not included	B	Most β-lactams including carbapenems	–	+	L1 from *Xanthomonas maltophilia*, Ccra from *Bacteroides fragilis*
4	4	Not included	Not included	ND	Penicillins	–	?	Penicillinase from *Pseudomonas cepacia*

Csase - cephalosporinase; PCase - penicillinase; CXase - cefuroxime-hydrolysing β-lactamase; CA - clavulanic acid; ND - not determined.

Extended Spectrum β-lactamases (ESBLs)

ESBLs are defined as β-lactamases capable of hydrolyzing oximino-cephalosporins that are inhibited by clavulanic acid and are placed into functional group 2be. ESBLs are not active against cephamycins, and most strains expressing ESBLs are susceptible to cefoxitin and cefotetan. ESBLs are not active against cephamycins, and most strains expressing ESBLs are susceptible to cefoxitin and cefotetan.

ESBL detection methods

The detection methods can be divided into:

a) Phenotypic methods

b) Molecular methods

a) Phenotypic methods:

They are based upon the resistance that ESBLs confer to oxyimino-beta-lactams (*e.g.* ceftriaxone, cefotaxime, ceftazidime and aztreonam) and the ability of a β-lactamase inhibitor, usually clavulanate, to block this resistance. Several tests have been proposed.

Double disk diffusion test: The Jarlier double disk approximation or double disk synergy (DDS) was the first detection test described in 1980's. DDS is a disk diffusion test in which 30 μg antibiotic disks of ceftazidime, ceftriaxone, cefotaxime and aztreonam are placed on the plate, 30 mm (center to center) from the amoxicillin/clavulanate (20μg/10μg) disk. A clear extension of the edge of the antibiotic's inhibition zone toward the disk containing clavulanate is interpreted as synergy, indicating the presence of an ESBL. This test remains a reliable method for the detection of ESBLs. However, it has been suggested that the sensitivity of this test can be increased by reducing the distance between the disks to 20 mm. The use of cefpodoxime as the expanded spectrum cephalosporin of choice has been suggested as evaluation of DDS has shown sensitivities and specificities ranging from 79% to 97% and 94% to 100% respectively.

Cephalosporin/clavulanate combination: The British Society for Antimicrobial Chemotherapy has recommended the disk diffusion method for phenotypic confirmation of ESBL presence using ceftazidime/clavulanate and cefotaxime/clavulanate combination disks with semi-confluent growth on screening test may be introduced into the routine Iso-Sensitest agar. The zone diameter of each combination is compared with zone diameter of cephalosporin alone and a ratio of cephalosporin/clavulanate zone size to cephalosporin zone size is calculated. A ratio of 1.5 or greater indicates the presence of ESBL.

Disk replacement method: Three amoxicillin/clavulanate disks are applied to a Mueller-Hinton plate inoculated with the test organism. After one hour at room temperature, these antibiotic disks are removed and replaced on the same spot by disks containing ceftazidime, cefotaxime and aztreonam. Control disks of these three antibiotics are simultaneously placed at least 30 mm from these locations. A positive test is indicated by a zone increase of 5 mm for the disks which have replaced the amoxicillin/clavulanate disks compared to the control disks.

Three dimensional test: The three dimensional test was described by Thomson and Sanders. It gives phenotypic evidence of ESBL-induced inactivation of extended-spectrum cephalosporins or aztreonam without relying on the demonstration of inactivation of the β-lactamases by a β-lactamase inhibitor[128]. The test depends on the ability of a culture of the test organism to distort the zone of inhibition around an oxyimino-beta lactam disk. This test was determined to be sensitive but it is more technically challenging and labor intensive than other methods.

E test for ESBL

The E test ESBL strip is a two-sided strip in which clavulanate is added to one side of a dual oxyimino-β-lactam gradient looking for a reduction in the MIC of cephalosporins in the presence of clavulanate. The availability of cefotaxime as well as ceftazidime strips improves the ability to detect ESBL types which preferentially hydrolyze cefotaxime such as CTX-M-types enzymes. This method is useful for both screening and phenotypic confirmation of ESBL production. The reported sensitivity as a phenotypic confirmatory test for ESBL is 87-100% and specificity is 95-100%.

Broth microdilution assay

Phenotypic confirmatory testing can also be performed by broth microdilution assays using ceftazidime (0.25 to 128 µg/ml), ceftazidime plus clavulanic acid (0.25/4, 128/4), cefotaxime (0.25 µg to 64 µg/ml) and cefotaxime plus clavulanic acid (0.25/4 to 64/4). The use of both antibiotics is recommended. The test is done using standard methods. Phenotypic confirmation is considered as 3-twofold-serial-dilution decrease in MIC of either cephalosporin in the presence of clavulanic acid to its MIC when used alone.

ESBL detection tests – Phenotypic Methods

Test	Advantages	Disadvantages
Standard CLSI interpretative criteria	Easy to use, performed in every laboratory	ESBLs not always "resistant"
CLSI ESBL confirmatory test	Easy to use and interpret	Sensitivity depends on choice of oxyimino-cephalosporins
Double Disc Approximation Test	Easy to use, easy to interpret	Distance of disc placement for optimal sensitivity not standardized
Three Dimensional Test	Sensitive, easy to interpret	Not specific for ESBLs, Labour intensive
E Test ESBL strips	Easy to use	Not always easy to interpret, not as sensitive as double-disc test
Vitek ESBL Test	Easy to use, easy to interpret	Reduced sensitivity

AmpC β-lactamases

AmpC β-lactamases are clinically important cephalosporinases encoded on the chromosomes of many of the *Enterobacteriaceae* and a few other organisms, where they mediate resistance to cephalosporins (cephalothin, cefazolin, cefoxitin), most penicillins and β-lactam/β-lactamase inhibitor combinations. In 1981, the sequence of the AmpC gene from *E. coli* was reported. In India, the first AmpC β-lactamase was detected from New Delhi in the year 2003.

AmpC detection methods

Screening Tests:

AmpC screening using dilution method – MICs of cefoxitin are determined for all study isolates by agar dilution following CLSI methods.

Doubling dilutions of cefoxitin are prepared in Mueller-Hinton agar, with and without the addition of a fixed concentration of cloxacillin (100 µg/ml), to provide two fold concentrations ranging from 0.125 µg/ml to 256 µg/ml. Isolates with a ≥ 4 fold reduction in cefoxitin MIC in the presence of cloxacillin are considered to be positive for AmpC β-lactamase.

AmpC screening using disk diffusion

The 30 µg cefoxitin disk is supplemented with either 400 µg of phenylboronic acid, 200 µg of cloxacillin, or both compounds. The 10 µg cefpodoxime disks are supplemented with either 400 µg of phenylboronic acid or 200 µg of cloxacillin. The turbidity of the test organism is matched with 0.5-McFarland-standard suspensions prepared from overnight cultures and it is swabbed on Mueller-Hinton agar and incubated at 35°C for 16-18 hrs after the application of above mentioned disks. Zones of inhibition are measured to the edge of obvious inhibition, ignoring any microcolonies present within a clear zone of inhibition. An organism that demonstrates a defined increase in zone diameter around the antibiotic disk with added inhibitor compound compared to that with the antibiotic-containing disk alone is considered to be an AmpC producer.

Disk approximation technique

Lawn culture of test isolate (0.5 Mc Farland) is put over Muller-Hinton agar plate (MHA). 10 µg imipenem, 30 µg cefoxitin, and 20/10 µg amoxicillin-clavulanate disks as the inducing substrates and 30 µg ceftazidime disks as the reporter substrate are used. These disks are applied by use of an applicator at a distance of 20 mm from each other. Any obvious blunting or flattening of the zone of inhibition between the ceftazidime disk and the inducing substrates is interpreted as a positive result for AmpC β-lactamase.

Disk Antagonism Test

Lawn culture of test isolate (0.5 McFarland) is put over Muller-Hinton agar (MHA) plate. Ceftazidime (30 µg) and Cefoxitin (30 µg) disc are placed 20mm apart from centre to centre. Plates are incubated for 18-20 hours at 37°C. Isolates showing blunting of ceftazidime zone of inhibition adjacent to cefoxitin disc are considered screen positive and AmpC β-lactamase inducibility is recognized.

Modified Double Disk approximation Method (MDDM)

A 0.5 McFarland of test isolate is swabbed on Mueller Hinton Agar plate and disk of cefotaxime (30 µg) and ceftazidime (30 µg) are placed adjacent to clavulanic acid (10 µg) and Cefoxitin (30 µg) disk at a distance of 20 mm from each. The plate is incubated at 37°C for 18-24 hours.

Isolates showing blunting of ceftazidime or cefotaxime zone of inhibition adjacent to cefoxitin disk or showing reduced susceptibility to either of the above test drugs (ceftazidime or cefotaxime) and cefoxitin are considered as "screen positive" and selected for detection of AmpC β-lactamases.

Confirmatory tests:

AmpC Disk Test

A lawn culture of E. coli ATCC 25922 is prepared on MHA plate. Sterile disks (6 mm) are moistened with sterile saline (20 µl) and inoculated with several colonies of test organism. The inoculated disk is then placed beside a cefoxitin disk (almost touching) on the inoculated plate. The plates are incubated overnight at 35°C. A positive test appears as a flattening or indentation of the cefoxitin inhibition zone in the vicinity of the test disk. A negative test has an undistorted zone.

Modified three dimensional test

Fresh overnight growth from Mueller Hinton agar (MHA) is transferred to a preweighed sterile micro centrifuge tube. The bacterial mass is suspended in peptone water and pelleted by centrifugation at 3000 rpm for 15 minutes. Crude enzyme extract is prepared by repeated freeze thawing of the bacterial pellet (approximately 10 cycles). Lawn culture of E. coli ATCC 25922 is prepared on MHA plates and cefoxitin (30 µg) disk is placed on the plates. Linear slits (3 cm) are cut using sterile surgical blade, 3 mm away from cefoxitin disk. At the other end of the slit a small circular well is made and the enzyme extract was loaded. A total of 30 to 40 µl of extract is loaded in the well at a 10 µl increment. The plates are kept upright for 5 to 10 minutes until the liquid dried and are incubated at 37°C for 24 h. Enhanced growth of the surface organism at the point where the slit inserted the zone of inhibition of cefoxitin is considered a positive three-dimentional test and is interpreted as evidence for the presence of AmpC β-lactamases. Three different types of results are recorded: Isolates showing clear distortion of the zone of inhibition of cefoxitin are taken as AmpC producers; Isolates with no distortion are recorded as non-AmpC producers; Isolates showing minimal distortion are considered as indeterminate strains.

Characteristics of screening and confirmatory tests for detection of AmpC β-lactamase

Tests	Characteristics
Screening tests Decreased susceptibility to expanded-spectrum cephalosporins and resistance to cefoxitin	Highly sensitive but nonspecific May rule in strains with decreased outer membrane production or chromosomal AmpC hyper producers in E. coli.
Decreased susceptibility to expanded-spectrum cephalosporins & lack of inhibition by β-lactamase inhibitors	Highly sensitive but nonspecific. May represent production of class B or D β-lactamase, IRTs or hyper production of ESBLs.
Decreased susceptibility to expanded spectrum cephalosporins but retained susceptibility to cefepime or cefpirome	Poorly sensitive and specificMay miss co-production of ESBLs.

Confirmatory tests	
Three-dimensional test	First phenotypic confirmatory test to be described. Technically demanding
AmpC disk test	Easier to perform than three-dimensional test, but slightly more complicated than inhibitor-based tests.
3-Aminophenylboronic acid (APB) - based disk test/ Micro dilution test	Easier to perform as analogous to ESBL confirmatory test. Potential to be incorporated into automated susceptibility testing systems.
Benzo(b)thiophene-2-boronic acid (BZBTH2B) - based disk test	Improved sensitivity when combined with clavulanic acid.

Metallo β-lactamases

Metallo β-lactamases (MBLs) are metalloenzymes of Ambler class B and are Clavulanic acid resistant enzymes. They require divalent cations of zinc as co-factors for enzymatic activity and are universally inhibited by EDTA, as well as other chelating agents of divalent cations. Most prevalent in non-fermentative gram negative bacilli, but also reported from Enterobacteriaceae. First plasmid-mediated MBL from *Pseudomonas aeruginosa* was reported from Japan in 1991. In India, the first Metallo β-lactamase was detected from Bangalore in the year 2001.

Tests for detection of MBLs

Screening tests:

- Hodge test
- Double-Disk Synergy Tests (DDST)
- EDTA-Disk Diffusion Synergy Test

Confirmatory tests:

- Minimum Inhibitory Concentration (MIC) detection by microdilution method
- MBL E-test

All of these tests are based upon the ability of chelating agents, EDTA and thiol-based compounds, to inhibit the MBL activity.

Hodge test

The test of Hodge et al is modified by substituting *Escherichia coli* ATCC 25922 for penicillin-susceptible *Staphylococcus aureus* ATCC 25923, and 10 μg imipenem disk for a 10-U penicillin disk. The surface of a Mueller–Hinton agar plate is inoculated evenly using a cotton swab with an overnight culture suspension of *E. coli*, which is adjusted to one-tenth turbidity of the McFarland no. 0.5 tube. After brief drying, an imipenem disk is placed at the center of the plate, and imipenem-resistant test strains from the overnight culture plates are streaked heavily from the edge of the disk to the periphery of the plate. The presence of a distorted inhibition zone after overnight incubation is interpreted as modified Hodge test positive.

In the original test of Hodge et al, a penicillin G disk and penicillin G-susceptible *Staphylococcus aureus* strain were used to differentiate penicillinase-producing gonococci. The test can be improved by using an IPM disk to which 10 ml of 50 Mm zinc sulfate (140 mg/disk) has been added or by using Mueller-Hinton agar to which zinc sulfate has been added to a final concentration of 70 mg/ml.

Double-disk synergy test (DDST)

Test organisms are inoculated on to plates with Mueller Hinton agar as recommended by the CLSI. A 0.5 *M* EDTA solution is prepared by dissolving 186.1 g of disodium EDTA.2H$_2$O in 1000 ml of distilled water and adjusting it to *p*H 8.0 by using NaOH. The mixture is sterilized by autoclaving. Two 10 mg imipenem discs and two 30 mg ceftazidime discs are placed on the surface of an agar plate. EDTA solution is added to one of them to obtain a desired concentration of 750 mg. The inhibition zones of imipenem, ceftazidime and imipenem EDTA and ceftazidime EDTA discs are compared after 16-18 hrs of incubation in air at 35°C. This test can also be done using meropenem (MEM) discs. Sensitivity and specificity of the latte test is 100% and 97% respectively, whereas sensitivity and specificity of the former test is 96% and 91% respectively.

EDTA Disk Diffusion Synergy test

An overnight broth culture of the test strain, (opacity adjusted to 0.5 McFarland opacity standard) is used to inoculate a plate of Mueller-Hinton agar. After drying, a 10 μg imipenem disc and a blank filter paper disk (6 mm in diameter, Whatman filter paper no. 2) are placed 10mm apart from edge to edge, 10 μl of 0.5 *M* EDTA solution is then applied to the blank disc, which results in approximately 1.5 mg/disc. After overnight incubation, the presence of an enlarged zone of inhibition is interpreted as EDTA disc synergy test positive.

Microdilution method

A simple microdilution method for the determination of the MIC with a combination of imipenem and EDTA can be performed. However, for performing the agar dilution method to determine the same, EDTA (1 ml) solution is added to 1 ml of the imipenem solution spanning similar concentrations as done for MIC to imipenem alone. EDTA and imipenem 2 ml each in graded concentrations is added to 18 ml of molten Mueller Hinton agar and poured on plates that are allowed to set. A fixed inoculum of the test strains are spot inoculated on these plates before incubation. The reading is taken after 18-24 hrs of incubation. The highest dilution inhibiting the growth of the organisms is taken as the MIC. A MIC reduction of 4-512 fold with imipenem EDTA combination compared with Imipenem alone is taken as positive.

E-Test

The E-test MBL is reliable for detecting the IMP-1- and VIM-2-producing *Pseudomonas* and *Acinetobacter* isolates. Several colonies from a 24 hrs culture plate are used to prepare the inoculum with a 0.5 McFarland standard density. Mueller-Hinton agar plates are streaked by using cotton swabs or a spiral inoculator. The E-test MBL strips containing graded concentration of Imipenem at one end and similar concentrations of Imipenem on the other end along with a fixed concentration of EDTA, are then applied, and the plates are incubated at 35°C in an incubator for 16-20 hrs. A ratio of the MICs of the imipenem (IP) to IP plus EDTA (IPI) of ≥ 8 or the presence of a phantom zone, i.e., an extra inhibition zone between the IP and IPI regions, or a deformation of the IP or IPI ellipses is interpreted as being positive for MBL production. The sensitivity and specificity of this test is 96% and 91% respectively.

Determination of MICs

MIC is determined by the two-fold serial broth microdilution method using Mueller–Hinton broth. A culture of test strains grown at 37°C for 6 hrs in Mueller–Hinton broth is diluted to 10^7 CFU/ ml, and these dilutions are inoculated into the drug-containing broth with an inoculation apparatus at the final inoculum size of 10^5 CFU/mL. The MIC is defined as the lowest antibiotic concentration that completely prevents visible growth after incubation at 37°C for 20 hrs.

Advantages of the phenotypic tests

- The test are simple to perform and to interpret and can easily be introduced into the workflow of a clinical laboratory. All IPM-nonsusceptible can be routinely screened for MBL production using the EDTA disk screen test and PCR confirmation can be performed at a regional laboratory.

- The tests are cheap to perform.

- The discs can be prepared at one time and stored for long periods ameliorating the needs for tedious reagent preparation, whenever the test has to be performed.

- Regular screening with these methods could help detect the new previously undocumented forms of MBL producing genes.

Disadvantages of the phenotypic tests

- Difficult and subjective to interpret.

- Technically demanding and time-consuming,

- 2-mercaptopropionic acid and 1, 10-phenanthroline are toxic for routine handling

- MBL E-test, which is widely available but rather expensive and gives variable results.

Antimicrobial stewardship

- Antimicrobial stewardship is a key component of a multifaceted approach in preventing emergence of antimicrobial resistance. Good antimicrobial stewardship involves selecting an appropriate drug and optimizing its dose and duration to cure an infection while minimizing toxicity and conditions for selection of resistant bacterial strains (giving maximum benefit and minimum of adverse events). Studies conducted over the years indicate that antibiotic use is unnecessary or inappropriate in as many as 50% of cases and this creates unnecessary pressure for the selection of resistant species.

- Because the pharmaceutical industry pipeline for new antibiotics has been curtailed in recent years, and it may be ≥ 10 years before important new antibiotics to treat certain resistant bacteria find their way to market, a premium has been set on maintaining the effectiveness of currently available agents.

- Several strategies, including prescriber education, formulary restriction, prior approval, streamlining and antibiotic cycling have been proposed to improve antibiotic use. Although rigorous clinical data in support of these strategies are lacking, the most effective means of improving antimicrobial stewardship will most likely involve a comprehensive program that incorporates multiple strategies and collaboration among various specialties within a given healthcare institution. Computer-assisted software programs may be especially useful in implementing these comprehensive programs.

- The antimicrobial stewardship program has shown to improve appropriateness of antibiotic use and cure rates, decrease failure rates and reduce healthcare-related costs. Presently, available data suggest that good antibiotic stewardship reduces rates of *Clostridium difficile*-associated diarrhea, resistant Gram-negative bacilli and Vancomycin Resistant Enterococci.

Good Antimicrobial Stewardship Strategies

- Education/guidelines for antimicrobial use: active methods work better than passive

- Formulary restriction: restrict dispensing of some antimicrobials to approved indications

- Review and feedback: daily review of targeted antimicrobials for appropriateness
- Computer assisted strategies: use IT to implement previous strategies
- Antimicrobial cycling: scheduled rotation of antimicrobials used in a hospital or unit.

Ongoing effort by hospitals is a must in order to optimize antimicrobial use to improve patient outcomes, ensure cost-effective therapy and reduce adverse sequelae of antimicrobial use, including antimicrobial resistance. There is a need for tracking pathogens and susceptibilities. Multi drug resistant isolates and the ESBL menace are on the rise and MBL's need attention. Antimicrobial resistance should be recognized as a serious problem and efforts taken to halt its growth. Every hospital should decide on appropriate antimicrobial stewardship programs.

REFERENCES

1. Topley & Wilson's, Microbiology and Microbial Infections. Bacteriology, Vol. 1, 10th Ed. Borriello SP, Murray PR, Funkay G. John Wiley & Sons Ltd., West Sussex, UK. 2009.

2. Performance Standards for Antimicrobial Susceptibility Testing; Twenty-Fifth Informational Supplement. Clinical Laboratory Standards Institute. M100-S25; January 2015.

3. De A. Practical and Applied Microbiology. 5th Ed. The National Book Depot, Mumbai. 2014.

4. Manchanda V, Singh NP. Occurrence and detection of AmpC β-lactamases among Gram-negative clinical isolates using a modified three-dimensional test at Guru Tegh Bahadur Hospital, Delhi, India. J AntimicrobChemother 2003; 51(2): 415-8.

5. Singhal S, Mathur T, Khan S, Upadhyay DJ, Chugh S, Gaind R, et al. Evaluation of methods for AmpC β-lactamase in Gram-negative clinical isolates from tertiary care hospitals. Indian J Med Microbiol 2005; 23(2): 120-4.

6. Black JA, Moland ES, Thomson KS. AmpC disk test for detection of plasmid-mediated AmpC β-lactamases in Enterobacteriaceaelacking chromosomal AmpC β-lactamases.J ClinMicrobiol 2005; 43(7): 3110-3.

7. Lee K, Chong Y, Shin HB, Kim YA, Yong D, Yum JH. Modified Hodge and EDTA-disk synergy tests to screen Metallo-β-lactamase strains of Pseudomonas and Acinetobacterspecies. ClinMicrobiol Infect Dis 2001;7:88-91.

8. Lee K, Lim YS, Yong D, Yum JH, Chong Y. Evaluation of the Hodge Test and the imipenem-EDTA Double Disk Synergy Test for differentiating Metallo-β-lactamase producing isolates of *Pseudomonas spp.* and *Acinetobacter spp.* J ClinMicrobiol 2003;41:4623-9.

9. http://www.idsociety.org/Stewardship_Policy.

METHICILLIN RESISTANT STAPHYLOCOCCUS AUREUS (MRSA) AND VANCOMYCIN INTERMEDIATE RESISTANT S. AUREUS (VISA)

9
Chapter

METHICILLIN RESISTANT *STAPHYLO-COCCUS AUREUS* (MRSA)

In 1958-59, penicillinase resistant penicillins (PRPs) were introduced. First case of hospitalized person having infection with MRSA was reported in 1961. During 1961-1980, MRSA was reported from almost all parts of the world. 1980-1990 had global epidemic of MRSA.

Incidence	Global	Indian
1961-1969	5%-10.5%	11.3%
1970-1979	40-45%	12-15%
1980-1989	60-70%	28-70%
1990-2000	70-80%	70-80%

Methicillin resistance in affected by various physical and chemical agents :

a) Temperature

b) Visible light

c) Osmolality

d) pH

e) Chelating agent, etc.

Mechanisms of resistance:

1. **Homogenons resistance** – 100% both at 30°C and 37°C.

   a) **Intrinsic resistance** is chromosomally mediated which occurs by transposition and site - specific integration MIC > 16μgml - **High level resistance**.

      Only 1 in 10^5 cells express resistance phenotypically.

   b) **Extrinsic resistance** is plasmid mediated. Inactivate PRPs slowly due to β-lactamase production, Resistance to β-lactams only.

      Drug inactivation due to β-lactam inactivating β-lactamase.

2. **Heterogenous resistance** – thermosensitive.

   - At 37°C variable

   - 100% at room temperature

PBP 2a or 2b produced in increasing amount at 30°C and at high salt concentration.

Abnormal PBP 2a production is induced by β-lactams, cephalosporins, imipenem, etc, PBP2b is a penicillin-resistant peptidoglycan transpeptidase, encoded by mec A gene. Mec A gene produces new PBPs-PBP2 and PBP2b that has low affinity for β-lactam antibiotics.

3. **Tolerance** to the killing action of penicillin and cephalosporins.

   All these 3 types of resistance are present in both *S. aureus* and Cogulase negative *S. aureus* (CONS).

Compariason of three types of resistance

		Extrinsic	Intrinsic	Tolerance
1.	MIC	Very high	High	Normal
2.	MBC	Very high	High	High
3.	Limited to β-lactam antibiotics	Yes	Yes	No
4.	Phenotypic expression	99.9%	1 in 105	1 in 102
5.	Rate of growth	Rapid	Slow	Rapid
6.	Stability	Stable	Stable	Unstable
7.	Rate of occurrence	80-90%	1-8%	45%
8.	Clinical importance	Yes	Yes	Yes
9.	Phage types	Many	Few	Many
10.	Protein A	Common	Low	-

Typical MRSA Infections

Wound infection, Burns infection, Indwelling venous catheter infection, Prosthetic device infection, urinary tract infection in catheterized patients, Multiple injection therapies, Respiratory infection, Infection of skin and muscle, Chronic ambulatory peritoneal dialysis (CAPD).

Underlying conditions

Malignancy, Cardiovascular diseases, Chronic congestion of lungs, Diabetes, Chronic renal failure, Liver disease, Gastrointestinal disease, Central nervous system disorders, Chronic joint diseases, Alcoholism and psychosis.

Characteristics of MRSA

- MIC $\geq$ 20 μg/ml.

- Slow growing.

- Variable size, usually larger.

- Heterogenicity in resistance.

- Higher resistance to lysis by lysostaphin.
- Usually penicillinase positive.
- Cross resistance to all β-lactam antibiotics.
- Phage typing difficult.
- Causes serious diseases with increased mortality.

Detection of methicillin resistance

1. Disc diffusion method
2. Agar dilution method
3. Mueller Hintor agar with 2-4% NaCl is the ideal medium.

   Time 40 - 48 hours.

   Inhibition zone size < 16 mm.
4. Broth dilution method – Dilutions from 0.003 μg/ml to 64 μg/ml in Mueller Hinton broth.

Epidemiological typing of MRSA

1. Bacteriophage typing – 88A, 83A, 84, 85 in MRSA strains.
2. Protein pattern profile by SDS - PAGE - 33 KDa band identified in MRSA strain.
3. Pulse field gel electrophoresis (PFGE)
4. Plasmid profile
5. PCR detection of mec A gene by **Extended PCR**. DNA fragment is amplified.

Radiolabelled fragment is used as DNA probe. Amplified PCR product detected enzymatically within 3-4 hours as electrophoresis is avoided. It is a sensitive test detecting 5×10^2 CFU/tube. Mec A gene could also be detected by same procedure from clinical materials like throat swab, sputum, urine, pus, swabs, etc.

Borderline oxacillin resistant *S. aureus* **(BORSA)** have also been detected now.

Treatment of MRSA infections

- Vancomycin or fusidic acid as monotherapy increases resistance.
- So combination of ciprofloxacin and rifampicin or clindamycin and rifampicin or fusidic acid rifampicin or vancomycin, rifampicin and co-trimoxazole is given.
- Linezolid is given in nosocomial pneumonia and complicated skin infections.

MRSA carrier. An individual found to be culture positure for MRSA from one or more biological sites, at entry, e.g. anterior nares, hands, axilla, perianal region, etc.

Treatment of carriers

- 2% mupirocin nasal ointment given twice daily for 5 days for nasal carries. Screening is done after 24 hours, 7 days and 14 days after completing treatment.

- 0.5% chlorhexidine bath once daily for 7 days for hand carriers.

VANCOMYCIN INTERMEDIATE RESISTANT *STAPHYLOCOCCUS AUREUS* (VISA)

First case of *S. aureus* with reduced susceptibility to vancomycin against *S. aureus* was reported in Japan in May 1996. MIC 8-16 μg/ml for intermediate susceptible strains of *S. aureus*. The vancomycin intermediate susceptible strains of *S. aureus* are known as **VISA**. Resistance is mainly due to overuse and improper use of vancomycin in MRSA patients. It is also due to cross resistance from other glycopeptides e.g. teicoplanin.

Mechanisms of Vancomycin resistance:

1. Production of cytoplasmic ~39-kilo Dalton (kDa) protein which alters the peptidoglycan structure.

2. Production of ~37-kilo Dalton (kDa) protein which is homologous to NA+ dependent, D-specific-2-hydroxyacid dehydrogenases. This includes both D-lactate dehydrogenases and the enterococcal vancomycin resistance protein Van H, so it is designated as ddH gene. In resistant mutants, there is production of increased autolytic enzymes. Cross resistance from other glycopeptides like teicoplanin can occur.

Detection of VISA

1. **Broth dilution method.** Dilutions ranging from 0.015 μg/ml to 16 μg/ml. Growth in tube 4 μg/ml istaken as a breakpoint for VISA.

2. **Agar dilution method.** Various dilution plates of vancomycin are used and a control plate without any antibiotic. The antibiotic concentration of the first plate showing > 99% inhibition is taken as MIC of *S. aureus*.

3. **Macro dilution E-test.**

4. **Modified population analysis profile (PAP)** method for screening of VISA.

5. Detection of increase in PBPs by **SDS-PAGE** and then blotting on a nitrocellulose membrane.

6. PBP2a and PBP2b detected in VISA is measured by :
   - Western blotting
   - HPLC
   - Densitometry
   - Reverse-phase (RP) HPLC

7. Increase in cell wall synthesis is detected by :
   - Radioactive 14C labeled N-acetylglucosamine incorporated and then radioactivity counted using liquid scintillation counter.
   - Transmission electron microscopy (TEM).

8. For screening of VISA – 5 μg/ml vancomycin incorporated in MHA can also be used.

Treatment of VISA infections

-   Quinupristin/dalfopristin, ansamycin, linezolid or broad spectrum quinolones like gatifloxacin, gemifloxacin, moxifloxacin and fourth generation cephalosporins.
-   Trovafloxacin and ampicillin-sulbactam alone or in combination are used effectively in few cases.
-   Daptomycin.
-   Teicoplanin is effective in few cases.

Prevention

-   Handwashing, wearing of gloves, isolation of patient, strict adherence to discipline and personal will for efficient patient care.
-   Strict following of universal safety precautions, hospital hygiene and appropriate waste disposal techniques play a role in control of MRSA and VISA cases.

Control of outbreak of MRSA and VISA infections

-   Early detection of colonised person/asymptomatic carrier by screening various sites for MRSA and treating them with mupirocin ointment.
-   Elimination of source.
-   Facility of isolation.
-   Creating awareness among staff and patient's relatives.
-   Scrupulous hand washing using 0.5% chlorhexidine in isopropyl alcohol.

REFERENCES

1. Koneman EW, Allen SD, Janda WM, Schreckenberger P, Winn Jr. WC. In Color Atlas & Textbook of Diagnostic Microbiology. 6th Ed. Lippincott, Philadelphia. 2006.

2. Tenover F. C., Lancaster M. V., Hill B. C., Steward C. D., et al. Characteristics of staphylococci with decreased susceptibilities to vancomycin and other glycopeptides. J ClinMicrobiol 1998;36:1020-7.

3. Cui L, Iwamoto A, Lian JQ, Neoh HM, Maruyama T, Horikawa Y, Hiramatsu K. Novel mechanism of antibiotic resistance originating in vancomycin-intermediate *Staphylococcus aureus*. Antimicrob Agents Chemother. 2006;50:428-38.

4. Howden BP, Davies JK, Johnson PD, Stinear TP, Grayson ML. Reduced vancomycin susceptibility in *Staphylococcus aureus*, including vancomycin-intermediate and heterogeneous vancomycin-intermediate strains: resistance mechanisms, laboratory detection, and clinical implications. ClinMicrobiol Rev. 2010;23:99-139.

BACTERIOCINS AND BACTERIAL PLASMIDS

Chapter

BACTERIOCINS

Jacob in 1953 coined the word bacteriocins.

They are substanes of high molecular weight - generally proteins, that are formed by bacteria and have strain-specific lethal activity on other bacteria, nearly always including some members of the same species as the producer. They are named according to the species or genus of origin:

- Colicin (family Enterobacteriaceae)
- Pyocin (*Pseudomonas aeruginosa*)
- Vibriocin (*Vibrio cholerae*)
- Megacin (*Bacillus megaterium*)
- Staphylococcin (*Staphylococcus spp.*)

Difference from phage

A specific receptor on sensitive cell is required for adsorption for both of them (may share the same receptor).

a) When bacteriocines destory bacterial cells, they do not multiply.

b) Non-particulate (some are rod shaped bodies resembling phage tail)

Action of Gram negative bacteria : Narrow range of activity confined to the species of the producer strains and closely related species within individual species, genera or groups of genera with some exceptions.

Action of Gram positive bacteria : Wider range of activity.

- Proteins of low mol. wt., the biosynthesis of which leads to death of producer cell, that are active at the intraspecies level, and that requires a specific receptor for adsorption.
- Their action is bactericidal.
- Producer is immune to its own bacteriocin.
- Genetic determinant is a plasmid – Col factors.
- Tagg et al. (1976) postulated that the term 'Bacteriocine - like substance' should be used.

Colicines. Produced by and act upon strains within the family Enterobacteriaceae. Range wider than pyocines. Formed spontaneously in liquid or in solid media.

Increased production is induced by ultraviolet light, Mitomycin C.

Colicins are classified into a series of groups according to their specificity for colicine receptors, e.g. E1, E2 & E3 though sharing the same receptor, differ in their mode of action.

Nature. Spontaneously released colicines were lipopolysaccharide - protein complexes with antigenic specificity identical with that of the O antigen of the producer strain (Goebel, 1955).

Activity is destroyed by trypsin.

Induced cultures - active preparations of protein free from lipopolysaccharide.

Mol. wt. 50,000 - 90,000.

Col Plasmids – Two groups:

- Group I – Mol. wt. 5×10^{6}, specifying colicines E1, E2, E3 & K. Presence of another self -transmissible plasmid is necessary.
- Group II – Larger plasmids ($6\text{-}9 \times 10^{7}$), present in fewer copies, specifying colicines I, B and V, self - transmissible, may belong to F or I incompatibility groups.

Functions

Colicins kill, but do not lyse sensitive strains.

- A, E_1 and K – uncouplers of oxidative phosphorylation.
- E_2 – inhibits DNA synthesis and degrades DNA (is a DNA endonuclaease)
- E_3 – inhibits protein synthesis by attacking the 30s ribosomal subunit.
- Ia and Ib – inhibit all macromolecular synthesis.

- It is a 2-stage process in which reversible attachment to the cell envelope lead to irreversible changes which were transferred to and amplified in the cell membrane causing widespread biochemical disturbances (Nomura 1967).

- Adsorption is not an essential prelude to the lethal action, e.g. in L-forms (cell envelope absent), action is enhanced rather than diminished.

- Some strain of a bacteria form a small mol. wt. protein ('immunity protein') that specifically neutralizes the colicin; the gene for its production is present on the corresponding Col plasmid, e.g. E_3 immunity protein is coded for on the Col E_3 plasmid.

- Preprations from which immunity protein has been removed have a broad range of activity, e.g. purified colicine E_2 degrades cell-free DNA from bacteria, phages and Simian virus 40.

Pyocins. Usually on members of same species of *Pseudomonas aeruginosa*. A few attack *P. fluorescens* and rough mutants of Enterobacteria (smooth state is resistant).

- S type - non particulate proteins resembling colicines.
- R type - particulate, contractile sheathed rods resembling headless phages, some act on gonococci and serologically ungroupable meningococci.
- S type - particulate, longer, thinner, non-contractile unsheathed rods.

Vibriocins. On members of genus Vibrio. Structures resembling contractile phage tails. One found to inhibit DNA synthesis and to degrade DNA.

Serratin. There are 2 groups : One attack on Enterobacteria (resembling colicins). Another attack only strains belonging to the same genus.

Pesticin kill *Y. pseudotuberculosis* (may be useful in identification of plague bacillus).

Bacteriocins of gram postive bacteria

In *Bacillus, Clostridium, Corynebacterium, Lactobacillus, Listeria, Mycobacterium, Staphylococcus* and *Streptococcus*.

Difference from Colicins

- Not inducible by UV light or mitomycin C.
- Wider range of activity - on same species, also some members of other genera.
  - Agents from *Staphylococcus* act on *Streptococcus* and *Corynebacteria*.
  - Agents from *Bacillus spp.* on micrococci.
  - Agents from *Clostridia* on strains of *Bacillus*.
  - Agents *Streptococcus, Staphylococcus* and *Corynebacteria* inhibit various Gram negative bacilli.
- Host-cell immunity to a number of them is only partial.
- Production often dependent on conditions of growth (presence of particular nutrients and O_2).
- Some formed only on solid media.
- Have a protein component, Mol. wt. vary over a wide range, some considerably lower (10,000).
- Some resistant to trypsin.
- In some protein has not been separated from lipid and carbohydrate components.
- Few have not be demonstrated in cell-free preparations.
- Ability to produce is often irreversibly lost in culture, and in some, rate of loss is accelerated by treatment with acridines or growth at a higher temperature (so plasmid-determined).

- *Staphylococcus aureus* (502A) interferes with growth of other strains of the species on skin and mucous membranes does not form an inhibitory agent in vitro.
- Inhibitory activity of resistant streptococcal flora of throat might prevent colonisation by Group A *Streptococcus*.
- Streptococcal bacteriocines are inactivated by proteases in saliva.
- Strains of *S. mutans* coexist with bacteriocines (to which they are susceptible) in dental plaque, being protected from their action by extracellular polysaccharide.

Bacteriocin Typing systems

- Active - A strain is characterised by the range of acitivity of its bacteriocines against a set of indicator strains.
- Passive - It is characterised by the pattern of its susceptibility to the bacteriocines of a set of indicator strains.
- Combined active - Passive typing is often used.

BACTERIAL PLASMIDS

Plasmid is an extrachromosal genetic element, autonomously replicating, cyclic, double stranded DNA molecules distinct from the cellular chromosome, present in the cytoplasm of the bacteria.

Plasmids carry genes that are not essential for the host cell growth, while the chromosome carries all the necessary genes. It is transferred to the progeny cells during multiplication.

Two main classes :

Large plasmids – F, R and certain bacteriocinogens.

- R factors (resistance to various antibiotics)
- Col factors (Bacteriocinogens)

   60-120 Kb long (i.e. 10-20 μm). 6Kb = 1 μm

   (1Kb = Length of an average gene)

   (0.5 μm = 1 mega dalton (M dal) = 10^6 daltons = 3 kilobase pair (Kbp)

Mostly conjugative (i.e. self transferable by contact)

Small plasmids – 1.5 - 15 Kb (0.25-2.5μ)

- Some bacteriocinogens.
- Some resistant determinants.
- Non conjugative, but can be mobilised for transfer by a conjugative plasmid in the same cell.

Transfection – uptaking of DNA from solution. Original strand and freshly synthesised complementary strand come together and then free end passes into recipient.

Control specificity:

Compatible plasmid.

Incompatible plamid – cannot persist for many generations with the same cell.

Transmissible plasmid – Plasmids that contain information for self transfer to another cell by conjugation.

Non transmissible plasmid – When conjugation does not occur and transfer is mediated by bacteriophage, e.g. Staphylococcal plasmid for penicillinase production.

F factor. First plasmid to be isolated in *E. coli* was F factor, but it is not a typical one. Its unusual character is that it can insert itself in bacterial chromosome.

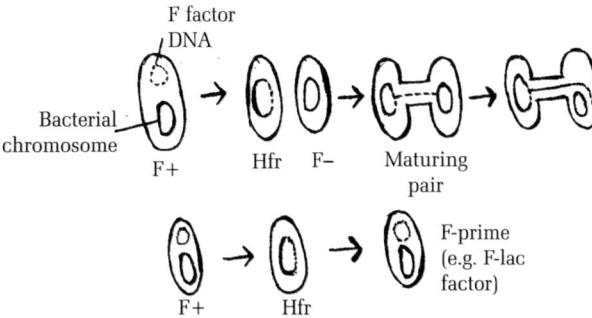

Curing of plasmids: By Acriflavine, Acridine orange, SDS, Ethidium bromide.

Phenotypic classification

R factor	Now not used because of various combination of genes in the same plasmid
Col-plasmid	
F - plasmid	

Observed Phenotypic effects of plasmids:

- Fertility (can produce sex pili for conjugation)
- Resistance to : Various antibiotics - widespread.
  Various heavy metals
  Ultraviolet irradiation (*E. coli*)
  Phages (*E. coli*)
  Serum bactericidal activity - widespread.
  Ethidium bromide (*S. aureus*)
- Production of :
  Bacteriocins (widespread)
  Proteases (*S. lacti*)
  Exotoxin (*C. botulinum*)
  Enterotoxin (*E. coli, S. aureus*)
  Exfoliative toxin
  Hemolysins (*E. coli*)
  Chloramphenicol (*Streptomyces*)

Adhesive factor (CAF I, CAF II, 987 P fimbriae)

Invasiveness (*S. sonnei*)

Skin reactive factor - increased capillary permeability.

Mitogen factor - stimulation and proliferation of normal unsensitised lymphocytes.

MIF - Platelet aggregation factor.

- Metabolism of various sugars, hydrocarbons.
- Tumorigenesis in plants - *Agrobacterium*.

Extraction of Plasmid

Bacteria grown
↓
Suspended in GET (glucose, EDTA, Tris)
↓
Centrifuged
↓
Deposit dissolved in 0.2 N NaOH + 1% SDS (Na-dioctyl sulphonate)
↓
DNA will come out
↓
Treated with 3N Sodium acetate
↓
All proteins precipitated and DNA (except plasmid, due to mol. wt.)
↓
Supernatant treated in absolute alcohol
(TRIS acetate can be added to get rid of rRNA)
↓
Bromophenol blue and ethidium bromide added
↓
Passed through agarose gel
↓
Examined through U-V light, when the dye has come to the opposite side
↓
Fluorescence band indicate presence of plasmid.

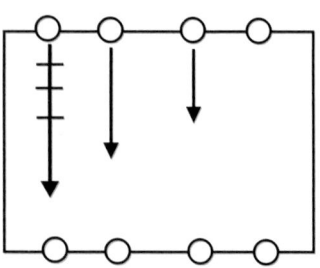

Extraction of Plasmid

REFERENCES

1. Topley & Wilson's, Microbiology and Microbial Infections. Bacteriology, Vol. 1. 10th Ed. Eds. Borriello SP, Murray PR, Funkay G. John Wiley & Sons Ltd., West Sussex, UK. 2009.

2. Summers D. The Biology of Plasmids. 1st Ed. Wiley-Blackwell. 1996; pp21-2. ISBN 978-0632034369.

3. Lederberg J. Cell genetics and hereditary symbiosis. PhysiolRev 1952; 32: 403-30. PMID 13003535.

4. Anderson ES, Threlfall EJ. The characterization of plasmids in the entertobacteria. J HygCamb 1974; 72: 471-87.

5. Shlaes DM, Currie-McCumber CA. Plasmid analysis in molecular epidemiology: a summary and future directions. Rev Infect Dis 1986; 8: 738-46.

6. Couturier M, Bex F, Bergquist PL, Maas WK. Identification and classification of bacterial plasmids. Microbiol Rev 1988;52: 375-95.

BACTEREMIA, SEPTICEMIA AND ENDOCARDITIS

11
Chapter

BACTEREMIA & SEPTICEMIA

Bacteremia - Presence of bacteria in blood as evidenced by positive blood culture.

Septicemia - Presence of mutiplying microbes or their toxins in blood.

Systemic Inflammatory Response Syndrome (SIRS)

Recent concept characterized by the presence of two or more of the following criteria:

– Fever or hypothermia ($> 38°C$ / $< 36°C$)

– Tachypnoea (> 20/min)

– Tachycardia (> 90/min)

– Leukocytosis (> 12000/mm3) or leukopenia (< 4000/mm^3) or > 10% band forms.

It may be infectious or noninfectious.

Sepsis - SIRS with proven or suspected microbial infection

Severe sepsis - SIRS with infection + organ dysfunction

Septic shock - Above with hypotension despite adequate fluid resuscitation

MODS - Multiple organ dysfunction syndrome.

SIRS Continuum

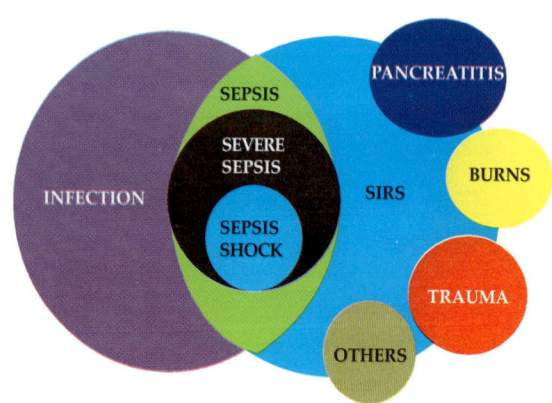

Mortality according to severity

Clinical stage	Mortality
SIRS	7%
Sepsis	16%
Severe sepsis	20%
Septic shock	46%

Classification of bacteremia

By site of acquision of infection:

- Nosocomial bacteremia – catheters, dental procedures, UTI, peritonitis, colorectal Ca, Surgery (Opx, GI, GU), antibiotics exposure, hemodialysis, parenteral fluids & nutrition.
- Community acquired bacteremia – pneumonia, meningitis, foreign bodies, septic arthritis, brain abscess, pericarditis, pressure ulcers, cellulitis.

By origin:

- Primary bacteremia – bacterial invasion of blood with no preceding infection
- Secondary bacteremia – bacterial invasion of blood following infection at other site.

By causative agent:

- Gram positive, Gram negative, Anaerobic and polymicrobial

By duration:

- Transient, Intermittent, Continuous.

Anaerobic Bacteremia

- Focus: Abdominal, skin and soft tissues, respiratory, gynecological, urological.
- Risk factors: > 65 years, malignancies and immuno-suppression (associated with greater frequency and mortality), diabetes and chronic liver disease.
- The most frequently isolated bacteria: Various species of *Bacteroides fragilis* group, *Clostridium spp.*, *Peptostreptococcus spp.*, *Prevotella spp.*
- Patients mostly at risk are the elderly with different comorbidities or oncological processes
- Mortality is high.

Bacteremia – Most common microorganisms

Brushing teeth	α-haemolytic streptococci from oral flora
Eating/chewing	
Dental work	
IV lines (colonised/infected)	*Staphylococcus aureus* from skin/nose, Gram negative bacilli
IV drug use	
Infected site/abscess	*Streptococcus pneumoniae, S. aureus*

Complications of bacteremia

- Sepsis
- Shock
- Endocarditis
- Osteomyelitis

ENDOCARDITIS

Endocarditid is infection of the endocardial surface of heart characterized by:

- Colonization or invasion of the heart valves (native or prosthetic) or the mural endocardium by a microbe
- Leading to formation of bulky, friable vegetation composed of thrombotic debris and organisms
- Often associated with destruction of underlying cardiac tissue.

Sites involved

- Heart valves
- Ventricular septum defects
- Mural endocardium
- Intracardiac devices

Predisposing factors

Cardiac and vascular abnormalities:

- Rheumatic heart disease (RHD)
- Myxomatous mitral valve
- Degenerative calcific valvular stenosis
- Bicuspid aortic valves
- Prosthetic valves

Host factors:

- Neutropenia
- Immunodeficiency
- Malignancy
- Therapeutic immunosuppression
- Diabetes mellitus
- Alcohol
- IV drug abuse

Modified Duke's Criteria for diagnosis of Infective Endocarditis

Definitive Endocarditis if,

- Two major or,
- One major and three minor or,
- Five minor

Possible Endocarditis if,

- One major and one minor or,
- Three minor

Major Criteria

- Positive blood culture
    - Typical organism from two cultures
    - Persistent positive blood cultures taken > 12 hours apart
    - Three or more cultures taken over more than 1 hour - positive.
- Endocardial involvement
    - Positive echocardiographic findings of vegetations
    - New valvular regurgitation

Minor Criteria

- Predisposing valvular or cardiac abnormality
- Intravenous drug misuse
- Pyrexia $\geq 38°C$ ($\geq 100.4°F$)
- Embolic phenomenon
- Vasculitic/immunologic phenomenon
- Blood cultures suggestive: organism grown but not achieving major criteria
- Suggestive echocardiographic findings.

Lesions in endocarditis

- Friable, bulky vegetation containing fibrin, inflammatory cells and microbes
- Aortic and mitral valves involved most commonly.
- Right side valve involvement in I. V. drug users.

Endothelial Injury

Uninfected Platelet-Fibrin thrombus

Transient bacteremia and attachment

Proliferation and pro-coagulant state

Infected, friable, bulky vegetation

Acute endocarditis	Subacute endocarditis
Destructive and tumultuous infection	Disease appear insidiously and pursue a protracted course of weeks to month
Frequently of a previously normal heart valve on deformed valves	Infection in a previously abnormal heart, particularly
A highly virulent organism - β-hemolytic streptococcus, *Staphylococcus aureus*	Organisms of low virulence - viridans streptococcus, enterococci, CONS, *Candida*, *Aspergillus*, HACEK, *Coxiella*
Hematogenous seeding in organs	Rare
If untreated, leads to death within weeks	Indolent, slow, rarely – embolus, ruptured aneurysms

Sub-acute Endocarditis
- Persistent fever
- Constitutional symptoms
- New signs of valve dysfunction
- Heart failure
- Embolic Stroke
- Peripheral arterial embolism

Etiology of endocarditis

Type	Etiology	Source
Native valve endocarditis	Viridans streptococci, staphylococci, HACEK group	Opx, skin
	S. gallolyticus	GIT (polyps, tumors)
	Enterococci	Genitourinary
	S. aureus, CONS	Catheters, community acquired
Prosthesis	S. aureus, CONS, GNB, diphtheroids, fungi	
IV drug Users	S. aureus, P. aeruginosa, Bacillus, lactobacillus, diphtheroids, polymicrobial, Candida	
Transvenous pacemakers	S. aureus, CONS	

HACEK Group
- Hemophilus aphrophilus
- Aggregatibacter actinomycetecomitans
- Cardiobacterium hominis
- Eikenella corrodens
- Kingella kingae

Infection may become evident 12 months after surgery

Other fastidious organisms: Bartonella, Abiotrophia, Granulicatella, Tropheryma whipplei.

Diagnosis of bacteremia, septicemia and endocarditis

Blood Culture

The isolation of bacteria in blood cultures provides valuable information to choose appropriate antimicrobial therapy. Low cost, no risk to the patient.

If patient is haemodynamically stable, withold antibiotics till antibiotic susceptibility results are available. Blood culture is critical for diagnosis.

When to do blood culture?
- Infective Endocarditis
- Sepsis
- FUO
- Unexplained leucocytosis or leucopenia
- Suspected meningitis, osteomyelitis, septic arthritis, pneumonia, other severe and deep seated suspected infections.
- Key diagnostic investigation
- Isolation of microorganisms from culture is important for diagnosis and also for treatment.
- At least 3 sets of samples should be taken from different venepuncture sites over 24 hours.
- Meticulous aseptic technique is required when taking blood cultures. Guidelines for best practice should be consulted.

- Chronic or subacute presentation – three sets of optimally filled blood cultures
- 1:10 ratio of blood: broth
- In adults, 20 ml blood is collected (10 ml per bottle)
- In children, 1-5 ml blood is collected
- Preferably not from IV lines
- Patient with CVC/arterial line – one each from peripheral site and invasive line
- Document date and time of specimen obtained and time of specimen collection
- Should be taken from peripheral sites with ≥ 6 hrs between them prior to commencing antimicrobial therapy.
- Taking blood cultures at different times is critical in identifying a constant bacteraemia, a hallmark of endocarditis.
- Blood culture bottle contains trypticase soy broth, Hartley's broth, soyabean casein digest broth, brain heart infusion broth, brucella broth, Columbia broth base. Anaerobic blood culture broth contains the same contents, along with addition of 0.5% cysteine.
- It is not always appropriate to withhold antimicrobial therapy while three sets of blood cultures are taken over a 12 h period.
- In cases of suspected infective endocarditis (IE) and severe sepsis or septic shock at the time of presentation, two sets at different times within 1 h prior to commencement of empirical therapy, to avoid undue delay in commencing empirical antimicrobial therapy.
- Positive Result
  - One set gives 90% sensitivity, remaining 2 sets add 8%
  - Multiple same cultures are important in confirming significance, especially for less typical organisms

- Negative Result
  - Prior antibiotic therapy
  - 'Culture negative endocarditis' – fastidous orgs/ non-culturable
  - May support a non-endocarditis patient diagnosis.
- Full identification of bacteria from positive blood cultures in suspected bacteremia/endocarditis is essential
- Full sensitivity testing is a must
- Liaison with microbiologist is required.
- Bacteremia is continuous in IE rather than intermittent, so positive results from only one set out of several blood cultures, should be regarded with caution.

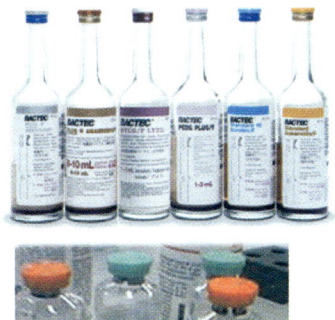

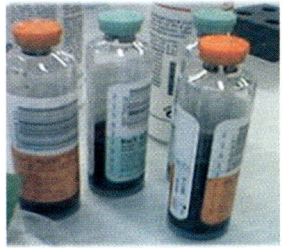

Newer blood culture systems

1) **Biphasic broth-slide system**
- Agar "paddles" attached to top of bottle
- Closed system

2) **Continuous monitoring blood culture systems**
- Bactec – measures $14CO_2$
- BacTAlert – measures change in pH
- VersaTrek ESP – measures consumption of gases manometrically.

Non culture diagnosis for microbes

- Serology: Can be done when the diagnosis is suspected and the cultures are negative. Aids in cases where the organisms will not grow in blood cultures, e.g. *Brucella, Coxiella, Legionella, Bartonella*, etc.
- In endocarditis – culture of valve tissue, special stains, DFA, PCR.

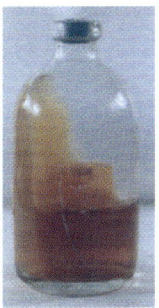

Castaneda's method of blood culture showing broth and agar slant

REFERENCES

1. Mims' Medical Microbiology. 5th Ed. Eds. Richard Goering, Hazel Dockrell, Mark Zuckerman, Ivan Roitt & Peter L. Chiodini. Saunders. 2013.

2. Topley & Wilson's, Microbiology and Microbial Infections. Bacteriology, Vol. 1. 10th Ed. Eds. Borriello SP, Murray PR, Funkay G. John Wiley & Sons Ltd., West Sussex, UK. 2009.

3. De A. Bacteremia and septicemia - Review article. The Indian Practitioner 1994; XLVII: 45-52.

4. De A. Practical and Applied Microbiology. 5th Ed. The National Book Depot, Mumbai. 2014.

5. Cunha BA. Bacterial Sepsis. Ed. Bronze MS. emedicine.medscape.com/article/234587-overview.

6. Everett ED, Hirschmann JV. Transient bacteremia and endocarditis prophylaxis. A review. Medicine 1977; 56: 61-77. PMID:834137.

7. Lee A, Mirrett S, Reller LB, Weinstein MP. Detection of bloodstream infections in adults: how many blood cultures are needed? J ClinMicrobiol. 2007;45:3546.

8. Brown DR, Kutler D, Rai B, Chan T, Cohen M. Bacterial concentration and blood volume required for a positive blood culture. J Perinatol. 1995;15:157.

9. Isaacman DJ, Karasic RB, Reynolds EA, Kost SI.

10. Effect of number of blood cultures and volume of blood on detection of bacteremia in children. J Pediatr. 1996;128:190.

11. Karen C. Carroll, Melvin P. Weinstein. Manual and Automated Systems for Detection and Identification of Microorganisms. In: Manual of Clinical Microbiology. 9th Ed.Eds. Patrick R. Murray. ASM Press, Washington, D.C. 2007; p.192.

IMMUNOLOGY

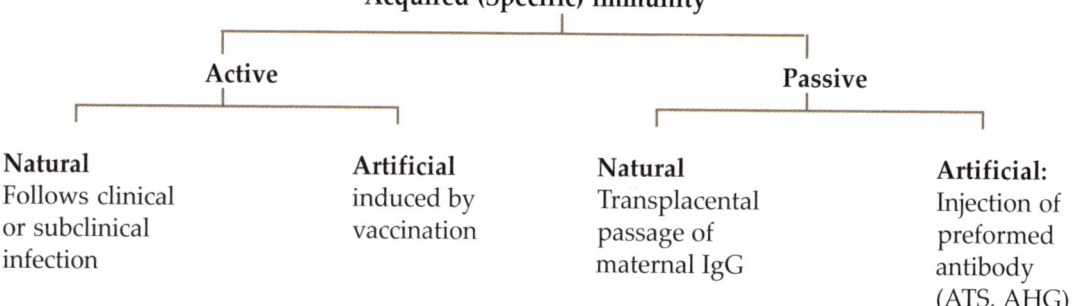

Chapter 12

IMMUNITY

Innate (Natural) immunity

1. Species : Birds immune to tetanus
2. Racial : Algerian sheep immune to anthrax
3. Individual : Seen in endemic area

Acquired (Specific) immunity

Active

Natural
Follows clinical
or subclinical
infection

Artificial
induced by
vaccination

Passive

Natural
Transplacental
passage of
maternal IgG

Artificial:
Injection of
preformed
antibody
(ATS, AHG)

Factors for Innate Immunity

1. Age
   - Two extremes of life
   - Neonates
   - Old persons

2. Hormonal disorders
   - Diabetes mellitus
   - Hypothyroidism
   - Adrenal dysfunction

3. Nutrition
   - Malnutrition predisposes to Gram negative bacterial sepsis, tuberculosis, herpes, measles and candidiasis.

Mechanisms of Innate Immunity

A) Epithelial Surfaces:

Skin;

Nose, nasopharynx and respiratory tract;

Mouth, stomach and intestinal tract;

Conjunctivae;

Genitourinary tract.

B) Tissue Factors :

a) Cellular Factors -
   - Microphages - Polymorphonuclear leucocytes
   - Macrophages - In blood (monocytes)
     - In tissues (histiocytes)

b) Humoral Factors
   - Properdin, Complement, Lysozyme
   - Other antibacterial substances (leukins, plakins, β-lysin)
   - Interferon (antiviral effect).

Active Immunity	Passive Immunity
Produced actively by the host's immune system	Received passively by the host
Induced by infection or by contact with immunogens	Introduction of readymade antibodies
Immune response-durable and effective	Short-lived and less effective
Immunity develop after a lag period	No lag period
Immunological memory present-booster doses effective	No immunological memory - 'immune elimination'
Negative phase may occur	No negative phase
Not applicable in immuno-deficient hosts	Applicable in immunodeficient hosts
Used for prophylaxis to increase body resistance	Used for treatment of acute infections

Vaccines

- Bacterial
    - Live - BCG for tuberculosis, anthrax, plague
    - Killed - TAB for enteric fever.

- Viral
    - Liver - Measles, mumps, rubella, influenza, Sabin Polio vaccine
    - Killed - Salk vaccine for polio.

- Bacterial products - Toxoids for tetanus and diphtheria

Local Immunity by IgA (mucosal or secretory)

- Inhibit attachment of bacteria to epithelial cells and neutralise viruses.

Herd Immunity - Collective resistance to the disease displayed by the community in its environmental setting.

Natural Defense Mechanisms of the Body

Respiratory tract

- Mucus
- Ciliated epithelium
- Cough reflex
- Antibody
- Phagocytosis

Digestive tract

- Stomach acidity
- Normal flora
- Intestine alkaline pH
- Mechanial flushing
- Enzymes e.g. Lysozyme
- Bacteriocines (Colon)

Eyes

- Washing by tears
- Lysozyme

Mouth

- Tongue
- Lysozyme in saliva

Skin

- Anatomic barrier; Sweat; sebum
- Antimicrobial secretions; Lactic acid; Free fatty acid
- Low pH
- Commensal microbes

Genitourinary tract

- Washing by urine
- Acidity of urine
- Lysozyme
- Vaginal lactic acid
- Semen

Phagocytosis

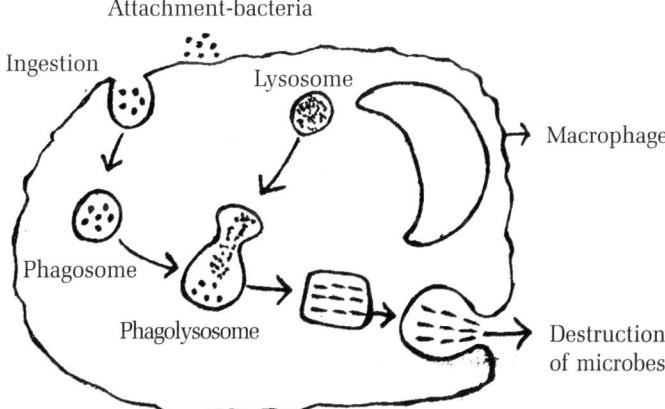

Various Non-specific Defense Mechanisms of our body against microorganisms

Biochemical	Chemical and physical
Lysozyme in tears, PMN, nasal secretions and saliva lyses mucopeptide of the cell wall of many Gram negative bacteria	Mucus of respiratory tract
Sebaceous gland secretions	Cilia lining trachea
Commensal organisms in gut and vagina	Acid in stomach
Spermine in semen	Skin
Properdin in serum causes lysis of Gram negative bacteria with the help of Mg++ and complement.	

Specific Defense Mechanisms

The mononuclear phagocyte system

Blood – monocytes

Brain – microglial cells

Lung – alveolar macrophages

Spleen – splenic macrophages

Liver – Kupffer cells

Kidney – mesangial phagocytes

Lymph node – resident & recirculating macrophages

Connective tissue – histiocytes

Bone marrow – precursors

Joint space – synovial A cells

Bone – osteoclasts

Serous cavities – serosal macrophages

Antigen (Antibody and generator)

Protein in nature, when introduced into living animal causes production of antibody which reacts specifically with those antibodies in an observable manner. It evokes immune response by producing specific antibody or specially sensitised T cells or both.

Types of antigens:

- Complete antigen or immunogen - high molecular wt. proteins (> 10,000) some are polysaccharides.

- Incomplete antigen or hapten - can react specifically with antibody, but unable to induce an immune response itself. Immunogenic when covalently linked to carrier proteins. Low molecular wt. (< 10,000), non protein substance.

    Hapten - (haptein = to grasp)

    Simple and Complex –

    1. Bacterial - Capsular polysaccharide of pneumococci, polysaccharide 'C' of hemolytic streptococci.

2. Drugs and chemicals - Agents causing allergic contact dermatitis and drug hypersensitivity.

3. Blood group (ABO) substances - glycoproteins.

4. Lipids - Forssman, cardiolipin and transplantation antigens.

Properties of Antigen :

1. Foreignness – Immune system distinguishes between self and nonself.

2. Size – Larger (macro) molecules like hemocyanin (M.W. 6.75 millions) are highly antigenic. Substances less than 10,000 daltons M.W. like insulin (5700) are non-antigenic or weakly antigenic. Low mol. wt. substances are rendered antigenic by absorbing them on large inert particles such as bentonite or kaolin.

3. Chemical nature – Proteins and polysaccharides. Lipids and nucleic acids are less antigenic, enhanced antigenicity on combing with proteins.

4. Susceptibility to tissue enzymes – Antigens introduced into the body are degraded by the host (by phagocytosis and intracellular enzymes) into fragments of appropriate size containing the antigenic determinants.

5. Antigenic specificity.

6. Species specificity.

7. Isospecificity – Alloantigens or isoantigens are present in some, but not all member of a species, which are able to produce alloantibodies or isoantibodies in individuals who are free from the antigens.

8. Organ specificity.

9. Autospecificity.

10. Heterogenetic specificity.

Different Antigens

Group	Antigens
Foreign	Microbial : cell walls, pili, enzymes, toxins, etc.
	Drug : Sedormid
	Environmental : Dust, pollens
Autoantigens	Self : Thyroglobulin, DNA, corneal tissue.
	Blood : Blood group antigens (ABO, RH)
	Tissue : HLA antigens
Heteroantigens	Heterophile antigens, cross-reacting microbial antigens (Streptococcal cell wall components with human cardiac and glomerular tissues)

Biological properties of major immunoglobulin classes in man

Immunoglobulin	IgG	IgA	IgM	IgD	IgE
Major characteristics	Protects body fluids	Protects body surfaces	Protects blood stream	Mainly lymphocyte receptor, role	Initiates inflammation, rises in parasitic infections, causes allergic symptoms. unknown
	LateAb		Early Ab		Reaginic Ab
Percent of total Ig	70-80	10-15	5-10	1	0.01
Intravascular distribution	45%	42%	80%	75%	0.01
Sedimentation coefficient	7S	7S & 11S	19S	7S	8S
Heavy chain	g	a	μ	d	e
Complement fixation - Classical	++	-	+++	+	-
- Alternative	-	++	-	-	rarely
Crosses placenta	++	-	-	-	-
Present in milk	+	+	-	-	-
Secretion by seromucous glands	-	++	-	-	-
Fixes to homologous mast cells and basophilss	-	-	-	-	++
Binds to macrophages & polymorphs	+++	+	-	-	+

IgA - Dimer in external secretions, carries secretory component. IgA dimer and IgM contain J chains. Serum IgA is a trimer. IgM is a pentamer. IgG, IgD and IgE are monomers. Aminoterminal of Ig forms the antigen-combining site. Carboxyterminal determines biological properties, e.g. complement fixation, placental transfer, skin fixation and catabolic rate.

Antimicrobial actions of antibodies

- Opsonizes for phagocytosis
- Complement activation, enhances phagocytosis & induces lysis
- Prevents attachment to host cells and prevents penetration of host cells
- Neutralizes toxins
- Inhibits motility of parasites and agglutinates parasites
- Inhibits microbial growth and metabolism.

Characteristics of Antigen-Antibody (Ag-Ab) reaction:

1. Reaction is highly specific, cross-reactions may occur between related antigens.
2. Entire molecules of Ag and Ab react during combination, where only surface antigens participate.
3. No denaturation of Ag and Ab during reaction.
4. Combination is firm, but reversible. Firmness depends on affinity and avidity.

   Affinity – intensity of attraction between Ag-Ab molecules.

   Avidity – strength of the bond after formation of Ag-Ab complexes.

5. Ags and Abs combine in varying proportions. Abs are generally bivalent (IgM have 5 or 10 combining sites). Ags have valencies upto hundreds.

Two parameters of serological tests are sensitivity and specificity which is in inverse proportion to each other.

Sensitivity - Ability of the test to detect very minute quantities of Ag or Ab.

Specificity - Ability of the test to detect reactions between homologous Ags and Abs only.

Antigens and Antibodies combine with each other specifically and in an observable manner.

Different types of Ag-Ab reactions :

A. Precipitation Reaction
B. Agglutination Reaction
C. Complement Fixation Test
D. Neutralization
E. Opsonisation
F. Labelled Antibody Assays
   1. Radioimmunoassay (RIA)
   2. Immunofluoroscence (IF)
   3. Enzyme-linked immunosorbent assay (ELISA)

A. Precipitation Reaction

When soluble Ag reacts with specific Ab in presence of electrolytes (NaCl) at optimal pH (7.4) and temperature (37°C), the Ag-Ab complex forms an insoluble precipitate (ppt.). When ppt. remains suspended instead of sedimenting, the reaction is called flocculation. Better reaction takes place with IgG Ab.

Mechanism of precipitation – Marrack's (1934) lattice hypothesis.

Applications – Very sensitive, can detect 1 µg protein Ag.

Qualitative test detects presence of Ag and helps in :

1. Identification of bacteria (grouping of streptococci)

2. Identification of bacterial components (Ascoli's thermoprecipitin test in anthrax)

3. Detection of unknown Ab (VDRL and Kahn test in syphilis)

4. Medicolegal identification of human blood or seminal fluid.

5. Standardization of toxins and antitoxins.

6. Testing for food adulteration.

7. Detecting serum proteins, human myeloma protein, etc.

8. Detecting hepatitis B antigens and antibodies, DNA antibodies in SLE, etc.

9. Pneumococcal, meningococcal and *H. influenzae* antigens by Counter immunoelectrophoresis (CIEP).

10. Measuring IgG, IgM & complement components by Radial immunodiffusion (RIA).

11. Identifying identity, partial identity and non-identity of antigens by Ouchterlony procedure.

B. Agglutination Reaction

An Ab combines with a particulate Ag in presence of electrolytes at optimal pH and temperature, resulting in visible clumping of the particles. Better reaction takes place with IgM Ab.

Principle – When Ag and Ab are mixed in optimal proportions – lattice formation.

Uses :

a) Identification of unknown culture

b) Blood grouping and cross-matching

c) Macroscopic slide agglutination test for leptospirosis

d) Tube agglutination test for serological diagnosis of enteric fever (Widal test), typhus fever (Weil-Felix reaction), brucellosis (Brucella agglutination test), infectious mononucleosis (Paul-Bunnel test) and leptospirosis (Microscopic agglutination test).

e) Streptococcus MG agglutination test and cold agglutination test for diagnosing primary atypical pneumonia.

Fallacies –

i) **Prozone Phenomenon** – Sera with high concentration of Ab fails to react with Ag in undiluted state and also inhibit agglutination. In great Ab excess, Ag cannot attach simultaneously to both sides of the Ab. Give positive agglutination only when it is diluted several times.

ii) **Blocking antibody** – Detected by performing test in hypertonic (5% saline) or in albumin saline. IgG (also IgA) acts as blocking or incomplete Ab (non-agglutinating), thus inhibiting agglutination reaction.

f) Coomb's test – Devised by Coomb's, Mourant and Race (1945).

i) **Direct** – Detects presence of incomplete Abs adsorbed onto RBC surface. Red cells are washed in saline 3 times by centrifugation, to remove serum proteins except globulin. Washed RBC's are mixed with Coomb's reagent (antihuman gammaglobulin) or rabbit AHG in presence of bovine albumin. Clumping of RBC's occur. Indicated in autoimmune hemolytic anemia, erythroblastosis foetalis.

ii) **Indirect** – Detects incomplete Ab present in serum. Patient's serum is mixed with saline. Washed Rh positive group O (or that of same group of patient) red cells are incubated at 37°C for 30 minutes. Red cells are washed in saline 3 times. Washed cells are then mixed with Coomb's serum – agglutination occurs in positive cases.

Used for detecting anti Rh Ab in serum of Rh negative pregnant women of Rh positive husband.

g) Passive agglutination test – Precipitation reaction is converted into agglutination reaction by carrier particles like latex particles, bentonite, fixed staphylococcal cells and RBC's by coating with soluble Ag. Serve as reagent for detection of specific Ab in serum by agglutination method. Polystyrene latex particles (0.8-1µ diameter) used for detection of HBs Ag, ASO, CRP, RA factor, HCG, bacterial typing and other organisms.

Latex Particle Agglutination Tests available commercially

Bacteria	Fungi
Campylobacter species	*C. albicans*
(coli, fetus, jejuni, laridis)	*C. immitis*
E. coli	*C. neoformans*
H. influenzae type b	*H. capsulatum*
M. pneumoniae	*S. schenckii*
N. meningitidis	
N. gonorrhoeae	**Parasites**
Proteus spp.	*T. gondii*
Rickettsia spp.	*T. spiralis*
Salmonella spp.	
Shigella spp.	**Viruses**
S. aureus	*Adenovirus*
S. pneumoniae	CMV
S. pyogenes	HSV
S. agalactiae	HIV
Other β-hemolytic streptococci	*Rotavirus*
Detects as little as 10 ng of Ab per ml.	

Reversed passive agglutination – When instead of Ag, Ab is adsorbed onto carrier particles for estimation of Ags.

h) Haemagglutination Test –

a) Rose-Waaler Test - detects auto-antibody (RA factor) which appears in serum of rheumatoid arthritis patients, which acts as an Ab to gammaglobulin (IgM to IgG) – agglutinates red cells coated with globulins.

b) TPHA – detects treponemal Ab in syphilis.

i) Coagglutination – Simple slide test described by Kron Vall in 1973, a modified agglutination reaction.

-   Based on presence of protein A on surface of some members of *S. aureus* (Cowan 1 strain).

-   Fc portion of IgG binds to protein A of *S. aureus* and Ag combining Fab terminal remains free.

-   Used for detection of bacterial Ags in blood, urine and CSF, e.g. Streptococcal grouping, typing of Gonococci, Mycobacterial grouping and detection of Meningococcal, Pneumococcal and *H. influenzae* Ags.

C. **Complement Fixation Test**

-   Ability of Ag-Ab complexes to 'fix' complement is made use of in this test.

-   Versatile and sensitive test, with various types of Ags and Abs and detect as little as 0.04 μg of Ab and 0.1 μg of Ag.

-   Two steps and five reagents - Ag, Ab, complement, sheep erythrocytes and amboceptor (rabbit Ab to sheep RBC), each reagent standardised separately.

Classical example is Wassermann reaction, formerly a routine serological test for syphilis.

Indirect CFT – Certain avian (duck, turkey, parrot) & mammalian (horse, cat) sera do not fix guineapig complement. Test is set in duplicate. After step 1, standard antiserum known to fix complement is added to one set. If test serum contained Ab, Ag is used up in step 1, so standard antiserum added subsequently would not be able to fix complement. So hemolysis means a positive result.

Other complement dependent serological tests:

1. Immobilisation Test - T. pallidum immobilisation test.

2. Immune Adherence - Ag-Ab complexes of some bacteria adhere to RBCs, platelets & macrophages in presence of complement e.g. *V. cholerae, T. pallidum*.

3. Cytolytic or Cytocidal Reaction - Live bacteria (*V. cholerae*) when mixed with its specific Ab in presence of complement, bacteriolysis occurs - 'Pfeiffer Phenomenon'.

D. **Neutralization** – When an antitoxin combines with a toxin, the toxin is neutralized. Toxin being an Ag in solution, is also precipitated.

Measured in vivo & in vitro.

1. Neutralization in vivo

a) Toxigenicity test - I/D inoculation of C.diphtheriae toxin in guineapig previously protected by anti diphtheritic serum (ADS) – unprotected animal dies.

b) Schick test - 0.2 ml diphtheria toxin (1/50 MLD) injected I/D in man. No reaction occurs at the site of injection if person contains circulating antitoxin.

2. Neutralization in vitro

a) Agar gel precipitation test – detects production of toxin by *C. diphtheriae* (Elek test).

b) Nagler reaction - *C. perfringens* toxin is neutralized by antitoxin when organism is grown in serum or egg yolk medium containing antitoxin.

c) Steptolysin O neutralization – Serum of patient infected with *S. pyogenes* contain antistreptolysin O, that neutralizes streptolysin O.

d) Virus neutralization – Done for typing viral isolates. Interferes with attachment of virions to cellular receptors which is a reversible process. In stable neutralisation, Ab molecules establish antigenic contact on two antigenic sites on a virion and make them susceptible to attack by nucleases and proteases on entering the cell or fail to transcribe their own mRNA. This test can be done in cell cultures, egg embryos and animals.

E. **Opsonization** – It is a process by which a particulate Ag (microbial cell) becomes more susceptible to phagocytosis by combination with oposonin (Ab-like substance or other component of serum).

- Opsonic index – ratio of phagocytic activity of patient's blood for a particular bacterium to that of a normal individual.

- Opsonophagocytic test – PMN leucocystes of patient is mixed with bacteria or yeasts in presence of serum. Rate of intracellular killing of bacteria or yeast is tested by subculture.

THE COMPLEMENT CASCADE

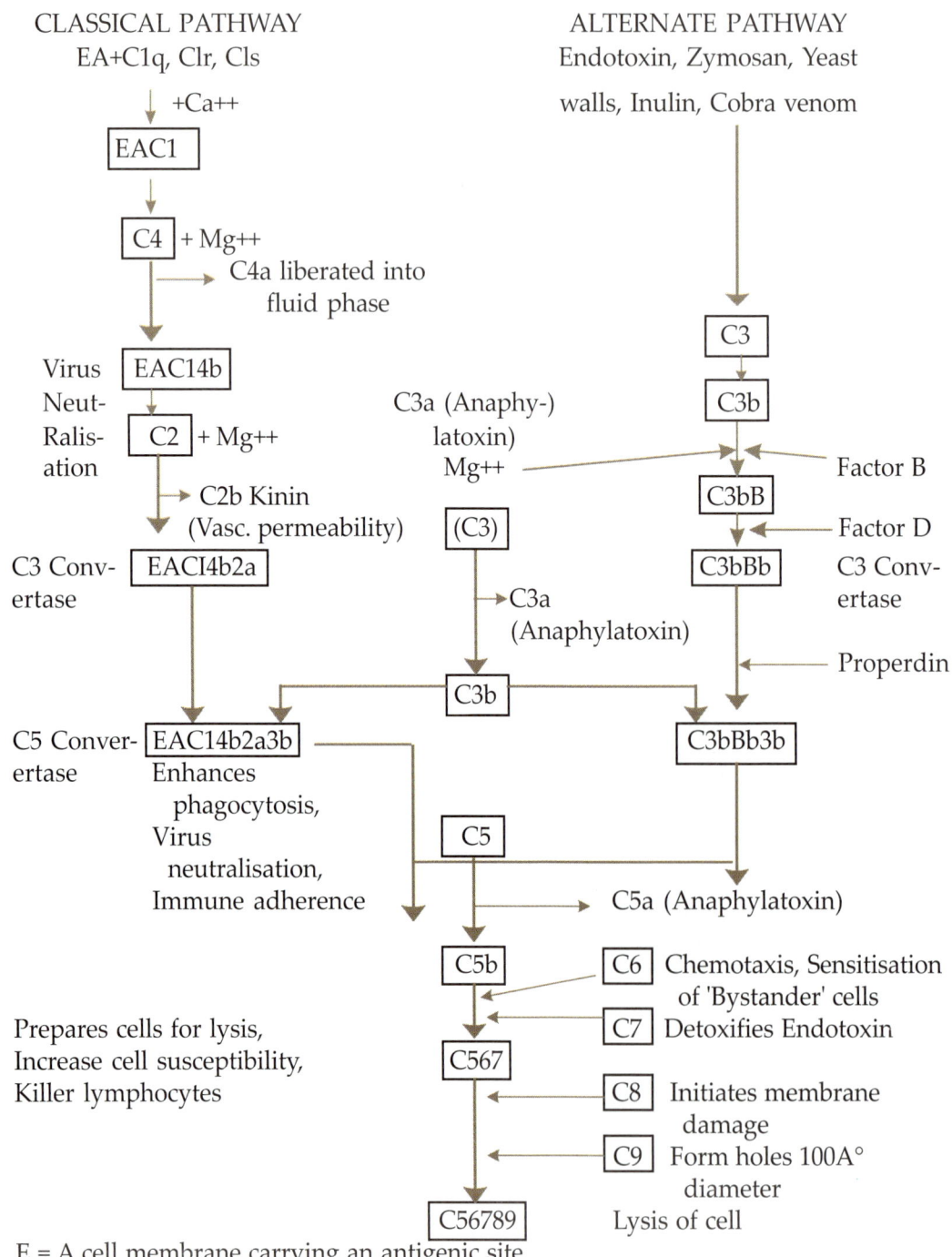

E = A cell membrane carrying an antigenic site
A = Antibody to that antigen

Biological Functions of Complement

1. Immune Adherence and Opsonization – Receptors for C3b facilitate adherence of C3b on Ab coated (Opsonized) microorganisms and promote phagocytosis.

2. Chemotaxis – C5a and C567 complex attract leucocytes.

3. Anaphylatoxin effect – C3a, C4a and C5a cause degranuution of mast cells with release of histamine and other mediators – cause increased vascular permeability and smooth muscle contraction.

4. Cytolysis – complex C56789 acts on Ag - Ab complexes on the membrane of many types of cells (erythrocytes, bacteria, tumour cells) resulting in killing or lysis of cells.

5. Hypersensitivity reactions – Complement participates in type II and type III hypersensitivity reactions.

6. Endotoxic shock – Endotoxins activate the Alternative pathway. Excessive C3 activation leads to tissue damage by DIC in endotoxic shock with gram-negative septicemia or Dengue haemorrhagic fever. Endotoxin gets coated with C3b and sticks to platelets by immune adherence. Complex of C567 causes lysis of platelets with release of clotting factors.

Complement Deficient States

Deficiency	Syndrome
C1 Inhibitor	Hereditary angioneurotic odema
Components of Classical pathway	SLE, other collagen diseases
C1, C2, C4, C3 and C3b	Recurrent pyogenic infections
Inactivator	Bacteremia mainly with gram-negative
C5 to C8	Diplococci (*N. meningitidis*), toxoplasmosis
Properdin	Severe meningococcal disease

Stem cell originate in yolk sac in 6th-8th week of gestation. Foetal liver bone marrow just before birth.

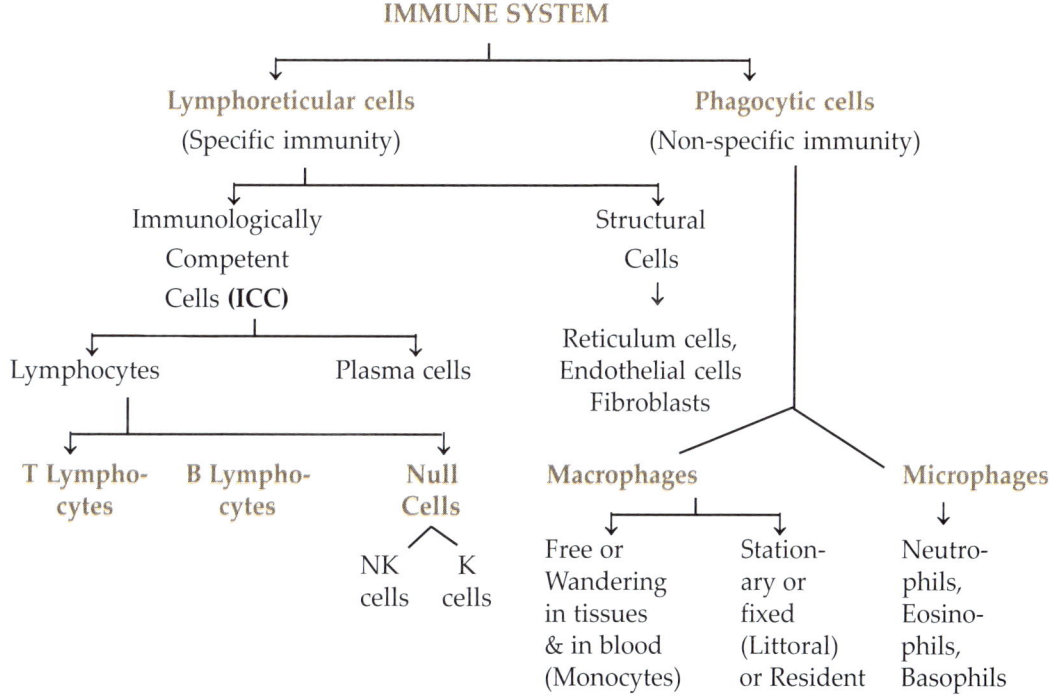

Central (Primary) Lymphoid Organs

- Thymus
- Bonemarrow

Peripheral (Secondary) Lymphoid organs

- Lymph node
- Spleen
- Mucosa associated
- Lymphoid tissues (Malt)

Nonspecific Effector Cells. Large Granular Lymphocytes (LGL) or Natural Killer (NK) Cells constitutes 3% of peripheral lymphocytes

1. **Natural Killer (NK) Cells** – Found in spleen and peripheral blood of men and animals. Action is nonspecific, do not need Ab for activity. Cytolytic for certain tumourlines, virally transformed target cells (Herpes, Mumps) and involved in allograft rejection. Interferon, IL-2 and agents that activate macrophages (BCG vaccine) enhance the activity of NK cells. Play an important role in controlling the

development of neoplastic cells (Antitumor immunity) and viral replication (Antiviral immunity) in the body.

2. **Null Cells (K Cells) :** A population of lymphocyte like cells that cooperate with Ab to destroy cellular targets, mediates Ab-dependent cell cytotoxicity (**ADCC**). They are neither T nor B cells, so referred to as null cells. ADCC cells do not require complement and are not MHC restricted, but kills target cells only in presence of Ab.

3. A population of cells that infiltrates tumours and has a greater tumour killing potency. These are more T cell-like.

4. Lymphokine Activated Killer (LAK) cell – Activated by IL-2 and is a potent tumoricidal Agent.

Functions of Macrophages

1. Active phagocytosis. Discovered by Metchnikoff in 1882.

2. Role in acquired immune response. They trap and process bacterial Ags and present them to lymphocytes to induce special immune response. T lymphocyte will accept the processed Ag only when it is presented by macrophage, carrying on its surface the self MHC determinant but not with different determinant (MHC Restriction).

3. They are also an important link between innate and acquired immune mechanism.

4. Activated macrophages secrete IL-1 which acts as endogenous pyrogen and induces T cell to synthesise IL-2.

Cellular Activation

Ag recognition + Binding of Ag to specific receptor of cell – Cellular activation, proliferation, differentiation and production of effector cells.

In B Cells – Effector cells are Ab producing plasma cells and memory cells.

In T Cells – Effector cells eliminate the Ag or synthesise molecules that help other cells to destroy the pathogen.

Ag Processing and Presentation

Ag must be associated with MHC class II molecules on the surface of Ag presenting cell **(APC)** → Processed Ag is then presented by APC to T helper cell (CD4+), so that latter recognise it and produce Ab response. Majority of APCs express MHC class II molecules. Role of each APC depends on type of immune response (I.R.) and on the location.

Ag delivers message directly to a B cell and after intracellular processing, to a T cell. Effective cooperation between B and T cells requires recognition of Ag on MHC II.

Antigen Presenting Cells (APCs)

Group	Type	Location Expression	MHC Class II
Phagocytic cells	Monocytes	Blood	+
	Macrophages	Tissues	+
	Follicular dendritic cells	Spleen & lymph nodes	+
	Microglia	Brain	++++
	Kupffer cells	Liver	
Lymphocytes	B Lymphocytes	Lymphoid tissue	+ to +++
	T Lymphocytes	Site of I.R.	0 to ++
Non-phagocytic	Langerhan's cells	Skin	++
- Constitutive presenters	Interdigitating cells	Lymphoid tissue	++
- Facultative presenters	Astrocytes	Brain	0
	Follicular cells	Thyroid	0
	Fibroblasts	Connective tissue	++
	Endothelium	Vascular & Lymphatic tissue	0

Signals for activation of B cells:

1. Binding of an antigenic determinant to B cell

2. An activated signal produced by T cell, IL-4.

3. Other lymphokines (IL-5 and IL-6) bring resting B cell (Go) to a state of full activation (G1).

T cell Activation :

1. Activation of CD4+ T cells require 2 signals:
   - Presentation of Ag by Ag-presenting cells in association with MHC II molecules.
   - IL-1 from an APC.

2. Activation of CD8+ T cells. Ag recognition restricted to MHC I molecules.

   Require 2 signals for activation:-
   - Ag fragment-MHC I.
   - IL-2 produced by a CD4+ helper T cell.

Major Histocompatibility Complex (MHC)

Gorer (1930) → H2 Ag in mice, located in chromosome 17 is the MHC Ag, related to allograft rejection and tumour immunity. Human counterpart of H2 Ag system → alloantigens present on surface of leucocyte called Human Leucocyte Ag (HLA) and cluster of genes encoding for them is HLA complex. HLA complex located on short arm of chromosome 6, has 8 genetic loci grouped into three classes.

- **Class I MHC Ags (A, B, C)**
  - Found on surface of all nucleated cells, abundantly on lymphoid cells and sparsely on cells of liver, lung and kidney.
  - They are essential for immune recognition by lymphocytes. The CD8 T cells are specific for MHC Class I Ags.
  - They are principal Ags responsible for graft rejection and destruction of viral infected cells.

- **Class II MHC Ags**
  - Principally found on surface of macrophages, monocytes, B lymphocytes and activated T lymphocytes (CD4).
  - Plays significant role in graft versus host (GVH) response and in mixed lymphocyte reaction (MLR). Immune response (IR) genes are identical to MHC Class II genes.

- **Class III MHC Ags**
  - These genes encode C2, C4 and Factor B of complement system.

Functions of MHC

- Class I and Class II genes provide a system for intercellular communication.

- MHC Ags are essential for recognition of Ag by T cells – T cells with CD4 molecules on their surface recognise Ag in association with MHC Class II molecules, while T cells with CD8 molecules are restricted by MHC Class I molecules.

- MHC Class III products are important complement system and are responsible for many cellular events.

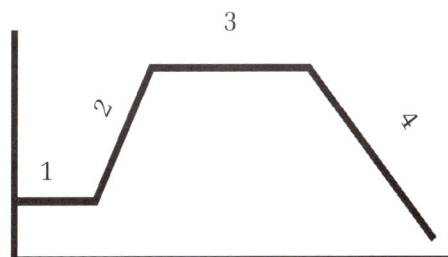

Antibody Production:

1. Lag phase

2. Log phase

3. Stationary phase

4. Phase of decline.

Humoral or Ab Mediated Immunity (AMI)

1. Primary defence against most extracellular bacterial pathogens.

2. Defence against viruses that infect through respiratory or intestinal tracts.

3. Prevents recurrence of virus infections.

4. Participates in pathogenesis of immediate hypersensitivity (Types I, II and III) and certain autoimmune diseases.

Humoral Immune Response – Production of Abs are in three steps:

1. Afferent limb - entry of Ag, its distribution and fate in tissues and its contact with appropriate I.C. cells.

2. Central functions - processing of Ag by cells and the control of Ab forming process.

3. Efferent limb - secretion of Ab, its distribution in tissues and body fluids and its effects.

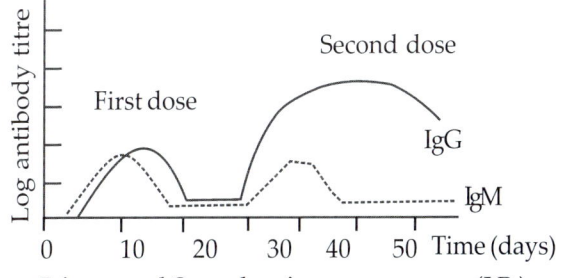

Primary and Secondary immune response (I.R.)

Primary I.R. – Slow, sluggish, shortlived, long lag phase (5-7 days), low titre of Abs (mainly IgM), do not persist long.

Secondary I.R. – Prompt, powerful and prolonged with a short or negligible lag phase (2-3 days) and much higher level of Abs (10 or more times greater mainly IgG) that lasts for long periods, negative phase present initially. Boosted to higher levels by further injection of Ag (booster dose).

Immunocompetent B lymphocytes

(Recognise Ags and initiate Ab synthesis) – undergoes clonal proliferation and blast transformation and converted into plasma cells (end cells that synthesise and secrete Abs). Ab formation by B cells is stimulated by helper T cells and inhibited by suppressor T cells.

Few B cells develop into 'Memory cells' which recognise the same Ag introduced subsequently. Long life span, disseminated throughout blood, lymph and tissues.

Characteristics of T and B cells

Characteristics	T cell	B cell
Peripheral blood lymphocytes	65-85%	15-25%
Thymus cells	96%	Negligible
Thymus specific antigens	+	-
Surface immunoglobulins	-	+
Receptor for Fc piece of IgG	-	+
EAC rosette (C3 receptor)	-	+
SRBC rosette	+	-
Numerous microvilli on surface	-	+
Blast transformation with		
- Phytohaemagglutinin	+	-
- Concanavalin A	+	-
- Endotoxins	-	+

Site	T cell area	B cell area
Lymph Node	Paracortical area	Cortical follicles.
Spleen	Lymphatic sheath lying adjacent to the central arterioles	Perifollicular region, mantle layer

Adjuvants

Substances which when injected together with antigen, enhances antibody production.

Types

a) Depot: Alum, Aluminium hydroxide, Al -phosphate, Freund's incomplete adjuvant (water in arachis oil).

b) Bacterial: Freund's complete adjuvant (water in arachis oil and dead mycobacteria).

c) Chemical: Silica particles, berrilium sulphate, bentonite, calcium alginate.

Actions

-   Sustained release of antigen from depot.

-   Lymphocyte stimulation activating factor

-   Antitumour effect by stimulating specific CMI.

To avoid autoagglutination of any culture

-   Repeated subculture on nutrient agar and kept at room temperature

-   Repeated passage in experimental animals

-   Incorporating acriflavine 0.5% in the culture media

-   Boiling broth culture for half an hour.

Significance of two serological tests for diagnosis of a particular disease

For a particular disease, ideally two serum samples should be collected – one in the acute phase of the disease and other after 10-15 days in the convalescent phase of the disease. Rise in titre of antibodies in the second serological test as compared to that of the first is diagnostic of that disease. Low titres of antibodies might be present even in normal individuals or those suffering from some other diseases e.g. anamnestic reaction or after vaccination in case of Widal test (*S. typhi* TO, TH, AH and BH antigens are present in low titres after typhoid vaccine is given).

Serum inactivation prior to a serological test

Serum is inactivated because it contains complement and many inhibitory substances which might interfere with the serological test by masking the antibody present in serum (of the disease we want to diagnose). So serum is kept at 56°C for 30 minutes in a water bath to inactivate the complement and the inhibitory substances.

Monoclonal Antibodies

-   Antibodies produced by a single antibody forming cell or clone directed against a single antigen or antigenic determinant is called monoclonal antibody, e.g. plasma cell tumor (myeloma).

-   Kohler and Milstein (1975) prepared a hybrid cell line (hybridoma) by fusion of mouse myeloma cell with an antibody producing lymphocyte from spleen (B cells) of the same inbred strain of mouse in HAT medium (hypoxanthine, aminopterin and thymidine). This can produce unlimited quantities of monoclonal Ab of any required specificity indefinitely in cell culture conditions. Hybridoma can be frozen for prolonged storage.

-   Kohler and Milstein received Nobel prize in 1984 for the same.

Characteristics of hybrid cells :

i) They no longer produce their own Ig but that of the normal B cell parent

ii) Lack the enzyme HGPRT (hypoxanthine- guanine phosphoribosyl transferase) necessary for nucleic acid synthesis which is provided by the B cell component of hybridoma.

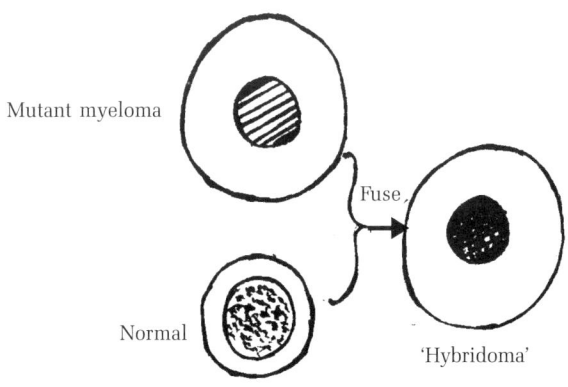

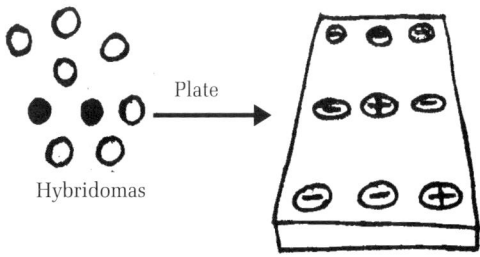

● Cells producing anti-X Abs
⊕ Wells in which anti-X Abs are produced

Production of Monoclonal antibodies

Properties of Hybridoma:

- Hybrid cells have capacity for limitless growth (immortality).

- They produce B cell's Ig, can synthesise unlimited amount of Ig (10,000 molecules/ cell/min).

Uses of monoclonal antibody

1. Diagnostic use: commercial diagnostic kits for identification of viral and other Ags. toxins produced by *Shigella spp., Campylobacter spp., E. coli, Vibrio spp.,* etc.

2. Test for vaccines : Identification and purification of microbial products, both for vaccine and for industrial use.

3. Pure antibody : Large amount of pure Ab of defined class and subclass can be prepared.

4. Future use: Hybridomas can be stored in frozen state and revived subsequently.

CELL MEDIATED IMMUNITY (CMI)

Mediated by sensitised T lymphocytes against An Ag– undergoes blast transformation and clonal proliferation in paracortical areas of lymph nodes - activated lymphocytes release biologically active products (lymphokines) which are responsible for manifestations of CMI. It is a specific I.R. that do not involve Abs.

1. Delayed hypersensitivity (Type IV).

2. Immunity in infectious diseases caused by obligate and facultative intracellular parasites, e.g. **Bacteria** (T.B., Leprosy, Listeriosis, Brucellosis); **Fungi** (Histoplasmosis, Coccidiodomycosis, Blastomycosis); **Protozoa** (Leishmaniasis, Trypanosomiasis); and **Viruses** (Small pox, Measles, Mumps).

3. Transplantation immunity and Graft versus host reaction

4. Immunological surveillance and immunity against cancer

5. Pathogenesis of certain autoimmune diseases (Thyroiditis, Encephalomyelitis).

Tests detecting Cell Mediated Immunity

1. Skin test showing delayed hypersensitivity, e.g. tuberculin test, lepromin test.

2. Test to detect competence to develop CMI by applying chemicals like 1% dinitrochlorobenzene or dinitrofluorobenzene to skin in normal individuals and then applying 0.03% to 0.1% of the same chemical after 10-14 days.

3. In vitro evaluations of lymphocytic response

   - Lymphocyte transformation test – blast transformation of lymphocytes on exposure to mitogens (phytohaemagglutinin or concanavalin A) and the incorporation of tritiated thymidine into DNA can be measured.

   - Macrophage migration inhibitory factor test – Macrophages in a glass capillary tube normally fan out. But in presence of sensitised T lymphocytes and antigen of guineapig, peritoneal macrophages is prevented out of the tube by the migration inhibition factor (MIF).

   - Rosette formation – T cells form rosettes on incubation with sheep red blood cells.

The Major Cytokines

Cytokine	Immune system	Other cells	Principal targets	Principal effects
IL-1a IL-1b	Macrophages LGLs, B cells	Endothelium, fibroblasts, astrocytes, etc.	T cells, B cells, macrophages, endothelium, tissue cells	Lymphocyte activation Macrophage stimulation, Reduced leucocyte/ endothelial adhesion, Pyrexia, Acute phase proteins
IL-2	T cells		T cells	T cell proliferation and differentiation, Activation of cytotoxic lymphocytes and macrophages
IL-3	T cells		Stem cells	Multilineage colony stimulating factor
IL-4	T cells		B cells	B cell growth factor, isotype selection, IgE, IgG1
IL-5	T cells		B cells	B cell growth and differntiation, IgA selection
IL-6	T cells, B cells	Fibroblasts	B cells, hepatocytes	B cell differentiation, induces acute phase proteins.
IL-7		Bone marrow, stromal cells	pre-B cells, T cells	B cell & T cell proliferation
IL-8	Monocytes		Neutrophils, Basophils	Chemotaxis
IL-10	T cells		TH1 cells	Inhibitor of cytokine synthesis
TNFα	Macrophages, lymphocytes,		Macrophages, granulocytes, tissue cells	Activation of macrophages, granulocytes & cytotoxic cells, Reduced leucocyte/endothelial cell adhesion, Cachexia, Pyrexia, Induction of stimulation of acute phase protein, Stimulation of angiogenesis,
TNFβ (LT)	T cells			Enhanced MHC Class I production
IFNα	Leucocytes	Epithelia, fibroblasts	Tissue cells	MHC Class I induction, Antiviral effect, Stimulation of NK cells
IFNβ	T cells,	Epithelia,	Leucocytes,	MHC class I & II induction,
IFNγ	NK cells	fibroblasts	Tissue cells, TH2 cells	Macrophage activation, Reduced endothelial cell/ lymphocyte adhesion, Cytokine synthesis
M-CSF	Monocytes	Endothelium, fibroblasts		Proliferation of macrophage precursors
G-CSF	Macrophages	Fibroblats	Stem cells	Stimulate division and differentiation
GM-CSF	T cells, macrophages	Endothelium, fibroblasts		Prolifern. of granulocyte & macrophage precursors & activators
MIF	T cells		Macrophages	Migration inhibition

Most cytokines act in connection with others to produce their biological effects in vivo.

Therapeutic cytokines

IFN-α, IFN-β	In chronic Hepatitis B infection, Hepatitis C infection, Herpes zoster, Papillomavinus infection, ? HIV
IFN-γ	In lepromatous leprosy, Leishmaniasis, Cerebral toxoplasmosis, Chronic granulomatons disease
IL-2	Local treatment of skin lesions in leprosy
Anti-TNF	In septic shock
IL-1	Receptor antagonist in septic shock
IL-10, TGF-β	In septic shock
CSFs	In bacterial infection due to neutropenia in irradiated patients

TGF = Transforming growth factor

Interleukin-1. It promotes short term proliferation.

Interleukin-1, formerly known as "lymphocyte activating factor" (LAF) is defined as a macrophage - derived molecule of 12K-16K mol. wt. (15,000) which carries hormone-like activities and affects T-lymphocytes by increasing their metabolism and potentiating their responses to other stimuli. Recent findings indicate that similar or possibly identical molecules are produced by cells other than macrophages and IL-1 affects a variety of nonlymphoid cells such as fibroblasts, hepatocytes and even brain cells.

There is no correlation between released IL-1 activity and total production of the mediator, the sum of extracellular and intracellular activities. A complex relationship was observed between these 2 activities by using a variety of stimulatory agents and unstimulated cultures of murine macrophages :

- Both IC and EC activities in response to high concentration of latex beads
- Marked increase in IC level of IL-1 with only a minimal EC activity in cultures incubated with LPS
- Increase in EC activity with much smaller increase of IC level by silica particles.

This synergy between different agents provides a practical approach to obtain exceedingly high levels of IL-1. Damage to the macrophages also play a major role in increasing IL-1 production and release by activated macrophages exposed to LPS in vitro or in vivo. Human monocytes are more susceptible to LPS than murine resident macrophages resulting in both IC and EC levels of IL-1 activity.

Site of production of IL-1 - like mediators :

Macrophages and other mononuclear phagocytes are most efficient producers of IL-1, production increased by stimulatory agents.

Non macrophage cells – polymorphonuclear cells, glia and glioma cells, glomerular measangial cells, and epithelial cells of the skin or cornea, similarly enhanced by LPS or silica particles produce IL-1- like mediators.

1. Release of IL-1 like molecules by epithelial cells is markedly enhanced by damaging treatments such as physical perturbation and UV irradiation.

2. A large production of IL-1 remains in the internal environment of the producing cells. IL-1 is formed and released in response to a variety of insults and that it mediates many of the physiological responses of the body to injurious events.

Functions of IL-1

1. Main cell target of IL-1 is lymphocyte affecting by enhancing differentiation, and short time proliferation.

2. Main function of IL-1 is to enahnce the release of IL-2.

3. In vivo, IL-1 exhibits adjuvanticity by increasing the immune response to a soluble antigen.

4. IL-1 augments the immune response in vivo by promoting the production and release of other mediators.

Functions of IL-1 - like molecules

1. Macrophage - derived factors affects fibroblasts to stimulate its profliferation as well as collagenase production by these cells.

2. Macrophage - derived factors stimulate collagenase and prostaglandin production by synovial cells in culture.

3. Monokines stimulate chondrocytes byincreasing their secretion of collagenase and neutral proteases, as well as by elevating the chondrocyte incorporation of proline and sulfate into collagen and proteoglycan.

4. Injection of macrophage - derived factors into mice or rat's liver, produced proteins known as 'acute phase proteins' (APP) are released into the circulation and they are closely related to inflammatory processes (Serum amyloid A increases several 100 fold). IL-1- like molecule is the mediator which activates hepatocytes.

5. EP (Endogenous pyrogen) having capacity to cause fever, have physical properties similar to those of IL-1 and the target cell for its effect are certain neurous in the hypothalamus. However, IL-1 activity is elicited by the butyl ester of muramyl dipeptide (MDP) without the usual parallel apperance of EP activity, so now regarded as 2 different monokines.

6. An IL-1 - like polypeptide significantly increases the synthesis of prostaglandin E2 and consequently the degradation of proteins by muscle cells in culture. These reactions in vitro simulate certain pathologic processes (loss of body protein and myalgia) which accompany fever in vivo. Thus IL-1 plays a major role in orchestrating the body's reaction to injury (wound healing), fibrosis, fever and release of APP, inflammation and infection.

Interleukin-2. It promotes long - term proliferation of Tcell lines in culture. The activation and participation of lymphocytes in immune responses requires the T cell growth factor (TCGF) or interleukin-2, and generation of specific TCGF receptors. IL-2 is secreted by lymphocytes in response to antigen or (mitogen) lectin, and macrophage derived IL-1. T cells also produce specific membrane receptors of TCGF de novo in response to Ag or lectin. While the Ag confers specificity on a given immune response, the amount and interaction of TCGF with TCGF receptors determine the magnitude.

Functions

- IL-2 is required for the development of cytotoxic T lymphocytes as well as proliferative responses to Ag or lectin.
- IL-2 promotes lymphocyte mitogenesis, either directly or in synergy with other stimuli.
- T lymphocyte expansion and subsequent cessation of expansion may occur through altertions in TCGF production or receptor expression or both.
- IL-2 may be useful as a general restorative of immune competence in immunosuppressed individuals.

Properties

It is a glycoprotein of mol. wt. 15,000 daltons. Uchiyama and co-workers reported a monochonal Ab which appeared to react specifically with activated T cells and termed it 'anti-Tac'. It also reacted with continuous T cell lines maintained in TCGF, but not with TCGF - independent Tcell lines. Anti - Tac have been used to investigate the role of TCGF in various immune responses and radiolabelled Ab has been used to investigate the presence and abundance of TCGF receptors on various cell preprations, and their ability to be regulated.

IL-2 is adsorbed by its receptor on IL-2 dependent lines. IL-2 was adsorbed from partially purified preprns. and the remaining cytokines were tested for LAK (lymphokine- activated killer cell) activation – total loss of capacity to stimulate LAK, when IL-2 adsorption was complete. Cytokines other then IL-2 are not involved in the development of LAK, and that IL-2 alone suffices for direct activation. They are nonspecific potentiators of immune responses. Interleukins are proteins that have been known since the early 1970s and were called LAF, TCGF, etc. Since 1980, they have been grouped under the generic title 'interleukins'.

Interleukin-3. It activates suppressor T cells.

Factor-dependent cell lines generated from fresh lymphoid tissue resembled T cells phenotypically in that they expressed Thy. 1 and significant amounts of Ly antigens. The factor was partially purified and designated Interleukin-3. In addition to the T-cell markers, the IL-3 dependent cell lines were shown to contain relatively high levels of an enzyme, 20 a-steroid dehydrogenase. This enzyme which hydrogenates the 20 a position of progesterone is strictly limited to T-lineage cells. The finding of high 20 a SDH levels in the IL-3-dependent lines therefore confirmed their T-lineage and so proposed that IL-3 was involved in the differentiation of T cells. IL-3 was involved in the differentiation of T cells. IL-3 also rapidly induced 20 a SDH in spleen cells and was related to cell proliferation as it was inhibited by drugs that prevented mitosis.

Much remains to be learned about IL-3. Because of its wide specificity, it will receive much attention by immunologists, hematologists, biochemists and molecular biologists. Allusion has already been made to its potential in transplantation and cancer therapy, and the effect of lymphokines in vivo during pathological processes.

HYPERSENSITIVITY

Excessive or exaggerated reactions leading to tissue damage. **Allergy** (von Pirquet 1905) – Altered reactivity of an animal to repeated contacts with a foreign Ag.

Sensitizing dose: sensitizes I.S. by sensitizing appropriate B or T cells.

Shocking dose: (subsequent contact with same Ag) leads to variety of abnormal reactions at an interval of 2-3 weeks between 2 doses.

	Immediate Hypersensitiivty	Delayed Hypersensitivity
Timing	Appears immediately within minutes and recedes rapidly usually in one hour	Appears slowly in 24-72 hrs. and lasts longer for days
Immune Response	Ab mediated reaction Passive transfer possible by serum	Cell mediated reaction Transfer possible only with lymphoid cells or their extracts
Cellular Response	Limited PMN infiltration	Predominantly mononuclear cell infiltration
Desensitization	Easy but short-lived	Difficult but long-lasting

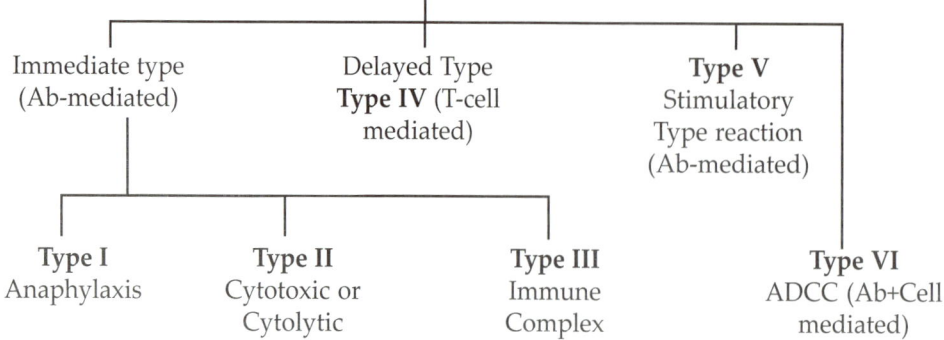

Coombs and Gell, 1963 classified Hypersensitivity

- Immediate type (Ab-mediated)
- Delayed Type **Type IV** (T-cell mediated)
- **Type V** Stimulatory Type reaction (Ab-mediated)

Type I Anaphylaxis — **Type II** Cytotoxic or Cytolytic — **Type III** Immune Complex — **Type VI** ADCC (Ab+Cell mediated)

A. Type I reaction - Anaphylaxis: (Ana - without; phylaxis - protection)

Systemic anaphylaxis in man observed in hypersensitive individuals by insect sting (Bees); injection of foreign serum (ATS, horse serum) or antibiotics (Penicillin).

Mechanism:- Minute sensitising dose (0.1 mg) of Ag by any route – **Waiting period** (IgE attaches to mast cells produced and basophils) – Massive shocking dose (0.1-10 mg) I/V – Ag combines with cell bound Ab on mast cell rapidly – Ag-Ab complex upsets adenylcyclase-cyclic AMP system in cell membrane – Degranulation of mast cells and basophils – Release of vasoactive amines (mediators).

Histamine receptors

– **H1** (mediates pharmacological effect on blood vessels, smooth muscles and mucosal surfaces);

– **H2** (mediates feedback effects via secondary messenger cAMP.

Primary mediators : From mast cell granules, basophils, platelets.

1. **Histamine** - Causes vasodilatation and vascular permeability, bronchospasm.
2. **Serotonin** - Causes capillary dilatation, vascular permeability and smooth muscle contraction.
3. **ECF-A** - Causes eosinophil chemotaxis.

Secondary mediators :

1. **Slow-reacting substance of anaphylaxis**

   (SRS-A) from leucocytes – mixture of leukotrienes, increased vascular permeability and smooth muscle contraction.

2. **Prostaglandins and thromboxanes** – Widely distributed in tissues, related to leukotrienes, derived from arachidonic acid via cyclooxygenase pathway. PGE_2 and A_2 cause vasodilatation; PGF2a cause airway constriction.

3. **Platelet activating factor (PAF)** from mast cells and monocytes – causes aggregation of platelets, neutrophil chemotaxis, vascular permeability, bronchospasm.

4. **Enzymes** (proteolytic and glycosidase) – Damage neighbouring cells.

   Target Organs – Organs predominantly affected in anaphylaxis.

   Guineapigs – Highly susceptible.
   Man, rabbit and dog – Intermediate susceptible.
   Rats – Resistant.

1. **Generalised or Systemic Anaphylaxis** – By I/V or I/M injection.

   In man, lung is the principal shock organ – causing bronchospasm, laryngeal oedema, respiratory distress, shock and death.

   Adrenaline injection is life saving.

2. **Local anaphylaxis** – By injection into skin/applied locally. Two types are there – mucosal and cutaneous.Transient redness and swelling (wheal and erythema), return to normal within 30 minutes.

3. **Passive cutaneous anaphylaxis** – Ag injected in passively sensitised individuals.

   In vivo – I/D Ab followed by I/V Ag and a dye.

   In vitro – Shultz-Dale phenomenon.

4. **Atopy** (Coca,1923) – out of place or strangeness. Occurs spontaneously in response to Ag encountered in environment in everyday life, e.g. hay fever and asthma.

   Familial distribution, inherited, IgE mediated.

   Inhalants – Pollens of ragweeds, house dust, grasses.

   Ingestants – Shell fish, prawn, milk, egg.

   Drug allergy – Penicillin, sulphonamides.

Prausnitz-Kustner (PK) reaction – Serum of Kustner injected I/D to Prausnitz, 24 hours later I/D injection of cooked fish extract – wheal and flare reaction in the sensitised area.

Desensitization

1. Acute desensitization – small amounts of Ag administered at 15 minutes interval for 1-2 hrs, short-lasting, hypersensitivity returns after days or weeks.

2. Chronic desensitization – long-term procedure, small amount of Ag administered at weekly intervals, IgG blocking Abs produced in serum which prevents Ag later from reaching IgE Ab, long-lasting.

Depot theray – injection of Ag with oil adjuvant.

Anaphylactoid reaction. By I/V injection of heavy metal salts, trypsin, peptone, starch or polysaccharides. It is a non-specific mechanism where alternate complement pathway is activated with release of anaphylatoxins.

B. Type II (Cytotoxic) reaction

   Ab (IgG & IgM) attaches to Ag (microbial product adsorbed on to a cell or a drug or a self molecule) via Fab region + complement – causes damage or lysis of cells.

Examples:

1. **Isoimmune Reactions** – ABO transfusion reactions, erythroblastosis foetalis.

2. **Autoimmune Reactions** – Autoimmune hemolytic anaemia, agranulocytosis or thrombocytopenia.

3. **Drug Reactions** – Penicillin, phenacetin, quinidine, sedormid (Abs produced against sedormid coated platelet Ag).

4. **Bacterial Reactions** – Salmonella and mycobacterial infections (immune reaction against a lipopolysaccharide bacterial endotoxin).

Demonstration

1. Direct antiglobulin test (Coombs test).
2. Agglutination test with tanned red cells, CFT, precipitation and immunofluorescence.

C. Type III (Immune Complex) Reaction

Ag-Ab complexes deposit in tissues (endothelial lining of blood vessels) – complement activation (C5a) – massive infiltration by PMN and platelets – inflammation and tissue damage, platelet aggregation leads to ischaemic necrosis of blood vessel.

Pathogenesis. Defect in phagocyte-complement system or when system is overloaded by Ag-Ab complex.

1. **Arthus reaction (Localized)** Arthus (1903). Repeated S/C injection of Ag into rabbits – High level of precipitating Ab appears in blood. Same Ag injected S/C or I/D – intense local oedema and haemorrhage within 3-6 hours.

2. **Serum Sickness (Generalized).** Single injection of high titre foreign serum (ATS), Ag slowly cleared from circulation and Ab production begins which reach high titres after 7-12 days. Some amount of excess Ag remaining in circulation combines with Ab forming small and soluble Ag-Ab complexes. These circulate or filter out in important organs and tissues (endothelial lining of capillaries of kidney, muscles, L.N. and joints). Once I.C. is formed, symptoms occur promptly – fever, urticaria, arthralgia, lymphadenopathy and splenomegaly.

Important I.C. diseases. Post-streptococcal glomerulonephritis, rheumatoid arthritis, serum hepatitis, dengue hemorrhagic fever.

Immune Complex detection:

Based on physical properties of complexes

1. Ultracentrifugation of serum.
2. Cryoprecipitation – Sera from patient with immune complex disease cryoprecipitate at 4°C.
3. Precipitation by polyethylene glycol (PEG), followed by estimation of IgG in the precipitate by single radial diffusion or laser nephelometry.

Detection of complexes or fixed Ig and/or complement components

1. Antiglobulin technique. Radiolabeled antigen and precipitate.
2. C1q binding technique.
   a) Plastic tube coated with C1q and serum. There is binding of immune complex. Estimation of amount and class of Ig in the complexes with radio or enzyme labeled class specific anti Ig.
   b) To serum, ^{125}I-C1q is added and then precipitated with PEG.

Binding of complexes to target cells with Fc and C3b receptors		
Fc receptors	- B lymphocytes - Raji cell line - K-cells	To the culture, serum and then radiolabeled anti IgG are added.
C3b receptors	- Macrophages - B lymphocytes - Raji cell line	

Tissue: Staining of tissue biopsies with fluorescent labeled anti Ig and anti C3.

Precipitative complexes – Elution of antibody from the complex by pepsin digestion or by pH shift to acid levels (pH3), when bound antibody tends to elute from the complex.

- Epstein-Barr virus demonstrated in complexes from patients with Burkitt's lymphoma.
- Schistosomiasis from circulating immune complex.
- Malaria antigen from acute glomerulonephritis (*P. falciparum*)
- DNA from Systemic Lupus Erythematosus (SLE)
- SLE, Rheumatoid Arthritis, Nephritis, Measles, Rubella, Influenza, HBV, Schistosomiasis.

D. Type IV (Delayed Hypersensitivity) Reaction – (Cell Mediated Immunity – CMI)

Mixed cellular reaction involving lymphocytes and macrophages, tissue damage mediated by T lymphocytes, not by Ab.

1. **Tuberculin (infection) type.** Tuberculin test – 1-5 TU of PPD injected I/D in a sensitised person – Erythema and swelling in skin after 48-72 hours. Infiltration of mononuclear cells (lymphocytes and 10-20% macrophages). Positive test indicates infection or vaccination in the past, but does not indicate current disease.

   Chronic infectious diseases caused by intracellular bacterial pathogens, fungi and protozoa – Lepromin test, Frei test, Histoplasmin and Toxoplasmin skin tests, herpes simplex and mumps.

2. **Contact Dermatitis Type.** Eczematous reaction (macule – papule – vesicle) with infiltration of lymphocytes and later macrophages after 48-72 hours of repeated contact with a sensitising material in a localised area of skin. A chemical in an oily base is applied on an inflamed area of skin.

   - **Drugs** – Penicillin or other antibiotics in ointment or creams
   - **Metals** – Ni, Cr
   - **Simple chemicals** – Hair dyes, picryl chloride, formaldehyde, dinitrochlorobenzene, soaps, cosmetics

- **Others** – Nylon, wool.

  Detected by **'Patch test'** of the offending agent on skin.

3. **Granulomatous Type.** Persistent Ag or Ag-Ab complexes within the macrophages shows granuloma containing epitheloid cells, giant cells and macrophages – hardening of skin after 4 weeks. **Examples** – T.B., leprosy, leishmaniasis, listeriosis, blastomycosis, schistosomiasis.

E. **Type V (Stimulatory Type) Reaction**

Modification of type II reaction – Ab+Ag on cell surface cause cell proliferation and differentiation instead of inhibition or killing. Abs are non-complement fixing, react with hormone receptor of cell surface, Ag-Ab combination enhances functional activity of cell, e.g. **Grave's Disease** – Thyroid stimulating Ab (LATS) combines with TSH receptor on thyroid cell surface and stimulates hormone activating adenyl cyclase causing excessive secretion of thyroid hormone. LATS can cross placenta causing hyperthyroidism in neonates which subsides few weeks after birth.

F. **Type VI Reaction (Antibody Dependant Cell mediated Cytotoxicity - ADCC)**

Mediated by NK cells, independent of complement. Ag-Ab complex is formed on target cells. NK cells combines via Fc fragment of Ab and cause lysis of cells. Small proportion bear T markers.

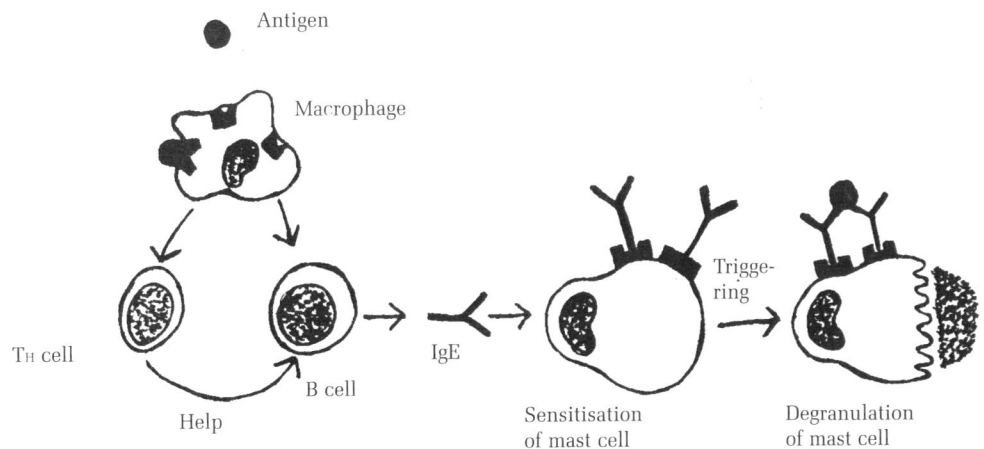

Mechanism of Type I hypersensitivity

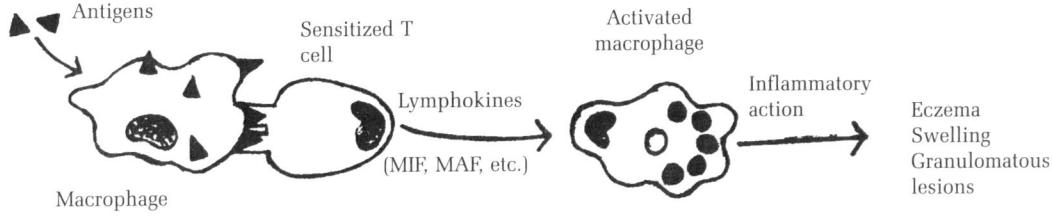

Mechanism of Type IV hypersensitivity

AUTOIMMUNITY

- Autoimmunity is a condition in which structural or functional damage is produced by immunologically competent cells or antibodies against the normal components of the body.

- It results from failure of the host's humoral and cellular immune systems to distinguish self from non self, resulting in attack on self-cells and organs by autoantibodies and self-reactive T cells.

- Incidence rates vary among the autoimmune diseases, with estimates ranging from less than one newly-diagnosed case of systemic sclerosis to more than 20 cases of adult-onset rheumatoid arthritis per 100,000 person-years.

- Prevalence rates range from less than 5 per 100,000 (e.g. chronic active hepatitis, uveitis) to more than 500 per 100,000 (Grave's disease, rheumatoid arthritis, thyroiditis).

- Until now, about 81 diseases are known as autoimmune diseases.

- More common in females as compared to males.

- Most common age group affected is 30-60 years.

- The overall estimated prevalence is 4.5%, with 2.7% for males and 6.4% for females.

What protects one from autoimmune diseases?

Tolerance

- Mechanism to protect an individual from potentially self-reactive lymphocytes

- A state of unresponsiveness to an antigen

- It can be central tolerance or peripheral tolerance

Central tolerance

- Deletion of lymphocyte clone during early maturation that may react with self-components

- Genetic rearrangements that gives rise to TCR or Ig receptors have role in this

- As these receptors could produce mature functional T or B cells that recognise self-antigen

- When these receptors are altered or edited, this clones will react with self-antigen may lead to disease

- So central tolerance process works to eliminate auto-reactive B cells in bone marrow and auto-reactive T cells in the thymus.

Peripheral tolerance

- Central tolerance is not a full proof process and does not eliminate all self-reactive lymphocytes

- So Peripheral tolerance inactivates these cells, and makes this process complete

- It is defined as inactivation of self-reactive T cells or B cells in periphery rendering them incapable of responding to self

- These can be by Anergy – unresponsiveness to antigenic stimulus

- It can be also induced by regulatory T cells (T reg cells), which acts at the sites of inflammation, down regulate autoimmune processes.

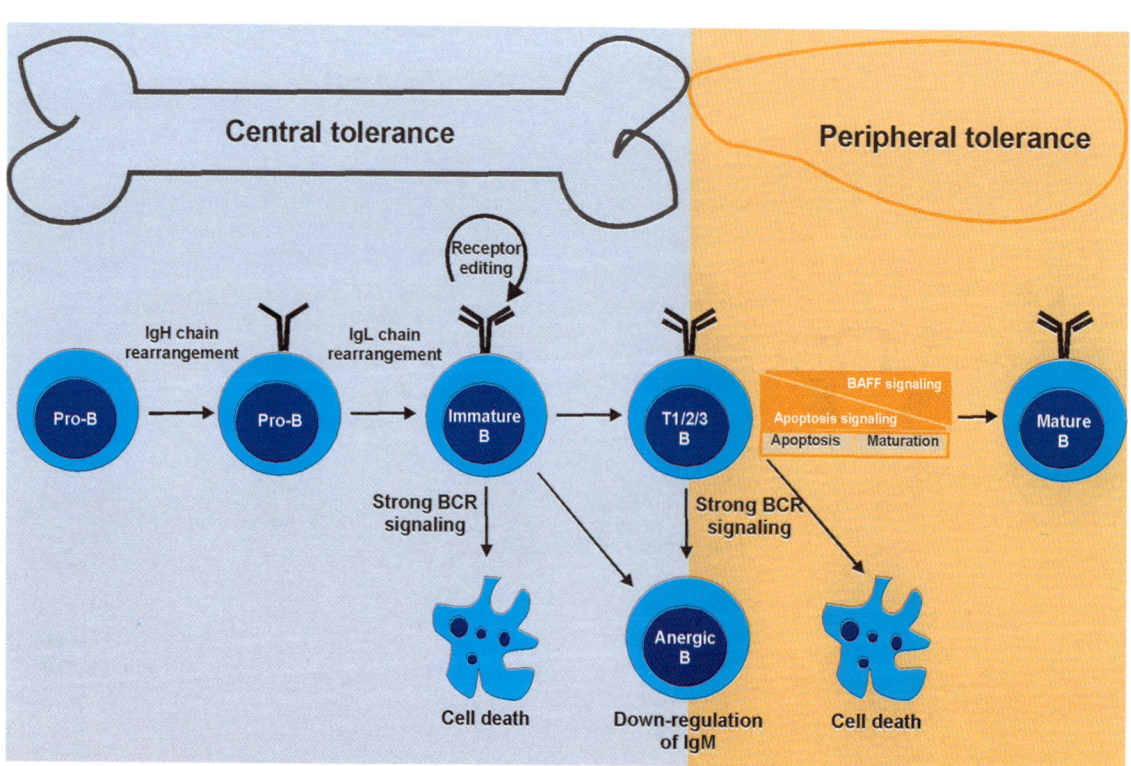

Mechanisms of autoimmunity

Derangements in these normal processes leads to autoimmunity these can be:

- **Antigenic alteration:** Cells or tissues may undergo antigenic alterations which can be
- Exogenous
- Endogenous

Exogenous

- Molecular mimicry:
  - Number of viruses and bacteria have been shown to possess antigenic determinants that are identical or similar to normal host cell components
  - Cross reactivity between microbial product and self-antigens, e.g. heart damage in Rheumatic fever which develop after streptococcal infection, in this case antibodies are to streptococcal antigens but they cross react with heart muscles.

Endogenous

- Loss of immunologic privilege (Sequestered antigens):
  - Many antigens are residing in immunologically privileged sites like lens protein anterior chamber of eye, sperms within testes
  - Sequestered antigens are not seen by developing T cells in thymus and will not induce tolerance
  - Damage to these barrier by trauma or inflammation can cause immunological reaction at protein expression sites.
- Alteration of self-antigen – e.g. Drug induced autoimmune disease, caused by medications e.g. hydralazine, isoniazid, procainamide
- Increased T cell stimulation
  - This bypasses antigen specific helper T cells and leads to polyclonal B cells activation with formation of multiple autoantibodies
  - A number of viruses and bacteria can induce non-specific polyclonal B cell activation, e.g. Cytomegalovirus, Epstein Barr virus
  - Super-antigenic stimulation, e.g. staphylococcal enterotoxin
- Inappropriate expression of class II MHC molecules
  - This can sensitise auto-reactive T cells
  - Certain agents, viral infections, trauma may induce localised inflammatory responses and thus increase IFN-α
  - If IFN-α induce class II MHC expression on non-antigen presenting cells, inappropriate and TH cell activation might follow with autoimmune consequences e.g. in IDDM, pancreatic beta cells expresses high levels of both class I and class II MHC molecules
  - Increased IFN-α causes inappropriate class II MHC expression in many autoimmune diseases like SLE

Genetic factors

Certain individuals are genetically susceptible to developing autoimmune diseases. This susceptibility is associated with multiple genes plus other risk factors.

Three main sets of genes are suspected in many autoimmune diseases. These genes are related to:

- Immunoglobulin
- T-cell receptors
- The major histocompatibility complexes (MHC).
- HLA DR2 is strongly positively correlated with Systemic Lupus Erythematous
- HLA DR3 is correlated strongly with Sjögren's syndrome, myasthenia gravis, SLE, and DM Type 1.
- HLA DR4 is correlated with the genesis of rheumatoid arthritis, Type 1 diabetes mellitus, and pemphigus vulgaris.

Gender differences in autoimmune diseases

- Females in general tend to mount more vigorous immune responses, this is particularly apparent in young women
- Sex hormones have role in this differences
- Estrogen play a significant role in aetiology of SLE, women exposed to oral contraceptives or hormone replacement therapy have an increased risk of developing SLE (approximately 2 fold)
- Estradiol binds to receptors on T cells and B lymphocytes, increasing activation and survival of those cells, favouring prolonged immune responses
- Prolactin has also role in autoimmune diseases, increased levels seen in SLE, rheumatoid arthritis, this hormone stimulates the release of both T helper 1 and T helper 2 cytokines, increases production of IFN-γ by lymphocytes and expression of IL 2, that stimulates autoantibody production.
- Many studies in mice indicates that Testosterone may be effective in ameliorating some autoimmune responses and so may be protective against several autoimmune diseases, including MS, Diabetes, SLE, Sjorgen's syndrome
- Exact mechanism is not understood but it is likely that these hormones, which circulate throughout body, alter immune responses by altering patterns of gene expression.

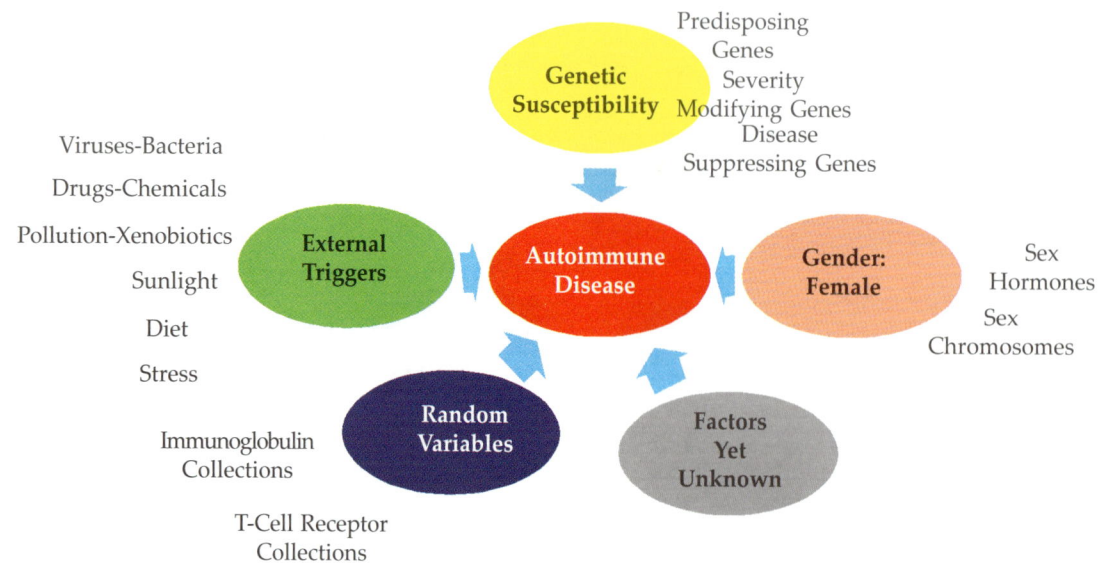

Etiology of Autoimmune diseases

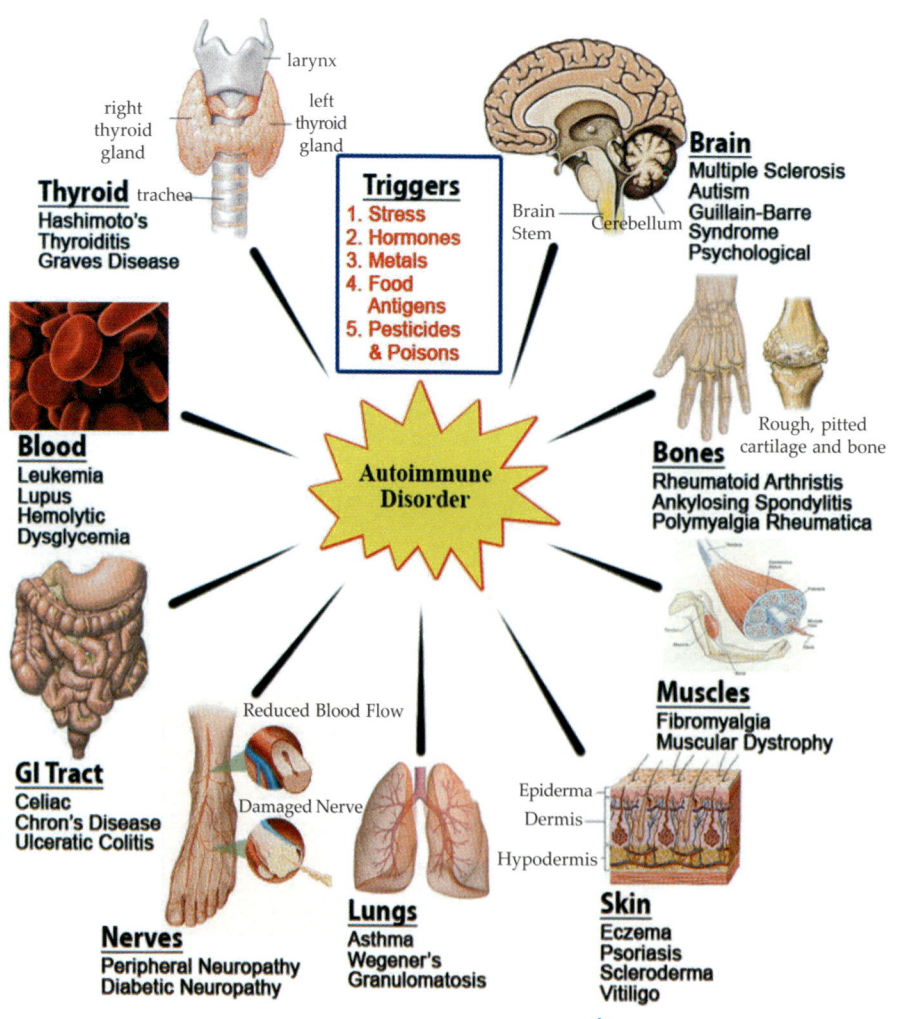

Tissues of the body affected by Autoimmune attack

Organ specific autoimmune diseases

Disease	Autoantibody
Myasthenia gravis	Anti-acetylcholine receptor
Grave's disease	Thyroid stimulating immunoglobulin or anti-TSH receptor autoantibody
Hashimoto's thyroiditis	Antibodies to thyroglobulin
Pernicious anaemia	Antibodies to intrinsic factor and gastric parietal cells
Addison's disease	Antibodies to adrenal cells
Primary biliary sclerosis	Antibodies to mitochondrial antigens
Autoimmune hemolytic anemia	Anti-red blood cell antibodies
Pemphigus	Antibodies to intracellular substances of skin and mucosa

Systemic autoimmune diseases

Disease	Autoantibodies
Systemic lupus erythematous	Antinuclear antibodiesAnti-ds DNA, anti-ss DNA antibodies, Anti-histone antibodies, Anti-ss antibodies
Rheumatoid arthritis	Anti γ-globulin antibodies
Goodpasture's syndrome	Anti-basement membrane antibodies
Sjogren's syndrome	Anti ss-A/Ro antibodies
Scleroderma	AntiScl-70, anti-centromere antibodies

Hashimoto's thyroiditis

- Hashimoto's thyroiditis is about 15-20 times more common in women than in men

- It frequently involves people between the ages of 30 and 50 years of age

- Some studies estimate that the current prevalence rate in the United States ranges between 0.3% - 1.2%. Other studies estimate the prevalence among the general population to be approximately 2%.

- Production of autoantibodies against thyroid protein including thyroglobulin and thyroid peroxidase and sensitisation of TH 1 cells specific for thyroid antigens

- Intense infiltration of the thyroid gland by lymphocytes, macrophages, plasma cells which forms lymphocytic follicles and germinal centres

- Visible enlargement of thyroid gland, a physiological response to hypothyroidism (decreased production of thyroid hormones)

- Symptoms of hypothyroidism:
  - Weight gain
  - Facial puffiness
  - Depression
  - Menstrual irregularities
  - Muscle joint pain

- **Laboratory diagnosis:**
  Detection of anti TPO antibodies by
  - Immunofluorescence assay
  - ELISA
  - Radio immune assay
  - Tanned erythrocyte agglutination
  - Complement fixation test

Sensitivity of anti TPO antibody ELISA with respect to IIFT is 81.5% and specificity is 100%.

Indirect immunofluorescence is considered as gold standard for detection for anti TPO antibodies by ELISA

- The Anti-TPO [^{125}I] RIA system provides a direct quantitative determination of auto-antibodies to thyroid peroxidase in human serum. Anti-TPO can be assayed in the range of 0-1900 IU/ml using 20µl serum samples.

- This determination is based on the competition between biotin labelled human polyclonal antibody and antibodies in the sample for the binding to 125 I-labelled TPO tracer.

Autoimmune Haemolytic anaemia

- The annual incidence of AIHA is estimated at 1/35,000-1/80,000 in North America and Western Europe.

- Autoantibody against red blood cell antigens, triggering complement mediated lysis or antibody mediated opsonisation and phagocytosis of the red blood cells.

- AIHA is classified as either warm autoimmune haemolytic anaemia or cold autoimmune haemolytic anaemia based on characteristics of autoantibodies involved in disease.

Warm type AIHA	Cold type AIHA
Chronic lymphocytic leukemia	Lympho-proliferative disorders
SLE	Scleroderma
RA	Mycoplasma, viral pneumonia
	Infectious pneumonia

Drug induced AIHA – caused by drugs for e.g. methyl dopa, penicillin, cephalosporin, dapsone this is type II immune response in which macromolecules on surface of RBCs and acts as an antigen.

Diagnosis

- Coomb's test: red cells are incubated with an anti-human IgG antiserum. If IgG autoantibodies are present on red cells, red cells are agglutinated by the antiserum
- This test is 98% sensitive.
- Evidences for haemolysis
- Peripheral smear - spherocytosis,
- Reticulocyte count - reticulocytosis
- Elevated serum bilirubin
- Raised serum LDH
- Raised urinary urobilinogen
- Hemosiderinuria,
- Serum haptoglobin-reduced

Goodpasture's Syndrome

- Autoantibodies specific for certain basement membrane antigens bind to the basement membranes of the kidney glomeruli and alveoli of the lungs
- Subsequent complement activation leads to direct cellular damage and an ensuing inflammatory response
- Damage to glomerular and alveolar basement membrane leads to progressive kidney damage and pulmonary haemorrhage
- **Diagnosis:**
  - Biopsy -The diagnosis of GPS is often difficult, The most accurate means of achieving the diagnosis is testing the affected tissues by means of a biopsy, especially the kidney as it is the best studied-organ for obtaining a sample, for the presence of anti-GBM antibodies.
  - Anti neutrophilic antibodies- About one in three of those affected also have cytoplasmic anti-neutrophilic antibodies in their bloodstream, which often pre-dates the anti-GBM antibodies by about a few months or even years.
  - Other – a urine analysis for protein or blood in urine, chest X-ray - abnormal white patches.

Pernicious anemia

- Antibody to gastric parietal cells and intrinsic factor - Vit B Pernicious anaemia (also known as Addison's anaemia) is one of many types of the larger family of megaloblastic anaemias.
- One way pernicious anaemia can develop is by loss of gastric parietal cells, which are responsible, in part, for the secretion of intrinsic factor, a protein essential for subsequent absorption of vitamin B12 in the ileum.
- Complete blood count - reveals anaemia, raised MCV, MCHC
- Peripheral smear:
  - Megaloblasts – large fragile immature erythrocytes
  - Ovalocytes typically seen
  - Hyper-segmented neutrophils
- The Schilling test, the classic test for PA, is no longer widely used, as safer and more efficient methods are available.
- Part one of the Schilling test consists of taking an oral dose of radiolabelled B12 and having the radioactivity of the urine measured over a 24-hour period.
- The second part of the test is a repeat of the first, with the addition of oral intrinsic factor.
- With lower than normal amounts of intrinsic factor produced in PA, the addition of intrinsic factor in the second test allows the body to absorb more B12, producing a higher urine radioactivity. This test can distinguish PA from other forms of B12 deficiency

Insulin dependent Diabetes Mellitus

- Caused autoimmune attack on the pancreas
- Antibodies formed against specialised insulin producing cells- beta cells, in islets of Langerhans
- Destruction of beta cells, resulting decreased production of insulin and increased blood glucose levels

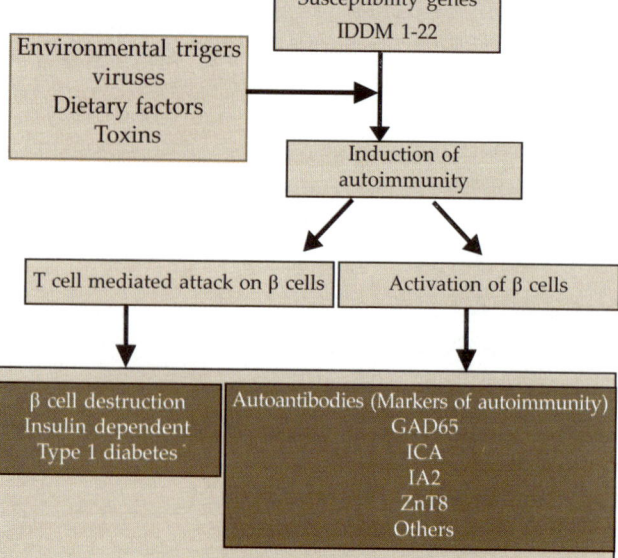

- Local cytokine production during this response includes IFN-γ, TNF-α and IL-1
- Beta cells destruction is thought to be mediated by cytokines released during DTH and lytic enzymes released from activated macrophages.
- Metabolic problems - ketoacidosis, increased urine production
- Late stages - atherosclerotic vascular lesions, renal failure, gangrene of extremities, blindness

Diagnosis:

- Glutamic acid decarboxylase autoantibody (anti GAD) detected by ELISA which has sensitivity 92% and specificity 98%,
- It can also detected by Radioimmuno assay with sensitivity 84% specificity 95%
- Islet cell antibody – detected by ELISA , RIA
- C- peptide level – measures level of insulin secretion, low levels.

Myasthenia gravis

- Prototype of autoimmune disease mediated by blocking antibodies
- Prevalence is about 5 per 1,00,000. In India around 53,000 affected with myasthenia gravis
- Production of autoantibodies that bind the acetylcholine receptor on the motor end plates of muscles, blocking the normal binding of acetylcholine and also inducing complement mediated lysis of the cells
- Antibodies causes destruction of cells bearing the receptors
- This results progressive weakening of the skeletal muscles leads to severe impairment
- **Method of detection:**
  - Immune-precipitation methodology – Human nicotinic ACh Rs from skeletal muscles labelled with I-α-bungarotoxin-conjugated acetylcholine receptors

  - Others - repetitive nerve stimulation, EMG, tensilon test.

Symptoms of Myasthenia gravis

Eye muscles

Drooping of one or both eyelids (ptosis).

Double vision (diplopia)

Face and throat muscles

Altered speaking (dyasarthria)

Difficulty swallowing (dysphagia)

Problems chewing

Limited facial expressions

Neck and limb muscles

Weakness in arms, legs, neck, fingers, etc.

Weakness in the chest muscles sometimes occurs. If this is severe, myasthenic crisis may result.

Systemic lupus erythematosus

- Prototype of systemic autoimmune diseases.
- The aetiology is not known as yet and the pathogenesis is complex, involving immunological, genetic, hormonal and environmental factors.
- It affects predominantly women in their reproductive years. The median age of onset in Indian SLE is 24.5 years
- The sex ratio (F:M) is 11:1.
- A prevalence study in India is 3 per 100,000. This is a much lower figure than reported from the west (varying from 12.5 per 100,000 adults in England to 39 per 100,000 in Finland and 124 per 100,000 in USA).
- Factors predisposing to SLE:
  - Genetic predisposition- HLA class II DR and DQ genes, alleles on chromosome 1 and 16
  - Female gender
  - Environmental factors, e.g. UV rays, viral infections, drugs

Antibodies in SLE

Antibody	Prevalence %	Antigen recognized
Anti- nuclear antibodies	98	Multiple nuclear
Anti-sm	70	DNA (double stranded)
Anti- RNP	40	Protein complexed to U1 RNA
Anti-histone	70	Histones associated with DNA
Anti-phospholipiid	50	Phospholipid, β2 glycoprotein, prothrombin
Anti-erythrocytes	60	Erythrocyte membrane
Anti-platelet	30	Surface and altered cytoplasmic antigens in platelets
Anti-neuronal	60	Neuronal and lymphocytes surface antigens

Pathogenesis

- When immune complexes of autoantibodies with various nuclear antigens are deposited along the walls of small blood vessels, type II hypersensitivity reaction develops, this results vasculitis and glomerulonephritis

- Complement mediated lysis results haemolytic anaemia and thrombocytopenia

- Excessive complement activation produces elevated levels the complement split products of C3a and C5a

- Neutrophil aggregation and attachment of neutrophil to endothelium of small blood vessels, this leads to neutropenia

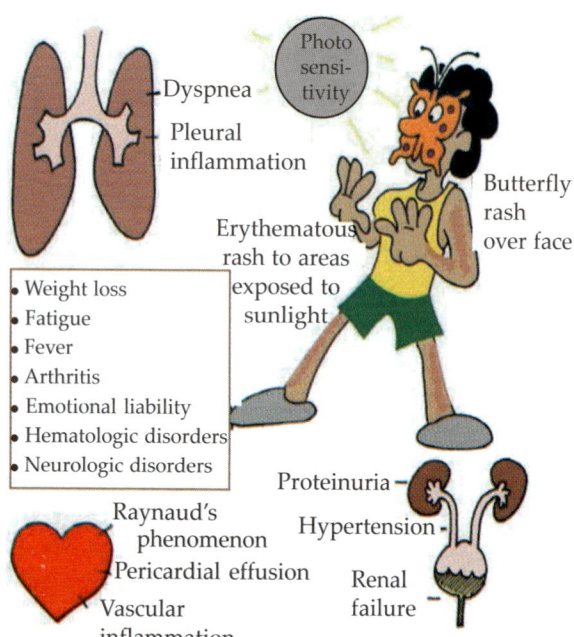

Symptoms of SLE:

Classification criteria for the diagnosis of SLE

- Malar rash – erythematous rash over malar eminences

- Discoid rash

- Photosensitivity

- Oral ulcers

- Arthritis – two or more peripheral joints,

- Serositis – pleuritis or peritonitis

- Renal disorders – proteinuria > 0.5 g/dl, cellular casts

- Hematologic disorders- haemolytic anaemia, leucopenia (< 4000/µl) or thrombocytopenia

- Antinuclear antibodies- abnormal titre of ANA

- If ≥ 4 of these criteria, well documented, are present at any time in a patient's history the diagnosis of SLE is likely to be.

- Specificity is ~95% and sensitivity is ~75%.

Laboratory diagnosis of SLE

Serological tests for detection of:

- Antinuclear antibodies – best screening test

- Anti-sm antibodies

- Anti-ds DNA antibodies

- Anti-histone antibodies by following methods:
  - ELISA
  - Indirect immunofluorescent staining with serum from SLE patient produces characteristic nuclear staining patterns
  - Immunofluoroscence assay
  - Hemagglutination assay
  - Radioimmunoassay

Antinuclear antibody detection tests

- The ANA test detects the autoantibodies present in an individual's blood serum.

- The common tests used for detecting and quantifying ANAs are indirect immunofluorescence and enzyme-linked immunosorbent assay (ELISA)

- Sensitivity 99% and specificity 49%.

Immunofluorescence

- In immunofluorescence, the level of autoantibodies is reported as a titre. This is the highest dilution of the serum at which autoantibodies are still detectable.

- Positive autoantibody titres at a dilution equal to or greater than 1:160 are usually considered as clinically significant.

- Although positive titres of 1:160 or higher are strongly associated with autoimmune disorders, they are also found in 5% of healthy individuals.

- Anti-ds DNA detected by ELISA and Immunofluoroscence

- Antigen used is ds-DNA in flagellate *Crithidia lucilliae*

- Farr assay - used to quantify amount of anti-ds DNA antibodies in serum , correlates with nephritis

- Polythelene Glycol (PEG) Assay

- Flow cytometry - detection of ANA multiplex polystyrene beads coated with autoantigens such as SS-A, SS-B, sm, histone

- Microarrays – newly emerging method for detection of ANA.

- Other Tests for following disease course:
  - Hemoglobin levels – anaemia
  - Platelet count – thrombocytopenia
  - Lymphopenia, leucopenia
  - Urine analysis –   proteinuria
    albumin > 500 mg/24 hrs
    cellular casts

Rheumatoid arthritis

- Rheumatoid arthritis is a chronic multisystem disease with characteristic feature of persistent inflammatory synovitis, usually involving peripheral joints in symmetric distribution
- RA affects between 0.5 and 1% of adults in the developed world with between 5 and 50 per 100,000 people newly developing the condition each year.
- Onset is most frequent during middle age, but people of any age can be affected.
- Women are affected approximately three times more often than men.

Etiology

- Infectious agents, e.g. Mycoplasma, Epstein Barr virus, cytomegalovirus, parvovirus, rubella virus
- Exact mechanism is controversial but it can be 'molecular mimicry'
- Microorganisms might induce immune response to components of the joints, in this reactivity to type II collagen and heat shock proteins has been demonstrated
- Produces group of autoantibodies called rheumatoid factors that are reactive with determinants in Fc region of IgG
- The classic rheumatoid factor is an IgM antibody
- This binds to normal circulating IgG antibody, forming IgM-IgG complexes that are deposited in joints.

Pathogenesis

- Micro vascular injury and increased number of synovial lining cells appear to be earliest lesion in rheumatoid arthritis
- Perivascular infiltrate appears initially predominantly of myeloid and during symptoms T cells also appears
- Synovium becomes oedematous and protrudes into the joint cavity as villous projections.
- The process involves an inflammatory and fibrosis of the capsule around the joints. It also affects the underlying bone and cartilage.
- This resulting pain, swelling, tenderness of joints, pain aggravated by movement
- RA can produce diffuse inflammation in the lungs resulting pleural disease, pleuro-pulmonary manifestations, pneumonitis,
- Inflammation in the membrane around the heart, and whites of the eye.
- It can also produce nodular lesions, most common within the skin
- Rheumatoid vasculitis

Symptoms

- Swelling in joints (specially in small joints)
- Red and Puffy Hands
  Joint Pain (early morning stiffness lasting more than one hour)
  Back Pain
  Tightning of skin
  Skin Rashes
  Weight Loss

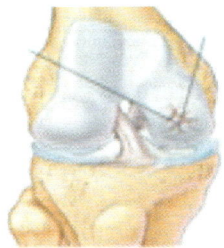

Causes

An autoimmune disorder, it occur when the white blood cells which generally help maintain immunity move from the blood stream into the membrane surrounding the joints, it results in inflammation blood cells play role in causing the synovium to become inflamed.

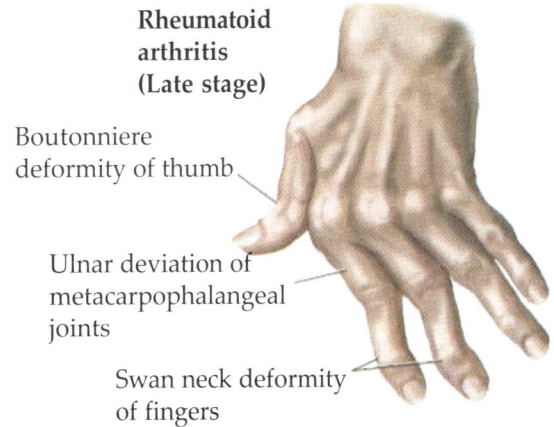

Rheumatoid arthritis (Late stage)

Boutonniere deformity of thumb

Ulnar deviation of metacarpophalangeal joints

Swan neck deformity of fingers

Diagnosis

Serological tests:

- R.A. tests (RA Factor test) - Rheumatoid factor (RF) is the autoantibody (antibody directed against an organism's own tissues) that was first found in rheumatoid arthritis. It is defined as an antibody against the Fc portion of IgG (an antibody against an antibody).
- Widely used test, detects IgM Rheumatoid factor
- Latex agglutination test
- Though not useful as screening test it has prognostic value in patients with severe and progressive disease.
- Other additional antibodies may be found in patients of RA:
  - Antibodies to citrullinated proteins
  - Anti-cyclic citrullinated peptide
- Detected by ELISA , in this CCP2 synthetic peptide antigen is used
- Similar sensitivity as RF, but higher specificity (95-98%)

- Rapid test for anti-CCP is also available now days, based on lateral flow immunochromatography

- Serum Anti-CCP levels are currently measured using an enzyme-linked immunosorbent assay (ELISA). The first generation of anti-CCP testing (CCP1) used citrullinated proteins derived from human filaggrin

- This method of testing was expensive and difficult to standardize, since it required purification of sufficient quantities of human antigen.

- The second generation of anti-CCP testing (CCP2) uses a synthetic peptide antigen, thus making the test cheaper and easy to standardize. CCP2 is currently the only commercially available method for testing for anti-CCP antibodies.

- Sensitivity 78%, Specificity 99%.

- Blood findings - normocytic, normochromic anaemia (ineffective erythropoiesis)

- Thrombocytopenia

- Raised ESR

- Raised CRP

- Synovial fluid analysis- increased protein content, reduced viscisity

- White cell count varies between 5-50,000/μl (polymorphonuclear predominate)

Multiple sclerosis

- Most people diagnosed at ages of 20 and 40

- Production of auto reactive T cells that participate in the formation of inflammatory lesions along myelin sheath of nerve fibres

- Cerebrospinal fluid of patients with active T lymphocytes , which infiltrate the brain tissue and cause characteristic inflammatory lesions, destroying the myelin

- Breakdown in myelin sheath leads to neurological dysfunction

- The Multiple Sclerosis Foundation estimates that more than 400,000 people in the United States and about 2.5 million people around the world have MS.

- Rates of MS are higher farther from the equator. It's estimated that in southern states (below the 37th parallel), the rate of MS is between 57 and 78 cases per 100,000 people. The rate is twice as high in northern states (above the 37th parallel), at about 110 to 140 cases per 100,000.

- Multiple Sclerosis International federation quotes a prevalence of 3/100,000 for India which may be an under estimation but it certainly is not as high as the rates found in high prevalence temperate zones (60-100/100,000) or higher.

- A study by Bharucha et al used a door-to-door survey of the Parsi community in Mumbai to arrive at the conclusion that the prevalence of clinically definite MS in the Parsi community living in Mumbai was 21/100,000.

- Wadia et al also arrived at a similar figure of 26/100,000 for the same community.

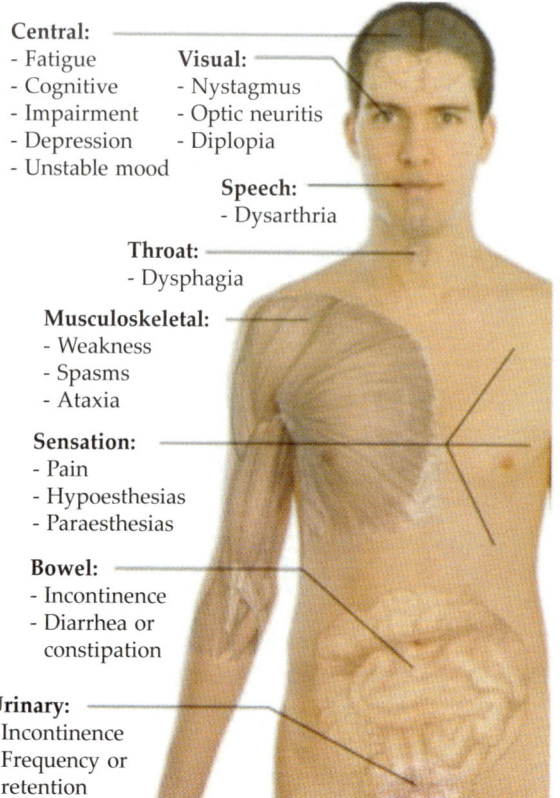

Central:
- Fatigue
- Cognitive
- Impairment
- Depression
- Unstable mood

Visual:
- Nystagmus
- Optic neuritis
- Diplopia

Speech:
- Dysarthria

Throat:
- Dysphagia

Musculoskeletal:
- Weakness
- Spasms
- Ataxia

Sensation:
- Pain
- Hypoesthesias
- Paraesthesias

Bowel:
- Incontinence
- Diarrhea or constipation

Urinary:
- Incontinence
- Frequency or retention

Main symptoms of Multiple sclerosis

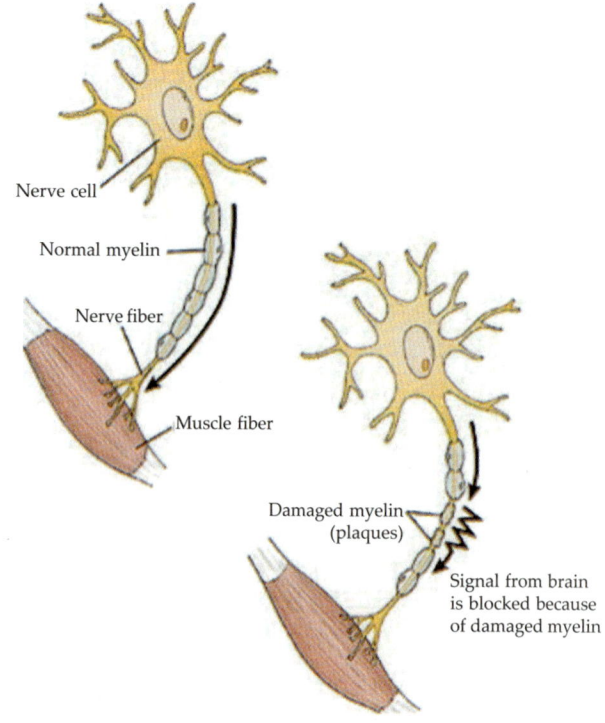

Nerve cell

Normal myelin

Nerve fiber

Muscle fiber

Damaged myelin (plaques)

Signal from brain is blocked because of damaged myelin

Damaged Myelin in Multiple Sclerosis

Diagnosis:

- Accurate diagnosis is mostly based mainly on neurological examination
- MRI - two or more lesions or abnormal areas in brain and spinal cord
- CSF examination- raised protein
- Evoked potentials
- Serological - anti neutrophil cytoplasmic antibodies, raised IgG antibodies in CSF.

Autoimmune disease	Antibodies detected	Common Method of detection
Hashimoto's thyroditis	Anti TPO antibodies	ELISA, immunofluorescence
Autoimmune haemolytic anaemia	Anti erythrocytes antibodies	Coomb's test
Goodpaster 's syndrome	Anti –GBM antibodies	biopsy, immunofluorescence
Pernicious anaemia	Anti parietal cell antibodies	immunofluorescence
IDDM	Antibodies against beta cells	Anti- GAD ELISA, RIA
Myasthenia gravis	Anti acetylcholine receptor antibodies	Immuno-precipitation
Systemic lupus erythematous	Anti-ANA, Anti ds DNA, anti-sm antibodies	ELISA, immunofluorescence
Rheumatoid arthritis	Antibodies against Fc portion of IgG antibodies	RA test, anti CCP ELISA
Multiple sclerosis	Antibodies against myelin sheath of neurons	MRI, CSF examination

Treatment

- Immunosupressive drugs e.g. Corticosteroids, Azathioprine, Cyclosporine A
  - These agents block signal transduction mediated by T cell receptor and inhibit antigen activated T cells
  - This causes non-specific suppression of the immune system, predisposes individual to infections
- Removal of thymus, e.g. in Myasthenia gravis often increases the likelihood remission of symptoms.
- Plasmapheresis - short term benefit in RA, SLE, Grave's disease

  Beneficial in patients with autoimmune diseases involving antigen – antibody complexes, which are removed with plasma
- Monoclonal antibody Rituxan , which kills B cells by targeting the surface marker CD20
- Blockers of TNFα - Enabrel, Remicade, Humira, widely used for RA, Psoariasis and Crohn's disease.

HLA (HUMAN LEUCOCYTE ANTIGEN) SYSTEM

- HLA forms part of the Major Histocompatibility Complex (MHC)
- Found on the short arm of chromosome 6
- MHC antigens are integral to the normal functioning of the immune response.
- Essential role of HLA antigens lies in the control of self-recognition and thus defence against micro-organisms and surveillance.
- β2-microglobulin on chromosome 15
- α and β chain = GeneA and GeneB
- Haplotype - combination of alleles inherited from Chr 6
- 2% meiotic recombination rate generates population diversity.

Characteristics of HLA

HLA comprises two classes: Class I and Class II

Class I – A,B,C most significant (other loci e.g. E,F,G,H, etc. are not so important in transplantation)

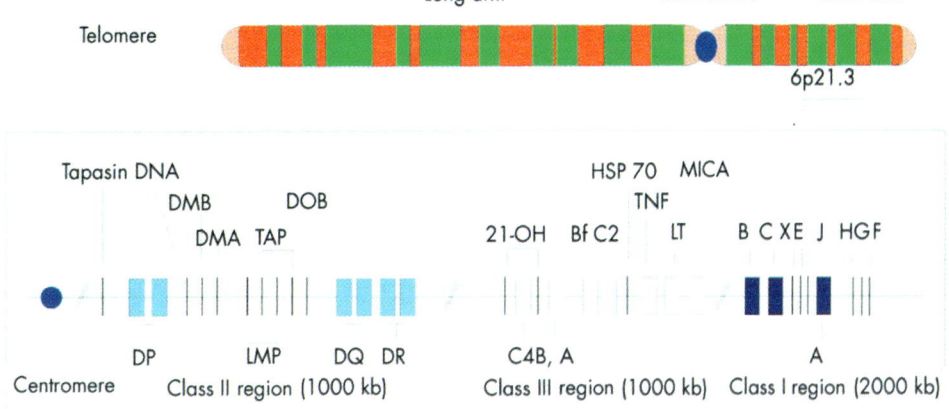

HLA gene

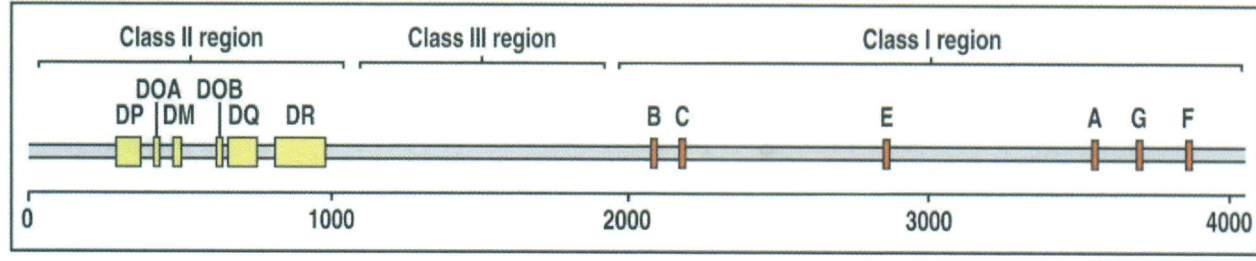

Chromosome Organization of HLA complex

- Expressed on most nucleated cells
- Have soluble form in plasma
- Are adsorbed onto platelets (some antigens more readily than others).
- Erythrocytes will adsorb some Class I antigens viz. Bg blood group system (B7, A28, B57....)
- HLA B most polymorphic system and studies have shown it is most significant followed by A and then C.
- 45Kd glycoprotein comprising three heavy chain domains, non-covalently associated with β2-microglobulin (coded on chromosome 15) which plays an important role in the structural support of the heavy chains.
- Class I molecules are assembled within the cell and ultimately sit on the cell surface with a section inserted into the lipid bilayer of the cell membrane and a short cytoplasmic tail where they present antigen in the form of peptide to cytotoxic T (CD8+) cells.
- HLA Class II five loci DR, DQ, DP, DM and DO
- HLA DR, DQ, DP most significant
- Expressed on B lymphocytes, activated T lymphocytes, macrophages, endothelial cells i.e. immune competent cells.
- Comprise 2 chains encoded by HLA genes, alpha and beta each with 2 domains.
- Hyper variable region is in the β1 domain.

- HLA Class II present peptide in the cleft to helper T (CD4+) cells. Thus Class II presentation involves the helper-function of setting up a general immune reaction involving cytokine, cellular and humoral defence.
- The role of Class II in initiating a general immune response is why they only need to be present on immunologically active cells.
- MHC I (single peptide binding chain a): 3 genes present antigen – HLA-A, HLA-B, HLA-C

- MHC II (two chains, a and b): 3 genes present antigen – HLA-DQ, HLA-DP, HLA-DR
- Each MHC II locus encodes a gene for the a chain and a gene for the b chain:
  - HLA-DQA, HLA-DQB   => MHC II isoforms
  - HLA-DPA, HLA-DPB   => MHC II isoforms
  - HLA-DRA, HLA-DRB   => MHC II isoforms

MHC isoform can bind multiple peptides.

- **HLA-A,B,C -** present peptide antigens to CD8 Tcells and interact with NK-cells
  - HLA-E, G interact with NK-cells
  - HLA-F ?
  - HLA-DP, DQ, DR present peptide antigens to CD4 Tcells
  - HLA-DM, DO regulate peptide loading of DP, DQ, DR.

Interallelic conversion

Recombination between alleles of the same gene

Generation of new MHC alleles

HLA B*5301 found in African populations and associated with resistance to severe malaria.

Gene conversion

HLA B*4601 found in South East Asian populations and associated with susceptibility to nasopharyngeal carcinoma.

MHC selection by Infectious Disease

- Pathogens adapt to avoid MHC - recent MHC isoform may provide a survival advantage (hence higher frequency level)
- Epidemic diseases place survival advantages on those who can best present pathogenic peptides
- Only a minority of HLA alleles are common to all humans - most are recent and specific to ethnic groups.

B C XE J HGF

A
Class I region (2000 kb)

Tapasin DNA
DMB   DOB
DMA TAP

Centromere

DP   LMP   DQ DR
Class II region (1000 kb)

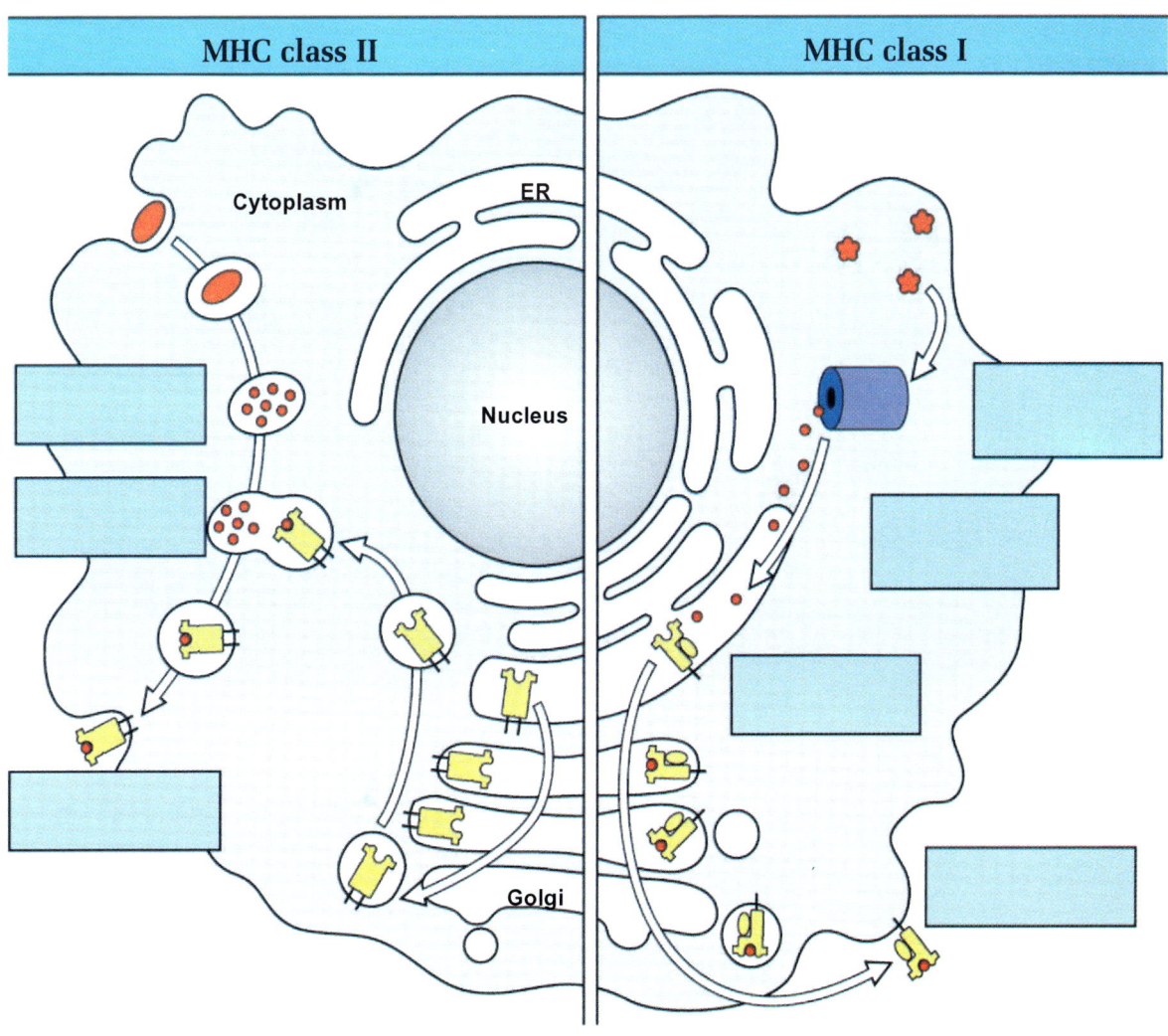

MHC class II | **MHC class I**

HLA Type and Disease Susceptibility

Ankylosing spondylitis	B27
IDDM	DR4/DR3
Multiple Sclerosis	DR2
Narcolepsy	DR2
Rheumatoid arthritis	DR4
Lupus (SLE)	DR3
AIDS (rapid)	HLA-A29, HLA-B22
	HLA-C16, HLA-DR11
AIDS (slow)	HLA-B14, B27, B57
	HLA-C8, C14

Heterozygous

Six different MHC I proteins on all cells

- Maternal: 3 MHC I genes - HLA-A_M, HLA-B_M, HLA-C_M
- Paternal: 3 MHC I genes - HLA-A_P, HLA-B_P, HLA-C_P

Six different MHC II proteins on all cells (some individuals have 8 due to two HLA-DRB genes)

Maternal: Three MHC II genes

- HLA-DPA_M, HLA-DPB_M
- HLA-DQA_M, HLA-DQB_M
- HLA-DRA_M, HLA-DRB_M

Paternal: Three MHC II genes

- HLA-DPA_P, HLA-DPB_P
- HLA-DQA_P, HLA-DQB_P
- HLA-DRA_P, HLA-DRB_P

MHC Polymorphism and Organ Transplants

- Developing T cells that recognize complexes of peptide and MHC molecules on HEALTHY tissue (self-peptides presented by self MHC) are DESTROYED.

- This results in the preservation of T cells that recognize non-self MHC (allogenic MHC). Called alloreactive T cells (1-10%) of total T-cell repertoire.

- It is primed for rejection of foreign organs that express allogenic MHC.

HLA Typing methods:

Serology

- Complement Dependent Cytotoxicity (CDC)

- Viable peripheral blood lymphocytes are obtained by discontinous density gradient centrifugation using Ficoll/Tryosil or Ficoll/Sodium Metrizoate at a density of 1.077 at 19°-22°C.
- Microlymphocytotoxic test: Three stages

Microlymphocyotototxic test

- Viable lymphocytes are incubated with HLA specific antibodies. If the specific antigen is present on the cell, the antibody is bound.
- Rabbit serum as a source of complement is added and incubated. If antibody is bound to the HLA antigen on the cell surface, it activates the complement which damages the cell membrane making it permeable to vital stains.
- Results are visualised by adding dye, usually a fluorochrome e.g. Ethidium Bromide although both Trypan Blue and Eosin have been used in the past.
- If the reaction has taken place the EB enters the cell and binds to the DNA.
- For ease, double staining is normally used – a cocktail of Ethidium Bromide and Acridine Orange, quenched using Bovine Haemoglobin to allow simultaneous visualisation of both living and dead cells.
- Test is left for 10 minutes and then read using an inverted fluorescent microscope.
- A mixture of T and B lymphocytes can be used for HLA Class I typing.
- B lymphocytes are required for HLA Class II typing by serology. (Normal population 85-90% T and 10-15% B cells)
- This can be achieved using a number of methods.
- In the past, neuraminidase treated sheep red blood cell rosetting and nylon wool have been used.
- Immunomagnetic bead separation is the current method of choice.
- It utilises polystyrene microspheres with a magnetisable core coated in monoclonal antibody for a HLA Class II β chain monomorphic epitope. Positive selection.
- **Pros**
    - Easily performed does not require expensive equipment.
    - Takes around three hours to perform.
    - Low level resolution, with good antisera reliable results.
- **Cons**
    - Requires large volumes of blood.
    - Requires viable lymphocytes.
    - Difficult to find good antisera for rarer antigens in different populations.

Cellular typing

- Not/rarely used by laboratories these days.
- Requires panels of homozygous typing cells.
- Cell culture method therefore takes a long time – labour intensive.
- Involves use of radioisotopes.

Molecular methods

- All commonly used molecular methods require good quality genomic DNA. There are numerous methods for extraction of DNA from whole blood.
- There are 'in house' methods based on Miller et al's 'Salting Out' which are cheap and easy but labour intensive.
- There are also numerous commercial kits available such as individual matrix capture columns, beads and semi- automated systems. This however can increase the cost per extraction from around 65p to £3.60p.
- All methods rely on DNA extraction from the nucleated cells following cell lysis and protein digestion.
- The application of molecular techniques to HLA typing began around 1987 when the Southern Blot technique was used to identify restriction fragment length polymorphisms (RFLP's) associated with known serological DR/DQ and cellular Dw defined specifities.
- Around 1992 polymerase chain reaction (PCR) methods were developed.
- Most methods currently used have a PCR element within the technique.

Polymerase Chain Reaction (PCR)

- Three steps per cycle – denaturation, annealing and extension. Amplification is exponential yielding 2 power n where n = number of cycles.
- The introduction of the programmable Thermal Cycler revolutionised the use of PCR within the routine laboratory.
- Amplification: DNA of interest is amplified by a single pair of biotinylated primers which flank the whole of exon e.g. exon 2 of the HLA DRB1 gene. PCR amplifies all the alleles in the exon.
- Hybridisation: PCR product is denatured and then added to a 'well' containing the nylon membrane with the bound probes and incubated with hybridisation buffer. PCR product hybridises to probes with complementary sequences.
- Excess product is washed away during a series of wash steps.
- Temperature is VERY important during these stages.
- Visualisation of results is achieved by incubating with a conjugate and enzyme, often streptavidin and horse radish peroxidase which binds to the biotin

of the PCR product and then adding a substrate. Band with PCR product turns blue.

- Strips will have internal control bands to show the test has worked.
- Interpretation is usually achieved by entering the band pattern into a computer programme.
- This is an excellent method for low resolution batch testing.
- Can be semi-automated.

Other Molecular Methods

- **PCR SSP** (Sequence Specific Priming)

  Can be used for HLA Class I and II typing using a panel of primer pairs either for low to medium resolution, whereby primers amplify groups of alleles or high resolution whereby primer pairs amplify specific alleles.

  Each PCR reaction takes place in a separate tube therefore the number of tubes depends on the level of resolution. Each tube also contains a pair of primers for part of the human growth hormone gene as an internal control. These are at a much lower concentration thus do not compete with specific primers.

- Electrophoresis is used following amplification. PCR product is run out on an agarose gel containing ethidium bromide. Each product moves according to its size and is compared to a molecular weight marker.

- Interpretation: Every tube should produce an identical sized product as internal control and either a specific band or dependent on whether the allele/s is/are present or not.

- Results are visualised using 312nm UV transillumination and recorded either by video imaging or polaroid photograghy.

- PCR SSOP **(Sequence Specific Oligonucleotide Probes)**

- 'Dot blot' in house method usually whereby one labels ones own probes with Digoxigenin.

- 'Reverse dot blot' normally commercial where specific oligonucleotide probes are attached to a nylon membrane. Dynal and Innotrans for example produce such kits.

- **Pros**
  - Does not require viable cells
  - Samples do not have to arrive in the lab the day they are taken
  - PCR SSOP good for batch testing
  - Can be semi automated

- **Cons**
  - Requires good quality DNA
  - Require a degree of redundancy within the primers used
  - Sequence of alleles must be known.

- **Other Molecular methods**
  - **Sequence Based Typing (SBT)**

    DNA sequencing is the determination of the sequence of a gene and thus is the highest resolution possible. Sequence based typing involves PCR amplification of the gene of interest e.g. HLA DRB1 followed by determination of the base sequence. The sequence is then compared with a database of DRB1 gene sequences to find comparable sequences and assign alleles. This method also allows for detection on new alleles.

  - **Reference Strand Conformational Analysis (RSCA)** offers sequence level typing without the need to sequence. Assigns HLA type on the basis of accurate measurement of conformation i.e. shape dependent on DNA mobility in polyacrylamide gel electrophoresis (PAGE). It is a complex and difficult technique, not taken up by laboratories for routine use.

  - **Luminex technology** – SSOP based. Just beginning to be introduced into laboratories for routine use on non-urgent samples.

REFERENCES

1. Ananthanarayan & Paniker's Textbook of Microbiology. 9th Ed. Kapil A. Universities Press, Hyderabad. 2013.
2. Chakraborty PA. Textbook of Microbiology. 4th Ed. New Central Book Agency (P) Ltd., Kolkata. 2009.
3. Roitt I, Brostoff J, Male D. Immunology. 5th Ed. Mosby - Year Book, Inc. USA. 1998.
4. Topley & Wilson's, Microbiology and Microbial Infections. Immunology. 10th Ed. Eds. Kaufmann SHE, Steward MW. John Wiley & Sons Ltd., West Sussex, UK. 2009.
5. Harrison's Principles of Internal Medicine. 18th Ed. Eds. Braunwald E, Fauci AS, Kasper DL, Hauser SL, Longo DL, Jameson JL. McGraw-Hill, New Delhi. 2011.
6. Kuby Immunology Paperback. 7th Ed. Owen J, Punt J, Stranford S. International Ed. 2013.
7. Daniel P. Stites. Basic & Clinical Immunology. Appleton & Lange, Internat.8r.e. Ed. 1994.
8. Kumar A. Indian guidelines on the management of SLE. J Indian Rheumatol Assoc 2002; 10: 80-96.
9. Cooper GS, Stroehla BC. The epidemiology of Autoimmune diseases. Autoimmun Rev. 2003; 2: 119-25.
10. http://www.healthline.com/health/multiple-sclerosis/facts-statistics-infographic#sthash.rIQIKawJ.dpuf.

RESPIRATORY TRACT INFECTIONS

13
Chapter

Site of infection	Specimen for culture	Bacteria associated with infections
Upper respiratory tract	**Acute** Nasopharyngeal swab Sinus washings	S. pneumoniae, S. pyogenes S. aureus
	Chronic Sinus washings Surgical biopsy specimen Swab of posterior pharynx Swab of tonsils (abscess) Nasopharyngeal swab	H. influenzae Klebsiella spp. and other Enterobacteriaceae Bacteroides spp. and other anaerobes S. pyogenes C. diphtheriae N. gonorrhoeae B. pertussis
Lower respiratory tract	Sputum Blood Bronchoscopy secretions Transtracheal or translaryngeal aspirate Lung aspirate or biopsy Bronchoalveolar lavage* Transbronchial and open-lung biopsies and brushings for P. jirovecii, fungi, tissue forms and mycobacteria.	S. pneumoniae H. influenzae S. aureus K. pneumoniae and other enterobacteriaceae Burkholderia cepacia Branhamella catarrhalis Legionella pneumophila** Mycobacterium tubercluosis, M. avium-intracellulare F. nucleatum, Prevotella melaninogenicus and other anaerobes Bordetella pertussis and other species, Brucella spp., Nocardia spp., Actinomyces spp., Chlamydia trachomatis, Mycoplasma pneumoniae, Coxiella burnetii, B. anthracis, Y. pestis, F. tularensis.

* In intubated patients undergoing ventilation.

** Scanty and watery sputum.

Colony count > 10^3/ml and intracellular bacteria in > 25% of inflammatory cells indicate pneumonia.

Diffuse pneumonia in I.C.P. patients - P. carinii, M. avium - intracellulare, Cytomegalovirus.

Fungi: Candida albicans, Cryptococcus neoformans, Histoplasma capsulatum, Coccidioides immitis, P. jirovecii.

Parasites: E. histolytica, A. lumbricoides, Strongyloides, P. westermani, Cryptosporidum.

Viruses: CMV, RSV, Influenza, Parainfluenza, Herpes simplex, Varicella zoster, Adenovirus.

Pneumonia by "HACEK" group

They are fastidious gram negative bacteria, forms part of normal flora of human oropharynx or urogenital tract. They are associated with pneumonia, endocarditis, bacterial and mixed flora wound infections. Organisms belonging to this group are :

H = *Hemophilus aphrophilus / paraaphrophilus*

A = *Actinobacillus actinomycetemcomitans*

C = *Cardiobacterium hominis*

E = *Eikenella corrodens*

K = *Kingella spp.*

They grow in blood agar or chocolate agar in CO_2 atmosphere within 48-72 hours.

In blood cultures, they take few days to 2 weeks for growth.

Collection and transport of respiratory specimens

I. **For upper respiratory tract** – Cotton-, Dacron-, or Calcium alginate - tipped swabs are suitable. May be toxic to some *Chlamydia species.*

Preferably cultured within 4 hours of collection. In case of delay, transported in Stuart's transport medium.

-   Nasopharyngeal swabs are better for RSV, parainfluenza virus, *Bordetella pertussis, Neisseria spp.* and the viruses causing rhinitis. Ideally should be transported in veal infusion broth.

-   For *Chlamydia,* transport medium is 2-sucrose phosphate (2SP).

-   For other organisms – throat swabs.

-   For *C. diphtheriae* – throat swabs and nasopharygeal swabs.

-   For *B. pertussis* – **'Cough plate'** method is ideal.

II. **For lower respiratory tract** – Early morning sputum samples in a sterile wide - mouthed container with tightly fitted screw cap. "Aerosol induced specimen", by allowing patient to breathe aerosolized droplets of a solution containing 15% NaCl and 10% glycerine for approx. 10 mins. or until a strong cough reflex is initiated.

-   For myocobacteria – Gastric aspirate in young children. Transported to laboratory immediately so that acidity can be neutralized.

-   Tracheal aspirates or tracheostomy suction.

-   Pleural fluid by thoracentesis for direct exmination and culture in case of pleural empyema.

-   Bronchial washings.

-   Blood culture from patients with pneumonia is positive in 20% cases.

-   Bronchoalveolar lavage (BAL) obtained by bronchoscopy (for *Pneumocystis* and fungal elements). Acute bacterial pneumonia correlates with $> 10^3$-10^4 bacterial colonies per ml of BAL fluid.

-   Bronchial brush specimen suspended in 1 ml of broth solution with vigorous vortexing and then inoculated into culture media.

-   Percutaneous transtracheal aspirate is good for anaerobic bacteria, including *Actinomyces.*

-   Thin needle aspirated material from lung in pneumonia (mainly used in children).

-   Biopsy specimens – direct examination for *Pneumocystis.*

-   Transbronchial biopsies more used for histological diagnosis.

-   Open lung biopsy for severe viral infections such as herpes pneumonia, *Pneumocystis,* etc.

Laboratory diagnosis of respiratory infections

I. **Microscopy:**

A.

1. Wet mount seen under microsocpe for counting epithelial cells and leucocytes per low power field and grading the sputum samples by Bartlett's grading or Murray and Washington's system.

Bartlett's Grading System for assessing the quality of sputum samples

	Grade
Number of neutrophils per 10X low-power field	
< 10	0
10-25	+1
> 25	+2
Presence of mucus	+1
No. of epithelial cells per 10X low power field	
10-25	- 1
> 25	- 2

Average number of epithelial cells and neutrophils counted in 20-30 separate 10X microscopic fields and total is calculated. A final score of 0 or less indicates lack of active inflammation or contamination with saliva. Repeat sputum specimens are advised.

Murray and Washington's Grading System for assessing the quality of sputum samples

	Epithelial cells per low-power field	Leucocytes per low-power field
Group 1	25	10
Group 2	25	10-25
Group 3	25	25
Group 4	10-25	25
Group 5	< 10	25

2. 10% KOH mount – yeast cells and pseudohyphae

3. Toluidine blue O for *P. carinii* and *N. asteroides.*

4. Saline and iodine mounts for ova, cysts and larva.

B. Staining methods

1. Gram stain

a) Pus cells with gram-positive lanceolate - shaped diplococci – *S. pneumoniae*

b) Pus cells and plenty of light staining Gram negative coccobacilli – *H. influenzae*

c) Pus cells and Gram negative short stout rods - *Klebsiella spp.*

d) Budding yeasts with pseudohyphae – *C. albicans*

e) Pus cells and Gram positive fine, filamentous, beaded and branching organisms – *Actinomyces* and *Nocardia spp.*

2. Albert's stain for *C. diphtheriae.*

3. Acid fast staining with
    - 20% H_2SO_4 for *M. tuberculosis*
    - 10% H_2SO_4 for *Cryptosporidium*
    - 1% H_2SO_4 for *Nocardia spp.*

4. Negative staining with India ink for capsule of *Klebsiella, S. pneumoniae, Hemophilus, Cryptococcus.*

5. Gomori methenamine silver staining for *Pneumocystis carinii.*

6. Auramine or auramine - rhodamine staining for mycobacteria.

7. Calcofluor white fluorescent stain or PAS stain for fungi.

8. Direct fluorescent assay (DFA) using fluorescein conjugated antibody for *Bordetella, Legionella,* RSV, HSV, CMV, VZV, Adenovirus.

9. Monoclonal and polyclonal fluorescent stains for *Chlamydia trachomatis*

II. Culture :

1. 5% sheep blood agar and chocolate agar.

2. MacConkey agar - Mucoid colonies of *Burkholderia cepacia* in cystic fibrosis.

3. Streptococcus selective agar.

4. Loeffler's serum slope, K-tellurite agar and L-J medium.

5. Modified Thayer Martin medium for *N. gonorrhoeae.*

6. Bordet Gengou medium for *B. pertussis* (Cough plate method).

7. Brain heart infusion agar and Sabouraud's dextrose agar for fungi.

8. Bacteroides bile esculin agar and Neomycin blood agar for anaerobes.

9. Buffered charcoal yeast extract (BCYE) agar with 3µg/ml lincomycin and 80µg/ml ansamycin for *Legionella.*

10. Brucella blood agar for *Brucella spp.*

III. Other tests :

1. Quellung test for *H. influenzae* type b, *S. pneumoniae* and *Klebsiella.*

2. Satellitism for *H. influenzae.*

3. PYR test for *S. pyogenes.*

4. Niacin test and MGIT for mycobacteria.

5. Bile solubility and optochin sensitivity for *S. pneumoniae.*

6. Paraffin bait technique for *Nocardia spp.*

IV. Serological tests

1. Fluorescent antibody test (FAT) for HSV, Adenovirus, Parainfluenza virus, RSV, *B. pertussis* (done directly on sample from nasopharynx).

2. Direct detection of group A streptococcal antigen nitrous acid extracts within 10 mintues by Coagglutination test.

3. Latex agglutination test (LAT) using latex beads coated with antibody for *S. pyogenes.*

4. Counterimmunoelectrophoresis (CIEP).

5. Enzyme immunoassays (EIA-based ICON strep A test).

6. ELISAs for HSV, CMV, etc.

7. Chemiluminiscent nucleic acid probe assay for direct detection of Group A streptococci in pharyngeal specimens (Group A Streptococcus Direct Test [GP-ST], Gen Probe, San Diego CA). Sensitivity 93.5%; Specificity 99.7%.

8. RIA for *Legionella,* from urine of patients

9. PCR for *B. pertussis* (100% sensitive), *Mycobacterium tuberculosis,* etc.

REFERENCES

1. Mims' Medical Microbiology. 5th Ed. Eds. Richard Goering, Hazel Dockrell, Mark Zuckerman, Ivan Roitt & Peter L. Chiodini. Saunders. 2013.

2. Koneman EW, Allen SD, Janda WM, Schreckenberger P, Winn Jr. WC. In Color Atlas & Textbook of Diagnostic Microbiology. 6th Ed. Lippincott, Philadelphia. 2006.

DIAGNOSIS OF TUBERCULOSIS

Diagnostic Dilemmas

- The chronic nature of the disease with an insiduous onset
- Wide spectrum of clinical manifestations
- Tendency to latent infection
- Paucibacillary distribution of the organisms
- Slow rate of growth
- Presence of environmental mycobacteria causing disease mimicking tuberculosis
- HIV co-infection
- Childhood tuberculosis
- Extrapulmonary tuberculosis.

Mycobacteria causing disease mimicking tuberculosis

- M. avium-intracellulare complex (MAIC)
- *M. fortuitum*
- *M. kansasii*
- *M. scrofulaceum*
- *M. chelonei*
- *M. abscessus*
- *M. genavense*
- *M. szulgai*
- *M. xenopi*

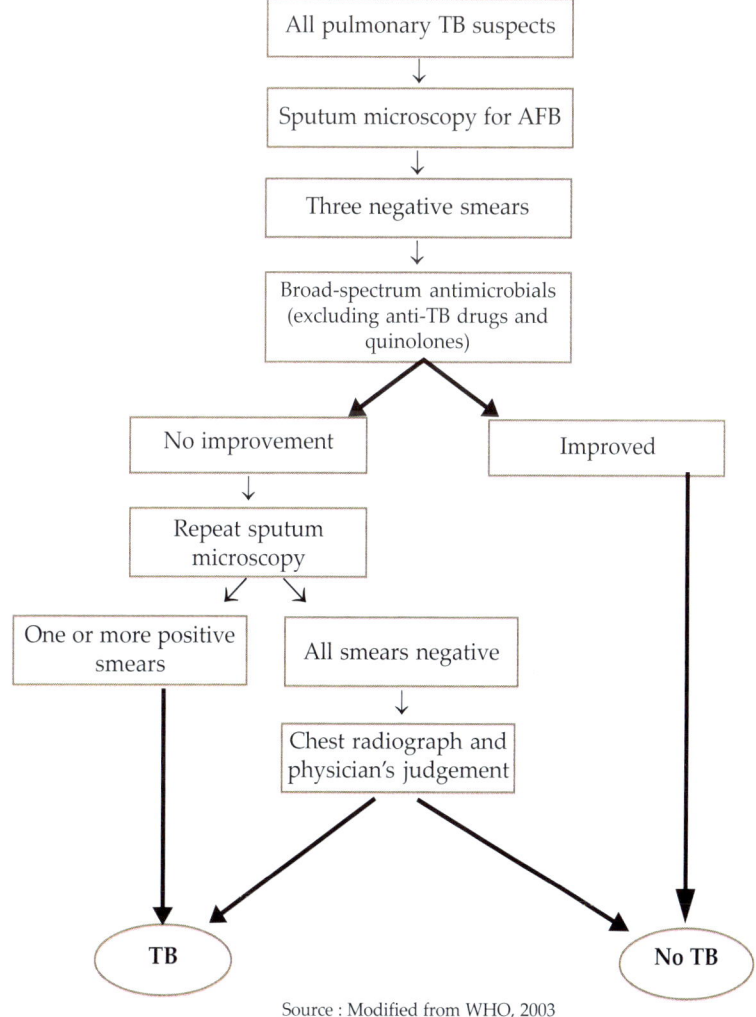

Source : Modified from WHO, 2003

Specimens

Pulmonary Tuberculosis	Extra Pulmonary TB
• Expectorated Sputum - at least 3 of which one should be EM • Induced sputum - especially in cases with non- • Quantity - 3 to 5 ml • Other specimens - Bronchoalveolar lavage (BAL), Biopsy, Gastric lavage.	• Representative of the site of infection • Blood and BM - in productive cough X 3suspected disseminated infections (especially immunocompromised) • Adequate quantity (10 ml - body fluids) • Swabs - Not recommended; Inhibits transfer from swab to culture medium. Should be rejected. If received, place tip directly onto solid culture medium or put whole in 5 ml broth. Colonies are formed in the fibers at the junction with culture media.

Decontaminate and liquefy, if required.

Concentrate the specimen by centrifugation [RCF = 1.12 x R max (in mm) x (rpm/1,000) x 2].

Microscopy

- Simplest and most rapid
- Acid fast (acid-alcohol fast)/Fluorescent stains
- All Non tuberculous mycobacteria (NTMs) are not fluorescent (*M. fortuitum*)
- Sensitivity – 5000 orgs/ml
- Observe direct smears and smears from centrifuged sediment
- If no digestion/decontamination done prior to sedimentation, observe smears prepared from supernatant also.

- Average observation time per slide – 2 minutes
- By examining for the recommended time (10 minutes), case detection doubles.
- Simplest and most rapid test for diagnosis
- Inexpensive
- Identifies infective patients
- Monitor response to treatment (Repeat sputum examination)
- Following treatment, culture becomes negative earlier than smear
- Fluorescin diacetate staining can be used to assess viability of mycobacteria
- Early and accurate identification of treatment failure.

No. of acid-fast bacilli observed in smears, concentrations of culturable bacilli in sputum specimens and probability of positive results

No. of bacilli observed	Estimated concentration of bacilli per ml of specimen	Probability of a positive result
0 in 100 or more fields	< 1000	< 10%
1-2 in 300 fields	5000 – 10,000	50%
1-9 in 100 fields	About 30000	80%
1-9 in 10 fields	About 50000	90%
1-9 per field	About 100000	96.2%
10 or more per field	About 500000	99.95%

False Positive

- Food particles (waxes and oils)
- Precipitates
- Other organisms (Environmental mycobacteria, Nocardia, Yeasts, spores of *B. subtilis*)
- Inorganic materials
- Artefacts (fibers and pollen, scratches on the slide)
- Carry over contamination.

Types of TB	Sensitivity	Specificity
Pulmonary	30-80% Lower in children and HIV co-infected	95% Beaded forms - MTBC Cross banding - *M. kansasii* Presence of AFB indicates only that Mycobacteria are present, not viable or dead and not MTB/MOTT The predictive value is high in high burden countries.
TB lymph node	60-80% (FNAC)	
Pleural effusion	> 10% + HIV – 20%	
TB meningitis	20-58%	

Peptide nucleic acids (PNAs)

- A potential advancement in microscopy
- A fluorescent stain format
- PNAs are DNA like molecules in which the sugar phosphate backbone is replaced with a peptide like structure.
- The binding of PNA to DNA or RNA is sequence specific and the interaction is stronger than that of a DNA–DNA interaction.
- Labelling the PNA with a fluorescent dye enables visualisation with a suitable microscope.
- PNAs specific for the *M. tuberculosis* complex and NTM have been formulated and tested against a panel of cultured mycobacterial species with some success.

Isolate the organism after Decontamination, Liquefaction and Concentration.

Culture

- Solid and liquid media
- Sensitivity – 10 to 100 bacilli /ml (viable)
- Egg based (Lowenstein Jensen) or Agar based (Middlebrook 7H10/11)
- Kirschner's/Middlebrook 7H9 broth
- Limitation – Slow growth.

Rapid Methods of Culture

A) Manual

- Microcolony detection on solid media
  - Plates poured with thin layer of Middlebrook 7H11 agar medium
  - Observed microscopically on alternate days
  - In less than 7 days, micro colonies can be detected.
  - Bird nest with cording – MTBC
  - Bird nest without cording – *M. kansasii*
  - Less expensive and requires about half the time needed for conventional culture
  - The recovery of mycobacteria is less efficient and labour intensive.
- Microscopic observation of broth cultures
  - Rapid and relatively inexpensive method
  - Compares very well with other well established systems in terms of both sensitivity and specificity, and speed when compared to solid media .
  - *M. tuberculosis* grows more rapidly in liquid medium forming strings and tangles, which can be observed under the inverted light microscope with 40X magnification.

- It requires P2 Bio-safety cabinets, relatively expensive Middlebrook 7H9 broth, oleic acid dextrose catalase (OADC) and anti-microbial supplements and high technical skill.

- Septichek AFB
  - Biphasic medium
  - Liquid phase (MB7H9 + CO2+ enrichment + antimicrobial cocktail)
  - The solid phase has 3 components:
    1 – Mycobacteria
    2 – MTB (MB7H11 + NAP)
    3 – Other bacteria (chocolate agar)
  - Consists of a capped bottle containing 30 ml of Middlebrook 7H9 broth under enhanced (5-8%) CO_2. A paddle with agar media enclosed in a plastic tube, and enrichment broth containing glucose, glycerine, oleic acid, pyridoxal, catalase, albumin, polyoxyethylene 40 stearate, azlocillin, nalidixic acid, trimethoprim, polymyxin B and amphotericin B. One side of the paddle is covered with non-selective Middlebrook 7H11 agar, the reverse side is divided into two sections: one contains 7H11 agar with *para-nitro-α*-acetylamino-*β*-hydroxypropiophenone (NAP) for differentiation of *M. tuberculosis* from other mycobacteria.The other section contains chocolate agar for detection of contaminants.

  - This method requires about 3 weeks of incubation.

  - The unique advantage of this technique is the simultaneous detection of *M. tuberculosis*, non-tuberculous mycobacteria (NTM), other respiratory pathogens and even contaminants.

  - The system gives a better culture result when compared to other methods including BACTEC 460 TB system.

Mycobacteria Growth Indicator Tube (MGIT) System (BBL, BD), uses 7H11 Middlebrook broth and an O_2-sensitive fluorescent sensor to indicator. Microbial growth is detected earlier in this method, than by conventional culture on L-J medium.

- **MGIT** consists of 16 x 100 mm round bottomed glass tube containing 4 ml of modified 7 H 11 broth base and 0.5 ml OADC (oleic acid, bovine albumin, dextrose and catalase) and 0.1 ml of PANTA solution (polymyxin B, amphotericin B, nalidixic acid, trimethoprim and azlocillin). 7H11 medium contains defined salts, vitamins, cofactors, oleic acid, albumin, catalase, glycerol, glucose and 0.1% casein hydrolysate.

- A fluorescent compound embedded in silicone is present at the bottom of the tube which is sensitive to dissolved O_2 in the broth.

- 0.5 ml of specimen is added, the tube is inverted for mixing and incubated at 37°C.

- As bacteria consume dissolved oxygen, fluorescence is unmasked and detected under long wave ultraviolet light by Wood's lamp. Tubes are read at day 2 and then alternate days for bright orange colour at the bottom of each tube with an orange reflection on the meniscus.

- Non-homogenous turbidity or small grains or flakes in medium within 7-10 days also indicate a positive sample

B) Automated systems

- Mycobacteria in clinical samples can be detected in half the time compared to conventional culture methods.

- There is a good correlation in drug susceptibility tests and most of these results can be obtained within 8-12 days.

- BACTEC 460 is the reference standard.

Different Automated Culture Systems

	BACTEC 460TB	MB/BacT	ESPII (VersaTrek)	MGIT 960
Principle	Detects radiolabeled CO_2 which is released when radiolabeled palmitic acid is utilised as carbohydrate substrate	Detects CO_2 by a gas permeable sensor	Detects gas pressure changes Reduction – early phase due to O_2 consumption; Increase - Release of gases in later phase	Detects decrease in oxygen
Detector	Scintillation counter	Gas permeable sensor which detects colour change from green to yellow when CO_2 produced	Sensor to detect pressure changes	Oxygen quenched fluorescent compound (ruthenium salt) embedded in silicone at bottom Observe under UV light
Other features	Non automated Needle aspiration assembly NAP – Differentiating MTB from MOTT	Automated Hourly monitoring	Automated Monitors every 10 minutes Shaker incubator Special cellulose sponge resembling alveoli	Automated Hourly monitoring Can monitor 960 tubes at a time
Advantages	Comparison of average time of detection between paired specimen showed that the BACTEC 460 and MGIT 960 systems are 87 versus 86 days for *M. avium* complex (MAC) & 13.4 versus 15.5 days for *M. avium* complex (MAC) and 13.4 versus 15.5 days for *M. tuberculosis* respectively	Comparison of the of detection between paired specimens showed that the BACTEC 460 showed that the mean time for detection of *M. tuberculosis* by the BACTEC system was 11.6 days vs 13.7 d. by the MB/BacT system. MB/BacT with the computerized data management system is an acceptable alternative for BACTEC 460 method.	The mean time for recovery of all mycobacteria, *M. tuberculosis* complex & MAC was found to be 13.1, 15.5 and 10.9 days respectively. A reliable nonradiometric, less labour-intensive less labour-intensive alternative to BACTEC 460 system for the growth and detection of mycobacteria.	MGIT 960 exhibits greater potential as a rapid, accurate & cost effective method for a high volume mycobacteriology laboratory.

Recovery rates of various mycobacteria with different media and their combinations

Organism	ESP (%)	BACTEC (%)	L-J medium (%)	ESP +L-J medium (%)	BACTEC + L-J medium (%)	ESP + BACTEC (%)
M. tuberculosis complex	85.27	97.67	82.94	96.12	99.22	99.22
MAC	94.59	75.67	56.76	100	78.38	97.30
M. gordonae	75.00	71.87	3.12	75.00	71.87	100
M. xenopi	0.00	92.31	61.53	61.53	100	92.31
Other MOTT	50.00	75.00	50.00	75.00	75.00	100

All liquid culture systems should be used in combination with a solid medium, and not as a stand-alone system.

Culture Performance

Site	Sensitivity (%)	Comments
Pulmonary	40-80 CN cavitary disease is unlikely to be TB	Highly specific.
TB lymph node	77-90	Rule out false positivity due to lab contamination.
Pleural Effusion	12-70 Improved sensitivity – bedside inoculation, collecting in anticoagulantBiopsy – 40-80	20-40% specimens will be smear negative and culture positive.
TB meningitis	50-70	
Genitourinary TB	< 40 patients CP for PTB, 5-8% are urine CP21% of patients with EPTB are UCP	CP in SN patients is better than PCR positivity in SN_{43}.

Phage based Assays

- Relatively easy to perform
- Infrastructure needed as for routine mycobacterial cultures
- The turnaround time is 2 days compared to about 2 hours (microscopy) or up to 2 months (culture)
- Two main phage-based approaches are used to detect *M. tuberculosis*
  - Amplification of phages after their infection of *M. tuberculosis*, followed by detection of progeny phages using helper cells (plaque formation)
  - Detection of light produced by luciferase reporter phages (LRP) by live *M. tuberculosis*.
- Two large-scale studies have shown that the test detected 65-83% of the confirmed TB cases within 48 hours.
- The specificity of the tests in each of the studies was > 95%
- Sensitivity and specificity of phage assay with respect to growth on L-J medium was 92.86% and 97.83% respectively.
- Phage-based assays are highly specific but not sensitive enough to be equivalent to culture.
- Phage based assays cannot replace conventional tests.
- Solid media allow a prompt recognition of contaminations and mixed cultures.

Molecular Methods

Method	Target	Detection method	Sensitivity in respiratory samples (%)	Sensitivity inextra respiratorysamples (%)	Overall specificity (%)
AMTD2	16S rRNA	Chemiluminometric	80-100	60-90	95-100
LCx	b antigenic protein	Fluorimetric	80-90	65-80	90-100
AMPLICOR	16S rRNA	Colorimetric	75-100	45-60	90-100
BD ProbeTec	IS6110 &16S rRNA	Fluorimetric	55-100	30-80	45-100
INNO-LIPA v2	IR16S-23S	Colorimetric	50-95	60-80	90-100
GenoType *Direct*	23S rRNA	Colorimetric	60-95	60-80	95-100
PCR real time	16S rRNA	Fluorimetric	70-90	65-85	85

In smear negative samples the sensitivity is reduced to 50%.

Nuclei acid amplification tests (NAATs)

AccuProbe DNA hybridization tests, the PCR-based Inno-LiPA Rif. TB (LiPA) assay, and a PCR-based DNA sequencing of the *rpoB* gene for the rapid identification of the *Mycobacterium tuberculosis* complex.

PRA (Polymorphism Restriction Amplification)

Amplification, by PCR of a fragment of the hsp65 gene, followed by a restriction with 2 restriction enzymes (*Bst*EII y *Hae*III).

Pyrosequencing

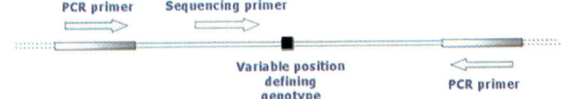

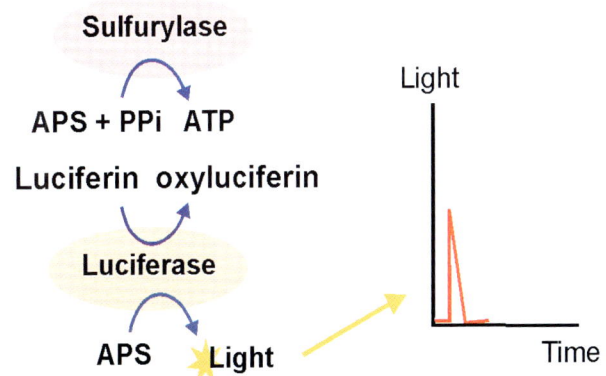

Nucleotide incorporation generates light seen as peak in the pyrogram

Restriction Fragment Length Polymorphism (RFLP)

- Insertion sequence present exclusively in the *M.tuberculosis* complex: IS*6110*
- High polymorphism between no related strains regarding the number of copies and their localization in the chromosome.
- Advantages: High discriminative power.
- Disadvantages: Slow, laborious and with certain complexity.

Spoligotyping

- The DR sequences (*direct repeat*) are repeated sequences of 36 bp in only one locus of the *M.tuberculosis* chromosome, separated by sequences of 34 to 41 bp.
- The technique is based on a PCR of the locus where the DR sequences are located. The amplification product is hybridized with oligos synthesized from the inter-DR spaces.
- The presence or absence of different DR allows a specific pattern for each strain.
- Advantages: Few DNA is required, easy interpretation
- Disadvantages: Lesser discriminative power than the RFLP.

Mycobacterial Interspersed Repetitive Units (MIRUs)

- Determine the number of repetitive units in 12 (15 or 20) different locus of one genetic sequence called "mycobacterial interspersed repetitive units (MIRUs)". The number of repetitions is detected by PCR.
- The number of repetitive units in each locus is calculated by the size of the fragment amplified with the specifics primers.

Transmembrane amplification (TMA):

Comparative Evaluation of TMA and PCR

- MIRU-VNTR is more discriminative than the *spoligotyping* and similar to the RFLP-IS6110.
- Advantages: rapid, simple and automatic.
- Disadvantages: In study.

Line Probe Assays for DST

- INNO-LiPA Rif TB
- GenoType MTBDR*plus*

Resistance	Sensitivity	Specificity
RMP	98.1%	98.7%
INH	84.3%	99.5%

Drug	Gene locus	Gene function	Resistance %
Fluoroquinolones Amikacin, Capreomycin,	gyrA	DNA-Gyrase A	Approx. 80-90%
Kanamycin	Rrs tlyA	16S rRna Methylase	Approx. 80%
Ethambutol	embB		Approx. 30-60%

Xpert MTB

The assay is fully automated with only 3 manual steps at the beginning. Addition of SR to raw sputum and 15 min later, after the sample has been inactivated and liquified, transfer to the Cartridge and in the instrument. All the rest is automated: Sample concentration, removal of inhibitors, ultrasonic lysis of cells and a nested real time PCR. Time to result less than 2 hours. Sensitivity and Specificity seems to be very good for the detection of TB and Rifampicin resistance.

Feature	TMA	PCR
Detects	Live Bacteria	Live and dead both
Nucleic Acid Detected	rRNA	DNA
False Positive Issues	No false positive issue because no contamination 1) The test picks up RNA and this is present only in live bacteria 2) No aerosol is formed due to a single tube assay format	Encountered very often because 1) DNA left behind by dead bacteria is picked up in PCR 2) Aerosols are formed because of pipetting from the PCR tubes.
Testing Time	3 hours	8-12 hours
Can be used as therapy monitoring tool	Yes (because it can detect only live bacteria)	No (because it can detect live and dead bacteria)
Positive Predictive Value	100%	< 75% due to problems of contamination.

Conclusion

- The sensitivity of the molecular tests vary, and is affected by the amount of bacteria present in the samples, and also by the clinical suspicion level. Low sensitivity is present in samples with low bacterial load, especially in extra respiratory samples

- At the moment, new molecular methods cannot substitute the conventional ones. The gold standard is the culture, and the other methods have to be considered and interpreted as complementary diagnostic methods.
- Communication between clinicians and microbiologists is imperative

Tuberculin Test. Intradermal injection of 5 TU (1 TU ≡ 0.02 µg of PPD). After 48-72 hours – induration of 10mm diameter or more is positive.

Positive test	False negative test
a) Confirms active infection in children under 5 years of age b) Signifies past infection by tubercle bacilli	a) Miliary tuberculosis b) Anergy due to overwhelming infection of measles, Hodgkin's disease, sarcoidosis, lepromatous leprosy, malnutrition, administration of immunosuppressive agents and corticosteroids.

False positive test

In presence of other mycobacteria like *M. avium* or other atypical mycobacteria.

BCG vaccine. It is an attenuated strain of *M. bovis* obtained by 231 repeated subcultures, every 3 weeks in glycerine bile potato medium. Lyophilised form is commonly used. Given 0.1 ml intradermally within 0-15 days of birth. Papule after 3 weeks, ultimately ulcer by 5 weeks which heals up leaving 4-8 mm diameter permanent scar. A person becomes tuberculin positive after 4-6 weeks of BCG vaccination.

a) Does not give absolute protection but a milder course in immunised children.

b) Prevents serious forms like miliary tuberculosis, tuberculous meningitis, skeletal tuberculosis, etc.

c) Stimulates T-lymphocytes which offer some protection against leprosy and leukemia.

Theoretically 1 vaccine should give life-long protection, but major trials have shown 7-10 years protection after the vaccination.

Old therapy

Triple therapy with streptomycin, INH and PAS for 2 years.

REVISED NATIONAL TUBERCULOSIS CONTROL PROGRAM (RNTCP) AND DIRECTLY OBSERVED TREATMENT SHORT COURSE CHEMOTHERAPY (DOTS)

Global Scenario

Tuberculosis (TB) remains one of the world's deadliest communicable diseases, second only to HIV causing deaths amongst communicable diseases. In 2013, 9 million people fell ill with TB and 1.5 million died from the disease.

WHO Global Tuberculosis Report 2014

- Over 95% of TB deaths occur in low- and middle-income countries, and it is among the top 5 causes of death for women aged 15 to 44.

- In 2013, an estimated 550,000 children became ill with TB and 80,000 HIV-negative children died of TB.

- TB is a leading killer of HIV-positive people causing one fourth of all HIV-related deaths.

- Globally in 2013, an estimated 480,000 people developed multidrug resistant TB (MDR-TB).

- The estimated number of people falling ill with TB each year is declining, although very slowly, which means that the world is on track to achieve the Millennium Development Goal to reverse the spread of TB by 2015.

- The TB death rate dropped 45% between 1990 and 2013.

- An estimated 37 million lives were saved through TB diagnosis and treatment between 2000 and 2013.

- India and China alone account for 24% and 11% of total cases, respectively. More than half (56%) were in the South-East Asia and Western Pacific Regions. A further one quarter were in the African Region, which also had the highest rates of cases and deaths relative to population.

- Globally, 3.5% (95% CI: 2.2–4.7%) of new TB cases are estimated to have MDR-TB. Globally, 5% of TB cases were estimated to have had MDR-TB in 2013 (3.5% of new and 20.5% of previously treated TB cases). Drug resistance surveillance data show that an estimated **480 000 people** developed **MDR-TB** in 2013 and 210 000 people died.

- **Extensively drug-resistant TB (XDR-TB)** has been reported by **100 countries** in 2013. On average, an estimated 9% of people with MDR-TB have XDR-TB.

- Globally, 20.5% (95% CI: 13.6–27.5%) of previously treated cases are estimated to have MDR-TB

- TB will cost the world's poorest countries upto $3 trillion.

Indian scenario (WHO Global Tuberculosis Report 2014 (updated Feb 2015)

- India is the second-most populous country in the world, one fourth of the global incident TB cases occurs in India annually.

- In 2013, out of the estimated global annual incidence of 9 million TB cases, 2.1 million were estimated to have occurred in India.

- 1,000 TB deaths every day.

Estimated TB Burden, 2013 in India

Population (millions)	Mortality (Excluding HIV)		Mortality HIV Positive		Prevalence		Incidence
	Number (Thousands)	Rate*	Number (Thousands)	Rate*	Number (Thousands)	Rate*	Number Thousands)
1,252	240 (150-350)	19 (12-28)	38 (31-44)	3 (2.5-3.5)	2,600 (1,800-3,700)	211 (143-294)	2,100 (2,000-2,300)

*Rates are per 100,000 population

Confirmed MDR	Estimated MDR amongst notified cases		
	Total	New Pulmonary cases	Previously treated
25244	62,000 (50,000-74,000)	20,000 (17,000-24,000)	41,000 (30,000-52,000)

Every year 9 million people get sick with TB and 3 million do not get the care they need.

World TB Day 24th March 2015: "Reach the 3 Million: Reach, Treat, Cure Everyone"

Stop TB Strategy

Vision	A TB-free world
Goal	To dramatically reduce the global burden of TB by 2015 in line with the Millenium Development Goals and the Stop TB Partnership targets.
Objectives	- Achieve universal access to high quality care for all people with TB
	- Reduce the human suffering and socioeconomic burden associated with TB
	- Protect vulnerable populations from TB, TB/HIV and Multidrug Resistant TB
	- Support development of new tools and enable their timely and effective use
	- Protect and promote human rights in TB prevention, care and control.
Targets	- MDG 6, Target 8*: Halt and begin to reverse the incidence of TB by 2015
	- Targets linked to the MDGs and endorsed by Stop TB Partnership:
	– 2015: reduce prevalence of and deaths due to TB by 50%
	– 2050: eliminate TB as a public health problem.

*Goal 6: "Combat HIV/AIDS, malaria and other diseases"

Target 8: "By 2015, to have halted and begun to reverse the incidence of malaria and other major diseases…"

- Indicator 23: between 1990 and 2015 to halve prevalence of TB disease and deaths due to TB

- Indicator 24: to detect 70% of new infectious cases and to successfully treat 85% of detected sputum positive patients.

Beyond 2015

- The end of 2015 marks a transition from the MDGs to a post-2015 development framework.

- Within this broader context, WHO has developed a post-2015 global TB strategy (the **End TB Strategy**): approved by all Member States at the May 2014 World Health Assembly.

- The overall goal of the strategy is to end the global TB epidemic, with corresponding 2035 targets of:
  - 95% reduction in TB deaths and
  - 90% reduction in TB incidence (both compared with 2015).

- The strategy also includes a target of zero catastrophic costs for TB affected families by 2020.

RNTCP Program

The Revised National Tuberculosis Control Program (RNTCP), based on the DOTS strategy, began as a pilot in 1993 and was launched as a national programme in 1997. Rapid RNTCP expansion began in late 1998. By the end of 2000, 30% of the country's population was covered, and by the end of 2002, 50% of the country's population was covered under the RNTCP. By the end of 2003, 778 million population was covered, and at the end of year 2004 the coverage reached to 997 million. By December 2005, around 97% (about 1080 million) of the population had been covered, and the entire country was covered under DOTS by 24th March 2006.

Objectives of RNTCP

- To achieve and maintain a cure rate of at least 85% among newly detected infectious (new sputum smear positive) cases

- To achieve and maintain detection of at least 70% of such cases in the population.

12th Five Year Plan

- RNTCP has entered in an ambitious National Strategic Plan (NSP) 2012-17 as part of the country's 12th Five year Plan.

- The theme of the NSP 2012-17 is "Universal Access for quality diagnosis and treatment for all TB patients in the community" with a target of "reaching the unreached".

- **Vision:** The vision of the Government of India is a "**TB-free India** - through achieving Universal Access by provision of quality diagnosis and treatment for all TB patients in the community".

Objectives

- Early detection and treatment of at least 90% of estimated all type of TB cases in the community, including Drug resistant and HIV associated TB.

- Successful treatment of at least 90% of new TB patients, and at least 85% of previously-treated TB patients

- Reduction in default rate of new TB cases to less than 5% and re-treatment TB cases to less than 10%

- Initial screening of all re-treatment smear-positive till 2015 and all Smear positive TB patients by year 2017 for drug-resistant TB and provision of treatment services for MDR-TB patients;

- Offer of HIV Counselling and testing for all TB patients and linking HIV-infected TB patients to HIV care and support

- Extend RNTCP services to patients diagnosed and treated in the private sector.

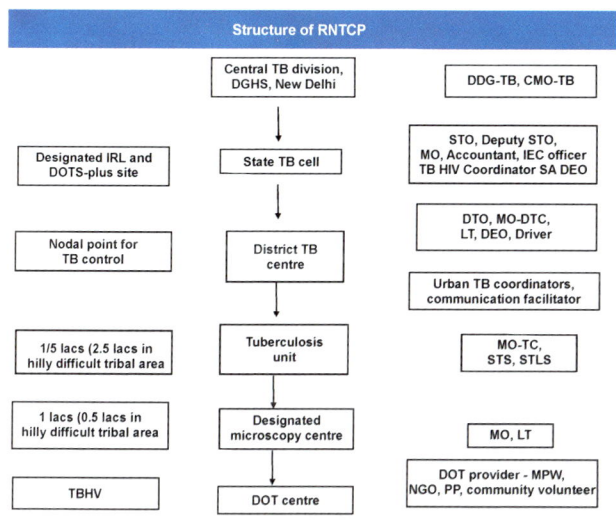

DGHS: Directorate General of Health services.

Structure of RNTCP Laboratory Network

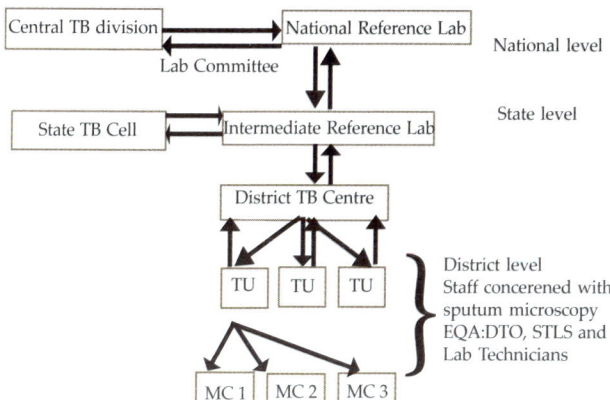

Supranational Reference Laboratories: National Institute of TB and Respiratory Diseases (NITRD) New Delhi (National Centre for excellence); National Institute for Research in Tuberculosis No.1, Chennai. TRC: TB Research Centre, Chennai.

National Reference Laboratories (NRLs): The six NRLs under the programme include National Institute for Research in Tuberculosis (NIRT), Chennai; National Tuberculosis Institute (NTI), Bangalore; National Institute of Tuberculosis & Respiratory Diseases (NITRD), Delhi; National Japanese Leprosy Mission for Asia (JALMA) Institute of Leprosy and other Mycobacterial Diseases, Agra; Regional Medical Research Centre (RMRC), Bhubaneswar; and Bhopal Memorial Hospital and Research Centre (BMHRC), Bhopal*. The NRLs work closely with the IRLs, supervise their activities and also undertake periodic training of the staff with respect to EQA and C & DST.

IRL: There are 8-11 IRLs under 1 NRL. One IRL has been designated in the STDC Public Health Laboratory/ Medical College of the respective state. The functions of IRL include supervision and monitoring of EQA activities, Mycobacterial culture and DST as well as Drug Resistance Surveys (DRS) in selected states. The IRL conducts regular trainings to ensure that the district and sub-district laboratory staff has the technical know-how to efficiently perform smear microscopy activities. Additionally, they undertake onsite evaluation and panel testing of each district in the state at least once a year. There are 27 IRLs in Guwahati, Hyderabad, Patna, Raipur, Delhi, Ahmedabad, Karnal, Dharampur, Ranchi, Bengaluru, Thiruvanthapuram, Indore, Bhopal, Pune, Nagpur, Cuttack, Puducherry, Ajmer, Chennai, Patiala, Dehradun, Lucknow, Agra and Kolkata.

Solid DST: IRL at Pune, Nagpur; Medical Colleges at GMC, Mumbai and MGIMS, Wardha.

Liquid DST: IRL Nagpur, NGO based Hinduja Hospital and SRL Mumbai.

LPA: IRL Pune, Nagpur, Medical College at GMC, Mumbai, NGO Hinduja Hospital.

2nd line DST: IRL Nagpur, Medical College at GMC, Mumbai, NGO Hinduja Hospital.

TB Surveillance: NIKSHAY

- Central TB Division (CTD) in collaboration with National Informatics Centre (NIC) undertook the initiative to develop a Case Based Web online (cloud) application named Nikshay.

- This software was launched in May 2012 and has following functional components:

  - Master management & User details

  - TB Patient registration & details of diagnosis, DOT Provider, HIV status, Follow-up, contact tracing, Outcomes

  - Details of solid and liquid culture & DST, LPA, CBNAAT details

- DR-TB patient registration with details
- Referral and transfer of patients
- Private health facility registration and TB Notification
- Mobile application for TB notification
- SMS alerts to patients on registration
- SMS alerts to programme officers
- Automated periodic Reports: Case Finding, Sputum conversion, Treatment outcome.

Municipal Corporaton of Greater Mumbai (MCGM) launched its **"TB Harega Desh Jeetega"** campaign in 2015, in a landmark moment for advocacy for TB prevention and control in India.

Basis of TB treatment

- Intermittent (thrice weekly) treatment regimens
- Treatment given under direct observation
- Standardized treatment regimens in two categories
- Regimen decided by MO on basis of
- Sputum smear results
- History of previous anti-TB treatment
- Disease classification (pulmonary/extra pulmonary)
- Severity of illness

Revised TB definitions

Existing term RNTCP	Existing RNTCP definitions	New term RNTCP	New definitions under RNTCP
Pulmonary TB suspects	Persons having cough of 2 weeks or more with or without other symptoms. Should have 2 sputum samples examined for AFB.	Presumptive TB	Person with any of the symptoms and signs suggestive of TB, including cough > 2 weeks, significant weight loss, hemoptysis, any abnormality in chest radiograph. In addition, bacteriologically confirmed TB patients, PLHIV, diabetics, malnourished, cancer patients, patients on immunosuppression or steroid hould be regularly screened for signs and symptoms of TB.
Disease Classification (Pulmonary or Extra pulmonary based on anatomical site)			
Pulmonary Tuberculosis	Smear positive and smear negative definitions together make PTB	Pulmonary Tuberculosis	Pulmonary Tuberculosis (PTB) refers to any bacteriologically confirmed or clinically diagnosed case of TB involving the lung parenchyma and the tracheobronchial tree N.B. Miliary TB is classified as PTB because there are lesions in the lungs. A patient with both pulmonary & extra pulmonary TB should be classified as a case of PTB
Extra Pulmonary Tuberculosis	Tuberculosis of organs other than the lungs such as pleura, lymph nodes, intestine, genitourinary tract, joint and bones, meninges of the brain, etc. is called as extra-pulmonary TB. Pleural TB is classified as extra pulmonary.	Extra Pulmonary Tuberculosis	**Extrapulmonary tuberculosis (EPTB)** refers to any bacteriologically confirmed or clinically diagnosed case of TB, involving organs other than the lungs such as pleura, lymph nodes, intestine, genitourinary tract, joint and bones, meninges of the brain, etc.
Smear-positive pulmonary TB	A person with one or two smears positive for AFB out of the two specimens subjected for smear examination by direct microscopy is classified as having Smear positive pulmonary TB	Not to be used	-
Smear-negative pulmonary TB	A patient with symptoms positive for TB with 2 smear examination negative for AFB, with evidence of pulmonary TB by microbiological methods (culture positive or by other approved molecular methods) or by Chest X-ray is classified as having smear negative pulmonary TB	Not to be used	-
-	Do not exist	Bacteriological confirmed TB case	Bacteriological confirmed TB case (Definitive TB case) refers to a presumptive TB patient from whom a biological sample is positive for acid fast bacilli, or positive for M. tuberculosis on culture, or positive for TB through Quality Assured Rapid Diagnostic Molecular test
-	Do not exist	Clinically diagnosed TB case	A clinically diagnosed TB case (Probable case) refers to a presumptive TB patient who is not bacteriologically confirmed but

			positive but has been diagnosed with active TB by a clinician on the basis of X-ray abnormalities, histopathology or clinical signs with a decision to treat the patient with a full course of Anti-TB treatment.
New	A TB patient who has never had treatment for TB or has taken Anti-TB drugs for less than one month is considered as a new case.	New	A TB patient who never received treatment for TB or received Anti-TB treatment for less than one month.
Previously treated	No such definition exists	Previously treated	A TB patient who received Anti-TB treatment for one month or more in the past
Relapse	A TB patient who was declared cure on treatment completed by a physician and who reports back to the health facility and is now found to be sputum smear positive is a relapse case	Recurrent TB	A TB patient previously declared as successfully treated (cured/treatment completed) and is subsequently found to be baceriologically confirmed TB case is a recurrent TB case
Failure	Any TB patient who is smear positive at 5 months or more after initiation of treatment is considered as failure	Treatment after failure	A TB patient whose Anti-TB treatment failed after end of the most recent course
Treatment after default	A patient who has received treatment of TB for a month or more from any source and returns for treatment after having defaulted, i.e. not taken Anti-TB drugs for consecutively 2 months or more and is found to be smear positive is case of treatment after default	Treatment after lost to follow up	A TB patient previously treated for TB for 1 month or more and was declared lost to follow up in their most recent course of treatment and subsequently found bacteriologically confirmed TB case
Others	A patient who does not fit into any of the types mentioned above. The reason for labeling a patient in this type must be specified in the Treatment card and TB Register	Other previously treated patients	A patient who has been previously treated for TB but whose outcome after their most recent course of treatment is unknown or undocumented
Transferred in	A TB patient who has been received for treatment in a Tuberculosis Unit, after starting treatment in another TB unit, where he has been registered is considered as a case of transferred in	Transferred in	A TB patient who is received for treatment in a Tuberculosis Unit, after registered for treatment in another TB unit is considered as a case of transferred in
Transferred out	A patient who has been transferred to another TU / district / state and whose treatment outcome is no available is considered as 'Transferred out'	Not evaluated	A TB patient for whom no treatment outcome is assigned, this includes former 'transfer-out'
Cured	Initially sputum smear positive patient who has completed treatment and had negative sputum smears on two occasions, one of which is at the end of the treatment is declared as cured.	Cured	Bacteriologically confirmed TB patient at the beginning of treatment who was smear or culture negative at the end of the complete treatment
Treatment completed	Initially sputum smear positive patients who have completed treatment with negative smears at the end of the intensive phase / two months in the continuation phase, but none at the end of treatment, the outcome is declared as treatment completed. Initially sputum smear negative patient who has received full course of treatment and has not become smear positive at the end of the treatment. EPTB patient who has received full course of treatment and has not become smear positive during or at the end of treatment is also declared as treatment completed	Treatment completed	A TB patient who completed treatment without evidence of failure but with no record to show that the smear or culture results of biological specimen in the last month of treatment was negative, either because test was not done or because result is unavailable
Failure	Any TB patient who is smear positive at 5 months or more after starting the treatment is considered as 'Failure'	Failure	A TB patient whose biological specimen is +ve for by smear or culture at end of treatmnt
Defaulted	A patient after treatment initiation has interrupted treatment consecutively for > 2 months	Lost to follow up	A TB patient whose treatment was

Components of DOTS

- **Political and administrative commitment.** TB is the leading infectious cause of death among adults. TB kills more men than women, yet more women die of TB than all causes associated with childbirth combined. Since TB can be cured and the epidemic reversed, it warrants the topmost priority, which it has been accorded by the Government of India. This priority must be continued and expanded at the state, district and local levels.

- **Good quality diagnosis.** Good quality microscopy allows health workers to see the tubercle bacilli and is essential to identify the infectious patients who need treatment the most.

- **Good quality drugs. An uninterrupted supply of good quality anti-TB drugs** must be available. In the RNTCP, a box of medications for the entire treatment is earmarked for every patient registered, ensuring the availability of the full course of treatment the moment the patient is initiated on treatment. Hence in DOTS, the treatment can never interrupt for lack of medicine.

- **Supervised treatment to ensure the right treatment**, given in the right way. The RNTCP uses the best anti-TB medications available. But unless treatment is made convenient for patients, it will fail. This is why the heart of the DOTS programme is "directly observed treatment" in which a health worker, or another trained person who is not a family member, watches as the patient swallows the anti-TB medicines in their presence.

- **Systematic monitoring and accountability.** The programme is accountable for the outcome of every patient treated. This is done using standard recording and reporting system, and the technique of 'cohort analysis'. The cure rate and other key indicators are monitored at every level of the health system, and if any area is not meeting expectations, supervision is intensified. The RNTCP shifts the responsibility for cure from the patient to the health system.

Case Detection

- **Sputum microscopy** is the primary tool for diagnosis
- **Advantages:**
- Provides definitive diagnosis
- Easy to perform
- Reliability and reproducibility
- Cost effective
- No harm to patient unlike X-rays.

Sputum Smear Interpretation

Examination finding	Result as recorded	Grading	No. of fields examined
>10 AFB per oil immersion field	Positive	+3	20
1-10 AFB per oil immersion field	Positive	+2	50
10-99 AFB per 100 oil immersion fields	Positive	+1	100
1-9 AFB per 100 oil immersion fields	Positive	Scanty*	100
No AFB in 100 oil immersion fields	Negative	Negative	100

*Record exact number seen in 100 fields

Diagnostic algorithm of TB

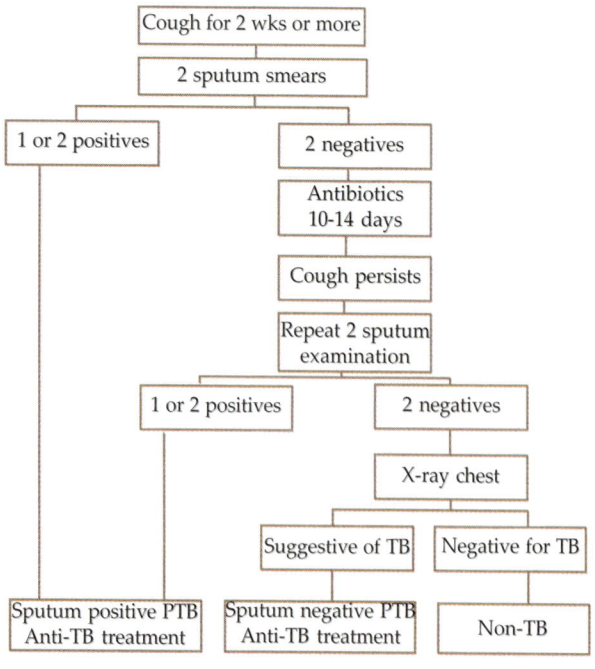

Treatment Regimens

New*	Previously treated**
New sputum smear-positive, New sputum smear-negative,	Sputum smear-positive relapse, Sputum smear-positive
New extrapulmonary tuberculosis, others	failure, Sputum smear-positive treatment after default, others#
$2H_3R_3Z_3E_3 + 4H_3R_3$	$2H_3R_3Z_3E_3S_3 + 1H_3R_3Z_3E_3 + 5H_3R_3E_3$
2 months Intensive phase + 4 mths continuation phase	3 mnths Intensive phase + 5 months continuation phase

H-Isoniazid; R-Rifampicin; Z-Pyrazinamide; E-Ethambutol; S-Streptomycin.

*New categories includes former Categories I & III

**Previously treated is former Category II

Others include patients who are Sputum Smear-Negative or who have Extra-pulmonary disease who can have recurrence or resonance.

Four drugs at Thrice-Weekly Schedule for 2 months Intensive phase & Two drugs at Thrice-Weekly Schedule for remaining 4 months continuation phase. Five drugs at Thrice-Weekly Schedule for initial 2 months followed by four drugs for next 1 month Intensive phase. Three drugs at Thrice-Weekly Schedule for remaining 5 months continuation phase.

Basis for Regimens

CAT I: New sputum smear Positive patients, high bacillary population, chances for naturally occurring resistant mutants higher, therefore 4 drugs in intensive phase

CAT II: Because of previous treatment, chances of harboring resistant bacilli are higher; hence 5 drugs in IP and total duration of treatment is 8 months. In continuation phase lower bacterial population; hence less chance of resistant organisms, therefore 3 drugs are enough.

Regimen for Non-DOTS treatment in RNTCP Areas

- Self-administered non rifampicin containing regimen
- Needed in few cases of adverse reaction to rifampicin and pyrazinamide
- Upto a maximum of 1% of patients may get Non-DOTS treatment in an RNTCP area.
- Tuberculosis treatment card to be filled for these patients as well
- Regimen: 2HSE + 10 HE.

Special situations

- **Hospitalization**
  - General policy is treatment on ambulatory basis.
  - Indoor treatment adviced if general condition of patient is serious
    Pneumothorax
    Massive haemoptysis
    Large pleural effusion
  - Treatment with prolongation pouches supplied by DTO of the district in which hospital is situated.

- **Pregnancy and post-natal period**
  - Streptomycin not to be given. Other drugs in RNTCP are safe
  - Breast feeding should continue
  - Chemoprophylaxis for baby if mother is smear positive

- **Renal failure**
  - Rifampicin, isoniazid and pyrazinamide can be given
  - Streptomycin and ethambutol require close monitoring.

Directly Observed Treatment

- Directly observed treatment (DOT) is one element of the DOTS strategy
- An observer watches and helps the patient swallow the tablets
- Direct observation ensures treatment for the entire course
  - with the right drugs
  - in the right doses
  - at the right intervals

DOTS Strategy

A strategy to ensure treatment completion in which

- Treatment observer (DOT provider) must be accessible and acceptable to the patient and accountable to the health system; NOT a family member
- Can be health care workers, ASHA, Anganwadi Workers, NGO workers, private practitioners, community volunteers, shop keepers, cured patients etc
- DOT provider administers the drugs in intensive phase.
- Ensures that the patient takes medicines correctly in continuation phase.
- Provides the necessary information and encouragement for completion of treatment.

Drug administration

- A suitable DOT provider and DOT centre is selected in consultation with patient
- Tuberculosis Treatment Card is accurately and completely filled after initial home visit
- Initial counselling at the health facility and at patients home is important to achieve treatment compliance
- Ensure that treatment is being directly observed for all doses of the intensive phase and the first of the thrice weekly dose in the continuation phase.

Drug dosages for adults in the blister packs* in RNTCP

Drugs	Dose (thrice a week)	No. of tablets in blister pack
Isoniazid (H)	600 mg	2 X 300 mg
Rifampicin (R)	450 mg**	1 X 450 mg
Pyrazinamide (Z)	1500 mg	2 X 750 mg
Ethambutol (E)	1200 mg	2 X 600 mg
Streptomycin (S)	0.75 gm***	-

*Adult patients who weigh < 30 kgs receive drugs in PWBs from the respective weight band suggested for pediatric patients.

**Patients who weigh > 60 kg at the start of treatment are given an extra 150 mg of Rifampicin.

***Patients over 50 years of age are given 0.5 gm of Streptomycin.

Drug administration

- Streptomycin injections should be given
  - After oral drugs are administered
  - With disposable syringes and needles
- Chemoprophylaxis to be given to children (under 6 years of age) of smear-positive patients
- Patients missing doses should be traced and put back on treatment
  - Within one day in intensive phase
  - Within one week in continuation phase.

Monitoring of Treatment

- Follow up sputum microscopy determines
  - Conversion rate
  - Cure rate

- Sputum smear microscopy schedule
  - Initial sputum examination
  - End of Intensive phase of treatment
  - 2 months into Continuation phase of treatment
  - End of treatment.

Schedule of follow-up sputum smear examination

Category of treatment	Pretreatment sputum	Test at month	If: Result	Then
Cat-I	+	2	- +	C.P. – sputum at 4 & 6 months I.P. for 1 month, Sp. at 3, 5 & 7
	-	2	- +	C.P. sputum at 6 months I.P. for 1 month, Sp. at 3, 5 & 7
Cat-II	+	3	- +	C.P. sputum at 5 & 8 months I.P. for 1 month, Sp. at 4, 6 & 9

DOTS-Plus

- The emergence of resistance to drugs used to treat tuberculosis (TB), and particularly multidrug-resistant TB (MDR-TB): significant public health problem in a number of countries and an obstacle to effective TB control.
- Traditionally, DOTS-Plus refers to DOTS programmes that add components for MDR-TB diagnosis, management and treatment.

MDR TB (Multi Drug Resistant TB)

Initial intensive phase: 6- 9 months	Continuation phase: 18 months
- Inj. Kanamycin - Tab. Ethionamide - Tab. Ofloxacin - Tab. Pyrazinamide - Tab. Ethambutol - Cap. Cycloserine	- Tab. Ethionamide - Tab. Ofloxacin - Tab. Ethambutol - Cap. Cycloserine

At least six months of Intensive Phase (IP) should be given, extended up to 9 months in patients who have a positive culture result taken at 4th month of treatment.

Follow up: Smear examination should be conducted monthly during IP and at least quarterly during CP Culture examination should be done at least at 4, 6, 12, 18 and 24 months of treatment.

CAT V- XDR TB (Extensively Drug Resistant TB

XDR TB – MDR TB + Resistant to Fluroquinolones and at least one Second line injectable Anti-TB drug [Aminoglycosides (Amikacin/Kanamycin) or Capreomycin or both].

On average, an estimated 9% of people with MDR-TB have XDR-TB.

- The **Intensive Phase (6-12 months) will consist of 7 drugs:**

  Capreomycin (Cm), PAS, Moxifloxacin (Mfx), High dose-INH, Clofazimine, Linezolid and Amoxyclav.

- The **Continuation Phase (18 months) will consist of 6 drugs:**

  PAS, Moxifloxacin (Mfx), High dose-INH, Clofazimine, Linezolid and Amoxyclav.

Points to remember

- TB is a notifiable disease.
- If you are not sure of individualized treatment regime, please do not start it. Instead you may register the patient under RNTCP.
- Do not start a fluroquinolone to a TB suspect.
- Please do simple sputum microscopy for AFB smear for all TB suspects, rather than directly starting from higher investigations like CT scan.
- Serological TB tests are banned in India eg. TB IgG and TB IgM.
- Do not even attempt to treat drug resistant TB, in absence of requisite training. Refer to specialist/RNTCP/PMDT.

Pediatric Guidelines

- All efforts should be made to demonstrate bacteriological evidence in the diagnosis of pediatric TB.
- Ideal sample: Sputum
- Alternative specimens: Gastric lavage, Induced sputum, broncho-alveolar lavage, should be

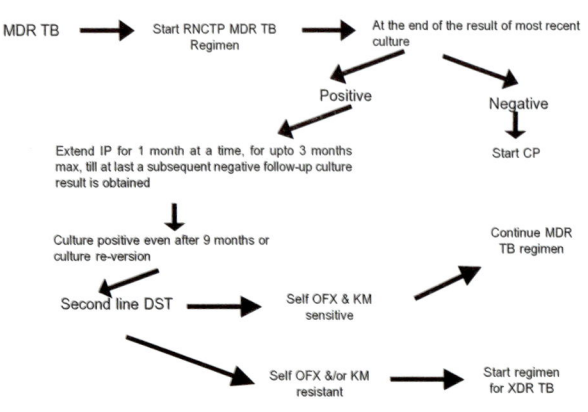

TREATMENT ALGORITHM OF DRUG RESISTANT TB

collected, depending upon the feasibility, under the supervision of a pediatrician.

- A positive Tuberculin skin test/Mantoux positive were defined as 10 mm or more induration. The optimal strength of tuberculin 2 TU (RT 23 or equivalent) to be used for diagnosis in children.

- There is no role for inaccurate/inconsistent diagnostics like serology (IgM, IgG, IgA antibodies against MTB antigens), various in-house or nonvalidated commercial PCR tests and BCG test.

- Loss of weight was defined as a loss of more than 5% of the highest weight recorded in the past three months.

Treatment Regimens

- There will be only two treatment categories – one for treating 'new' cases and another for treating 'previously treated cases' as in adults.

- **Extending intensive and continuation phase:**

    a. Children who show poor or no response at 8 weeks of intensive phase should be given benefit of extension of IP for one more month.

    b. In patients with TB Meningitis, spinal TB, miliary/disseminated TB and osteoarticular TB, the continuation phase shall be extended by 3 months making the total duration of treatment to a total of 9 months.

A further extension may be done for 3 more months in continuation phase (making the total duration of treatment to 12 months) on a case to case basis in case of delayed response and as per the discretion of the treating physician/ pediatrician.

TB Preventive Therapy

- The dose of INH for chemoprophylaxis is 10 mg/kg (instead of currently recommended dosage of 5 mg/kg) administered daily for 6 months.

- TB preventive therapy should be provided to:

    a. All asymptomatic contacts (under 6 years of age) of a smear positive case, after ruling out active disease and irrespective of their BCG or nutritional status.

    b. Chemoprophylaxis is also recommended for all HIV infected children who either had a known exposure to an infectious TB case or are Tuberculin skin test (TST) positive ($\geq$ 5mm induration) but have no active TB disease.

    c. All TST positive children who are receiving immunosuppressive therapy (*e.g.* Children with nephrotic syndrome, acute leukemia, *etc.*).

    d. A child born to mother who was diagnosed to have TB in pregnancy should receive prophylaxis for 6 months, provided congenital TB has been ruled out.

BCG vaccination can be given at birth even if INH chemoprophylaxis is planned.

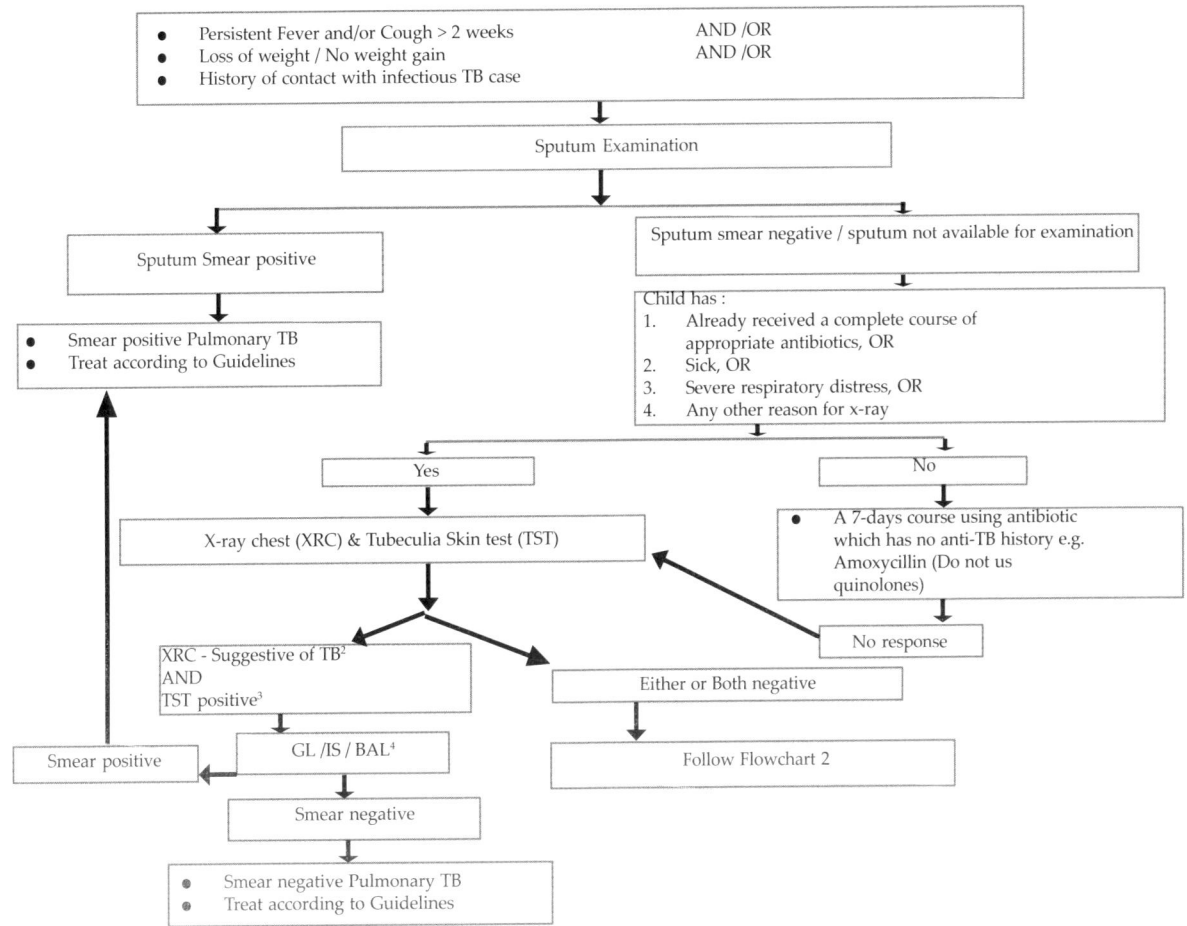

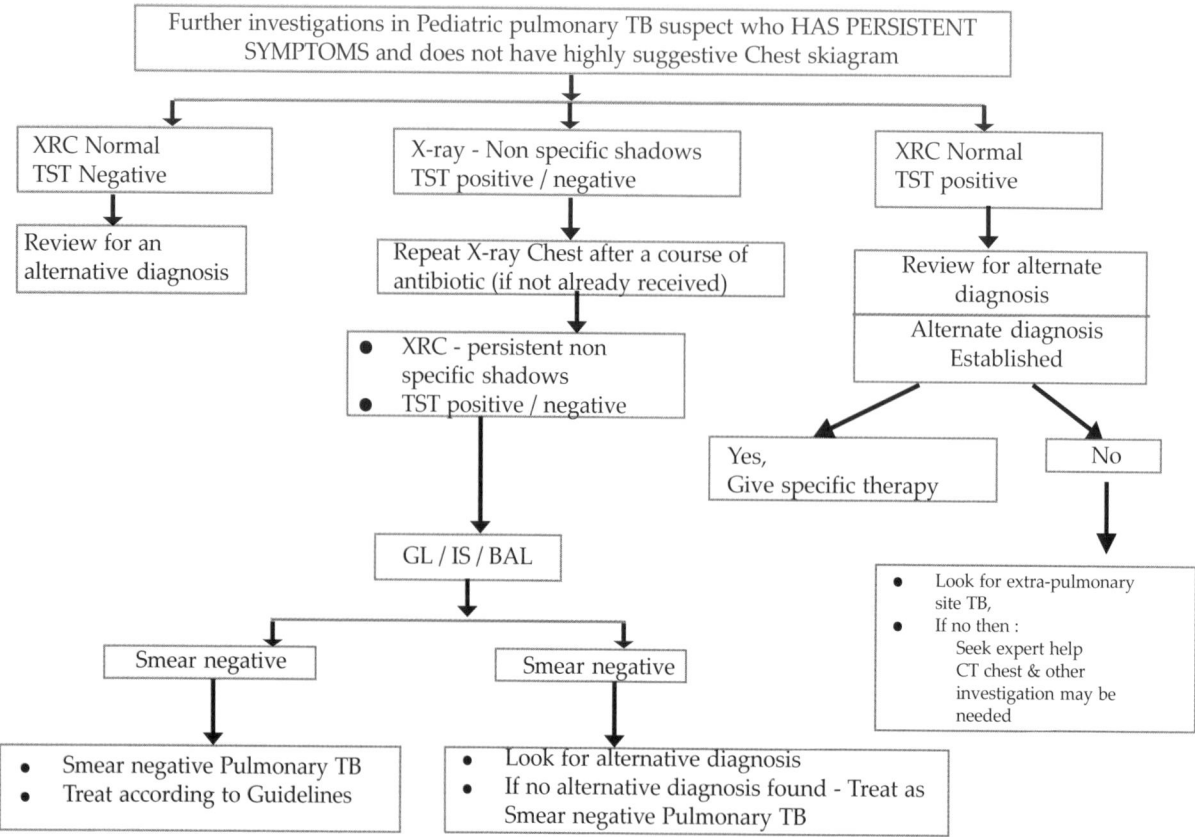

GeneXpert and DST Update

- The rapid TB test, known as Xpert MTB/RIF is a fully-automated, closed system that performs real-time PCR.

- It has the potential to revolutionize and transform TB care and control.

- The test:
  - Simultaneously detects TB and rifampicin drug resistance (detects rpoB gene)
  - Provides accurate results < 2 hours so that patients can be offered proper treatment on the same day
  - Minimal bio-safety requirements and training needs, and can be housed in non-conventional laboratories.

WHO Recommendations: Pulmonary TB

- Xpert MTB/RIF should be used rather than conventional microscopy, culture and DST as the initial diagnostic test in adults suspected of having MDR-TB or HIV-associated TB (strong recommendation, high-quality evidence).

- Xpert MTB/RIF should be used rather than conventional microscopy, culture and DST as the initial diagnostic test in children suspected of having MDR-TB or HIV-associated TB (strong recommendation, very low-quality evidence).

- Xpert MTB/RIF may be used rather than conventional microscopy and culture as the initial diagnostic test in all adults suspected of having TB (conditional recommendation acknowledging resource implications, high-quality evidence).

- Xpert MTB/RIF may be used rather than conventional microscopy and culture as the initial diagnostic test in all children suspected of having TB (conditional recommendation acknowledging resource implications, very low-quality evidence).

- Xpert MTB/RIF may be used as a follow-on test to microscopy in adults suspected of having TB but not at risk of MDR-TB or HIV-associated TB, especially when further testing of smear-negative specimens is necessary (conditional recommendation acknowledging resource implications, high-quality evidence).

WHO Recommendations: Extra-Pulmonary TB

- Xpert MTB/RIF should be used in preference to conventional microscopy and culture as the initial diagnostic test for CSF specimens from patients suspected of having TB meningitis (strong recommendation given the urgency for rapid diagnosis, very low-quality evidence).

- Xpert MTB/RIF may be used as a replacement test for usual practice (including conventional microscopy, culture or histopathology) for testing specific nonrespiratory specimens (lymph nodes and other tissues) from patients suspected of having extrapulmonary TB (conditional recommendation, very low-quality evidence).

Procurement of Xpert MTB/RIF

Since the time of the initial WHO recommendation of Xpert MTB/RIF in December 2010, 110 high-burden and low/middle-income countries (see dark blue countries in map) have procured 3,553 GeneXpert instruments and 8.8 million Xpert MTB/RIF cartridges in the public sector under concessional pricing, as of 30 September 2014.

Specimen type	Comparison (No. of studies, No. of samples)	Median (%) pooled sensitivity (pooled 95% CI)	Median (%) pooled sensitivity (pooled 95% CI)
Lymph node tissue and aspirate	Xpert MTB/RIF compared against culture (14 studies, 849 samples)	84.9 (72-92)	92.5 (80-97)
	Xpert MTB/RIF compared against a composite reference standard (5 studies, 1 unpublished)	83.7 (74-90)	99.2 (88-100)
Corobrospinal fluid	Xpert MTB/RIF compared against culture (16 studies, 709 samples)	79.5 (62-90)	98.6 (96-100)
	Xpert MTB/RIF compared against a composite reference standard (6 studies, 512 samples)	55.5 (51-81)	98.8 (95-100)
Pleural fluid	Xpert MTB/RIF compared against culture (17 studies, 1385 samples)	43.7 (25-65)	98.1 (95-99)
	Xpert MTB/RIF compared against a composite reference standard (7 studies, 698 samples)	17 (8-34)	99.9 (94-100)
Gastric lavage and aspirate	Xpert MTB/RIF compared against culture (12 studies, 1258 samples)	83.8 (66-93)	98.1 (92-100)
Other tissue sample	Xpert MTB/RIF compared against culture (12 studies, 699 samples)	81.2 (68-90)	98.1 (87-100)

Meta-analysis of Xpert MTB/RIF in diagnosing EPTB and RIF resistance

Burden of DR TB in India

- India - world's highest MDR-TB burden country
  - 64,000 emerging annually in notified PTB cases
- Sub-national drug-resistance surveys (2005-09)
  - 2.4% MDR in new, 15% MDR in previously treated cases
  - 21-24% Ofx resistance, 4-7% XDR in MDR isolates
- National drug resistance survey underway
- World's largest PMDT service expansion
  - Started in 2007, dramatically accelerated from 2011 to achieve
  - Nationwide rapid scale up in March 2013.

Key Features of RNTCP PMDT

- Decentralized lab diagnosis and DST
- Specimen transport to LPA (45) or Xpert MTB/Rif (89) site
- Xpert MTB/Rif or LPA – is preferred DST method and available across India
- Treatment with standardized regimen for M/XDR TB – largely ambulatory
- Scope of strengthening MDR-TB regimen in baseline Ofx / Km resistance
- Base line Second Line DST started in 6 states (KA, TN, MH, KE, GJ, DL).

Introduction of Rapid Diagnostics in India

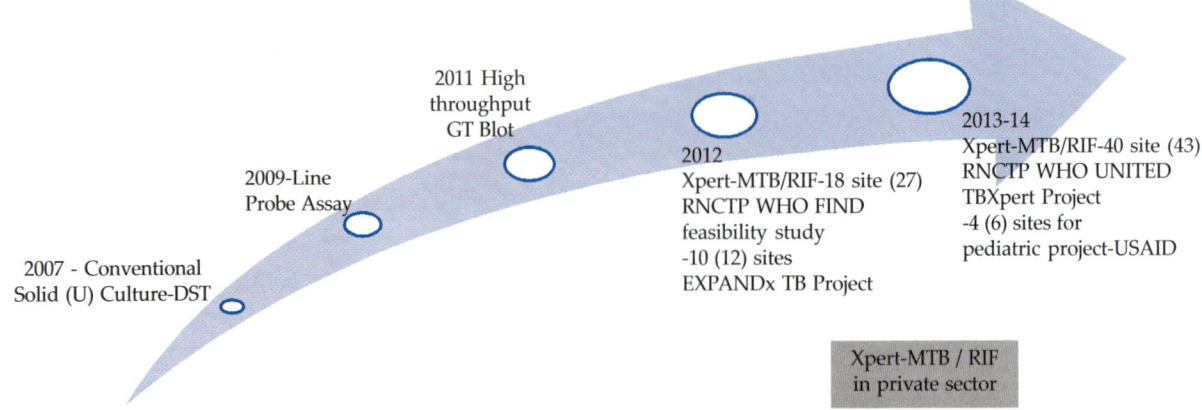

2011 High
throughput
GT Blot

2009-Line
Probe Assay

2007 - Conventional
Solid (U) Culture-DST

2012
Xpert-MTB/RIF-18 site (27)
RNCTP WHO FIND
feasibility study
-10 (12) sites
EXPANDx TB Project

2013-14
Xpert-MTB/RIF-40 site (43)
RNCTP WHO UNITED
TBXpert Project
-4 (6) sites for
pediatric project-USAID

Xpert-MTB / RIF
in private sector

Foundation for New Innovative Diagnostics, launched in World Health assembly 2003. Initial 5 year grant from BMGF.

Policy for use of Xpert-MTB/Rif in India

Use of Xpert-MTB/Rif in programmatic settings

- For Diagnosis of Rif Resistance
  - Presumptive MDR TB cases

- Prioritize Xpert-MTB/Rif to detect MTB in Presumptive TB cases among
  - People living with HIV / AIDS
  - Pediatric cases.

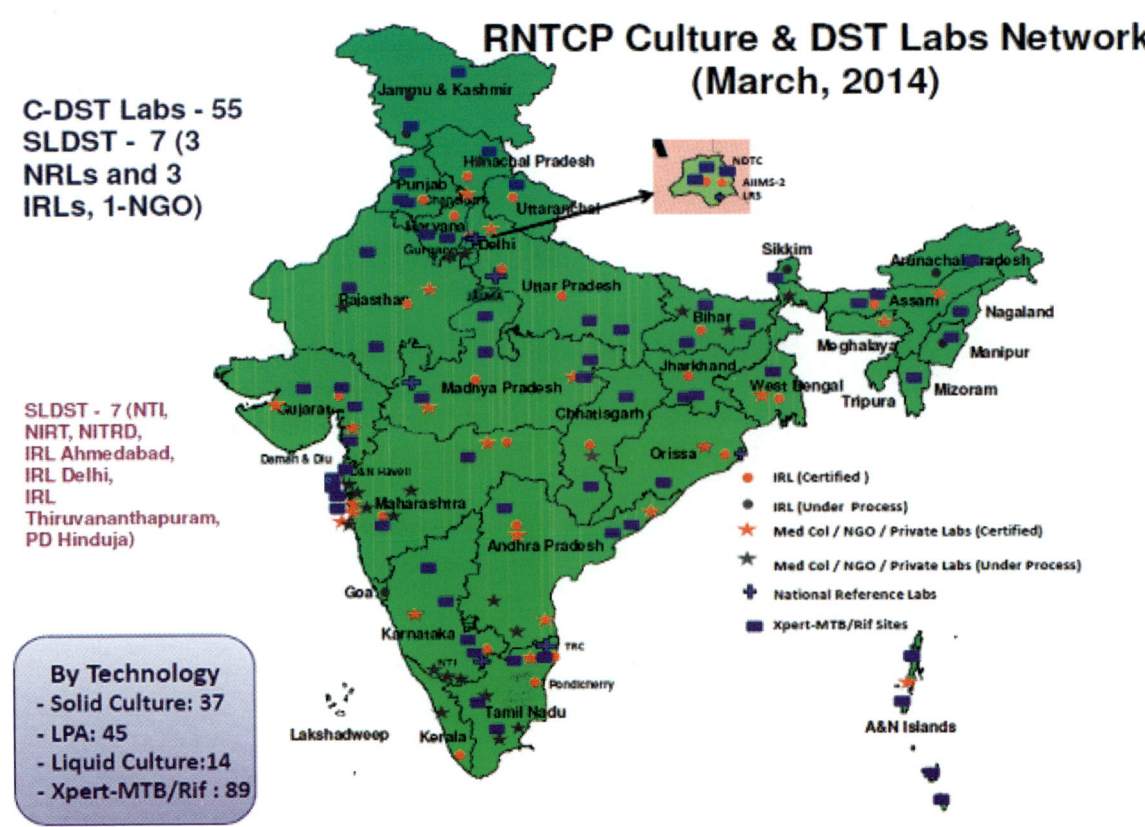

**RNTCP Culture & DST Labs Network
(March, 2014)**

C-DST Labs - 55
SLDST - 7 (3
NRLs and 3
IRLs, 1-NGO)

SLDST - 7 (NTI,
NIRT, NITRD,
IRL Ahmedabad,
IRL Delhi,
IRL
Thiruvananthapuram,
PD Hinduja)

By Technology
- Solid Culture: 37
- LPA: 45
- Liquid Culture:14
- Xpert-MTB/Rif : 89

● IRL (Certified)
● IRL (Under Process)
★ Med Col / NGO / Private Labs (Certified)
★ Med Col / NGO / Private Labs (Under Process)
✚ National Reference Labs
■ Xpert-MTB/Rif Sites

Xpert-MTB/Rif projects in India

- 18 sites under TB Unit feasibility and impact study (March 2012 - December 2013)
  - conducted to create in-country evidence on feasibility for rapid adoption and guide future scale-up of the technology to **diagnose TB and DR TB**
- 10 sites under EXPANDx TB (till December 2014)
  - for further accelerating services of PMDT to **diagnose DR TB**
- 40 sites under UNITAID (till December 2015)
  - for scale-up and deployment of Xpert-MTB/Rif as a decentralized tool and innovative **PPM** to **diagnose TB and DR TB**
- 4 sites under USAID (till June 2014)
  - Accelerating access to quality **TB and DR TB diagnosis** for **paediatric cases** in 4 major cities in India
- 30 sites proposed under PEPFAR, USAID (till September 2015)
  - Innovative intensified TB case finding and appropriate treatment at **high burden ART centres** in 4 states.

Priority areas for deployment of Xpert-MTB/Rif

- Medical Colleges with DR TB centres
- ART Centres
- Urban TB Districts
- High workload districts
- CBNAAT Laboratories under RNTCP in Mumbai.

TDR TB

- The term "totally drug-resistant TB" (TDR-TB) refers to an isolate of *M. tuberculosis* resistant to all locally tested medications
- However, the published studies initially describing TDR-TB did not including susceptibility testing for less frequently used agents with activity against TB, including cycloserine, terizidone, clofazimine, linezolid, or carbapenems.

TB drugs used to treat drug resistant TB according to group (class)

Group 1 TB drugs : First Line Oral Agents

- Isoniazid
- Rifampicin/ rifabutin
- Pyrazinamide
- Ethambutol

Group 2 TB drugs : Injectable Agents

- Streptomycin
- Kanamycin
- Amikacin
- Capreomycin

Group 3 TB drugs : Fluoroquinolones

- levofloxacin
- moxifloxacin
- Ofloxacin
- Gatifloxacin
- Ciplrofloxacin

Group 4 TB drugs : Oral Bacteriostatic Second Line Agents

- Para–aminosalicylic acid
- Cycloserine
- Terizidone
- Thionamide
- Protionamide

Group 5 TB drugs: Agents with an unclear role in the treatment of drug resistant TB

- Clofazimine
- Linezolid
- Amoxicillin/clavulanate
- Thioacetazone
- Imipenem/cilastatin
- High dose isoniazid
- Clarithromycin

PAS powder, which is a regular TB drug, I put her on another four non-TB drugs - Amoxyclav, Clarithromycin, Linezolid and Cycloserine.

Impact of HIV on TB

- Increases rate of TB re-activation and progression
- Increases TB morbidity
- Increases TB mortality (5-14 fold)
- Alters clinical manifestations of TB
- Creates diagnostic challenges
- Complicates treatment
- HIV increases risk of developing active tuberculosis
- 5 -10% chance per year of re-activation
- 9 times greater risk compared to HIV negative people
- 50% probability per lifetime of re-activation

Prospective cohort studies have documented high rates of TB re-activation among patients with latent infection. Not only is re-activation a concern, but exogenous re-infection with a different strain of *M. tuberculosis* has been documented in patients with advanced HIV-1 disease.

Clinical Manifestations

- Clinical presentation of TB in HIV patients is variable, depending on CD4
- Extra-pulmonary disease is more likely as CD4 count declines
- Reported in up to 70% when CD4 < 200
- Atypical clinical and radiographic manifestations

Treatment of TB & HIV

- Treatment: Directly observed treatment, Short-course (DOTS) regimens.
- Category I: 2(HRZE)3 ; 4(HR)3
- Category II: 2(SEHRZ)3 ; 1(HERZ)3 ; 5(HER)3
  - H-Isoniazide
  - R-Rifampicin
  - Z-Pyrazinamide
  - E-ethambutol
  - S-Streptomycin

Anti retroviral therapy and Anti Tubercular treatment

- **Nucleoside reverse transcriptase inhibitors (NRTIs)**

  No major interactions with rifampicin or rifabutin.

- **Non nucleoside reverse transcriptase inhibitors (NNRTIs)**

  NNRTI may be used with rifabutin, but the rifabutin dose is increased to 450 mg/day when used with efavirenz

- **Protease inhibitors (PIs)**

  The combination of rifampicin and Darunavir/r is **contraindicated,** Rifabutin may be used

- **Integrase Inhibitors**

  Doubling the dose of raltegravir when used with Rifampicin or use with Rifabutin.

DOTS in the context of HIV

DOTS can:

- Prolong and improve the quality of life.
- Prevent emergence of MDRTB.
- Stop the spread of TB.
- Reverse the trend of MDRTB.
- In the context of HIV, failure to use DOTS can result in rapid spread of disease, tripling of cases and increased drug resistance.

ATYPICAL MYCOBACTERIA (Runyon's classification, 1959)

Group	Species	Importance
I - Photochromogens	M. kansasii	Lung infection simulating tuberculosis
	M. simiae	
	M. marinum	Swimming pool granuloma
II - Scotochromogens	M. scrofulaceum	Cervical adenitis in children
	M. szulgai	
III - Nonphotochromogens	M. avium	Pulmonary disease in
	M. intracellulare	immunocompromised patient
	M. ulcerans	Buruli ulcer
IV - Rapid growers	M. chelonae	Post-injection abscesses and wound infections.
	M. fortuitum	

Niacin positive: *M. tuberculosis, M. simiae,* some strains of *M. chelonae.*

Nitrate reduction test

Group	Positive	Negative
Slow-growing nonchromogens	M. tuberculosis	M. bovis
		M. simiae
Slow-growing scotochromogens	M. szulgai (weak)	M. scrofulaceum
	M. flavescens	M. gordonae
Rapid - growing nonchromogens	M. fortuitum	M. chelonae
Photochromogens	M. kansasii	M. marinum
		M. asiaticum
		M. simiae

Tween 80 hydrolysis test

	Positive	Negative
Photochromogens	M. kansasii	M. simiae
	M. marinum	
	M. asiaticum	
Nonphotochromogens	M. gastri	M. bovis
	M. malmoense	M. avium complex
		M. xenopi, M. simiae
		M. hemophilum
Scotochromogens	M. szulgai (slow)	M. scrofulaceum
	M. gordonae	M. xenopi
	M. flavescens	

Mycobacterium avium

Amongst the atypical mycobacteria, *M.avium* and *M.fortuitum* have been recently reported to cause gastrointestinal disturbances in immunocompromised persons, specially in HIV infected persons and those suffering from AIDS. After 1982, there is a dramatic increase in MAl complex (MAC due to HIV and AIDS).

M.avium is a non-photochromogen and *M. fortuitum* is a rapid grower. MAl complex consist of 25 serovars, of which types 1-6, 8 and 11 are *M. avium.* They are mostly opportunistic, colonise gut and respiratory tract. They are found in the environment in water, soil, mammals, reptiles, birds, etc. From the environment, they are ingested via contaminated food and water. They colonize the human gut and cause infections in immunocompromised patients.

Laboratory diagnosis of Atypical mycobacteria

1. Ziehl-Neelsen staining of stool smear using 25% H_2SO_4 – acid fast bacilli seen in smear which are typically short and cocoobacillary, staining is uniform usually, without beading.

2. Auramine-rhodamine staining and Acridine orange staining showing fluorescence.

3. Culture in L-J medium and in liquid media like 7H9 and Czapek broth.

4. The Mycobacteria Growth Indicator Tube (MGIT) system (BBL, Becton Dickinson Microbiology Systems).

5. Middlebrook 7H10 agar plate after 6 days incubation at 35°C in 10% CO_2 — small, smooth, grey-white colonies without pigment.

   Flat, thin, translucent microcolonies with a central darker umbonate elevation, suggestive of a fried egg are characteristics of the smooth strains of *M. avium*, seen in immunocompromised patients. The rhizoid-like peripheral extensions can be seen in a close-in photomicrograph of a single microcolony.

6. Dubos broth base with pyrazinamide — development of a pink-red band at the reagent layer (ferrous ammonium sulphate reagent) indicates ability of *M. avium* to deaminate pyrazinamide.

7. It produces heat stable catalase (after heating the culture at 68°C for 20 minutes). Grows on Thiophene-2-carboxylic hydrazide (T2H) 1µg/ml.

8. Paraffin slide culture technique can be done, as non-tuberculous mycobacteria is paraffinophilic like *Nocardia spp.*

9. Molecular studies
   - PCR for direct identification of amplified nucleic acid targets of *M. avium.*
   - Rapid diagnosis by nucleic acid probe for culture confirmation of *M. avium*

Treatment

Usually a combination of ethambutol with one of the drugs (INH/Rifabutin/Clofazimine) and one of the macrolides.

LEPROSY

It is a slow chronic progressive granulomatous disease. Incubation period varies from 3-30 years.

Prevalent in Tropics, Southern USA, Mediterranean countries of Europe.

Worldwide – 11million cases.

India – 4 million cases.

Prevalance rate – 6.7 per 10,000 population.

Active leprosy cases – 0.61 million (in March 1996)

< 15 years of age – 15%

Infectious cases – 15-20%

Annually – 0.4 million new cases detected

Over 70% cases in Orissa, Bihar, West Bengal, Nagaland and Daman and Diu.

Prevalence per 1000 population

Orissa – 1.6

Bihar – 1.02

West Bengal – 1.23

Nagaland – 1.8

Daman and Diu – 1.87

Ridley and Jopling's classification

Five groups :

- Tuberculoid (TT)
- Borderline tuberculoid (BT)
- Borderline (BB)
- Borderline Lepromatous (BL)
- Lepromatous (LL)

	Lepromatous leprosy	Tuberculoid leprosy
Lesion	Patches or nodules in skin, face and ear lobes, subcutaneous, tissue, mucous membranes of nose, extremities, with numerous lepra bacilli in tissues	Asymmetrical anaesthetic patches in face, gluteal region and limbs with thickened peripheral nerves with scanty or no lepra bacilli.
Infectivity	Highly infectious	Usually noninfective
CMI	Deficient / absent	Adequate
Anti bodies to *M. leprae*	+++	+
Complication	Testicular atrophy Necrosis of bones and cartilage, loss of upper incisor teeth, leonine facies	Peripheral neuropathy, corneal ulcer and lagophthalmos, claw hand, foot drop, trophic ulcers, loss of digits.
Lepromin test	Negative	Positive
Prognosis	Bad	Good

Laboratory diagnosis

Samples taken – Slit skin smear by scrapings from patches on the skin, ear lobes and nasal mucosa.

1. **Acid fast stain** of smear done by using 5% H_2SO_4 as a decolouriser.

   Bacteriological index **(BI)** = No. of viable bacilli in a smear.

   Negative = No bacilli in 100 fields.

   1 + = 1-10 bacilli in 100 fields

   2 + = 1-10 bacilli in 10 fields

   3 + = 1-10 bacilli per field

   4 + = 10-100 bacilli per field

   5 + = 100-1000 bacilli per field

   6 + > 1000 bacilli per field + clumps and groups in every field.

   Morphological index **(MI)** = Percentage of uniformly stained bacilli out of total bacteria present in tissue. This is done for assessing prognosis and response to treatment.

2. **Skin and nerve biopsy** for histology.

3. **Green fluorescence protein (GFP)** gene detect by immunofluorescence.

4. **Animal inoculation** in foot pad of mice or in nine-banded armadillos.

5. **Lepromin test**
   - Early reaction of Fernandez after 24-48 hours
   - Delayed reaction of Mitsuda after 3-5 weeks

It assesses prognosis, but has limited diagnostic value.

6. **Serological tests** – detection of antiphenolic glycolipid 1 (anti PGL-1) antibodies by

   a) Latex agglutination test

   b) *M. leprae* particle agglutination (MLPA) test

      In lepromatous type, positivity rate is 78-80%

      In tuberculoid type, positivity rate is 20-22%

   c) ELISA

   d) RIA

   e) Monoclonal antibodies against antigens of *M. leprae* through hybridoma technique.

   f) DNA-DNA hybridization using pieces of defined DNA probes to identify DNA of *M. leprae* and application of PCR.

Treatment

- **Dapsone monotherapy (DDS)**. Resistance to dapsone has been reported since 1981.

- **Multi-drug therapy (MDT)** – Dapsone, rifampicin, clofazimine.

- **ROM therapy since 1994** – Rifampicin, ofloxacin and minocycline.

Lepromatous type – treated for 2 years (99.9% bacteria killed)

Tuberculoid type – treated for 1 year.

REFERENCES

1. Cambanis A, Ramsay A, Wirkom V, Tata E, Cuevas LE. Investing time in microscopy: an opportunity to optimise smear-based case detection of tuberculosis. Int J Tuberc Lung Dis. 2007;11: 40-5.

2. Tortoli E, Cichero P, Chirillo MG, et al. Multicenter comparison of ESP Culture System II with BACTEC 460 TB and with LowensteinJensen medium for recovery of mycobacteria from different clinical specimens, including blood. J ClinMicrobiol 1998; 36:1378-81.

3. Barman P, Gadre D. A study of phage based diagnostic technique for tuberculosis. Indian J Tuberc 2007;54:36-40.

4. Akos Somoskov, Qunfeng Song, Judit Mester, Charise Tanner, Yvonne M. Hale, Linda M. Parsons.Use of Molecular Methods To Identify the Mycobacterium tuberculosis Complex (MTBC) and Other Mycobacterial Species and To Detect Rifampin Resistance in MTBC Isolates following Growth Detection with the BACTEC MGIT 960 System. J Clin Microbiol 2003; 41: 2822-6.

5. WHO Global Tuberculosis Report 2014 (updated Feb 2015)

6. http://www.stoptb.org/events/world_tb_day/2015

7. Mumbai Mirror 29th March, 2015; http://www.mumbaimirror.com/others/sunday-read/3-years-after-scare-citys-TB-fightback/articleshow/46729652.cms

8. http://www.who.int/tb/EndTBStrategy_infographic.jpg?ua=1

9. TB India 2014 Annual Status Report http://www.tbcindia.nic.in (last accessed on 27-Mar-2014)

10. Key indicators for the WHO South-East Asia Region, www.who.int/tb/data/ Accessed on 27-Mar-2014.

11. RNTCP Technical and Operational Guidelines for Tuberculosis Control. October 2005. Central TB Division, Directorate General of Health Services, New Delhi. http://tbcindia.nic.in/pdfs Technical%20&%20Operational% 20guidelines%20for%20TB%20Control.

12. National Guidelines on Diagnosis & Treatment of Pediatric Tuberculosis, 2012.http://tbcindia.nic.in/ Paediatric%20guidelines_New.pdf

13. Xpert MTB/RIF for the diagnosis of pulmonary and extrapulmonary TB - WHO policy update 2014.

14. http://www.newtbdrugs.org/pipeline.php

GASTROINTESTINAL INFECTIONS

15
Chapter

Defense Mechanism of the Gastrointestinal tract

Mouth: Flow of liquids, saliva, lysozyme, normal bacterial flora.

Oesophagus: Flow of liquids, peristalsis.

Stomach: Acid pH.

Small intestine: Flow of gut contents, peristalsis, mucus, bile, secretory IgA, lymphoid tissue (Peyer's patches), shredding and replacement of epithelium, normal flora.

Large intestine: Normal flora, peristatsis, shedding and replication of epithelium, mucus.

Infections of the G. I. tract

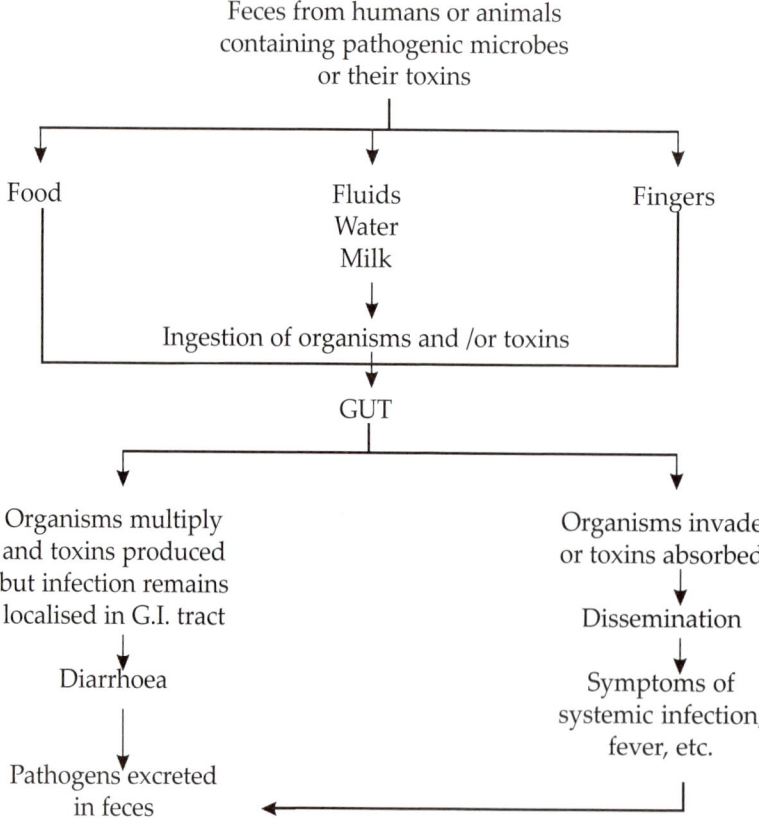

Damages resulting from infection of the Gastrointestinal tract

1. Pharmacologic action of bacterial toxins, local or distant to site of infection e.g. cholera, staphylococcal food poisoning.
2. Local inflammation in response to superficial microbial invasion e.g. shigellosis, amebiasis.
3. Deep invasion to blood or lymphatics, dissemination to other body sites e.g. hepatitis A, enteric fever.
4. Perforation of mucosal epithelium after infection, surgery or accidental trauma e.g. peritonitis, intra-abdominal abscesses.

Organisms causing diarrhoea

I.	**Bacteria**		**Incubation period**

A. Noninvasive Bacteria (Toxin type)

1. Enterotoxigenic *E. coli* (ETEC) — 24-72 hours
06, 08, 015, 025, 0115, 0148, 0167.

2. Enteropathogenic *E. coli* (EPEC)
026, 055, 086, 0111, 0126, 0128, 0142.

3. *V. cholerae* 01 — 8-24 hours

4. NAG vibrios

5. *C. perfringens* types A and C. — 8-20 hours

6. *S.* aureus (Gr. III 6/47, Gr. IV 42 D) — 2-6 hours

7. *Bacillus cereus*
 - Diarrhoeal — 8-16 hours
 - Emetic — 1-5 hours

8. *C. difficile* (AAC or PMC)

9. *Klebsiella, Pseudomonas, M. morganii*

B. Invasive Bacteria (Non toxin type)

1. *Salmonella spp.* — 6-48 hours

2. *Campylobacter spp.* — 3-10 days

3. *Shigella spp.*

4. *Enteroinvasive E. coli. (EIEC)*
0124, 028ac, 0112ac.

5. *Enterohemorrhagic E. coli (EHEC)*
or VTEC – 0157

6. *V. parahemolyticus, V. mimicus* — 2-48 hours

7. *Y. enterocolitica*

C. Toxi - invasive Bacteria

Shigella dysenteriae type 1

Produces neurotoxin and cytotoxin.

II. Viruses

Rotavirus, Norwalk virus, Adenovirus (types 40 and 41), Astrovirus, Calicivirus, Coronavirus, Small Round Virus-like objects (SRV)

III. Parasites

Protozoa: *E. histolytica, G. lamblia, Cryptosporidium, Isospora, Balantidium coli, Cyclospora, Microsporidia*

Helminths: *S. stercoralis, T. trichiura, T.saginata, T. solium, A. duodenale, N. americanus, A. lumbricoides, H. nana.*

IV. Fungi

Candida albicans (after prolonged antibiotic therapy)

Differences between Cholera Toxin (CT) and Heat Labile Toxin (LT)

CT	LT
Chromosomal mediated	Plasmid mediated
Antigenically similar	Not identical (difference in mol. structure lie in amino acid sequence)
GM1 ganglioside receptor (glycolipid)	GM1 + another glycoprotein receptor
Largely cell associated and reduced	Attaches only by oligosaccharide moiety.
CT cannot be expressed in *E. coli.*	LT can be expressed in *V. cholerae* 569 B

Differences between heat labile (LT) and heat stable (ST) toxins

LT	ST
Heat labile (60°C for 30 minutes)	Heat stable (100°C for 30 minutes)
Secretory response rapid.	Secretory response slow
Fluid accumulation in rabbit ileal loop:	
At 4-6 hrs +	At 4-6 hrs +
At 18 hrs +	At 18 hrs -
Rabbit skin test positive for vascular permeability factor	Rabbit skin test negative
Non dialysable toxin	Dialysable toxin
Suckling mouse assay is negative	Suckling mouse assay positive
Morphological changes in CHO and Y1 adrenal cells	No changes in CHO and Y1 adrenal cells
Detected by Biken test, ELISA, modified HRPGM1, ELISA and RIA	Detected by Biken test, but not by ELISA and RIA.

Some clinical and epidemiological features of Bacterial Food Poisoning

Organism	Incubation period (hours)	Duration and symptoms	Sources of organisms
Salmonella species	6-48 (12-36 hrs)	1-7 days Diarrhoea, abdominal pain, vomiting, fever nearly always present	Human and animal excreta, raw meat and poultry, etc.
Clostridium perfringens	8-20	12-24 hours Diarrhoea, abdominal pain, nausea but rarely vomiting, no fever	Dust, soil, human and animal excreta, raw meat and poultry, dried foods, herbs and spices.
Staphylococcus aureus	2-6	6-24 hours Nausea, vomiting, diarrhoea and abdominal pain, but no fever, collapse and dehydration in severe cases	Anterior nares and skin of man and animals, septic lesions (boils, carbuncles, whitlows). Raw milk of cows, sheep and goats, cream and cheese from raw milk.
Vibrio parahemolyticus	2-48 (12-18 hrs)	2-5 days Profuse diarrhoea often leading to dehydration, abdominal pain, vomiting and fever.	Raw and cooked sea foods, e.g. fish, prawns, crabs and other shell fishes.
Campylobacter jejuni	3-10 days	3 days-3 weeks Watery diarrhoea +++, sometimes with blood & mucus. Abdominal pain always present ++, Fever ++	Poultry, meat, milk.
Bacillus cereus	a) Diarrhoeal syndrome (8-16 hrs) b) Vomiting syndrome (1-5 hrs)	a) 12-14 hrs, abdominal pain, diarrhoea and sometimes nausea b) 6-24 hrs nausea and vomiting, sometimes diarrhoea	Common in soil and vegetation a) Meat products, soups, vegetables, puddings and sauces b) Cooked rice.
Yersinia enterocolitica		Fever, diarrhoea, abdominal cramps, vomiting	Milk or dairy products, raw pork. raw pork.
Clostridium botulinum	12-36 (8 hrs to 8 days)	Vomiting, constipation, blurred vision, ocular paresis, thirst, pharyngeal paralysis, aphonia, subnormal temperature (35.5°-36.7°C)	**Type A:** Home canned or preserved vegetables (U.S.); meats, fish, preserved bean curd (China). **Type B:** Home preserved pork, vegetable and fish (Poland, Germany, France). **Type E:** Uncooked products of fish and marine mammals (Japan, Canada, Alaska).

Bacterial Food Poisoning

A. Infection type

- *Salmonella spp.* e.g. *S. typhimurium, S. enteritidis*
- *Campylobacter jejuni*
- *Vibrio parahemolyticus*
- *V. mimicus, V. vulnificus*
- *Yersinia enterocolitica*

B. Toxin type

- *Staphylococcus aureus*
- *Bacillus cereus*
- *B. subtilis, B. licheniformis*

C. Intermediate type

- *Clostridium perfringens* type A
- *Clostridium botulinum*

Investigation of Outbreaks of Food Poisoning

Objectives :

- To verify that there is an outbreak of illness and the causative agent was food borne
- To determine the nature of the agent and the food stuff(s) by which it was transmitted
- To determine the way in which the food was contaminated
- To ensure that all cases or carriers of the agent are identified
- To stop the outbreak if it is continuing (Dept. of Health and Social Security).

Procedures :

- To secure a complete list of sufferers with clinical histories and a full list of foods consumed in the previous 2-3 days
- To record details of the origin and the mode of preparation and storage of the suspected foods
- To collect specimens for laboratory examination
  i)   The actual food consumed
  ii) The vomit and feces of patients
  iii) Blood, spleen, liver and intestine of fatal cases
- To endeavour to find out how the food(s) were contaminated and to find out the reservoir of the causative organism
- As soon as sufficient evidence has been obtained about the probable vehicle and the causative agent — further specimens are decided:
  iv) Specimens from contacts or food handlers
  v) Samples of food ingredients
  vi) Swabs and washings from the premises in which the food was prepared.

Laboratory Diagnosis

Macroscopic Examination

1. Consistency – formed, liquid, watery.
2. Colour – white, yellow, greenish, brown, black.
3. Others – mucus, blood, segments of parasites, adult worms.

Normal feces – Brown and formed or semi-formed

Infant feces – Yellow - green and semi-formed

Microscopic Examination

1. Direct examination
   a) Hanging drop preparation for darting motility of vibrios.
   b) Saline and iodine preparation for larva, ova, cysts, RBCs and pus cells.
      > 30 polymorphs/h.p.f. – Shigellosis.
      < 10 polymorphs/h.p.f. – Salmonellosis, Invasive *E. coli,* Amoebic dysentery.
      Segmented neutrophils – Campylobacteriosis.
   c) Dark field microscopy for *Campylobacter spp.* – corkscrew molility.
   d) Concentration methods – Floatation and Sedimentation.
   e) Electron microscopy for rotavirus.

2. Staining Methods
   a) Modified Gram stain for campylobacters.
   b) Modified Acid fast stain for *Cryptosporidium, Isospora,* etc.
   c) Trichrome staining (Chromotope 2 R) — for Microsporidia.
   d) Fluorescent staining – *Yersinia enterocolitica, Campylobacter spp.*

3. Culture in selective media and enrichment media.

4. Identification by biochemical test and slide agglutination with group and species specific antisera.

5. Toxin detection – LT, ST, VT, *C. difficile* toxin A and B, etc.
   Stool filtrates for toxin detection stored preferably at – 70°C, or at least at – 20°C until tested.
   Stools for virus culture must be refrigerated, if they are not invoculated within 2 hours, or frozen at – 20°C.

6. Serological tests – LAT (Latex agglutination test), ELISA, CIEP, FAT (Fluorescent antibody test), RIA, etc.

7. Molecular methods – PCR (Polymerase Chain Reaction), Ribotyping, etc.

Assays for LT

A. Whole Animals

1. Adult rabbit ligated ileal loop after 18 hours.
2. Infant rabbit bowel.
3. Rat perfusion test – Single jejunal segment perfused with enterotoxins along with control + antitoxin – measures net transport of water.
4. Canine model (Chronic Thiryvella loop)
5. Skin permeability assay in adult rabbit skin.

B. Tissue Culture Assays

1. Y1 mouse adrenal cells – steroid production.
2. CHO cells – morphological changes (number of elongated cells in each 100 cells).
3. Vero cells (for VT) – morphological changes (rounding of cells).
4. Henle intestinal cells – Human embryonic interstitial cells in monolayers – treated with LT – increased adenylate cyclase.

C. Enzymatic Assays

1. Pegion erythrocyte lysate assay (PEL)
2. Cat heart myocardial assay
3. NAD - glycohydrolase and ADP - ribsyltransferase activity.

D. Immunological Assays

1. Solid phase RIA
2. ELISA and Modified HRP-GM1 ELISA
3. Passive immune hemolysis of sheep red blood cells
4. Latex particle agglutination assay
5. Staphylococcus Coagglutination test (Cowan 1 strain NCTC 8530).

Assays for ST

1. Suckling mouse assay (infant mouse intragastric)
2. Adult rabbit ligated ileal loop after 6 hours.
3. Infant rabbit bowel.

Assays for both LT and ST

1. DNA probes (detection of individual colonies producing LT or ST.
2. Biken Test.

Adult Rabbit Ileal Loop Test (For LT and ST)

De and Chatterjee, 1953

1. Albino rabbits (2kg) are starved for 24 hours
2. Intestines exposed under light ether anaesthesia
3. Starting from ileocaecal junction, loops are made (10 cm in length), with gap of 5 cm between each loop — Total 8-10 loops.

4. Each strain was tested in 3 rabbits
5. First loop is positive control, last loop is negative control.
6. Abdomen is closed, rabbits are allowed only water to drink.
7. Sacrificed after 6 hours and 18 hours
   - I/V phenobarbital
   - I/V chloroform
   - 5 ml air I/V

- Accumulation of fluid per cm of gut = volume of fluid/length of loop
- Average value for 3 rabbits is calculated :
   - In 18 hr test > 1 is positive (for LT)
   - In 6 hr test > 0.4 is positive (for ST).

Suckling Mouse Assay (For ST)

Dean et al, 1972

1. Swiss albino mice (1-3 days old) are separated from mothers, randomly into groups of four.
2. 0.1 ml quantity of test strain and 2 drops of 2% Evans blue dye per ml is inoculated directly into milk filled stomach with No. 27 needle.
3. Mice are kept at 28°C for 4 hrs., then killed with chloroform.
4. Abdomen opened and intestine examined for distension.
5. Intestines from 4 mice weighed together, ratio of gut weight to remaining body wt. calculated.
6. Ratio of 0.09 or greater is positive.

Biken Test (For LT and ST)

Principle : Specific antitoxin reacts with toxin liberated by an actively growing organism on a special medium and produces a line of precipitation at the sites where they meet in optimal proportion.

For LT assay

1. Lincomycin 2.7 mg/ml is heated in water bath to 50-60°C and 0.5 ml poured in petri dish (85-90 mm size).
2. Biken agar No. 2 at 50-60°C poured into same petri dish (15 ml) by rotating the plate (10 times).
3. Culture is inoculated to produce a confluent growth around the site where central well will be punched. Around each central well, 3 test cultures and 1 positive control is inoculated.
4. After 48 hours, polymyxin B disc (500 IU/disc) is put on top of growth of each strain.
5. A well is punched at the centre, incubated for 5-6 hrs., 20μl of anti-LT antiserum is added and incubated for 20-24 hrs.

   Lines of precipitation in the zone between the growth and central well — Positive.

For ST assay

After step 4, four pieces of agar (7 mm diameter) is punched out from just outside the periphery of each inoculation site, put in 0.5 ml PBS, left overnight at 4°C to extract the ST (0.3-0.4 ml of overlying fluid).

Modified HRP-GM1 ELISA (For LT)

1. Bacteria are grown overnight in CAYE medium with shaking (200 rpm). Polymyxin B (200IU/ml) added to culture. 30 minutes later centrifuged at 4000g for 5 mins. Supernatant is tested for LT.
2. ELISA microtitre plates are coated with GM1. Incubated at 37°C for 4 hrs.
3. Plate is incubated with PBS (+1% bovine serum albumin) at 37°C for 30 mins.
4. Bacterial culture is added. Incubated at room temperature (RT) for 1 hr.
5. Rabbit anti-LT serum is diluted 1:100 in PBS Tween is added. Incubated at RT for 2 hrs.
6. Anti-rabbit Ig-horseradish peroxidase (HRP), 1:300 in PBS Tween and kept at RT for 32 hrs.
7. H_2O_2 is added, then substrate paraphenylene diamine. Incubated at RT for 20 mins. Reaction is stopped with 3M NaOH.
8. Visually brown or light brown colour is positive. Spectrophotometrically it is read at 450nm, an absorbance value of ≥ 0.1 is positive.

Advantages of modified ELISA

- Completed in 1½ days.
- Reagents are in stable form.
- Results are clearly read by naked eye.
- Shake cultures or polymyxin B is not essential, but sensitivity is improved by their use.

Filtrate

1. Formed stool is mixed with equal volume of HBSS or PBS.
2. Liquid stool should not be diluted.
3. 5 ml stool is centrifuged at 2500g x 20 minutes.
4. 2 ml supernatant is passed through sterile, disposable 0.45 µm membrane filter.
5. Filtrate is frozen at - 20°C until tested.
6. Freezing is done at - 70°C, if there is delay in toxin assay.

Cytotoxin Neutralization Assay

1. Growth medium - Eagle's MEM with Hank's BSS + FBS + L-glutamine + Penicillin, Streptomycin and Amphotericin B
2. Maintenance medium
3. Trypsin solution (2.5%)
4. Controls – neutralization, toxin, antitoxin, cell culture.

5. 0.1 ml of suspension + 0.1 ml from antitoxin/0.1 ml from toxin/each control.
6. Overnight incubation is done at 37°C in humid atmosphere.

Results

1. Toxin control - 100% cytotoxicity.
2. All controls and negative filtrates - confluent monlayers.
3. Positive filtrates - rounding of cells in 'T', not in 'A'.
4. If cytotoxicity is seen both in A and T wells - autocytotoxic. This is removed by
   - Diluting stool 1:10
   - Freezing and thawing filtrate once again.

Preparation of Toxin

1. The test strains are plated on non selective medium (NA).
2. 50 ml TSB with 0.6% yeast extract in 250ml conical flask is inoculated with 3-4 hour broth culture, incubated at 37°C in shaking water bath for 18 hours.
3. Culture is centrifuged at 4°C at 22,000 g for 30 mins.
4. Supernate is precipitated slowly by solid $(NH_4)_2SO_4$ to 90% saturation at 4°C. Kept overnight at 4°C, centriguged at 22,000g for 30 mins.
5. Supernate is discarded, ppt. dissolved in 10 ml 0.04 M PBS pH 7.2. Dialysed against, at 4°C with 5-6 changes of PBS.
6. Pass through membrane filter (0.45 µm size).
7. Filtrate is used as toxin (LT), stored at - 20°C. Filtrate is heated at 100°C for 30 mins. to destroy LT. Can be used for ST assay.

Tissue Culture Assay

1. **CHO cell culture method** (Guerrant et al 1974)
   - CHO cells grown in Ham's F 12 medium + 10% FBS in humid atmosphere with 5% CO_2 at 37°C, passaged by trypsinisation.
   - 5×10^3 cells in 0.25 ml medium + 1% FBS added to each chamber of an 8-chamber culture slide (Lab-Tek Products).
   - 0.2 ml cell free culture filtrates of test cultures are added. Incubated at 37°C for 18-24 hours.
   - Cells are fixed with absolute alcohol and stained with Giemsa (1:40) for 20 mins.

2. **Y1 mouse adrenal cell culture method** (Donta et al 1974)
   - Y1 cells are grown in Ham's F 10 medium + 15% horse serum + 2.5% FBS at 37°C in humid atmosphere with 5% CO_2.
   - A suspension containing 10 cells in 2 ml medium + supplements are added to each chamber of an 8-chamber culture slide.
   - Remaining is same as CHO cell culture method.

3. **Vero cell assay**

    -   MEM with Eagle's salts non-essential amino acids + 5% FBS + Penicillin, Streptomycin. pH 7.2.

    -   Vero cells are diluted to final concentration of 0.25 million cells/ml. 1.5 ml is taken in each Leighton tube (15 x 125 mm), incubated at 37°C for 2-3 days.

    -   0.1 ml culture supernate is added, incubated at 37°C for 1 day.

    -   Number of elongated cells in each 100 cells is counted.

        No alteration – Negative

Rounding and peeling off of :

-   25% cell sheet - Grade I
-   50% cell sheet - Grade II
-   75% cell sheet - Grade III
-   100% cell sheet - Grade IV

Grade II and above – Positive.

DNA probe tests for VTEC

Probe fragments are prepared from recombinant plasmids and labelled with digoxigenin and stored at -20°C.

Target DNA is prepared. Membranes are prepared for hybridization and dried. DNA is bound by taking for 2 hours at 80°C, then it is placed on an ultraviolet transilluminator of wavelength 302 nm for 4 minutes.

Diarrheagenic *Escherichia coli*

	Pathogenic group and common serogroups	Epidemiology	Laboratory diagnosis
1.	Enterotoxigenic *E. coli* (ETEC) 06, 08, 015, 025, 027, 063, 078, 0115, 0148, 0153, 0159, 0167	Most important casue of bacterial diarrhoea in children in developing counteries most common cause of traveller's diarrhoea	Organisms isolated from feces. Test for production of LT by ELISA and ligated rabbit ileal loop. For ST by suckling mouse assay (not in routine use) gene probes specific for LT and ST genes for detrection in feces, in food and water samples.
2.	Enteroinvasive *E. coli* (EIEC) 028ac, 0112ac, 0124, 0136, 0143, 0144, 0152, 0164	Important cause of diarrhoea in areas of poor hygiene. Infectious, usually from food. No evidence of animal or environmental reservoir.	Organism isolated from faces. Test for entero -invasive potential in tissue culture cells. Keratoconjunctivitis in rabbits.
3.	Enterohaemorrhagic *E. coli* (EHEC) 0157.	Serotype 0157 is most important in human infections. Outbreak and sporadic cases occur worldwide. Food and unpasteurised milk is important in spread.	Organisms isolated from feces. Proportion in fecal sample may be very low (< 1% of E.coli colonies) Usually sorbital non-fermenters. DNA probes using in colony hybridization tests.
4.	Enteropathogenic *E. coli* (EPEC) 026, 055, 086, 0111, 0114, 0119, 0125, 0126, 0127, 0128, 0142.	Belong to particular serotypes. Cause sporadic cases and outbreaks of infection in babies and young children – Infantile diarrhoea. Importance in adults is not known.	Organisms iolated from feces. Determine serotype of several colonies with poly-valent antisera for known EPEC types. Adhesion to tissue culture cells is demons-trated by a fluorescence Actin staining test.

Rotaviral Diarrhoea

Usually affects male children between 6 months - 2 years of age. Common during winter.

Pathogenesis: Ingested virus infects cell at tip of villi in small intestine, spreads to infect large number of

these cells (cells in crypt not infected)

↓

Release of virus particles into lumen

↓

Infected cells damaged and lost, leaving immature cells with reduced absorptive capacity for sugar, water, salts, infiltrating mononuclear cells, villi shorten (2-4 days)

↓

Fluid accumulation in lumen

↓

Diarrhoea, dehydration

↓

Viral replication ceases due to antibody/ interferon/infection of all susceptible cells

↓

Crypt cells repopulate villi

↓

Normal appearances regained (6-7 days).

Laboratory diagnosis of rotaviral dirrhoea

1. Electron microscopy — Centrifiged at 10,000 g for 20 minutes, then ultracentrifuged at 100,000 g.

2. Immune electron microscopy — specific antigen against rotavirus group antigen.

3. Serology for antigen detection
   a) Latex agglutination test –
      Sensitivity 62.5%, Specificity 97.4%
   b) ELISA – Sensitivity 97-100%.

4. Rotavirus ds RNA genome separated by polyacrylamide gel electrophoresis (PAGE) Specificity 100%.

5. Dot hybridisation technique using labelled ssRNA.

Oral Rehydration Solution (ORS) WHO

NaCl 3.5 g

KCl 1.5 g

Na-citrate 2.9 g

Glucose 20 g

Dissolved in 1 litre of drinking water.

Antimicrobials given in severe cases

- Tetracycline in cholera
- Metronidazole or vancomycin in antibiotic associated diarrhoea (AAD) or withdrawal of antibiotics
- Norfloxacin or ciprofloxacin
- Tinidazole

REFERENCES

1. Mims' Medical Microbiology. 5th Ed. Eds. Richard Goering, Hazel Dockrell, Mark Zuckerman, Ivan Roitt & Peter L. Chiodini. Saunders. 2013.

2. Koneman EW, Allen SD, Janda WM, Schreckenberger P, Winn Jr. WC. In Color Atlas & Textbook of Diagnostic Microbiology. 6th Ed. Lippincott, Philadelphia. 2006.

3. Saraswathi K, De A. Manual of National Workshop on Laboratory Diagnosis of Diarrheal Diseases - Conventional and newer techniques. 1994.

PROCESSING OF FECAL SAMPLES FOR BACTERIAL ENTERIC PATHOGENS

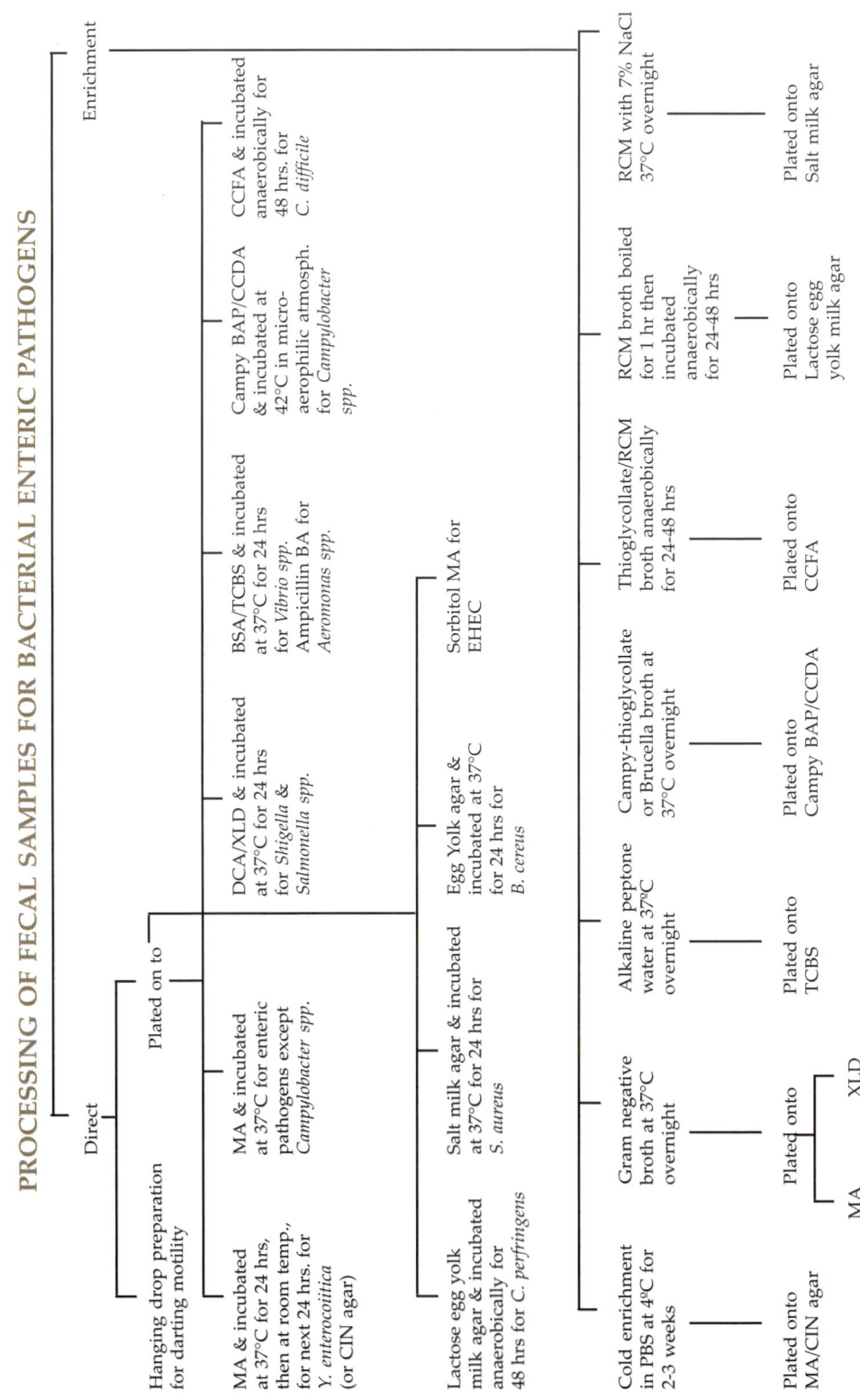

16
Chapter

ANTHRAX

Anthrax is a zoonotic disease caused by *Bacillus anthracis*. It is a non-motile gram positive rectangular bacilli having a polypeptide capsule and oval central spores. They occur singly or in short chains. On nutrient agar, **'Medusa head'** colonies are seen. On blood agar, colonies are non-hemolytic. On gelatin stabculture, it produces **'inverted fir tree'** appearance. Spores are destroyed by heating at 121°C for 15 minutes or by using 4% KMnO4 and 10% formaldehyde within half an hour.

Antigens

a) Capsular antigen polypeptide, D-glutamic acid. It inhibits opsonization and phagocytosis.

b) Cell wall antigen – poly laccharide

c) Somatic protein antigen

Toxin

Extracellular toxin has 3 components :

a) Oedema factor (EF or Factor I)

b) Protective antigen (PA or Factor II)

c) Lethal factor (LF or Factor III)

Three collectively produces **'toxic complex'** which is lethal, produces intensive vascular permeability and oedema. Toxin production is controlled by a plasmid.

Humans are secondarily infected from diseased animals by occupational means (85% cases) by contact with meat, hide or hair of infected animals and non occupational means (15% cases) due to handling of infected shaving brushes, leather goods, etc.

Three clinical types of anthrax are seen:

1. **Cutaneous anthrax** (malignant pustules) in face, neck, arms and back. Central necrotic area (black eschar) is surrounded by satellite vesicles containing serosanguinous fluid. 90% resolve spontaneously and 10% develop generalised infection and septicemia.

2. **Pulmonary anthrax** by inhalation of dust or filaments of wool from infected animals containing spores, specially in wool factories (Wool Sorter's disease). Hemorrhagic bronchopneumonia is seen, often followed by pleuro-pericardial effusion and septicemia. Hemorrhagic pneumonia has a high fatality rate.

3. **Intestinal anthrax** is seen in aborigines who consume improperly cooked infected meat. Severe enteritis is seen accompanied by bloody diarrhoea with high fatality.

4. **Septicemic anthrax** - the above three types, if not treated in time, progresses to septicemia.

Laboratory diagnosis

Specimens – Swabs, serous fluid of vesicles, pus from malignant pustules or material beneath the edge of black eschar, blood from septicemia, sputum from pulmonary anthrax, gastric aspirate, feces or food from intestinal anthrax.

Processing should be done under biological safety cabinet. After use, gloves, gowns and masks should be autoclaved or discarded. After work, cabinets and work benches should be decontaminated with 5% Na-hypochlorite or 5% phenol. All instruments should be autoclaved. Laboratory personnel should be immunized with anthrax vaccine. Nasal swab cultures are taken from patients to detect anthrax spores.

McFadyean's reaction – Capsular material appear amorphous and purplish around the bacilli, when stained with polychrome methylene blue. Microscopic examination of tissues should be done.

- Cultures of blood and spinal fluids before starting antibiotic therapy.

- Cultures of tissue or fluids from affected areas.

Selective medium – PLET (Heart infusion agar with polymyxin, lysozyme, EDTA and thallous acetate).

Biochemical tests – Catalase positive. Glucose, sucrose and maltose fermented with acid production only. Nitrates reduced to nitrites. Lecithinase postive.

Animal inoculation – I/P injection into white mouse or guineapig kills the animal in 36-48 hours.

Serological tests

1. Ascoli's thermoprecipitation test - filtered infected tissues are layered on top of anthrax antiserum in a test tube. A ring of precipitate appears at the junction within 15 minutes, for positive cases.

2. Direct fluorescent assay (DFA) to detect key bacterial proteins.

3. Strip test for antigen detection.

4. *B. anthracis* DNA detected by PCR from specimens collected from affected tissue or site.

5. Direct demonstration from clinical specimens by immunohistochemical staining.

6. API test and Phage sensitivity test.

Treatment

Penicillin is the drug of choice.

In septicemia – Scalvo's immune serum (by active immunisation in asses) along with antibiotics are given.

Optical fiber sensing device – a prototype biosensor which detects endotoxin at a level 20 times lower than previously achieved by other devices.

DIPHTHERIA

Commonly seen in children of 2-10 years of age. Infection is usually by droplet spread. Formation of a **pseudomembrane** (necrosed epithelium, fibrinous exudate, leucocytes, red blood cells and bacteria), fever and toxicity.

Common clinical types: Faucial, laryngeal, nasal.

Other types: Conjunctival, vulvovaginal, otitic, cutaneous.

Produces diphtheria toxin which inhibits protein synthesis.

A) **Local effect of the toxin** – pseudomembrane formation on tonsils, pharynx, larynx and nasal mucous membrane; **"Bull-neck"** appearance due to enlargement of cervical lymph nodes in severe cases.

B) **General effect of the toxin** in the blood stream, the toxin causes toxemia, has affinity for myocardium, nerve endings and adrenals.

Complications of diphtheria

1. **Local** – Pseudomembrane extending to larynx → laryngeal obstruction → asphyxia → death.

2. **Systemic**
   - Diphtheria myocarditis → heart failure → death.
   - Polyneuropathy and post-diphtheritic paralysis of palatine and ciliary muscles.
   - Degenerative changes in adrenal glands, liver and kidney.

There main biotypes of *C. diphtheriae*

a) **Gravis** – Severe infection, short rods, uniform staining, **"daisy head"** clonies on K-tellurite agar.

b) **Intermedius** – Moderate infection, long rods, irregular staining, **"frog's egg"** colonies.

c) **Mitis** – Mild infection, long curved rods, pleomorphic, **"poached egg"** colonies.

Laboratory diagnosis

1. Microscopy
   a) Gram stain of throat swab from area over pseudomembrane – Gram positive thin bacilli with metachromatic granules at both ends, arranged in Chinese - letter pattern.

   b) Albert's stain – Bacillary body stain green and granules bluish purple.

2. Culture
   a) Loeffler's serum slope – *C. diphtheriae* grows within 6-8 hours.

   b) McLeod's K-tellurite agar – Three biotypes of *C. diphtheriae* can be differentiated in this medium, but growth takes place after 36-48 hours.

3. **Toxigenicity tests**
   a) **In vivo tests** – Animal inoculation (guineapigs and rabbits)
      - Subcutaneous test
      - Intracutaneous test (better, as animal does not die).

   b) **In vitro tests**
      - Elek's agar gel precipitation test
      - Tissue culture test.

Teatment

1. Penicillin, Erythromycin or other antibiotics can be given.

2. Antidiphtheritic serum (ADS) – 20,000 units in moderately ill cases and 50,000 to 1,00,000 units in severely ill cases are given intramuscularly.

Prevention

One attack of diphtheria provides life long immunity

1. **Active immunity** by DPT vaccine – 3 doses at 4 weeks interval at 6th week, 10th week and 14th week. Boosters at 1½ years and 4½ years, given intramuscularly.

2. **Passive immunity** by antitoxin 500 to 1,000 units, given subcutaneously after skin test.

3. **Combined immunity** – diphtheria toxoid and antitoxin.

LABORATORY DIAGNOSIS OF GROUP A STREPTOCOCCI IN RHEUMATIC HEART DISEASE

1. **Serodiagnosis of rheumatic fever**

   ASO titre > 200 Todd U/ml. Normal value ≤ 200 Todd U/ml.

Rapid Mehods

a) Group A Str. Ag on throat swabs

b) Latex particle kit (Rapitex ASL)

c) Fluorescence AT for detection of Group specific Streptococcal polysaccharide

d) Antistreptozyme test (ASTZ) - Screening test

e) Anti-DNase B (> 300-350 units/ml)

f) Anti-streptokinase (ASK)

g) Anti-hyaluronidase (ADH)

h) Anti-NADase

2. **Indirect Methods**

   a) WBC count 12,000/cu. mm.

   b) Throat swab from tonsils, posterior pharynx (yellow grey exudate site)

   Should not be treated with antibiotics or given antiseptic mouth washes (gargles) for at least 8 hours prior to collection of throat swab.

   Sterile cotton or Alginate wool swab.

   c) Pyridoxine hydrochloride supplemented blood culture broth. Three blood cultures are collected within 3-6 hours, after which the treatment is started. Blood culture should be done for both aerobic and anaerobic bacteria. Blind Gram smear is done from the bottle and then subcultured onto solid media.

REFERENCES

1. Ananthanarayan & Paniker's Textbook of Microbiology. 9th Ed. Ed. Kapil A. Universities Press, Hyderabad. 2013.

2. Chakraborty PA. Textbook of Microbiology. 4th Ed. New Central Book Agency (P) Ltd., Kolkata. 2009.

VIRAL HEPATITIS AND LEPTOSPIROSIS

17

Chapter

VIRAL HEPATITIS

Hepatitis implies injury to the liver characterized by the presence of inflammatory cells.

Causes of Viral Hepatitis:

- Hepatitis A to E (more than 95% of viral cause)
- Herpes simplex virus
- Cytomegalovirus
- Epstein-Barr virus
- Yellow Fever virus
- Adenovirus.

Types of Hepatitis:

- Hepatitis A
- Hepatitis B
- Hepatitis D
- Hepatitis C
- Hepatitis E
- Hepatitis G

Hepatitis B

History

- First description of Hepatitis B dates back to 1885 when Lerman, a public health officer Berman, Germany gave a detailed report of an outbreak of jaundice that developed among workers of local company vaccinated against smallpox with glycerinised human serum.
- Blumberg in 1965 discovered Australia antigen, now called Hepatitis B surface antigen.
- Prince - recognition of its specific relation to Hepatitis B.
- Dane - identification of mature Hepatitis B virion.
- Magnius - recognition of antigenic complexity.
- Sherlock & Lancet - recognition of oncogenic potential.

Global distribution of Chronic HBV infection

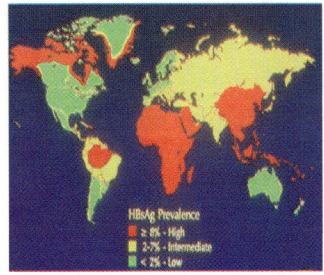

Epidemiology

- **High epidemicity** (> 8% carrier rate) – Equatorial Africa, South East Asia, China, part of South America
- **Low epidemicity** (< 2% carrier rate) – Western Europe, North America, Australia.
- **Intermediate epidemicity** (2-7% carrier rate) – Eastern Europe, Middle East, South Asia, parts of South America. India falls under the intermediate zone.

Classification

- FAMILY: Hepadna virus
- GENUS: Orthohepadna
- SPECIES: HBV

Structure of Hepatitis B virus

- HBV is spherical small DNA virus with an envelope containing the hepatitis B surface Ag and a nucleocapsid containing core Ag, i.e. HBcAg with 42 nm outer envelope and inner core of 27 nm.

  Under electron microscope sera from hepatitis B patient shows 3 types of particles:

- Spherical particle, 22 nm diameter
- Filamentous or tubular with diameter of 22 nm and of varying length
- Double walled spherical structure with 42 nm diameter - **Dane particle**.

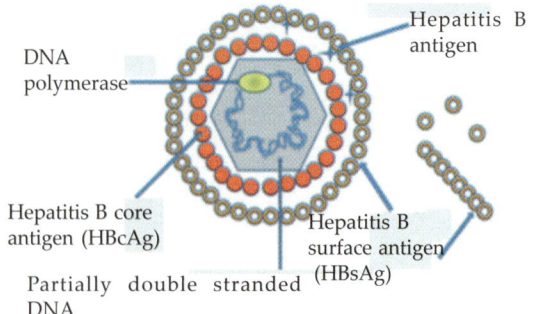

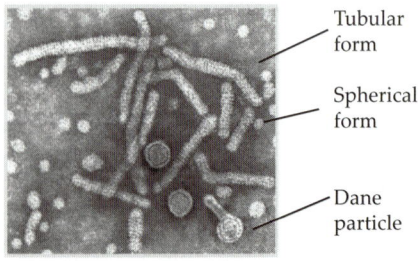

Genes	Regions	Gene products
S	S	Major protein (S)
	S + Pre S2	Middle protein (M)
	S + Pre S1 & S2	Large protein (L)
C	C	HBc Ag
	C + Pre C	Hbe Ag
P	-	DNA polymerase
X	-	HBx Ag

Antigenic diversity

HbsAg exhibits antigenic diversity. It contains 2 different antigenic components - the common group reactive antigen a & two small pairs of type specific antigen d-y and w-r. Only one member of each pair being present at a time. Thus HbsAg can be divided into four major subtypes – adw, adr, ayw and ayr.

Routes of transmission

- Parenteral
- Sexual
- Perinatal
- Blood Transfusion
- Direct contact with open skin lesions
- Other infectious materials – urine, saliva, bile
- Certain groups and occupations carry high risks of infections
- HBV is highly infectious more than HIV, as little as 0.00001 ml can be infectious.

Risk factors associated with Hepatitis B

Risk factors	%
Multiple sexual partners	24
I.V. drug users	20
MSMs	17
Sexual contact	13
Household contact	03
Others	23

Pathogenesis

Two phases:

- **Proliferative phase** – HBV DNA is present in episomal form with formation of complete virion and all associated antigens. Cellular expressions of viral HBsAg and HBcAg in association with MHC class I molecule leads to activation of cytotoxic CD8 lymphocytes
- **Integrative phase** – the incorporaion of viral DNA into genome of host cell. With disappearance of viral replication and appearance of antiviral antibodies infectivity ends and liver damage subsides but risk of HCC persists.

Course of Hepatitis B

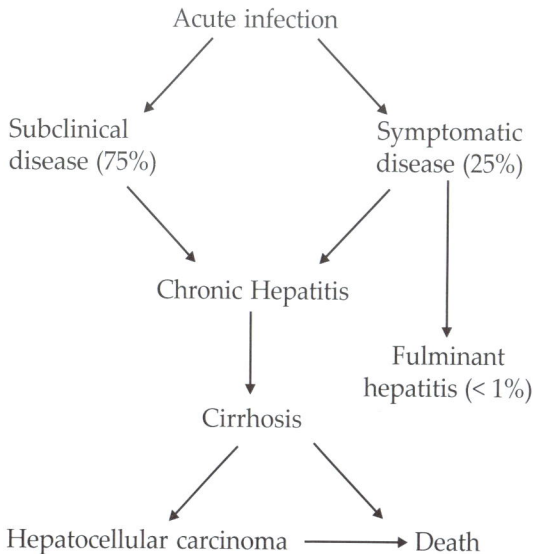

Acute hepatitis

- **Pre-icteric phase** marked by nonspecific constitutional symptoms like malaise, general fatiguability, nausea, loss of weight, muscle and joint ache, diarrhea; 10% patients develop serum sickness like symptoms.
- **Icteric phase** is due to conjugated hyperbilirubinemia with icterus, darker urine, pruritus, hepatomegaly, increased prothrombin time.

Chronic hepatitis

- Symptomatic, biochemical, serological evidence of continuing or relapsing hepatitis for > 6 months optimally with histological documented inflammation and necrosis
- Histological classification – Chronic persistent hepatitis; inflammation confined to portal tract
- Chronic active hepatitis – Portal tract inflammation spills into parenchymal and surrounds regions of necrosis of necrotic hepatocyte.
- Chronic lobular hepatitis – Persistant inflammation confined to the lobules.

Carriers

Individual without manifesting symptoms who harbors and therefore can transmit an organism.

- **Super carrier**: The carrier with high titre of HBsAg, HBeAg, DNA polytmerase, HBV in circulation; Some of them have enormous antigenemia and viremia; Upto 500 microgram of protein/ml of blood.
- **Simple carrier**: Low titre of HBsAg, negative HBeAg, HBV, DNA polymerase. Many super carrier in time becomes simple carrier.

Serological Markers

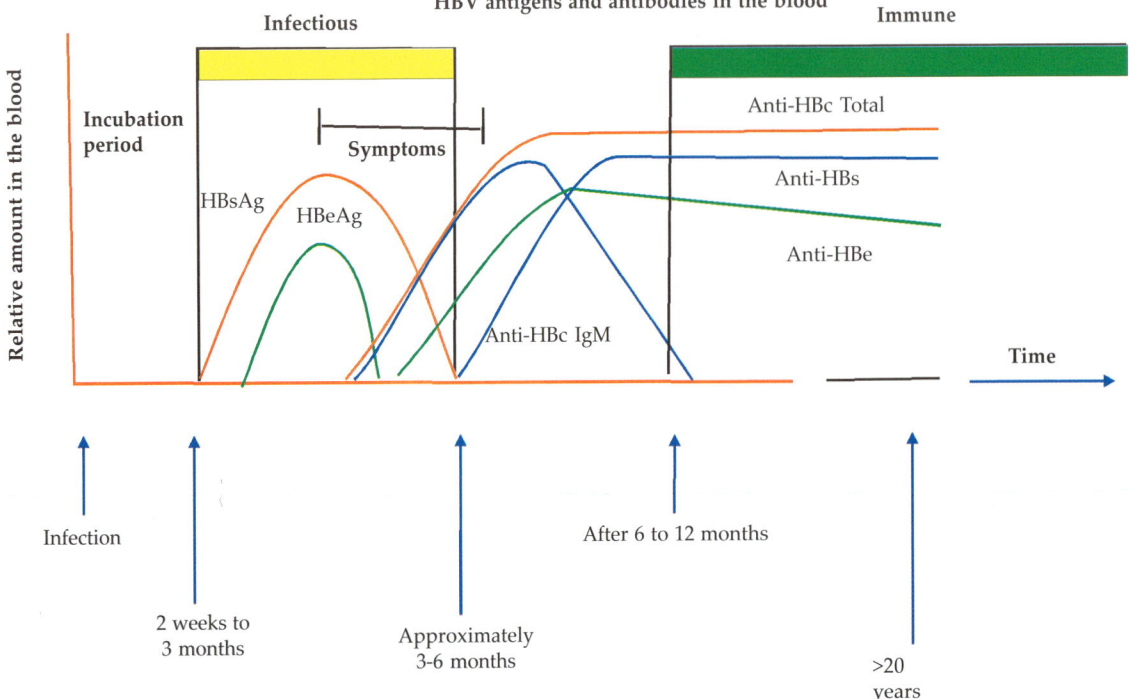

HBV antigens and antibodies in the blood

Interpretation of common serological patterns in HBV infection

HBsAg	HBeAg	AntiHBc	AntiHBs	AntiHBe	Interpretation
+	+	Ig M	-	-	Acute, highly infectious
+	+	Ig G	-	-	Chronic/late, highly infectious
+	-	IgG	-	+/-	Chronic/late/carrier, low infectivity
-	+/-	Ig M	-	+/-	Acute, infectious
-	-	Ig G	+/-	+/-	Remote infection
-	-	-	+	-	Vaccination

Laboratory diagnosis

- **Detection of antigens**
  - Rapid test
  - ELISA
  - Reverse Passive Hemagglutination Test - This test is based on the reverse passive hemagglutination assay (RPHA) using chicken-erythrocytes coated with highly purified antibodies to HBs antigens (IgG obtained from immunized guinea pigs). These coated chicken-erythrocytes are agglutinated in the presence of HBs antigens in human serum/plasma.

- **Detection of antibodies**
  - Anti HBc antibody detection; Microcapture antibody test
  - ELISA
  - Counterimmunoelectrophoresis
  - Complement fixation test.

HBV Rapid Test Kits

- HBV-5 Panel Test for the qualitative assessment of the markers of hepatitis B virus infection in human serum, plasma and whole blood.

- Intended use – This HBV Panel Test is an immunochromatographic assay to quickly detect five major markers of HBV infections – HBsAg, Anti-HBs (HBsAb), Anti-HBc (HBcAb), HBeAg and Anti-HBe (HBeAb) in human blood specimens.

Treatment

A) Active immunisation

- **Plasma derived vaccines** - Introduced in 1982, was prepared from purified, non-infectious 22 nm spherical forms of HBsAg, derived from healthy carriers.

- **Recombinant vaccines** - In 1987, plasma derived vaccine was supplanted by a genetically engineered vaccine derived from recombinant yeasts, consists of non- glycosylated HBsAg particles.

B) Passive immunization – Hepatitis B immuno-globulin.

C) Pre exposure prophylaxis

- Three IM Inj. of Hepatitis B vaccines at 0, 1 and 6 months (in deltoid or gluteal region)

- Pregnancy is not a contraindication.

RECOMBIVAX-HB [MERCK], 10 μg HBs Ag

Target group	No. of doses	Dose
Infant, children < 11 years	3	0.5 ml
Adolescent 11-19	3	0.5ml
Adults	3	1 ml
Hemodialysis patients	3	1 ml

ENGERIX -B [GLAXOSMITHKLINE] 1ml = 20 μg HBsAg

Target group	No. of doses	Dose
Infants, children < 10 years	3	0.5 ml
Adolescents 10-19 years	3	0.5 ml
Adults > 20 years	3	1 ml
Hemodialysis patients	3	1 ml

Duration of protection of vaccine:

- 80-90% – 5 years
- 60- 80% – 10 years

A) Post exposure prophylaxis

- Combination of HB immunoglobulins and vaccine.
- Perinatal exposure of infant born to HBsAg positive mother – Single dose of 0.5 ml HBs Ig I.M. in thigh, followed by complete course of HB vaccine within first 12 hours of life.
- Direct percutaneous inoculation or transmucosal exposure to HBsAg positive blood or body fluids – single I.M. 0.06 ml/kg HB Ig followed by complete course of HB vaccine to begin within 1st week.
- Exposed by sexual route: single IM 0.06 ml/kg within 14 days of exposure followed by complete course of HB vaccine.
- Hemodialysis patient – annual anti HBs testing after vaccinatimg antiHBs < 10 ml, booster dose is recommended.
- For people at risk of both Hepatitis A and B, combined vaccine, 720 ELU of inactivated HAV and 20 micro gm recombinant HBsAg at 0, 1, 6 months
- Other modes of treatment: Interferon alpha; Antiviral drug – lamivudine.

Hepatitis D

'D' in the virus designation is derived from the original term delta.

Pathogenesis

- In 1977, a discrepancy in immunofluorescence testing for HBV antigens in liver led RRIZZETTO and his associates to the iscovery of delta Ag which subsequently shown to be related to the new RNA virus HEPATITIS D.
- It was first seen in patients suffering from chronic Hep B.
- It is a defective virus that needs HBV as helper to acquire an envelope and infectivity.

Classification

- Family: Single stranded RNA satellites
- Genus: Delta virus
- Species: Hepatitis D

It is 36-43 nm double shelled partical that by electron microscope resembles DANE particles of HBV. The external coat Ag of HBsAg surrounds an internal 24 kd polypeptide delta antigen, associated with delta antigen is small circular molecule of single stranded RNA.

Genome replication

The genomic strand is transcribed by RNA polymerase 2 to short m RNA and long continuously growing strand which cleaves itself by ribozyme at 5′ end and thereafter a second time resulting in genome i.e. long linear RNA. This antigenome is ready to recircularise and now to repeat cycle with genomic transcript.

Genotype and variability

Three genotypes have been identified which differ by as much as 40% in their nucleiotide sequence.

- Genotype 1: most frequent and has variable pathogenicity
- Genotype 2: has found in Japan, causing relatively mild disease.
- Genotype 3: associated with HBV genotype f and fulminant hepatitis in South America.

Routes of transmission: Same as HBV. In endemic areas, commonly by percutaneous/close personal contacts. In non-endemic area infection is more often through blood and blood products, drug addicts, hemophiliacs. Incubation period is 30 to 180 days.

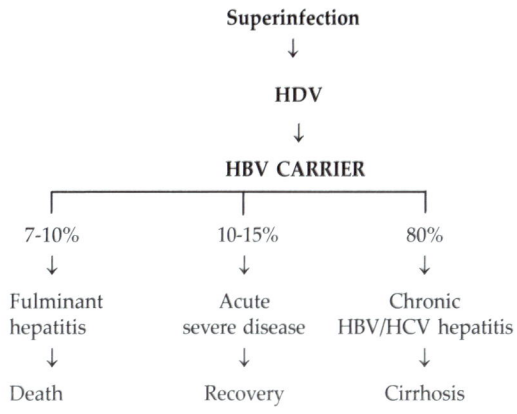

Laboratory Diagnosis

HDV infection occurs together with HBV. Thus detection of HDV and anti-HD Abs is indicated only for individuals with markers of active HBV infection, exception seen in patients who have received a liver transplant in which liver can be infected, inspite of seemingly absent HBV.

Detection of hepatitis D virus and Ag

- First assayed by immunoflorescence using chronic liver biopsies
- HD Ag detected by sandwich immunoassay

Antibody detection

- IgM anti HD - Standard Microcapture Assay

Treatment

- No specific prophylaxis
- Immunization with HBV vaccines effective as HDV cannot infect persons immune to HBV.

Hepatitis C

History

- When testing for hepatitis A and B were made available, it became clear that the two viruses did not explain all causes of viral hepatitisas many patients lacked markers of either infection. Thus concept of Non A Non B came. The diagnosis was based on exclusion criteria, not on specific positive markers.
- Houghten and associates identified a clone that met the criteria for being new viral genome and provided characterisation of HCV.
- For almost 20-25 years, it is known to cause 90-95% of cases of transfusion associated hepatitis.
- It is the most common cause of chronic liver disease and also common cause of patients waiting liver transplantation.
- Close contact with patient is important risk in India.

Classification

- Family: Flaviviridae
- Genus: Hepavirus
- Species: Hepatitis virus C

Structure

It is a small enveloped, single stranded positive sense RNA virus with 30-60 nm diameter, enveloped with glycoprotein spikes.

Genome

It has linear, single stranded open reading frame which is subsequently processed into functional proteins. The highly conserved region at 5' end encodes for nucleocapsid protein. A hypervariable region encodes for envelope E1, E2/NS1 and five less conserved non-structural regions NS1 to NS5 towards 3' end. The envelope proteins varies from isolate to isolate. At least 6 distinct genotypes as well as subtypes within genotype of HCV have been identified by nucleotide sequencing.

C	E1	E2/NS1	NS2	NS3	NS4	NS5
5' end		3' end				
Structural part			Non-structural part			

- Genotype 1 is important for prognosis, 1b has grave prognosis.
- Genotype 2 is found in Europe.
- Genotype 4, 5 found in Africa
- Genotype 3, 6 found in Southern and Eastern Asia.
- **Quasispecies**: Because of divergence of HCV isolates within a genotype or subtype, and within same host, may vary insufficiently to define a distinct genotype, these genotypic differences are referred as quasispecies and differ in sequence homology by only few %.
- Because of genotypic and quasispecies diversity of HCV resulting from high mutation rate interferes with effective humoral immunity.
- Since HCV does not replicate via DNA intermediate, it does not integrate into host genome.

Replication

Viral receptor CD 8 on hepatocyte
↓
Virus attaches Hepatocyte cytoplasm
↓
Virus RNA translation
↓
Polyprotein [RNA polymerase realeased by NS-3 protease]
↓
Replication by RNA dependent RNA synthetase
↓
Negative stranded RNA
↓
It acts as template for positive sense genomic RNA
↓
Translation of viral protein occur membrane associated ribosome
↓
Glycosylation of envelope protein occur in golgi apparatus
↓
Release by budding (enveloped)

Culture

- Although in vitro replication has been difficult, hepatocellular ca derived cell line have been described supports replication of genetically manipulated, truncated
- Or full length HCV RNA (but not intact virion)
- HCV replication has been documented in immunodeficient mouse model containing explants of human liver

Pathogenesis

It depends on:

- Viral factors: Replication efficiency; Genotype; Immune reactiveness; Cytopathic effects
- Host factor: Competence of immune system; Cytokine production
- Environmental: Alcohol intake

Modes of transmission:

A. Parenteral

Sporadic

- I.V. drug users
- Unscreened blood transfusion (Previously incidence was > 16%, now it is < 4%)
- Organ transplantation
- Blood product transfusion
- Nosocomial infection
- Occupational exposure 0-7%
- Colonoscopy
- Dialysis
- Perinatal < 10%
- Sexual < 20%
- Intranasal coccaine use

B. Non-parenteral

- Body fluids
- Household contacts

Sources of infection for persons with Hepatitis C

Risk factors	%
Injecting drug use	60
Sexual	15
Transfusion (before screening)	10
Others (hemodialysis, health care work, perinatal)	05
Unknown	10

Clinical features

Acute hepatitis C

- Acute asymptomatic (60-70%) → Recovery (15-25%)

- Acute icteric hepatitis (20-30%) → Chronic
- Acute fulminant hepatitis (rare)

Chronic hepatitis

80 -85% from acute asymptomatic icteric hepatitis.

Types:

- Chronic silent carrier - asymptomatic
- Chronic subclinical hepatitis - LFT raised
- Chronic persistent hepatitis
- Chronic autoimmune hepatitis
- Chronic active hepatitis

  Cirrhosis of liver

  Hepatocellular carcinoma

Causes of persistence of HCV infection

- Genetic variability
- HCV type
- Genotype
- Quasispecies
- Non hepatocellular reservoir
- Mononuclear cells
- Billiary cells
- Immune genetics
- HLA
- Production of TNF-α
- Cytokinin production

Acute infection

- Incubation period 6-7 weeks
- Fatigue, nausea, anorexia, weight loss, myalgia, arthralgia, fever
- Jaundice is in < 1/3rd of acute cases
- Resolves in 1-3 months.

Chronic infection

- Asymptomatic elevation of aminotransferase
- Fatigue is commonest, others are anorexia, vomiting, pruritus, arthralgia, myalgia
- Cirrhosis – on palpation, firm liver, ascitis, splenomegaly
- Decreased albumin, thrombocytopenia.

Laboratory diagnosis
Serological markers of Hepatitis C

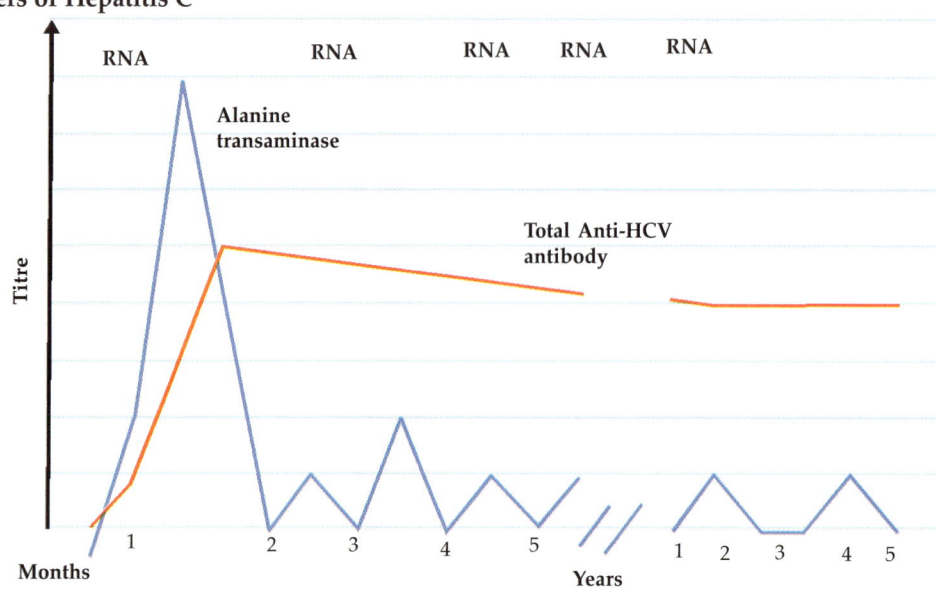

Conventional

Takes 3-4 hours, microtitre plates required.

ELISA kit	Sensitivity	Specificity	Comment
First generation	70-80%	Increased false positive	Ab detected after 7-8 weeks of window period
Second generation	80-90%	Few false positive	Window period shortened to 5-6 weeks
Third generation	90-95%	99%	Window period still shortened to 2 weeks

Assay	HCV Polyprotein[a]						
	Core	E1	E2/NS1	NS2	NS3	NS4	NS5
ORTHO ELISA[b]							
1st generation						c100-3 (a.a.1569-1931)	
2nd generation	c22-3 (a.a.2-120)				c200 (a.a.1182-1931)		
3rd generation	c22-3 (a.a.2-120)				c200 (a.a.1182-1931)		NS5 (a.a.2054-2995)
CHIRON SIA[c]							
3rd generation	c22p[d] (a.a.10-53)				c33c (a.a.1192-1457)	5-1-1p[d] (a.a.1694-1735) c100p[d] (a.a.1920-1935)	NS5 (a.a.2054-2995)
ABBOTT EIA[e]							
2nd generation	HC-34 (a.a.1-150)				HC-31 (a.a.1192-1457 and 1676-1931) c100-3 (a.a.1569-1931)		

a =   b =   c =   d =

Rapid HCV TRIDOT Test

- T1 and T2: two dots having antigen from non-structural gene NS1, NS4, NS5
- C is antigen from core
- Appearance of 2/3 dots including C: Positive
- It indicates past, present or resolved infection but cannot differentiate between the two.

Recombinant Immunoblot Assay (RIBA)

- It is a strip immunoblot assay which utilises recombinant HCV antigen
- And synthetic peptides are made as fusion protein between human superoxide desmutase and immobilized as individual bands on nitrocellulose strip.
- First generation: ag c 100-3, 5-1-1
- Second generation: c -22, ns -3, c100-3, 5-h2; c-22-3
- Third generation: c-200, c100-3, ns-5

Interpretation

- > 2 bands: positive
- One band: intermediate
- No band: negative

Uses

- Healthy blood donors
- Anti HCV positive patients with normal ALT levels
- ELISA negative, ALT high, risk factor present

Implications

- Antibody to structural antigen: viral replication
- Antibody to envelope antigen: past infection
- Antibody to core antigen: presence of infection without replication.

Particle-Agglutination Test for the Detection of Antibodies to Hepatitis C virus (HCV)

This test is based on Fujirebio's particle-agglutination assay using gelatin particles coated with recombinant HCV antigens, c22-3 and c200. These coated gelatin particles are agglutinated in the presence of antibodiesto HCV in human serum/plasma.

Molecular methods

- RT-PCR
- TMA (Transcription Mediated Amplification)
- bDNA (Branched DNA)

RT-PCR

- Both qualitative and quantitative
- Ampliclor HCV kits available
- Most specific test
- Can detect low level of RNA

- RT converts RNA to cDNA template → amplification → hybridization to HCV specific nucleotide probe.

Qualitative

- Amplicor HCV test: lower limit 50 IU/ml
- Ampliscreen HCV test: < 50 IU/ml

Quantitative

- Amplicor HCV MONITOR: 600-5,00,000 IU/ml.

TMA

- T-7 RNA polymerase and RT reacts under isothermal conditions to form detectable levels of RNA. From RT cDNA becomes a template for T-7 RNA polymerase enzyme which form no. of RNA next set by entering into cycle, e.g. VERSANT HCV RNA assay.
- Procleix HCV assay [Chiron] can detect < 50 IU/ml.

bDNA (Branched DNA signal amplification assay)

- Uses solid phase oligonucleotide probe → captures target RNA → hybridization of branched bDNA → secondary probe → amplification → binding to enzyme conjugated tertiary probe → substrate → chemiluminescence proportional to target RNA, e.g. VERSANT HCV RNA assay.
- Lower limit of detection is 615 IU/ml.
- Quantitation: 520-8,300,000 IU/ml.

Genotyping

- Single stranded conformational polymorphism (SSCP)
- In India, Type 3 is most common and 1b has the worst prognosis.

Immunostaining

- Poly/monoclonal Ab used to detect HCV Ag in liver, not commercially available.

Management

- Evaluation of progression of disease
- Counselling
- Palliative treatment
- Anti-retroviral therapy, Monotherapy or Combined therapy
- INF α2b – 3 million units S/C weekly + daily Ribavarin 1 gm/day
- For genotypes 2 & 3 – 24 weeks
- For genotype 1
  - > 8,00,000 IU/ml – 48 weeks
  - < 8,00,000 IU/ml – 24 weeks

Prevention

- Mandatory screening with more sensitive and cost effective test
- HBV and HIV should be under control
- Health education
- Discouragement of use of blood transfusion and other fractions

- Vaccination against HBV and HAV should be encouraged in high risk populations
- HCV vaccine: Because of genetic variability of the virus, vaccine is difficult to invent.
- Vaccine trials are on – Enveloped GP HCV; Naked DNA based; T lymphocyte based.

LEPTOSPIROSIS

Incubation period – 7 days (2-29 days)

Signs and symptoms

Fever, myalgia, conjunctival suffusion, albuminuria, jaundice, renal or central nervous system involvement, respiratory symptoms, other bleeding diathesis.

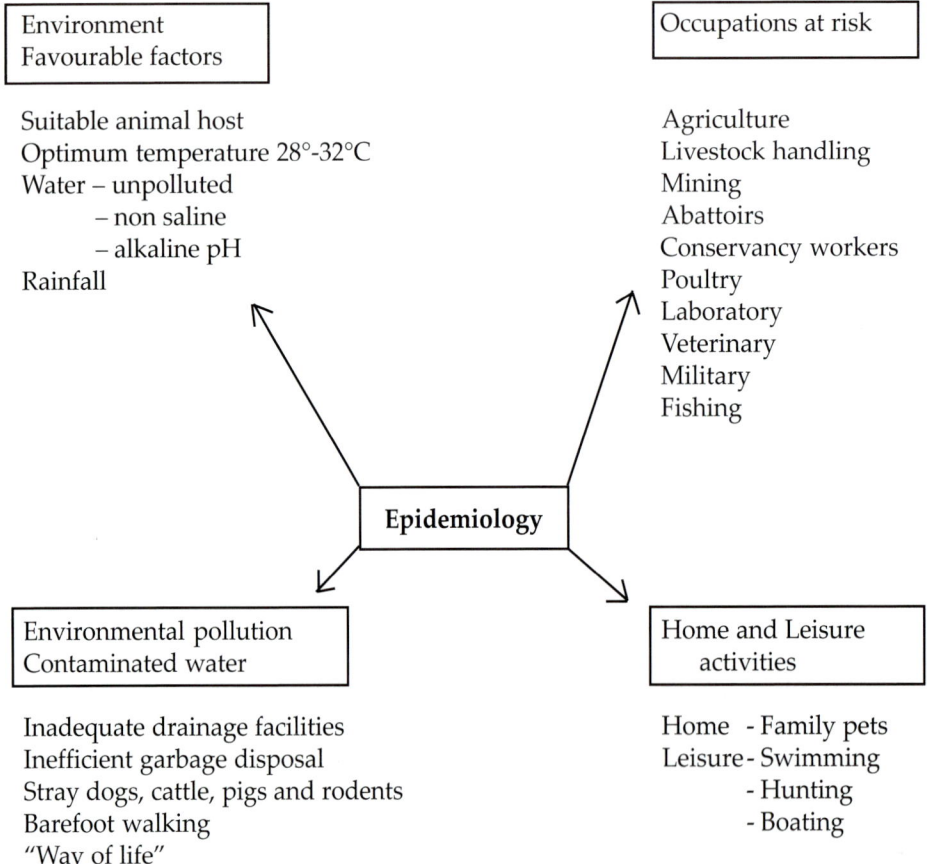

Environment Favourable factors

Suitable animal host
Optimum temperature 28°-32°C
Water – unpolluted
– non saline
– alkaline pH
Rainfall

Occupations at risk

Agriculture
Livestock handling
Mining
Abattoirs
Conservancy workers
Poultry
Laboratory
Veterinary
Military
Fishing

Epidemiology

Environmental pollution Contaminated water

Inadequate drainage facilities
Inefficient garbage disposal
Stray dogs, cattle, pigs and rodents
Barefoot walking
"Way of life"

Home and Leisure activities

Home   - Family pets
Leisure - Swimming
            - Hunting
            - Boating

Case fatality rate – 1.5%

Laboratory diagnosis

Leptospira, a spirochaete with hooked ends was discovered by Adolf Weil in 1886. They are broadly classified into *L. interrogans* which is pathogenic and *L. biflexa* which is saprophytic. Leptospirosis shows increased incidence in monsoon (rainfall > 100 inches; humidity 70-80%; optimum temperature 28°-32°C). It is also an occupational disease seen in agricultural workers, abattoir workers, veterinarians, sewage/drainage workers, dairy workers, sweepers and plumbers.

The samples should not be refrigerated. It is to be kept at room temperature. Storage for longer periods is done at -20°C.

Direct evidence of organism by Dark Ground Microscopy and Staining (Levaditi's and Fontana's).

Cultivation

First week from blood and CSF; Second week onwards from urine (PBS pH 7.2). Leptospira can be grown on semi-solid media supplemented with rabbit serum or bovine serum albumin and tween 80. Incubation may be for as long as 28 days.

EMJH (Ellinghausen McCollough Johnson and Harris) medium – Incubated at 30°C for 2 to 3 weeks → Dinger's ring 1cm below surface is diagnostic. It is visualized by darkfield microscopy.

Animal inoculation is done in Guineapigs intraperitoneally.

Serodiagnosis of leptospirosis

- Genus specific tests – detectable for months.
    - ELISA: IgM & IgG, Dot ELISA
    - MSAT(Macroscopic slide agglutination test)
    - Micro capsule agglutination test
    - Rapid tests:

        Lepto Dipstick assay

        Lepto Dri Dot test

        Lepto lateral flow test

        Latex agglutination test

    - Indirect fluorescent antibody (IFA) test

Observed under high power using fluorescent microscope (495 nm interference filter). Positive – apple green fluorescence. Sensitivity 89.2 %; Specificity 95.1%.

    - Indirect hemagglutination assay (IHA)
    - Complement fixation test (CFT)
    - Counter immunoelectrophoresis (CIEP)
- Serovar specific tests - detectable for years.
    - Microscopic Agglutination Test (MAT)
    - Serovar specific ELISA

Types of ELISA

- **IgM & IgG ELISA**

    - Biolisa IgM ELISA (Bios GmbH, Labordiagnostik)

    - Leptospira IgM ELISA (PanBio Pvt. Ltd., Brisbane, Australia)

Sensitivity:

- First acute phase – 52%
- Second acute phase and Convalescent phase – 89 to 93%

Specificity: 94% in all specimens.

- Serion ELISA Classic Leptospira (Institute Virion Serion GmbH, Würzburg, Germany)

    According to kit (Virion/Serion IgM ELISA) - Sensitivity: 96%; Specificity: 90%

- **Dot ELISA** - 95% positivity corresponding with MAT.

Microscopic Agglutination Test (MAT)

- Gold Standard test (performed only in Reference Centres).
- Performed with battery of antigens, covering the range of serovars that are likely to be circulating in a particular geographical area.
- A saprophytic serovar (Patoc 1) strain should also be included along with pathogenic serovars (it tends to behave as genus specific antigen & would therefore detect antibodies against serovar not yet known to exist in a particular geographic area.
- **Positive result** – The highest dilution of the serum showing approximately 50% agglutination of leptospires.
- If agglutination with more than one serovar is observed, then the serovar that gives the highest titre is considered as infective serovar.
- Seroconversion or four fold rise in titre in paired serum sample is definite evidence of infection.
- Sensitivity – First acute phase 30%; Second acute phase and Convalescent phase 60-76%.
- Specificity – 97%.

Molecular methods – Nucleic acid based techniques

- PCR (in vitro DNA amplification) in first week of disease.
- DNA hybridization (species specific).

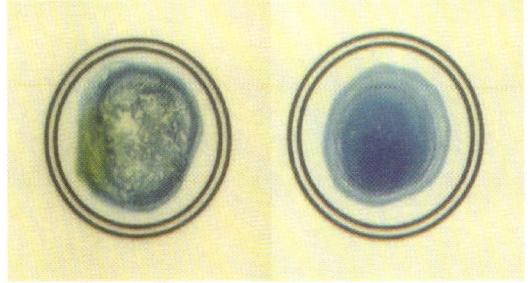

Lepto Tek Dri-dot test for detection of antibodies to leptospira antigen – positive test (left) and negative test (right)

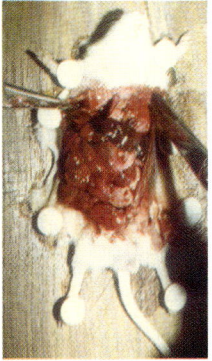

Post-mortem sample: Demonstration of leptospires in heart blood and peritoneal cavity of Guineapig

REFERENCES

1. Topley & Wilson's, Microbiology and Microbial Infections. Bacteriology, Vol. 2. 10th Ed. Eds. Borriello SP, Murray PR, Funkay G. John Wiley & Sons Ltd., West Sussex, UK. 2009.

2. Pungpapong S, Kim WR, Poterucha JJ. Natural History of Hepatitis B Virus Infection: an Update for Clinicians. Mayo Clinic Proceedings 2007; 82: 967-75. doi:10.4065/82.8.967. PMID 17673066.

3. Liaw YF, Brunetto MR, Hadziyannis S. The natural history of chronic HBV infection and geographical differences. Antiviral Therapy 2010; 15: 25-33.doi:10.3851/IMP1621. PMID 21041901.

4. Aspinall EJ, Hawkins G, Fraser A, Hutchinson SJ, Goldberg D. Hepatitis B prevention, diagnosis, treatment and care: a review. Occupational medicine. Oxford, England. 2011; 61: 531-40. doi:10.1093/occmed/kqr136.PMID 22114089.

5. Weinbaum CM, Williams I, Mast EE, Wang SA, Finelli L, Wasley A, et al. Recommendations for Identification and Public Health Management of Persons with Chronic Hepatitis B Virus Infection. National Center for HIV/ AIDS, Viral Hepatitis, STD, and TB Prevention, Division of Viral Hepatitis. MMWR 2008; 57(RR-8): 1-20.

6. A Comprehensive Immunization Strategy to Eliminate Transmission of Hepatitis B Virus Infection in the United States: Recommendations of the Advisory Committee on Immunization Practices. Part I: Immunization of Infants, Children, and Adolescents. MMWR 2005; 54(RR-16).

7. Recommendations for Prevention and Control of Hepatitis C Virus (HCV) Infection and HCV-Related Chronic Disease. U.S. Department of Health and Human Services, Centers for Disease Control and Prevention (CDC) Atlanta, Georgia. MMWR 1998; 47(RR-19): 1-54.

8. Recommended Testing Sequence for Identifying Current Hepatitis C Virus (HCV) Infection. Testing for HCV infection: An update of guidance for clinicians and laboratorians. U.S. Department of Health and Human Services, CDC. MMWR 2013; 62.

9. Recommendations for Prevention and Control of Hepatitis C Virus (HCV) Infection and HCV-Related Chronic Disease. U.S. Department of Health and Human Services, CDC. MMWR 2010; 59(RR-12).

10. Ratnam S. A manual on leptospirosis. 1st Ed. S. R. Publications, Madras. 1994.

11. De A, Varaiya A, Pujari A, Mathur M, Bhat M, Karande S, Yeolekar ME. An outbreak of leptospirosis in Mumbai. Indian J Med Microbiol 2002; 20: 153-5.

12. Human leptospirosis: Guidance for diagnosis, surveillance and control. WHO ILS.www.who.int/csr/don/en/ WHO_CDS_CSR_EPH_2002.23: pp1-109.

13. Clinical laboratory diagnosis of leptospirosis. WHO ILS. http://www.med.monash.edu.au/microbiology/staff/ adler/clinical-laboratory-diagnosis-of-leptospirosis.

ENTERIC FEVER

18
Chapter

Enteric fever is endemic in the Asia Pacific region, the Indian subcontinent, Central Asia, Africa and South America. WHO has estimated that there are 16.6 million cases of typhoid annually with about 600,000 deaths.

Laboratory diagnosis of Enteric fever

Infective dose is 10^5 organisms.

1. Blood culture is done in first week of fever

   5-10 ml of blood is inoculated in 50-100 ml of bile broth.

2. Widal test is done from second week of fever. H and O antigens are prepared from *Salmonella typhi* 901 strain are used. Local titre of H and O antigens are important.

3. Isolation from stool and urine is done in third and fourth weeks of fever respectively.

4. Bone marrow culture, culture of scrapings from rose spots.

5. Blood count shows leucoperia, i.e. W.B.C. count < 5000/cu. mm with relative lymphocytosis.

Bed side investigations:

6. Diazo test – Urine + equal amount of diazo reagent + 10% NH3 – red or pink froth developed. Positive in 80-90% cases.

7. Stool – pea soup appearance.

8. Urine – albuminuria.

Newer diagnostic tests:

9. Automated blood culture system detect *Salmonella spp.* within 6 hours – BACTEC/BacT Alert.

10. Staphylococcus Coagglutination test detects Salmonella antigen in blood during first week of fever which uses *Staphylococcus aureus* protein A (Cowan 1 strain) coated wtih antibody.

11. CIEP and ELISA detect porin antigen of *Salmonella spp.*

12. ELISA using monoclonal antibodies against *S. typhi* flagella.

13. Typhi dot (EIA dot test) detects separately both IgM and IgG antibodies to outer membrane protein of *S. typhi* within 48 hours. Only 10µl of serum is needed. It has a sensitivity and specificity of > 95%. It is a simple, easy and convenient test, result is obtained within 60 minutes. This tests indicates the stages of typhoid infection.

- IgM positive only - Acute typhoid fever.
- IgM and IgG positive - Acute typhoid fever (in the middle stage of infection).
- IgG positive - relapse or reinfection or previous infection. Confirmation can be obtained from Typhidot - M results.
- IgM and IgG negative - Probably not typhoid. The test may be negative when tested first time. Repeat testing 3-5 days later may show positive results.

This test reduces investigations and hospital stay.

14. PCR by using DNA probes.

Diagnosis of typhoid carriers

1. Bile or duodenal aspirate culture

2. Vi agglutination test

3. Stool culture specially after a cholagogue purgative, might grow *S.typhi*

4. "Sewer swab technique".

Stage of disease	Examination	Result (% positive)
1st week	Blood culture	95%
	Blood picture	Leucopenia with relative lymphocytosis
2nd week	Blood culture	60%
	Widal test	Low titre antibody
3rd week	Widal test	100%
	Blood culture	30%
	Stool and urine culture	50%
4th week	Widal test	100%
	Stool and urine culture	80%
	Blood culture	10%

Treatment

Chloramphenicol, Co-trimoxazole, Ciprofloxacin

For carriers

Ampicillin 6 gms daily for 3 months, cholecystectomy.

Control

- TAB vaccine – Killed vaccine given subcutaneously offers 60-80% protection.
- S. *typhi* Ty21a vaccine (Typhoral) – Live vaccine, 3 doses given orally on day 1, 3 and 5 to children above 6 years of age offer 65-96% protection.
- Vi Polysaccharide vaccine – given intramuscularly as a single dose offer 75% protection.

Procedure for TYPHIDOT

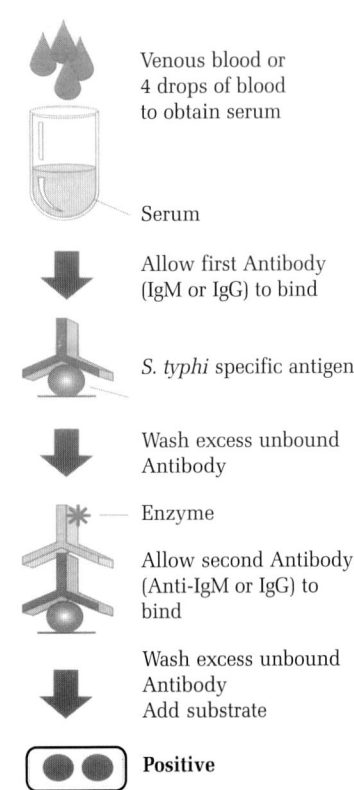

Venous blood or 4 drops of blood to obtain serum

Serum

Allow first Antibody (IgM or IgG) to bind

S. typhi specific antigen

Wash excess unbound Antibody

Enzyme

Allow second Antibody (Anti-IgM or IgG) to bind

Wash excess unbound Antibody
Add substrate

Positive

REFERENCES

1. Ananthanarayan & Paniker's Textbook of Microbiology. 9th Ed. Ed. Kapil A. Universities Press, Hyderabad. 2013.

2. Chakraborty PA. Textbook of Microbiology. 4th Ed. New Central Book Agency (P) Ltd., Kolkata. 2009.

3. Jesudason M, Esther E, Mathai E. Typhidot test to detect IgG & IgM antibodies in typhoid fever. Indian J Med Res 2002; 116: 70-2.

4. Parry CM, Hien TT, Dougan G, et al. Typhoid fever. New Eng J Med 2002; 347: 1770-82.

URINARY TRACT INFECTIONS

Chapter 19

Definitions

- **Bacteriuria** – Multiplication of organisms in urinary tract.
- **Significant bacteriuria** – Presence of $> 10^5$ organisms per ml in mid-stream sample of urine.
- **Pyuria** – Presence of pus cells in urine, which most often accompanies UTI.

Transient flora is found in distal urethra, most of which are derived from fecal flora.

Types of UTI

1) Lower UTI – Urethritis, Cystitis, Prostatitis.
2) Upper UTI – Acute pyelitis, Acute pyelonephritis.

Clinical features

1) **Asymptomatic** – "Covert bacteriuria" seen in 6% pregnant women, 0.6% men, 1-3% girls and 0.3% boys. It is detected only by urine culture.

2) **Symptomatic** – More frequent in women that men. (50% women suffer from UTI sometime during their life). In men usually in first year of life and over 60 years.

Symptoms of UTI

Dysuria, frequency, suprapubic pain, loin pain, tenderness.

Fever and rigors are seen usually in pyelonephritis.

Predisposing factors

1) Gender:

   More common in females due to
   a) Proximity of female urethra to anus
   b) Short length of female urethra (4cm)

2) Pregnancy:

   Predisposes to upper UTI due to
   a) Dilatation of ureters and pelvis
   b) Stasis in ureters
   c) Atony – reduced tone in ureteric musculature
   d) Temporary incompetence of vesico – urethral valves.

3) Other factors:

   a) Obstruction to flow of urine by tumour, stone, stricture, prostatic hypertrophy.
   b) Neurogenic bladder dysfunction (spinal cord injury, multiple sclerosis)
   c) Vesico ureteral reflux
   d) Bacterial virulence
   e) Genetic factors (receptors on uroepithelial cells).

Routes of infection

1) Ascending route – from perineum upwards.

   Causes of periurethral colonization by enteric bacilli:

   a) Alteration of normal perineal flora by antibiotics or contraceptives.
   b) Fecal incontinuance in infants.
   c) Entrance of periurethral bacteria inside bladder.
   d) Poor hygienic habbits in adults.

2) Hematogenous route – Rare.

   a) Renal infection of newborn.
   b) Bacteremia
      - Abscess formation in renal parenchyma
      - Excretion of organisms in urine

Causative organisms

Organism	Domiciliary (%)	Hospital-acquired (%)
Escherichia coli	80	50
Proteus mirabilis	08	12-15
Klebsiella species	08	12-20
Staphylococcus saprophyticus	0-1	1-2
Staphylococcus epidermidis	0-1	1-2
Enterococci	0-1	4-5
Other Gram-negative bacilli	< 1	< 1
Pseudomonas aeruginosa	0-1	8 - 12
Staphylococcus aureus	0	< 1

E. coli: Serogroups 02, 04, 06, 07, 08, 075 cause UTI.

- K antigens inhibit phagocytosis.
- P fimbriae adhere to receptors on uroepithelium.

Proteus spp.: produce urease, making urine alkaline. This causes necrosis of renal tubular epithelium leading to calculi formation and obstruction.

Klebsiella spp.: produce extracellular slime and polysaccharide leading to stone formation.

Incidence of Coagulase negative *Staphylococcus aureus* (CONS) is 14-44%.

S. saprophyticus: Novobiocin resistant CONS is responsible for 2-3% cases of urethritis and cystitis in sexually active otherwise healthy young females.

Incidence is about 46% in women of 16-25 years age group and in pregnant women it is 15.3%.

Chronic or complicated UTI

Infections associated with renal calculi, obstruction, instrumentation, secondary to prostatectomy or catheterization.

Organisms causing UTI

Gram negative bacilli :

E. coli, Klebsiella, Proteus, Pseudomonas, Enterobacter, Citrobacter, Serratia, Acinetobacter.

Gram positive cocci:

Enterococci, S. aureus, Streptococcus Groups A and B, S. saprophyticus.

Acid fast bacilli: *Mycobacterium tuberculosis.*

Fungus: *Candida albicans.*

Hospital-acquired UTI – 75% is seen after catheterization or instrumentation.

UTI may be Endogenous or Exogenous (cross-infection from patients, nurses, doctors, etc.).

Laboratory diagnosis

A) Specimen collection

Adequate urine sample, free from urethral or genital tract contamination.

- Mid-stream specimen of urine (MSU) in a sterile wide-mouthed container after proper genital toilet.
- Catheter specimen of urine (CSU) directly from the catheter and not from collection bag.
- Suprapubic aspiration (simple procedure, provides accurate specimen) – from infants and adult women.

B) Transport

- Transported to the laboratory within 2 hours of collection.
- Should be preserved by refrigeration at 4°C or in 1.8% boric acid solution, if there is delay in transport.

C) Processing

1) **Microscopy**

a) Direct examination of centrifuged deposit of urine
- Pus cells > 5/h.p.f.
- RBCs, bacteria, epithelial cells, casts, crystals
- Pyuria without bacteriuria – renal tuberculosis
- Sterile pyuria – Gram stained smear helps to diagnose
- Gram negative coccobacilli and polymicrobial flora – *H. influenzae* and anaerobes
- Painless sterile haematuria – *S. haematobium*

b) Gram stained smear of uncentrifuged urine ≥ 1 bacteria/oil immersion field.

Sensitivity 94%; Specificity 90%.

2) **Culture**

a) Media - Blood agar, MacConkey agar.

b) Semi-quantitative culture by

   i) Standard loop method (0.01 ml urine delivered by loop with 4mm external diameter, made of 23 SWG wire).

   ii) Dip slide method (200 colonies ≡ 10^5 organisms/ml). This is a commercially available plastic slide coated with CLED agar on one side and MacConkey agar on the other side.

   - Avoids transporting problem as it is done at bed side.
   - It is very useful for screening in busy OPDs.

Interpretation

- Significant bacteriuria > 10^5 organisms/ml
- Doubtful significance 10^4-10^5 organisms/ml
- Not significant < 10^4 organisms/ml

Cases when 10^4-10^5 organisms/ml are considered significant:

- Patient already on antibiotics
- Excessive water intake before culture
- Slow growing organisms, fastidious organisms
- Infection occurring above an obstruction
- Chronic UTI.

3) **Identification of organisms** by colony characters, biochemical tests and/or serological tests.

4) **Antibiotic sensitivity test** using the antibiotics amikacin, augmentin, nalidixic acid, norfloxacin, nitrofurantoin, cefotaxime, cefuroxime, ceftriaxone.

D) Chemical Screening Tests

a) Catalase test

b) Griess nitrite test

c) Glucose oxidase test

d) Triphenyl tetrazolium chloride (TTC) reduction test.

E) Commercial Screening Systems

1. **LE strip** impregnated with buffered indoxyl carbolic acid ester and a diazonium salt. Indirectly estimate number of segmented neutrophils. Sensitivity 88%; Specificity 94%. Reflects pyuria ($\geq$ 10 pus cells/h.p.f).

2. Combination of nitrate reduction and leukocyte - esterase activity – **Chemostrip LN**.

3. **Bac-T-Screen** uses vacuum section in which organisms in diluted urine sample are trapped in a filter paper before staining. This has 98% accuracy which is comparable with Gram stain.

   Disadvantages: Interfering chromogens are present in urine and clogging of filter can occur.

4. **BACTEC urine** is a screening system. This consist of a blood culture vialse with trypticase soy broth in which an aliquot of urine sample is inoculated. Sensitivity 97.6% at a breakpoint of 10^5 CFU/ml. Specificity 100%.

5. **Uriscreen system** (2-minute catalase tube test) to detect bacteriuria and pyuria.

   It is a rapid, manual, easy to perform enzymatic test. Detects 93% samples with > 10^5 CFU/ml.

6. **Urine screening system** employing bioluminescence
   - Lumac System   Sensitivity 92.4%
   - UTI Screen   Specificity 79.4%

7. **Bacterial ATP** by measuring light emitted by reaction of luciferin-luciferase.

8. **Antibody-coated** bacteria by using fluorescein - conjugated anti-human gamma globulin.

9. **Autobac** by light scatter photometry. It has 99% correlation with culture.

10. **DNA probe assay**

    More sensitive than UTI-screen tests.

    Disadvantages: High cost, complex protocol and time taken is 2½ hours.

For tuberculosis of kidney and urinary tract

Symptoms – Frequency, painless haematuria.

Microscopy – Numerous pus cells, RBCs are present but no pathogen is detected in routine culture.

Specimen

Early morning urine samples on 3 consecutive days.

Processing

Centrifuged deposit of urine is concentrated by Petroff's method, stored at 4°C and pooled. Then AFB staining and culture of the pooled sample are done on L-J medium.

Catheter-associated UTI

Plastic I.V. catheters and central venous catheters are associated with bacteremia – CDC and Prevention of National Nosocomial Infection Surveillance System (PNNISS) reported increased incidence of *S. epidermidis*.

Prevention of catheter-associated UTI

1. Insertion with sterile precautions

2. Modification in structure of catheter:

a) Physicochemical alteration

$\downarrow$ hydroxyethyl methacrylate

polyurethane surface

$\downarrow$

hydrophilic

$\downarrow$

reduce adhesin (for adhesion, hydrophobic nature is required)

b) Use of hydromes

c) Modification of Polyethylene tetrapthalate films using polyethylene oxide

d) Antiseptic hubs – protects entire length

e) Coating of catheter with antiseptic and antimicrobial compound
   - 2,4,4/trichloro-2-hydroxyphenyl ether
   - Benzalkonium chloride
   - Heparin
   - Benzalkonium chloride and heparin
   - Iodine
   - Silver sulphadiazine and chlorhexidine

f) Coating of catheter with antibiotic
   - Incorporation of antibiotic into polymer
   - Antibiotics like clindamycin, fusidic acid, cephalosporins, glycopeptides, quinolones and combination of rifampicin and minocycline.

REFERENCES

1. Mims' Medical Microbiology. 5th Ed. Eds. Richard Goering, Hazel Dockrell, Mark Zuckerman, Ivan Roitt & Peter L. Chiodini. Saunders. 2013.

2. Koneman EW, Allen SD, Janda WM, Schreckenberger P, Winn Jr. WC. In Color Atlas & Textbook of Diagnostic Microbiology. 6th Ed. Lippincott, Philadelphia. 2006.

3. De A. Urinary tract infections in men and women - Review article. Indian J Med Microbiol 1993; 11: 121-3.

CENTRAL NERVOUS SYSTEM INFECTIONS

20
Chapter

Acute Bacterial meningitis

Age group	Bacteria
In neonates	Group B streptococci
	Escherichia coli
	Listeria monocytogens
In children	*Hemophilus influenzae* type b
	Neisseria meningitidis
	Streptococcus pneumoniae
In adults	*S. pneumoniae*
	N. meningitidis
	Enteric gram negative bacilli
	Staphyloccus aureus
	H. influenzae
	Listeria moncytogenes

Chronic meningitis

Infectious	Non-infectious
Mycobacterium tuberculosis	Carcinoma
Cryptococcus neoformans	Sarcoid
Treponema pallidum	Granulomatous angitis
Candida allbicans	Systemic lupus
Brucella spp.	erythematosus
Leptospira canicola	Behcet's disease
Toxoplasma gondii	Subarachnoid
Nocardia spp.	hemorrhage
Borrelia burgdorferi	Drug/chemical toxicity
Neurocysticercosis	Post-infectious allergic
Actinomyces spp.	encephalomyelitis
Sporothix schenckii	
Partially treated bacterial meningitis	
Aspergillus spp.	
Zygomycoses	

Clinical Presentations

1. In neonates and children < 18 months – fever, irritability, poor feeding, vomiting.
2. In older children and adults – triad of headache, fever and nuchal rigidity. Seizures in 30-40% children.
3. CSF rhinorrhoea – fistula, neural tube defects.

Pathogenesis

Nasopharyngeal colonization by pathogen
↓
Transient immunosuppression
↓
Local invasion
↓
Bacteremia
↓
Meningeal invasion
↓
Bacterial replication
↓
Further bacteria and recruitment of inflammatory cells with production of purulent exudates
↓
Disruption of blood brain barrier
↓
Vasogenic cerebral oedema
↓
Release of toxins and free radicals
↓
Exudate in subarachnoid space
↓
Blocking of arachnoid granulations
↓
Obstruction to CSF drainage
↓
Backflow of CSF into brain interstitial space
↓
Periventricular white matter odema
↓
Infection in subarachnoid space
↓
Septic arteriolitis and phlebitis
↓
Infarcts and neurologic deficits
↓
Hemiparesis (following subdural effusion or vasculitis or infarcts)
↓
Purulent exudates around cranial nerves
↓
Cranial nerve palsies

Brain abscess

Predisposing condition	Likely pathogen
Otitis media/mastoiditis	Streptococci (anaerobic and aerobic) *Bacteroides fragilis* Enterobacteriaceae
Paranasal sinusitis	Streptococci *Bacteroides spp.* Enterobacteriaceae, *S. aureus*
Dental infection	Streptococci, *Fusobacterium spp.*, *Bacteroides spp.*
Trauma/Neurosurgery	*S. aureus*, Enterobacteriaceae Streptococci, *Pseudomonas aeruginosa*
Meningitis	*Listeria monocytogenes* *Citrobacter diversus*
Cyanotic heart disease	Streptococci, *Hemophilus spp.*
Pyogenic lung disease	Streptococci, *Nocardia asteroides*, *Actinomyces spp.*, *Fusobacterium spp.*, *Bacteroides spp.*,
Bacterial endocarditis	*S. viridans*, *S. aureus*, Enterococci, *Hemophilus spp.*
T-cell deficiency	*T. gondii*, *Nocardia spp.*, *L. monocytogenes*

Brain absecess develop in four clinical situations

- in association with contiguous suppurative focus
- hematogenous spread from a distant focus
- following trauma
- cryptogenic

Laboratory diagnosis

CT scan/MRI, followed by lumbar puncture (L.P.)

Samples – CSF, blood (when L.P. is contraindicated), subdural fluid, scrapings from septic spots on skin.

1. **CSF routine examination**

   a) Appearance – clear, cloudy or purulent, with/without blood

   b) Biochemicals – Glucose, total protein, chlorides

   c) Number of cells/litre – Number x 10^6/l, % of polymorphs and lymphocytes

Uncentrifuged CSF is used for cell counts and biochemical tests.

Characteristics of CSF

	Normal CSF	Pyogenic meningitis	Aseptic meningitis
Appearance	Clear	Turbid	Clear to opalescent
Pressure	Normal	Highly increased	Normal
Total protein	30-45 mg%	100-600 mg%	50-100 mg%, usually high in T.B. meningitis
Sugar	40-80 mg%	10-20 mg% or absent	Normal (< 40 mg% in T.B., leptospirosis, cryptococcosis)
Cell count	1-3 / cu.mm.	500-10,000/cu.mm.	100-1,000 / cu.mm.
Type of cell	Lymphocytes	Neutrophils - 95% Lymphocytes - 05%	Mainly lymphocytes

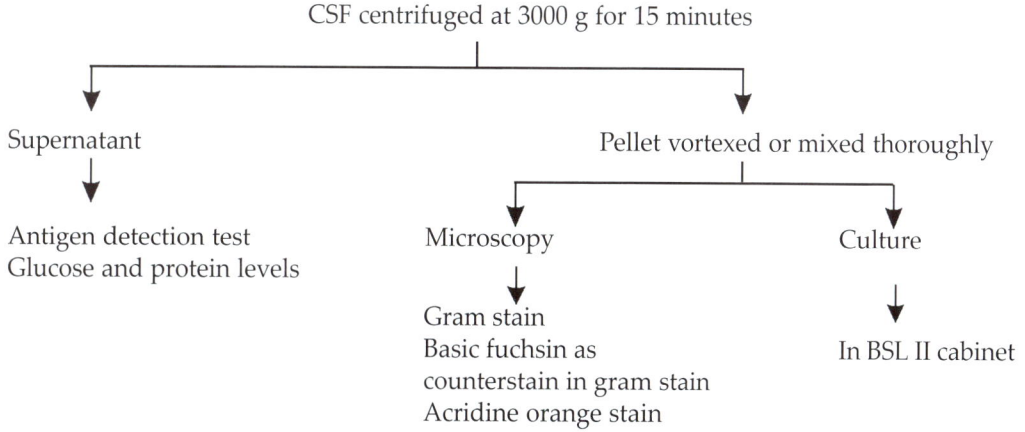

CSF centrifuged at 3000 g for 15 minutes

Supernatant

Antigen detection test
Glucose and protein levels

Pellet vortexed or mixed thoroughly

Microscopy

Gram stain
Basic fuchsin as
counterstain in gram stain
Acridine orange stain

Culture

In BSL II cabinet

2. **Microscopic examination**

   Gram stain:

   - Pus cells and gram positive diplococci or in chains
   - Pus cells and gram negative rods, coccobacilli or filamentous
   - Pus cells and gram negative intracellular diplococci

3. Culture

5% sheep Blood Agar, Chocoloate Agar	MacConkey Agar	BHI/thioglycollate broth/NBA/BBE for anaerobes
↓	↓	L-J for mycobacteria
S. pneumoniae	E. coli	SDA for fungi
H. influenzae	and other	
N. meningitidis	gram	
S. agalactiae	negative	
S. aureus	bacilli	
L. monocytogenes		

BA and CA are incubated at 37°C in candle jar. They are inspected daily and discarded as negative if there is no growth upto 72 hours.

4. Identified by standard biochemical tests.
5. Antibiotic sensitivity pattern including β-lactamase production.
6. BACTEC/BacTAlert culture of CSF
7. CSF VDRL
8. Lactate determination.
9. C-reactive protein estimation.
10. Latex agglutination test – Sensitivity 100%; Specificity 96.6%.
11. CIEP, ELISA.
12. Antimycobacterial, anticysticercal, antitoxoplasmal antibodies, etc.
13. Molecular diagnosis is done in cases of:
    - prior antibiotic therapy
    - when number of organisms are few
    - non cultiviable organisms.
    a) Nucleic acid probes
    b) PCR for H. influenzae – Sensitivity 90%; Specificity 99%.
    c) PCR-EIA for S. pneumoniae DNA in CSF – Sensitivity 94%; Specificity 90%.
    d) PCR for B. anthracis.
    e) Genotyping.

Quality control and quality assurance are done by Central or Reference Laboratory.

Treatment: Supportive

- Antioedema measures
- Nursing care

Bacteria	Antibiotics given for 3-6 weeks
S. pnemoniae	Ampicillin or penicillin (erythromycin, if hypersensitive to latter).
S. aureus	Cloxacillin, augmentin.
H. influenzae	Ampicillin, augmentin or cefuroxime.
K. pnemoniae	Gentamicin or ciprofloxacin.
P. aeruginosa	Piperacillin, amikacin.
L. monocytogenes	Ampicillin or penicillin

CONVENTIONAL VIRAL INFECTIONS OF THE CENTRAL NERVOUS SYSTEM

Acute	Subacute or Chronic
Poliomyelitis (Polio virus types 1-3)	Progressive multifocal leucoencephalopathy (PML)
Herpes simplex	Herpes zoster
Herpes zoster	Progressive Rubella
Measles panencephalitis	Subacute sclerosing panencephalitis (SSPE - a complication of measles)
Rabies	Cytomegatovirus infection
Echovirus	Human immunodeficiency virus 1
Coxsackie A and B	and 2
Arboviral encephalitis	Human T lymphotropic virus 1
- Japanese encephalitis	
- West Nile fever	
- Dengue fever	

Intrauterine/Neonatal – Hydrocephalus, Microcephaly by CMV, Rubella, HSV-2.

Virus related disorders – Reye's syndrome, Guillain-Barre syndrome.

Viral inclusion bodies

Intranuclear	Intracytoplasmic	No inclusions
Herpes simplex (Cowdry A)	Rabies (Negri bodies)	Japanese encephalitis
Herpes zoster	SSPE	HIV (AIDS)
SSPE	CMV	HTLV1
CMV		Poliomyelitis
		PML

Unconventional viral infections of the CNS

No viral particle is identified, no RNA or DNA, has long incubation period, no inflammation, no immune response, no inclusions, transmissible. Caused by abnormal Prion protein Prpsc in fibrillar form in presynaptic terminals and is normally present in neurons.

Types of Prion diseases

Kuru	Human
Creutzfeldt Jakob Disease (CJD)	Human
Variant CJD (mad cow disease)	
Scrapie	Sheep
Bovine spongiform encephalopathy	Cattle
Mink encephalopathy	Mink
Mule Deer encephalopathy	Mule Deer

Laboratory diagnosis

1. Electron microscopy

2. Inclusion bodies – Negri bodies, Cowdry A, etc.

3. Detection of viral antigen immunohistochemistry using monospecific antibodies to viral proteins.

4. Japnese Encephalitis (JE) Antigen (Ag) by RPHA and IFA, HSV Ag by FA

5. Immune electron microsocpy by using specific antibodies

6. IgM/IgG antibodies detection in CSF/serum by ELISA, by rising titre – JE antibodies, measles, rubella, herpes simplex antibodies, etc.

7. Viral antigen detection by PCR.

8. Coculture of brain and other infected tissues with permissible Vero cells for Subacute Sclerosing Pan Encephalitis (SSPE).

9. Corneal smear and nuchal skin biopsy for rabies

10. Neutralizing antibodies in CSF in rabies, enteroviruses.

FUNGAL INFECTIONS OF THE CNS

Yeast form – *Candida, Cryptococcus, Histoplasma, Coccidioides, Sporothrix, Blastomyces*

Branching hyphae:

- Septate – *Aspergillus, Cladosporium*
- Non-septate – Zygomycosis (Mucor)
- Non-septate pseudohyphae – *Candida, Absidia*

Diagnosis of fungal infections of CNS

1. Radiology – Chest X-ray, CT and MRI.

2. Microscopy:
   - Staining
     - PAS – fungal wall magenta coloured.
     - Methenamine silver – fungal wall black.
     - Gram stain
     - AFB stain with 1% H_2SO_4.
   - Wet mount, KOH mount.
   - India ink preparation.

3. Culture on SDA with/without antibiotics – further identification by sugar assimilation and other biochemical tests.

4. Serology
   - Antigen detection from CSF by LAT, ELISA; Cryptococcus LAT from CSF – Sensitivity 100%; Median EIA Specificity 98%.
   - Coagglutination test for *C. neoformans;*
   - For *Aspergillus, Candida, Histoplasma* (RIA/ELISA)
   - Antibody detection for *Coccidioides* by CFA.

5. Metabolite detection
   - D-arabinitol for invasive candidiasis.
   - D-mannitol for cryptococcosis and invasive aspergillosis.

6. Cell wall product detection

   (1-3) βD glucan is a characteristic cell wall component of fungi including *Candida, Cryptococcosis, Aspergillus,* etc. Evaluation of b-Dg in blood, CSF or other normally sterile body fluids is an useful marker to detect deep mycosis or fungal sepsis.

7. Nucleic acid detection from blood and CSF.

PARASITIC INFECTIONS OF THE CNS

Protozoa

Plasmodium falciparum	Acute encephalopathy
Trypanosoma gambiense ⎫	Chronic encephalitis
T. rhodesiense ⎭	
T. cruzi	Acute meningoecephalitis
Entamoeba histolytica	Brain bascess (rare)
Naegleria	Acute meningoencephalitis
Acanthamoeba	Subacute or chronic meningoencephalitis
Toxoplasma gondii	Diffuse, focal or multifocal encephalitis, chorioretinitis

Nematodes

Trichinella spiralis	Acute meningoencephalitis, myositis
Angiostrongylus cantonensis	Acute eosinophilic meningitis
Gnathostoma	Hemorrhages, infarcts (rare)
Strongyloides	Meningitis (rare), paralytic ileus
Toxocara	Small granuloma (rare), ocular granuloma
Loa loa	Acute cerebral oedema, subacute encephalitis (rare)
Onchoreca volvulus	Chorioretinal lesions

Trematodes

Schistosoma japonicum	Cerebral granulomas
Pargonimus westermani	
S. mansoni ⎫	Myelitis (rare)
S. hematobium ⎭	

Cestodes

Cysticercosis by larva of *Taenia solium*	Small cysts or basilar arachnoiditis with hydrocephalus, ocular lesions
E. granulosus	Large cysts in CNS, orbit
Multiceps multiceps	Budding cysts (rare)

REFERENCES

1. Mims' Medical Microbiology. 5th Ed. Eds. Richard Goering, Hazel Dockrell, Mark Zuckerman, Ivan Roitt & Peter L. Chiodini. Saunders. 2013.

2. Koneman EW, Allen SD, Janda WM, Schreckenberger P, Winn Jr. WC. In Color Atlas & Textbook of Diagnostic Microbiology. 6th Ed. Lippincott, Philadelphia. 2006.

3. Central Nervous System Infections. Neurologic Clinics, 14. Ed. Marra C. M. Philadelphia, W.B. Saunders & Co. 1999.

4. De A, Gogate A. Brain abscess – a review of 74 cases. Indian Med Gazette 1998; CXXXII: 240-4.

5. Dunbar SA, Eason RA, Musher DM, Clarridge JE III. Microscopic examination and broth culture of cerebrospinal fluid in diagnosis of meningitis, J ClinMicrobiol 1998; 36: 1617-20.

NONSPORING ANAEROBIC INFECTIONS

21

Chapter

Distinguishing Properties of Family Bacteroidaceae

Properties	Bacteroides	Fusobacterium	Leptotrichia
Morphology	1-4 μm x 0.4-0.8 μm GNB & CB, often pleomorphic; singly, in pairs or short chains; rounded ends, few short filaments.	4-6 μm x 0.5-1 μm, GNB, tapered or pointed ends, filamentous; L-forms and sphero-plasts common.	5-15 μm x 1 μm, fusiform GNB with pointed ends, straight or slightly curved; bulbous swelling and spheroplasts, or long septate filaments (200 μm).
Type	B. fragilis	F. nucleatum	L. buccalis
G + C content of DNA (mol%)	28-61	26-34	34
Bile salts	Resistant	Inhibitory	Inhibitory
Biochemical reactions			
a. Glucose & Maltose	Acid & Gas	Acid & Gas	Acid only, no gas
b. Indole & H$_2$S production	–	+	–
c. Nitrate reduction	–/+	–	–
Metabolic products	Saccharolytic sp. – acetic & succinic acids. Asaccharolytic sp. – n-butyric acid always, IB and IV acids.	n-butyric acid, scanty acetic & propionic. acids.	Lactic acid mainly.
Threonine to propionine	Not deaminated	Deaminated	–
Antibiotic susceptibility			
a. Penicillin	R/S	S	S
b. Neomycin	V	S	S
c. Metronidazole	S	S	S
d. Cefoxitin	S	V	V
e. Chloramphenicol	V	S	S
f. Rifampicin	S	R/S	S

Different properties of Bacteroides species

Properties	Fragilis Group	Melaninogenicus-oralis Group	Asaccharolytic Group
Morphology	Small NM, GNB and CB, Pleomorphic; PS capsule	Short GNB or CB, many pleomorphic; oral strains CB in short chains	Many CB, few slightly longer
G + C content of DNA. (mol %)	40-44	40-42	50-54
Tolerance tests:			
a. Bile salts (20%)	Tolerant	Inhibitory	Inhibitory
b. Victoria blue 4R (1 in 80,000)	+	–/+	–
c. Gentian & ethyl violet (1 in 100,000)	Inhibitory	–	–

Biochemical reactions:			
a. Indole & H_2S	–	–	+
b. Glucose	Acid & gas	Acid	–
c. Xylose	Acid only	–	–
d. Esculin hydrolysis	+	–	–
e. Catalase	–/+	–	–
f. Superoxide dismutase	+	–	–
g. Glutamic acid	Decarboxylated	–	–
h. ALN disc test	+	+	–
i. Charcoal gelatin disc digestion	slowly or negative	within few days	within 24-48 hours
Metabolic end products	Acetic and succinic acids (B. splanchnicus n-butyric acid)	n-butyric acid not formed except P. levii; Isovaleric and isobutyric acids	n-butyric acid
Antibiotic susceptibility:			
a. Penicillin (2 units)	R	S/R	S
b. Neomycin (1 mg)	R	S	S
c. Kanamycin (1 mg)	R	R	R
d. Vancomycin (5 µg)	R	R	R/S
e. Colistin (10 µg)	R	S	S
f. Rifampicin (15 µg)	S	S	S

NM - Nonmotile; GNB - Gram negative bacilli; CB - Coccobacilli; PS - Polysaccharide.

Different Properties of *Fusobacterium species* and *Leptotrichia buccalis*

Properties	F. necrophorum	F. nucleatum	L. buccalis
First isolated	Loeffler in 1884 as agent of calf diphtheria.	Veillon and Zuber in 1898. (F. fusiforme)	Knorr in 1922 (F. plauti-vincenti Bergey's manual in 1957)
Morphology	Cocci to bacilli & long threads, one end narrow and pointed, other end thicker, spheroplasts and beaded forms common, irregular staining	Regular bacillus, tapered or pointed ends, navicular and spindle forms common	Fusiform GNB with pointed ends
G + C content of DNA (mol%)	33	27-28	34
Tolerance tests:			
a. Bile & bile salts	Inhibitory	Inhibitory	Inhibitory
b. Brillian green, Victoria blue 4R, gentian & crystal violet	Tolerant	Tolerant	Tolerant
Biochemical reactions:			
a. Glucose fermentation	+ (weak)	+ (weak)	+
b. Indole production	+	+	–
c. Lipase	+	–	–
Major end products	n-butyric and propionic acids	n-butyric and acetic acids	Lactic acid with trace acetic acid
Infections associated	Liver abscess, genital infections in man, putrid infns. like necrotizing tonsillitis with bacteremia and metastatic abscesses, meningitis	AUG* or Vincent's angina, brain abscess	Normal oral flora, periurethral area of women, Vincent's angina
Antibiotic susceptibility:			
a. Kanamycin & Colistin	S	S	S
b. Vancomycin	R	R	R
c. Rifampicin	many R	S	S

* Acute ulcerative gingivostomatitis.

NONSPORING ANAEROBIC INFECTIONS

- Anaerobic organisms are found throughout human body – on skin, mucosal surfaces and high concentrations in GI tract.
- They do not grow in presence of O_2 and are killed by O_2 or toxic O_2 radicals.
- Contain flavoprotein, so in the presence of oxygen produce H_2O_2, which is toxic.
- They grow at low or negative oxidation - reduction potential.
- No oxidative phosphorylation and fermentation.
- Obligate anaerobes lack certain enzymes:
  - Superoxide dismutase $O_2^- + 2H^+ \rightarrow H_2O_2$
  - Catalase $H_2O_2 \rightarrow H_2O + O_2$
  - Peroxidase $H_2O_2 + NADH + H^+ \rightarrow 2H_2O + NAD$

Obligate anaerobes

Strict obligate – No growth on agar surface exposed to O_2 levels higher than 0.5%, *e.g. Veilonella* species, *Eubacterium* species.

Moderate obligate – Grow when exposed to O_2 levels between 2-8%, *e.g. Bacteroides fragilis* group, *Prevotella* species.

- Polymicrobial anaerobic infection – Simultaneous infection with facultative anaerobes
  - Diminishes O_2 supply further
  - Aids growth of obligate anaerobes
- Two sources:
  - Normal human flora – Endogenous
  - Environment (e.g. soil) – Exogenous

Habitat of nonsporing anaerobes

- Normal flora of skin, mouth, mucous surfaces, respiratory tract, gastrointestinal tract, genital tract
- Outnumber aerobes in many habitats
  - Mouth and skin – 10 to 30 times > aerobes
  - Intestines – 1000 times > aerobes
- Estimated no. of anaerobes in:
  - Saliva – 10^8/ml
  - Small intestine – 10^5/ml
  - Colon – 10^{11}/gm

Normal anaerobic flora of human body

Organisms	Mouth and nasopharynx	Gut	Vagina	Skin
Clostridium	-	++	-	-
Lactobacillus	+	+++	+	-
Actinomyces	+	-	-	-
Bifidobacterium	+	++	+	-
Propionobacterium	-	-	-	++
Bacteroides fragilis	++	+	-	-
Prevotella	++	+	++	-
Fusobacterium	++	+	-	-
Gram positive cocci	++	++	++	+
Gram negative cocci	++	++	++	-
Spirochaetes	+	-	-	-

Classification of nonsporing anaerobes

Gram positive cocci

- *Peptostreptococcus spp.* - 22%
- *Peptococcus spp.* - < 1%
- *Anaerococcus spp.*

Gram positive bacilli

- *Lactobacillus spp.* - 3%
- *Eubacterium spp.* - 5%
- *Bifidobacterium spp.* - 1%
- *Propionibacterium spp.* - 13%
- *Actinomyces spp.* - 1%
- *Mobiluncus spp.* < 1%
- *Atapobium spp.*
- *Collinsella spp.*
- *Eggerthella spp.*

Gram negative cocci

- *Veillonella spp.* - 3%
- *Acidaminococcus spp.*
- *Megasphaera spp.*

Gram negative bacilli

- *Bacteroides spp.* - 25% (*B. fragilis* group 20%)
- *Fusobacterium spp.* - 4%
- *Prevotella spp.* - 6%
- *Porphyromonas spp.* - 1%
- *Leptotrichia spp.* - < 1%

Other Nonsporing anaerobes

- *Treponema, Borrelia*
- *Anaerovibrio*
- *Anaerobiospirillum*
- *Bilophila*
- *Butyrivibrio*
- *Capnocytophaga*
- *Centipeda, Cristispira*
- *Desulfomonas, Desulfovibrio*
- *Dialister*
- *Mitsuokella*
- *Selenomonas*
- *Succinimonas, Succinivibrio*
- *Campylobacter* species (*C. gracilis, C. curvus, C. rectus*)
- *Sutterella, Tissierella*

Important anaerobic bacteria

- Above the diaphragm – *Peptostreptococcus, Bacteroides* spp., *Fusobacterium, Prevotella, Porphyromonas*
- Below the diaphragm – *Bacteroides fragilis* group (multiple species including *B. fragilis*), Other *Bacteroides* spp.

- *B. fragilis* in the gut
- *P. melaninogenicus* and *P. oralis* in the mouth and oropharynx.

Predisposing factors for anaerobic infections due to Nonsporing anaerobes

- Trauma – Dead tissue
- Impaired blood supply
- Presence of other organisms – *E. coli* in abdominal wound may encourage growth of *B. fragilis*
- Presence of foreign bodies
- In conditions where host resistance is decreased due to trauma, tissue necrosis, impaired circulation, corticosteroids and cytototoxic agents

- Diabetes mellitus, Malnutrition, Malignancy
- Prolonged antibiotic therapy
- Synergistic infection – Meleny's gangrene (*Staphylococcus aureus* + Anaerobic streptococci).
- Production of leukotoxins (by *Fusobacterium* spp.)
- Phagocytosis intracellular killing impairments (often caused by encapsulated anaerobes) and by succinic acid (produced by *Bacteroides* spp.)
- Chemotaxis inhibition (by *Fusobacterium, Prevotella* and *Porphyromonas* spp.)
- Protease degradation of serum proteins (by *Bacteroides* spp.).

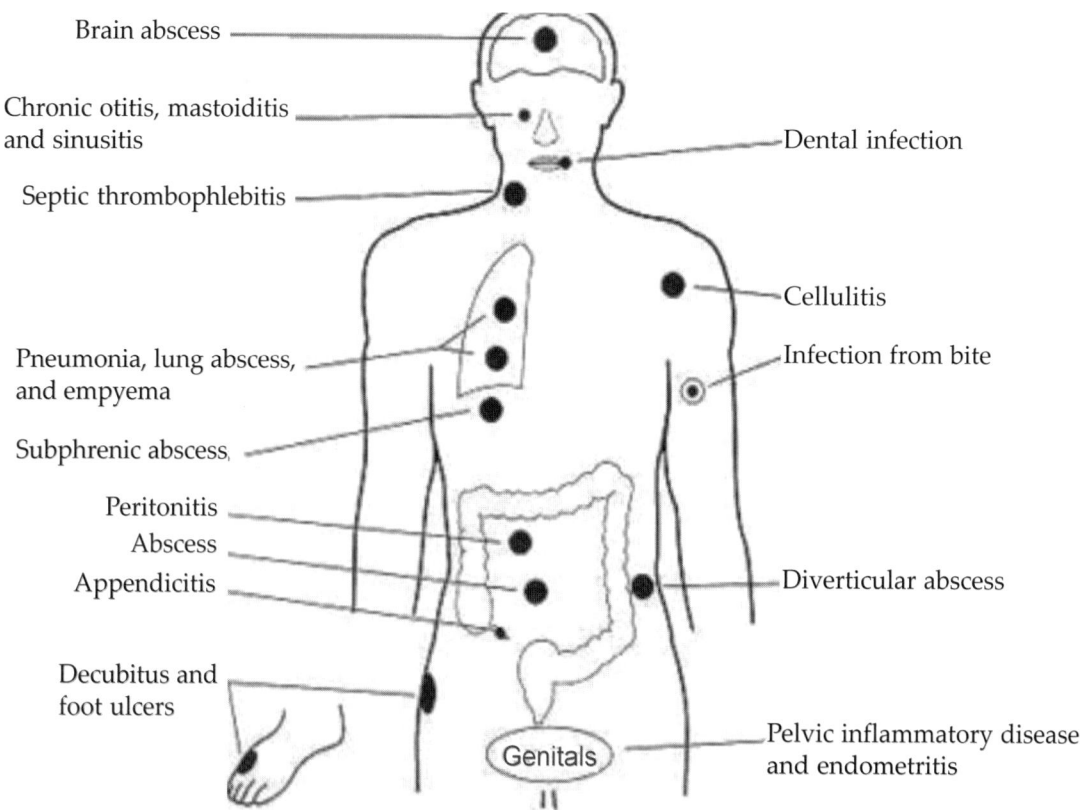

Common anaerobic infections

Exogenous infections

- Myonecrosis
- Crepitant cellulitis
- Benign superficial infections
- Infections following animal bites
- Septic abortion

Endogenous infections

- Abscess of any organ
- Actinomycosis
- Aspiration pneumonia, Bronchiectasis, Empyema

- Complications of appendicitis, cholecystitis
- Crepitant and non crepitant cellulitis, Wound infections
- Dental sepsis
- Endocarditis
- Osteomyelitis
- Chronic Suppurative Otitis Media (CSOM)
- Peritonitis, Diverticulitis
- Septic arthritis
- Gingivitis, Sinusitis
- Pelvic inflammatory disease

Infections produced by nonsporing anaerobes

Anaerobe	Infection
Gram negative bacilli	
Bacteroides fragilis	Brain abscess, intra abdominal abscess, infections of female genitalia, cellulitis, diabetic ulcer, septicemia
Prevotella melaninogenicus	Lung or liver abscess, empyema, aspiration pneumonia, pelvic infections in females, breast abscess, wound infections
Porphyromonas species	Dental root canal infections, periodontal disease, CSOM, mastoiditis
Fusobacterium necrophorum	Aspiration pneumonia, lung/ liver abscess, oral infections, chronic sinusitis, abdominal infections
Fusobacterium nucleatum	Periodontal disease
Leptotrichia	Acute necrotizing ulcerative gingivitis together with spirochetes and *F. nucleatum*
Gram positive bacilli	
Lactobacillus	Commensal in intestine, synthesizes biotin, vit. B12, flora of vagina - ferments glycogen – lactic acid - acidic ph of vagina; non pathogenic
Bifidobacterium	Non pathogenic; common in feces of infants
Eubacterium	Isolated from mixed infections of abdomen, pelvis or genitourinary tract
Propionibacterium	Inflammatory process in acne, anaerobic contaminant of blood cultures
Mobiluncus	Commensal in vagina
Actinomyces	Causes human actinomycosis – oral, cervicofacial, thoracic, pelvic and abdominal infections; *A. viscosus* and *A. naeslundii* commonly cause peridontal infections and dental caries.
Gram positive cocci	
Peptococcus	Pyogenic infections of wounds, puerperal sepsis
Peptostreptococcus	Pyogenic infections of wounds, puerperal sepsis, UTI, pleuropulmonary diseases, brain abscess.
Gram negative cocci	
Veillonella	Involved in mixed anaerobic infections

Clues to anaerobic infection

- Foul smelling discharge
- Location of infection in proximity to mucosal surface
- Infection secondary to human or animal bite
- Gas formation in tissues
- Black discoloration of blood containing exudates
- Necrotic tissue
- Presence of sulphur granules in discharge
- Failure of organisms seen on Gram stain of primary smear to grow aerobically.

Laboratory diagnosis

- Specimens free of contamination with normal flora should be collected.
- Anaerobiosis should be maintained throughout collection, transport and processing of samples.

Clinical samples suitable for anaerobic culture

- Bile
- Blood
- Bone marrow
- Bronchial washings obtained with double lumen plugged catheter
- Cerebrospinal fluid
- Culdocentesis aspirate
- Decubitus ulcer from base of lesion through debridement of surface debris
- Fluid from normally sterile sites like joint.
- Percutaneous lung aspirate or biopsy.
- Pleural aspirate/Transtracheal aspirate
- Tissue obtained at biopsy or autopsy
- Sulphur granules from draining fistula
- Peritoneal/ascitic fluid
- Uterine contents, if collected using a protected swab
- Suprapubic bladder aspirate
- Biopsy of endometrial tissue obtained with an endometrial suction curette.

Clinical samples unsuitable for anaerobic culture

- Bronchial washing or brush (except if collected with double lumen plugged catheter)
- Coughed out (expectorated) sputum.
- Feces/Rectal swab
- Gastric or small bowel contents (except if blind loop syndrome)
- Ileostomy or colostomy drainage
- Nasopharyngeal swab/Throat swab
- Secretions obtained by nasotracheal or orotracheal suction
- Swab of superficial (open) skin lesion.
- Urethral swab
- Vaginal or cervical swab
- Voided or catheterized urine

Recommended specimen collection

Source	Procedure
Pulmonary	Percutaneous transtracheal aspiration or direct lung puncture
Pleural	Thoracocentesis
Urinary tract	Suprapubic percutaneous bladder aspiration, nephrostomy specimens
Abscess	Needle and syringe aspiration of closed abscess
Female genital tract	Culdocentesis to obtain specimens
Sinus tracts or draining wounds	Aspiration by syringe, small plastic catheter introduced through decontaminated skin orifice, tissue biopsies, curettings
Joint fluids	Aspiration

Collection of samples

- **If material is plenty:** Specimen collected with syringe, air bubble expelled and needle end is sealed with sterile rubber stopper. Transported to laboratory within ten minutes.
- **If material is scanty:** Three swabs taken
    - In Stuart's transport medium (anaerobic culture).
    - In glucose broth (aerobic culture).
    - Sterile test tube (smear examination).
- **For fluids:** Sampleis filled upto the brim.

Anaerobic transport media

- Syringe and needle (sterile rubber stopper attached to needle).
- Hungate tube – It contains O_2 free gas and an agar or broth indicator system. Used when specimen is processed within 30 minutes.
- Sealed with a screw cap with a butyl rubber diaphragm.
- Pre-reduced and anaerobically sterilised (PRAS) semisolid Cary and Blair agar (BBL).
- Bio-Bag – It is used for transporting tissue, fluids and swabs.
- Vacutainer anaerobic transporter of BECTON DICKINSON (B-D):
    - Cotton swab is attached to a plunger in the stopper.
    - Specimen collected with the swab and placed in the inner tube.
    - Stopper is replaced and the plunger pushed down to release the central tube with the specimen into the outer tube and the hole in the rubber stopper is closed.

Common anaerobic media (Solid and Liquid)

Medium	Components / Comments	Primary purpose
Anaerobic blood agar (Ana BA)	May be prepared with Columbia, Schaedler, CDC, Brucella or brain heart infusion base, suppplemented with 5% sheep blood, 0.5% yeast extract, hemin, L-cystine, vitamin K_1	Nonselective medium for isolation of anaerobes and facultative anaerobes.
Bacteroides bile esculin agar (BBE)	Trypticase soy agar base with ferric ammonium citrate and hemin ; bile salts and gentamycin act as inhibitors	Selective and differential for *Bacteroides fragilis* group; good for presumptive identification.
Laked kanamycin/ paromomycin vancomycin blood agar	Brucella blood agar with kanamycin (75 µg/ml)/ paromomycin (100 µg/ml), vancomycin (7.5 µg/ml), vitamin K_1 (10 µg/ml), 5% laked blood	Selective for isolation of *Prevotella* and *Bacteroides* species
Anaerobic phenylethyl alcohol agar (PEA)	Nutrient agar base, 5% blood, phenylethyl alcohol	Selective for inhibition of enteric Gram negative rods
Egg yolk agar (EYA)	Egg yolk base	Nonselective for determination of lecithinase and lipase production by *Fusobacterium spp.*
Thioglycollate broth	Pancreatic digest of casein, soy broth and glucose enrich growth of most bacteria. Thioglycollate and agar reduce Eh; may be supplemented with hemin and Vit K_1	Nonselective for cultivation anaerobes, as well as facultative anaerobes and anaerobes
Cooked meat (chopped meat) broth	Solid meat particles initiate growth of bacteria; reducing substances lower oxidation reduction potential (Eh)	Nonselective for cultivation of anaerobic organisms; with addition of glucose, can be used for gas liquid chromatography
Peptone yeast extract glucose broth (PYG)	Peptone base, yeast extract, glucose, cysteine (reducing agent), resazurin (oxygen tension indicator), salts	Nonselective for cultivation of anaerobic bacteria for gas liquid chromatography
VL (Viande Levure) broth	Tryptone, NaCl, meat extract, yeast extract, cysteine hydrochloride	Conventional fermentation reactions for agar plate media, G.L.C. analysis

Methods of anaerobiosis

- McIntosh Fildes' jar – Works with evacuation replacement procedure. Air inside the jar evacuated and filled with 85%N_2, 10% H_2 and 5% CO_2.

- Gaspak system – Commercial gaspak foil sachet contains 3 tablets – one each of citric acid, sodium bicarbonate and sodium borohydride. Water is added just before use and kept inside the jar. H_2 and CO_2 are generated.

- Anaerobic chamber or glove box – It is a self contained anaerobic system with provisions of H_2, CO_2 and N_2 within it and catalytic conversion of residual O_2 to water; Flexible plastic bags or rigid glass tight cabinets are used; Has cuffs around middle arm and leaves hands unencumbered by gloves; The atmosphere is evacuated and replaced 3 times with gas mixture.

- Pre reduced anaerobically sterilised culture (PRAS) and Roll-tube techniques.

- Robertson's cooked meat medium.

- Thioglycollate broth.

Levels of identification

Three levels at which laboratories operate for anaerobic work up:

Level I

Presumptive identification from: Gram stain, Colony morphology, Aerotolerance testing, e.g. Gram negative bacillus with pointed ends, bread crumb colonies – presumptive diagnosis of *Fusobacterium nucleatum*.

Anaerobes should be maintained in pure culture in case it is needed to send to reference laboratories for complete identification and antibiotic susceptibility testing.

Level II

Simple tests to further group the anaerobes.

Speciation is based on: Colony morphology, Gram stain, Susceptibility to special potency antibiotic discs (vancomycin, colistin, kanamycin), Nitrate reduction test, Catalase test, Growth in 20% bile.

Level III

In addition to tests used in Level II, speciation may require additional biochemical tests and metabolic end product analysis, i.e. gas liquid chromatography (GLC).

Antibiotic identification disks

Most anaerobes have a characteristic susceptibility pattern to colistin (10 µg), vancomycin (5µg), and kanamycin (1 mg) disks. Disks used as aid in Gram reaction.

- Observe for a zone of inhibition of growth
- A zone of 10 mm or less indicates resistance
- Zone > 10 mm indicates susceptibility

- Gram positive organisms are sensitive to vancomycin, resistant to colistin.
- Gram negative are resistant to vancomycin.
- *Fusobacterium* spp. susceptible to kanamycin and colistin.
- *Bacteroides spp.* resistant to kanamycin and variable susceptibility to colistin.

Presumpto plate 1

Media	Characteristics
LD agar	Indole, growth on LD medium, catalase
LD esculin agar	Esculin hydrolysis, H_2S, Catalase
LD egg yolk agar	Lipase, lecithinase, proteolysis
LD bile agar	Growth in 20% bile, insoluble precipitate under and immediately surrounding growth

Presumpto plate 2

Media	Characteristics
LD glucose agar	Glucose fermentation, stimulation of growth by fermentable carbohydrates
LD starch agar	Starch hydrolysis
LD milk agar	Casein hydrolysis
LD DNA agar	Deoxyribonuclease activity

Presumpto plate 3

Media	Characteristics
LD mannitol agar	Mannitol fermentation
LD lactose agar	Lactose fermentation
LD rhamnose agar	Rhamnose fermentation
LD gelatin	Gelatin hydrolysis

Rapid and new methods for identification of anaerobes

Commercial packaged kits:

- API-20A (Bio Merieux Vitek)
- API-ZYM (Bio Merieux Vitek)
- Vitek anaerobe identification (ANI) Card (Bio Merieux Vitek)
- Crystal anaerobe identification system (Becton Dickinson microbiological system)
- Microscan rapid anaerobe panel
- Rapid IA-ANA II (Remel).

Serological Tests for *Bacteroides species*

1. **Agglutination tests** with 7 absorbed monospecific antisera – 21 serogroups and 45 serological patterns of *B. fragilis*.

2. **Precipitaion tests (ID and CIEP)** – Sonicate preparation of *B. fragilis* NCTC 2553.

3. **Direct haemagglutination and HI tests** – Sheep erythrocytes sensitized by phenol- extracted LPS. Indirect HAT using rabbit antisera against whole cells. Cross-reactivity is more.

4. **Coagglutination test** – Colony suspension of the organisms are used. It is a specific test.

5. **ELISA** – Protein antigens extracted from outer cell membrane by ultrasonic disintegration. Cross-reactivity is less.

6. **Indirect IF test** – Uses polyvalent antibody conjugates.

7. **DNA probes** – Dot-Blot hybridization with chromosomal DNA probes for pigmented *Bacteroides spp.*

8. **SDS-PAGE** with polypeptides extracted from *B. fragilis* and *P. melaninogenicus.*

Gas Liquid Chromatography (GLC)

- Metabolic products serve as markers for an organism's identity.

- The enzymes involved are genetically stable and the end products produce a fingerprint that is typical and useful for identification.

- The temperature of the column can be varied from 50°C to 250°C.

- Types of GLC:
  - Thermal conduction detector
  - Flame ionisation detector (commonly used).

- The **mobile phase** (or "moving phase") is a carrier gas, usually an inert gas such as helium or an unreactive gas such as nitrogen.

- The **stationary phase** is a microscopic layer of liquid or polymer on an inert solid support, inside a piece of glass or metal tubing called a column.

- The instrument is called a **gas chromatograph** (aerograph/gas separator).

- The gaseous compounds being analyzed interact with the walls of the column, which is coated with a stationary phase. This causes each compound to elute at a different time, known as the **retention time** of the compound.

Anaerobic blood culture isolates are presumptively identified by the qualitative analysis of volatile fatty acids:

- *Bacteroides fragilis* group with acetic and propionic acids

- *Fusobacterium* with acetic, butyric, and usually propionic acids

- *Veilonella* with acetic and propionic acids

- *Propionibacterium species* typically produces large amount of propionic acid

- Gram-positive cocci with acetic and butyric acids.

Molecular methods for identification of anaerobes

- Nuclear magnetic resonance (NMR) spectroscopy – NMR spectroscopy uses Mercury plus Varian 300 MHz (7.05 T) provides a fingerprint within the proton spectrum of six genera of anaerobes

(*Bacteroides fragilis, Prevotella melaninogenica, Prevotella denticola, Fusobacterium necrophorum, Peptococcus niger* and *Peptostreptococcus spp.*) reflecting their characteristic metabolites.

- After the NMR analysis (256/512 scans), the different peaks are noted.

- MR-based identification is of value in the identification of anaerobes. However, a larger database of the peaks produced by anaerobes needs to be created for identification of all genera and species.

- PCR-Restriction fragment length polymorphism (*Bacteroides spp.*)

- Pyrolysis mass spectroscopy

- Metronidazole-resistant *B. fragilis* strains are analyzed by pulsed-field gel electrophoresis.

- Nitrocellulose filter blots (*Mobiluncus spp.*).

Antimicrobial susceptibility testing

Why is susceptibility testing of anaerobes not generally used in clinical decision-making?

- Technical issues with testing
  - Slow growth
  - Lack of consensus regarding agar/method

- Anaerobic susceptibility patterns
  - Predictable
  - Unchanging over the years

- Limited data to correlate in vitro results with outcome
  - Infections often polymicrobial
  - Outcome affected by multiple factors (e.g. surgery).

Susceptibility testing of anaerobes should be done in 4 settings (Finegold, 1999):
- Determine patterns of susceptibility to new agents
- Monitor susceptibility patterns nationally and locally
- To study the changing pattern of resistance
- Assist in the management of individual patients
- Persistence of infection/failure of usual regimes/ difficulty in making decisions based on precedent
- Brain abscess/endocarditis/osteomyelitis/ prosthetic device infection/ septic arthritis.

Recommended methods:

- Agar dilution – Brucella agar/Wilkins Chalgren agar with hemin, Vit. K and laked 5% sheep blood

- Broth microdilution (only for *B. fragilis* group) – Brucella broth with with hemin, Vit. K and lysed 5% horse blood

- Inoculum – Agar (10^5 CFU per spot)
                      Broth (10^6 CFU/ml)

- Incubation –
  - For Broth microdilution at 37°C for 46-48 hours
  - For Agar dilution at 37°C for 42-43 hours
- Standard strains:
  - *B. fragilis* ATCC 25285;
  - *B. thetaiotaomicron* ATCC 29741;
  - *E. lentum* ATCC 43055 (for Gram positive anaerobes).

Antibiotics	MIC (µg/ml)
Penicillin	≤ 0.5
Pip/Taz	≤ 32/4
Tic/Clav	≤ 32/2
Cefoxitin	≤ 16
Ceftizoxime	≤ 32
Ceftriaxone	≤ 16
Imipenem/Meropenem	≤ 4
Tetracycline	≤ 4
Moxifloxacin	≤ 2
Clindamycin	≤ 2
Chloramphenicol	≤ 8
Metronidazole	≤ 8

Metronidazole – the wonder drug

- Metronidazole is an antimicrobial agent that has been used in clinical medicine for > 45 years.
- Metronidazole is still the drug of choice for treatment of anaerobic infections, as was described by Tally and colleagues 35 years ago.
- Metronidazole is considered to be a cost-effective drug, good activity against pathogenic anaerobic bacteria, favorable pharmacokinetic & pharmacodynamic properties, and minor adverse effects

Susceptibility of anaerobes

- *Besides the* wellknown high rate of resistance of *B. fragilis* group organisms to clindamycin, the emergence of resistance of *Peptostreptococcus species* isolates to β-lactam drugs has become obvious in the recent years.
- Almost all of the *Clostridium perfringens* isolates are susceptible to all of the agents tested, except tetracycline.
- **Resistance to metronidazole:** *Peptostreptococcus* (32%), *Fusobacterium* (25%), *Veillonella* (20%), *Prevotella* (13%) and *Bacteroides* species (5%).

B. fragilis group

- *Bacteroides fragilis group* organisms are more frequently resistant to various antimicrobial agents, including clindamycin.
- 6.7% were resistant to ampicillin-sulbactam, 20.2% to cefotetan, 30.3% to piperacillin, 48.3% to cefotaxime, and 42.7% to clindamycin.

- None of the *B. fragilis* group organisms are resistant to imipenem, cefoxitin, chloramphenicol, or metronidazole.

Peptostreptococcus species

- Few reports of increased resistance of *Peptostreptococcus species* are also there.
- The isolates resistant to penicillin G, cefotetan and metronidazole were identified as *Peptostreptococcus anaerobius.*
- Other *Peptostreptococcus spp.* were susceptible *to* piperacillin, cefotaxime and imipenem, while 7.4% were resistant to penicillin G, cefotetan and metronidazole. 25.9% were resistant to clindamycin.

Anti-anaerobic antibiotics

- An aminoglycoside **or** an anti-pseudomonal cephalosporin (cefepime) are generally added to metronidazole.
- Clindamycin should not be used as a single agent as empiric therapy for abdominal infections.
- Penicillin can be added to metronidazole in treating intracranial, pulmonary and dental infections to provide coverage against microaerophilic streptococci and *Actinomyces.*
- A trend towards tetracycline resistance is seen especially in *Bacteroides* and *Fusobacterium.*
- Increased resistance of *P. melaninogenicus* strains to penicillin.
- Resistance to meropenem is low (< 1%).
- For beta-lactam/lactamase inhibitors, piperacillin-tazobactam is most active for all species (resistance < 6%).
- *Veillonella* is susceptible to piperacillin, piperacillin-tazobactam, ticarcillin, ticarcillin-clavulanic acid.

A few more anaerobic antibody pearls

- Moxifloxacin is a potentially useful antibiotic against anaerobes (2nd line)
- Ampicillin and penicillin G – both active against *Actinomyces* and *Peptostreptococcus.*
- Doxycycline – activity against *Actinomyces.*
- Vancomycin – only against Gram positive anaerobes, but not against Gram negative.
- Chloramphenicol – Excellent activity against all nonsporing anaerobes.

Choice of antibiotics

- Above diaphragm – **Clindamycin** preferred (Metronidazole misses actinomyces and microaerophilic streptococci)
- Below diaphragm – **Metronidazole** prefered (Clindamycin misses some *B. fragilis*)
- Everywhere – β-lactam/β-lactamase combinations, carbapenems, moxifloxacin (2nd line).

PROCESSING OF ANAEROBIC SPECIMENS

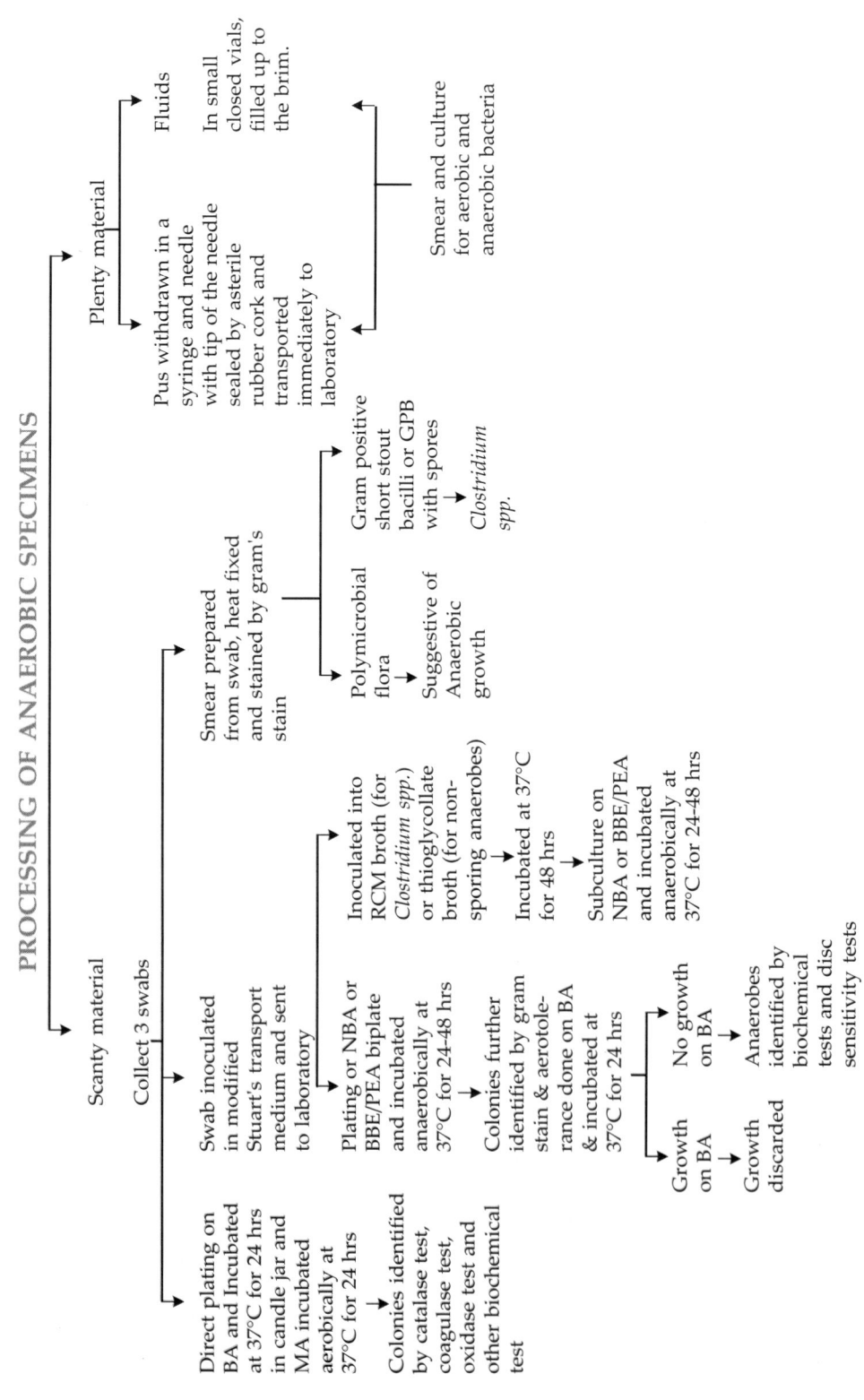

NBA = neomycin blood agar
BBE = Bacteroides bile esculin agar
PEA = Phenyl ethyl alcohol anaerobic blood agar
RCM = Robertson's cooked meat medium

Prevention and Treatment

- Surgical drainage – remove dead tissue, debris, foreign bodies
- Antimicrobial therapy – penicillin, metronidazole, etc.
- Hyperbaric oxygen
- Treatment of underlying disease.

REFERENCES

1. Wadsworth Anaerobic Bacteriology Manual Paperback. 5th Ed. Eds. SummanenP & Baron EJ. Star Publishing Co., California. 1993.

2. Koneman EW, Allen SD, Janda WM, Schreckenberger P, Winn Jr. WC. In Color Atlas & Textbook of Diagnostic Microbiology. 6th Ed. Lippincott, Philadelphia. 2006.

3. Forbes BA, Sahm DF, Weissfeld AS. Baily and Scott's Diagnostic Microbiology. 12th Ed. Mosby, Inc., an affiliate of Elsevier Inc. 2007.

4. Rathi N, Rathi A. Non spore forming anaerobic infections in children. Pediatr Infect Dis 2009; 1:3-8.

5. Seligman SA, Willis AT. Infection with non-sporing anaerobes in obstetrics and gynaecology. British J ObstrGynaec 1980;87:846-55.

6. I Phillips, E Taylor, S Eykyn. The rapid laboratory diagnosis of anaerobic infection. J Infect 1980;8:S155-8.

7. Sondag JE, Mariam Ali, Murray PR. Rapid Presumptive Identification of Anaerobes in Blood Cultures by Gas Liquid Chromatography. J ClinMicrobiol, 1980;11:274-7.

8. Appelbaum PC, Chatterton SA. Susceptibility of Anaerobic Bacteria to Ten Antimicrobial Agents. Antimicrob Agents Chemother. 1978;14:371-6.

9. Löfmark S, Edlund C, Nord CE. Metronidazole is still the drug of choice for treatment of anaerobic infections. Clin Infect Dis. 2010;50:S16-23.

10. A De, J Biswas, K Saraswathi, A Gogate. Microbiological features of necrotizing fasciitis. Indian J Med Microbiol 1998;17:18-21.

SEXUALLY TRANSMITTED INFECTIONS (STIS)

Chapter

Causative organisms of STIs

Bacteria	Viruses
Chlamydia trachomatis serotypes D–K, L1, L2, L3, *Gardnerella vaginalis*, *Hemophilus ducreyi*, *Myoplasma hominis*, *M. genitalium*, *N. gonorrhoeae*, *N. meningitidis* (oral-genital practices), *Ureaplasma urealyticum*, *Calymmatobacterium granulomatis*, *Treponema pallidum*, *Mobiluncus* (non-specific vaginitis)	Cytomegalovirus, HBV, HCV, HDV, HIV, HTLV-I, Herpes simplex virus type 2, Human papilloma virus (types 6 and 11, 16 and 18), Molluscum contagiosum
	Protozoa *Trichomonas vaginalis*
Fungi	**In homosexual population**
Candida albicans Other yeasts	*E. histolytica*, *G. lamblia*, *Salmonella*, *Shigella*, Camypylobacter, Cryptosporidium, Microsporidium.

STIs. Syphilis, gonorrhoea, chancroid, lymphogranuloma veneruem, condyloma acuminata, non-gonococcal urethritis, non-specific vaginitis, genital warts and ulcers, hepatitis B infection, HCV infection, HIV and AIDS.

Laboratory Diagnosis of STIs

Specimens to be collected:

- Vaginal swab
- Cervical swab
- Prostatic secretions
- Urethral discharge
- Swab from genital ulcers or warts
- Aspiration from bubo.

A. Microscopy

1. **Gram stain**
   i) Gram-negative intracellular reniform diplococci – gonococci. Predominance of lactobacilli – healthy vagina.
   ii) Vaginal epitheloid cells studded with gram variable bacilli "Clue cells" of Gardner and Dukes suggestive of bacterial vaginosis.
   iii) Gram-negative short ovoid bacilli in pairs or short chains resembling a school of red fish, sometimes gram-positive – *H. ducreyi.*
   iv) Gram-negative plemorphic rods with rounded ends and 'safety-pin' appearance – *C. granulomatis.*
   v) Gram-negative tiny comma-shaped bacilli – *Mobiluncus spp.*
   vi) Gram-positive budding yeast cells – *Candida spp.*

2. **Giemsa or Wright's stain:**
   i) Granules and filaments, beads and swollen forms – *Mycoplasma.*
   ii) Halberstaedter – Prowazek bodies – large reniform basophilic inclusion bodies surrounding the nucleus – *C. trachomatis* serotypes D-K.
   iii) Miyagawa's granulocorpuscles – *C. trachomatis* L1, L2, L3.
   iv) Blue bacillary body with well defined pink capsule on Wright's stain – *C. granulomatis.*
   v) Intranuclear Cowdry type A inclusion bodies on Wright's stain – HSV type 2.
   vi) Giemsa stain on centrifuged deposit of urine or saliva – cytomegalic cells (owl's eye appearance) in CMV infection.

3. **Tzanck smear** – 1% aqueous solution of toluidine blue O for 15 seconds. Multinucleated giant cells with faceted nuclei and homogenously stained ground glass chromatin - Herpes simplex virus type 2.

4. Wet mount of vaginal discharge – *Trichomonas vaginalis.*

5. Electron microscopy – HIV, HBV, HSV, *Mycoplasma, Chlamydia,* etc.

6. Dark ground miscroscopy – Motility of *Treponema.*

7. Antigen demonstration by fluorescent antibody technique by using fluorescent miscroscope – *Herpes simplex virus, T. vaginalis, G. vaginalis,* etc.

8. Direct fluorescent antibody test for *C. trachomatis.*

B. **Culture**

1. Chocolate agar enriched with isovitalex, fetal calf serum, vancomycin at 35°C in 10% CO_2 atmosphere and high humidity for 2-8 days – *H. ducreyi.*

2. Chocolate agar or Thayer Martin medium (with vancomycin, colistin and nystatin) – *Neisseria.*

3. Media enriched with 20% horse or human serum, yeast extract, penicillin, thallium acetate – "fried-egg" appearance of colonies, stained by Dienes method – *Mycoplasma.*

4. Sabouraud's dextrose agar – *Candida spp.*

5. Corn meal agar with Tween 80 – chlamydospore formation.

6. Diamond's medium – *T. vaginalis.*

7. i) On chorioallantoic membrane of chick embryo – herpes viruses type 2 produces large pocks.

   ii) Yolk sac – *C. trachomatis* L1, L2 and L3, *C.granulomatis.*

8. Tissue culture medium

   i) McCoy cells and HeLa cells – *C.trachomatis.*

   ii) Primary human embryonic kidney, human ammion – heaped up cells with focal degeneration – herpes virus.

   iii) Human fibroblast cultures – CMV.

C. **Serological Tests**

1. Standard tests and specific tests for syphilis.

2. CFT for herpes, Chlamydia, CMV.

3. Neutralization test for herpes.

4. Detection of IgM antibody by gel diffusion or virus agglutination – *C. granulomatis.*

5. Haemagglutination test – *G. vaginalis.*

6. Detection of various serological markers in blood in HBV infection – CIEP, RIA, ELISA, etc. HBcAg is detected in liver cells by immunofluorescence.

7. For HIV – ELISA, Indirect IF, RIA to detect anti-p24 antibody, Western Blot test, PCR.

8. For CMV – ELISA for IgM and IgG antibodies, rapid shell vial centrifugation, pp65 antigenemia assay, PCR, hybrid capture DNA assay, bDNA assay.

Causes of false negative ELISA – In immuno-compromised patients, in window period, in asssays detecting IgG antibody.

Causes of false positive ELISA – Technical error, intrinsic specificity of test, lipemic and hemolysed sera, hemophilia, dialysis, systemic disease.

Interpretation of Western Blot is same for HIV-1 and HIV-2

- Positive – Two envelop bands with or without gag or pol bands.

- Negative – No bands or presence of bands that do not correpond to structural HIV-1/ HIV-2 proteins.

- Indeterminate – Other profiles not considered positive or negative.

D. **Other tests**

1. Absolute CD4 count

2. T4 : T8 ratio

3. Increased IgG and IgA levels.

E. **Indirect tests:** Blood count for leucopenia, lymphopenia and thrombocytopenia.

F. **Skin test:** Frei test – Lygranum antigen injected intradermally for LGV.

GONORRHOEA

Clinical spectrum of gonococcal infection

- Asymptomatic infection – endocervix, pharynx, rectum, urethra.

- Symptomatic infection – bartholinitis, cervicitis, conjunctivitis, pharyngitis, proctitis, urethritis, vulvovaginitis.

- Local complications – bartholin abscess, epididymitis, lymphangitis, penile oedema, periurethral abscess, prostatitis, salpingitis (pelvic inflammatory disease).

- Systemic complications – disseminated gonococcal infection (arthritis, dermatitis, tenosynovitis).

Specimen collection

Sterile cotton, calcium alginate or polyethylene terephthalate (PET) swabs can be used.

From Endocervix. Vaginal speculum is moistened with warm water. Endocervix is cleaned with gauze to remove mucus. Swab is inserted 2 cm into the cervical canal, rotated and moved gently from side to side for 5-10 seconds to allow absorption of the exudate. Use of antiseptics, analgesics and lubricants should be avoided as these inhibit gonococci.

From Urethra. Should be taken at least 1 hour after patient has urinated. If no exudate is obtained, a thin swab is inserted 2-3 cm into the urethra and mucosa is scraped gently by rotating the swab for 5-10 seconds. In women, urethra is massaged against pubic symphysis and then collected as above.

From Rectum. Cotton swab is inserted 3 cm into the anal canal and rotated for 10 seconds, avoiding fecal contamination.

From Vagina. Recommended for women who have had a hysterectomy and for prepubertal girls. With a speculum, posterior fornix is swabbed for few seconds. If discharge in there, it can be collected without a speculum.

From Oropharynx. Region of tonsillar crypts and posterior pharynx should be swabbed.

Sensitivity and specificity of gram stain is 95% and 97% respectively. In females, secretions detect 40-60% of culture positive specimens, specificity 80-95%.

Carbohydrate and enzyme activities

	Glucose	Maltose	ONPG	GLU-AMP	PRO-AMP
N. gonorrhoeae	+	0	0	0	+
N. meningitis	+	+	0	+	0 (+)
N. lactamica (lactose+)	+	+	+	0	+

GLU-AMP = g-glutamyl aminopeptidase; PRO-AMP = Hydroxyprolyl aminopeptidase

Non-culture detection methods

- LAL – useful only in cases of purulent urethritis in men.
- DNA hybridisation – no higher sensitivity than ELISAs in asymptomatic females.
- ELISA – less reliable in women than in men.
- Spot DNA – expensive, less efficient for specimen with small number of organisms, not reliable for extra genital specimens, cannot be used to determine short-term test-of-cure, preclude antibiotic susceptibility testing.
- Serology – does not differentiate current from past infection, so not useful.
  CFT, LA, IF, anti-surface pili assays using HA, RIA, ELISA and immunoblotting.

Scoring vaginal Gram stain for bacterial vaginosis

Organism morphotype	No./Oil immersion field	Score
Lactobacillus-like	> 30	0
(parallel-sided GPR)	5-30	1
	1-4	2
	< 1	3
	0	4
Mobiluncus-like	> 5	2
(curved GNR)	< 1-4	1
	0	0
Gardnerella/Bacteroides-	> 30	4
like (tiny, gram-variable	5-30	3
coccobacilli/rounded	1-4	2
plemorphic GNR	< 1	1
with vacuoles)	0	0

Total score is added up and interpreted as follows:

Score	Interpretation
0-3	Normal
4-6	Intermediate, repeat test later
7-10	Bacterial vaginosis

Preservation of *N. gonorrhoeae*

- Overnight culture on chocolate agar or Columbia blood agar placed in small vial or tube with 0.5ml sterile skim milk or nutritive broth with 20% glycerol and kept at -25°C.
- For longer periods, suspensions in polypropylene tubes with skim milk and is kept at -70°C.
- By lyophilization or in a cryoprotective medium (nutrient broth to + 20% glycerol) in cryovials in liquid nitrogen.

HERPES SIMPLEX VIRUSES

IF (immunofluorescence) – Sensitivity for detection in genital specimens varies between 70% and 90% of culture positive specimens.

IP (immunoperoxidase) staining – sensitivity and specificity are same as IF.

HEMOPHILUS DUCREYI

Gram negative bacilli as 'schools of fish'.

Mueller Hinton (MH) agar enriched with 5% horse blood, 1% IsoVitaleX and 5% fetal calf serum – rate of isolation is 93-100%.

Gonococci (GC) agar base enriched with 1% Hgb, 1% IsoVitaleX and 5% fetal calf serum – rate of isolation is 72-89%.

Antibiotics tested on blood agar alone and in combination – Tetracycline, Chloramphenicol, Erythromycin, Kanamycin or Streptomycin, Ciprofloxacin or Fleroxacin, Ceftriaxone or Cefotaxime.

DONOVAN BODIES

A crushed preparation facilitates microscopic interpretation.

Leishman's stain – coccobacilli within large vacuoles (20-90 μm in diameter) in cytoplasm of large histiocytes and occasionally in plasma cells and PMN leucocytes. Blue - purple in colour and often surrounded by a prominent clear to acidophilic pink capsule. Typical bacteria resemble closed safety pins.

It can be cultivated in yolk sac of 5-day old embryonated chicken eggs. Organism is detectable after 72 hours incubation.

CF antibodies can be demonstrated if lesions persist for > 3 months.

Indirect IF for antibody detection – sensitivity 100% and specificity 98%.

VULVOVAGINAL CANDIDIASIS

Caused by *C. albicans* 85%, *C. glabrata* 15%, *C. krusei*, *C. tropicalis* and *C. stellatoidea* rarely. Found in genital tract of 25% of asymptomatic healthy women of childbearing age. Important predisposing factors for colonization and inflammation include:

a) Changes in reproductive hormone levels associated with premenstrual periods, pregnancy and oral contraceptives

b) Use of antibiotics eliminating the protective vaginal bacterial flora

c) Diabetes mellitus

TRICHOMONAS VAGINALIS

20% or more **'clue cells'** are seen. Frothy grey to green - yellow discharge, vaginal itching or irritation. Vaginal malodour and dysuria can also occur.

Sensitivity of direct microscopy is 60%, sensitivity of culture is 95%.

Diamond's medium – growth after 2-4 days in moderate anaerobic conditions and grow best at bottom of relatively long culture tubes filled with medium. Some may be recovered after prolonged incubation for 7 days.

GARDNERELLA VAGINALIS

Semi-selective human blood – Tween bilayer agar (HBT) is used and incubated at 36°C in 5% CO_2 atmosphere for 48 hours. Small, transluent to white, surrounded by a diffuse small zone of b-hemolysis. Gram variable coccobacilli on gram stain. Hippurate hydrolysis, starch hydrolysis, α-glucosidase, acid from glucose, maltose, starch, zone of inhibition with metronidazole disc (50µg) are given by 92% isolates, trimethoprim (5µg) 100% and sulfonamide (1µg) 0%.

MOBILUNCUS SPECIES

Large gram-negative curved rods – *M. mulieris* (2.9 µm) and *M. curtisii* (1.7 µm, gram variable).

	M. curtisii subsp. curtisii	*M. curtisii subsp. holmesii*	*M. mulieris*
NH_4 from arginine	+	+	–
Hippurate hydrolysis	+	+	–
Nitrate reduction	–	+	– (occasional +)
Acid (pH < 5.5) from glycogen	–	–	+

HUMAN PAPILLOMA VIRUSES

Lesion	Types
Condyloma acuminatum	6, 11, 42, 44, 51, 53, 55, 67
Intraepithelial neoplasia	6, 11, 16, 18, 30, 31, 33, 34, 35, 39, 40, 42, 43, 45, 51, 52, 56, 57, 59, 61, 62, 64
Carcinoma	6, 11, 16, 18, 31, 33, 35, 39, 45, 51, 52, 54, 56, 66

Gene	HIV-1 gene product	HIV-2 gene product
env	gp 160 precursor	gp 140 precursor
	gp 120 external	gp 125 external
	gp 41 transmembrane	gp 80 (gp105) – Dimeric form of transmembrane gp
gag	p 55 precursor	gp 36 transmembrane
	p 40 precursor	p 56 precursor
	p 24 core	p 26 core
	p 17 matrix	p 16 matrix
pol	p 66 Rev. tr.	p 68 Rev. tr.
	p 51 Rev. tr.	p 53 Rev. tr. (provisional)
	p 32 endonuclease	p 34 endonuclease.

GENITAL ULCERATIVE DISEASES (GUDs)

Ulcerative, erosive, pustular or vesicular genital lesion(s), with or without regional lymphadenopathy, caused by a number of sexually transmitted infections (STIs) and non–STI-related conditions.

Etiology

A) Sexually transmitted infections (STIs)

- Genital herpes: HSV1: 15% of all GUD and HSV2: 85% of all GUD
- Primary syphilis: *Treponema pallidum*
- Chancroid: *Haemophilus ducreyi*
- Lymphogranuloma venereum (LGV): *Chlamydia trachomatis* serovars L1, L2, L3
- Granuloma inguinale (Donovanosis): *Klebsiella* (formerly *Calymmatobacterium) granulomatis.*

B) Infectious non-STI causes

Fungal	Viral	Bacterial	Parasitic
Candida Deep fungi (rare)	*Cytomegalovirus* (rare) *Varicella* or *Herpes zoster* virus (rare) *Epstein-Barr* virus (rare)	*Staphylococcus* spp. *Streptococcus* spp. *Salmonella* spp. *Pseudomonas* spp. *Mycobacterium spp.*	Scabies (*Sarcoptes scabiei*)

C) Non-infectious causes

Bullous dermatoses	Non-bullous dermatoses	Malignancy

• Non-autoimmune
   - Contact dermatitis
   - Erythema multiforme (almost always HSV-related)
   - Toxic epidermolysis
 • Auto-immune
   - Pemphigus
   - Cicatricial pemphigoid
	• Nonspecific vulvitis/balanitis
	 • Aphthae or aphthous ulcers, aphthosis
	 • Lichen planus, erosive lichen planus
	 • Lichen sclerosus
	 • Behçet's disease
	 • Pyoderma gangrenosum
	 • Fixed drug eruption
	 • Lupus erythematosus
	 • Crohn's disease
	 • Vasculitis
		• Squamous-cell carcinoma
		 • Vulvar intraepithelial neoplasia
		 • Less common:
		   - Extramammary Paget's disease
		   - Basal-cell carcinoma
		   - Lymphoma/leukemia
		   - Histiocytosis X

Epidemiology

- Globally, the most frequent cause of STD-related GUD is genital herpes, followed by syphilis, then chancroid.
- In Asia and Africa, chancroid was once considered the most common type of genital ulcer, followed in frequency by primary syphilis and then genital herpes.
- But now genital herpes is the most common cause of genital ulceration in most developing countries.
- The majority of genital and perianal herpetic outbreaks in the U.S. are caused by HSV-2, though 10-50% of first episodes are due to HSV-1.
- LGV and donovanosis (granuloma inguinale) cause genital ulceration in developing countries.
- LGV outbreaks are occurring in Europe North America, and Australia causing anal and rectal disease in homosexual men, very often in association with HIV and/or hepatitis C virus infections.

Pathogenesis of Genital Herpes

Common sites involved:

Males: penis, urethra, rectal and perianal area.

Females: cervix, vagina, vulva, perineum.

The virus gains entry through contact with mucus membranes or small abrasions in the epidermis. Replication begins at the port of entry.

The virus then travels through the axon to the dorsal root ganglion of nerves of the spinal cord where it enters a latent state.

Viral activation results in replication and release from the neuron into surrounding epithelial cells.

Vesicles → pustules → ulcers → crusting → scabbing → healed skin

Pathogenesis of Donovanosis

Low propensity; direct inoculation in skin or mucous membrane

Incubation period: ill defined 2-3 years (3-40 days generally)

Subcutaneous nodule at inoculation site

↓

Ulcerates

↓

Painless beefy red granulomatous lesion

↓

Local spread; daughter lesions

↓

Local tissue loss

↓

Extensive scarring and pelvic fibrosis

↓

Inguinal region spread – periadenitis (pseudobubo)

Pathogenesis of Lymphogranuloma venerum

Incubation period 1-3 weeks

↓

Vesicle/ulcer on penis or fourchette

↓

The ulcer heals

↓

After 2-16 weeks

Swollen, painful, lymph nodes suppurate through multiple sinuses (bubo)

↓

Inguinal lymph nodes: 'groove sign'

↓

In women iliac lymph nodes are also involved

Complications

- Elephantiasis of the genital region
- Rectal strictures
- Rectal/ rectovaginal fistulae
- Pneumonitis
- CNS involvement

Pathogenesis of chancroid

Incubation period: 4-7 days

↓

Tender erythematous papules

↓

Pustular

↓

Erosions/ ulcers within 48-72 hours covered with yellow grey exudates overlying beefy bleeding base

↓

Painful unilateral inguinal lymphadenopathy

↓

Suppurate to form abscesses, fistulas or sinus tracts

Pathogenesis of syphilis

Incubation period: 9-90 days

↓

Primary syphilis:

↓

Papule on genital areas

↓

Ulcerates

↓

Classical hard chancre "Hunterian chancre"

↓

Flat red dull painless indurated exuding thick serous fluid

↓

Spontaneous healing in 10-40 days

↓

Thin scar

Most common site

Males: prepuce, corona, penis

Females: labia, vagina, cervix

Secondary syphilis (most infectious)

- Widespread mucocutaneous rash
- Coalesce at mucocutaneous junction → wart like condyloma lata
- Painless generalised lymphadenopathy
- Retinitis, Ophthalmitis
- Periosteitis, arthritis
- Meningitis

Latent syphilis

- 1/3 recover spontaneously
- 1/3 dormant / latent (asymptomatic but seropositive)
- 1/3 progress to tertiary stage.

Tertiary syphilis

- Cardiovascular: aneurysms, aortitis
- Coronary artery stenosis
- Neurosyphilis: General paralysis of insane
- Tabes dorsalis

Clinical Features of Genital Ulcers

Feature	Syphilis	Herpes	Chancroid	Lymphogranuloma venereum	Donovanosis
Incubation period	9–90 days	2–7 days	1–14 days	3 days–6 weeks	1–4 weeks (up to 6 mths)
Early primary lesions	Papule	Vesicle	Pustule	Papule, pustule, or vesicle	Papule
No. of lesions	Usually one	Multiple	Usually multiple, may coalesce	Usually one; often not detected, despite lymphadenopathy	Variable
Diameter	5–15 mm	1–2 mm	Variable	2–10 mm	Variable
Edges	Sharply demarcated, elevated, round, or oval	Erythematous	Undermined, ragged, irregular	Elevated, round, or oval	Elevated, irregular
Depth	Superficial or deep	Superficial	Excavated	Superficial or deep	Elevated
Base	Smooth, nonpurulent, relatively nonvascular	Serous, erythematous, nonvascular	Purulent, bleeds easily	Variable, nonvascular	Red and velvety, bleeds readily
Induration	Firm	None	Soft	Occasionally firm	Firm
Pain	Uncommon	Frequently tender	Usually very tender	Variable	Uncommon
Lymph-adenopathy	Firm, nontender, bilateral	Firm, tender, often bilateral with initial episode	Tender, may suppurate, loculated, usually unilateral	Tender, may suppurate, loculated, usually unilateral	None; pseudobubos

Laboratory Diagnosis of GUDs

Collection and transport of specimens

Herpes virus	Klebsiella granulomatis	Chlamydia trachomatis	Hemophilus ducreyi	Treponema pallidum
Mucocutaneous genital lesions, Swabs/aspirate from fresh vesicles/pustules, Cervical/rectal swabs Transport: viral transport medium; 4°C vials	Edge of granulomatous red tissue Punch biopsy/Tissue - crushed on glass slide	Swabs from ulcer (Rayon, dacron) Avoid cotton fibres/wooden swabs, Calcium alginate Cytobrush – cervical specimens Rectal swabs Punch biopsies Bubo pus: fluctuant – aspirate; non-fluctuant – inject sterile saline and then aspirate Preferably the third or second swab should be used.	Exudate from ulcer base/margin Transport: 4°C Thioglycolate hemin based medium with L-glutamine, bovine albumin. Or inoculate immediately on culture media.	Serous fluid from ulcer base/margin. Lymph node aspirate Serum

Transport medium for Chlamydia: In cryovials

- Sucrose phosphate saline (2SP)
- Storage: 4°C for 24 hrs or -60°C or liquid nitrogen for longer
- Swabs in transport medium should be shaken for 1-2 mins on a vortex mixer or sonicated and swab removed before storage.
- Tissues should be homogenized in a tissue grinder or shaken with glass beads in a vortex mixer and centrifuged to remove cellular debris.

Microscopy

Microscopy	T. pallidum	HSV	H. ducreyi	C. trachomatis	K. granulomatis
Dark field	Cork screw motility/ forward-backward movement with rotation about longitudinal axis, rigid, uniform, tightly wound spirals Sensitivity: 73-100%	-	-	EB appear yellow (natural autofluorescence)	-
Electron microscopy	-	+	-	-	-
Gram stain	-	-	Gram negative short plump rod/ coccobacilli. Shoal of fish Railroad appearance	Gram negative small spherical/ovoid	Gram negative pleomorphic short rod/coccobacilli/ curved/dumb bell shaped
Bipolar staining	-	-	-	-	Closed safety pin appearance
Histo-pathology/ cytology	Brain, GIT, placenta, umbilical cord or skin	Tzanck smear		Lymph node aspirate/biopsy EB (Miyagawa's granulo-corpuscles)	Donovan bodies (organisms contained within the phagosomes of large macrophages – Pund/Greenblatt cells.
Special stains	Silver impregnation (Fontana's Levaditi's Steine's Warthin starry) Giemsa	Giemsa Cowdry Type A Intranuclear inclusion bodies (Lipschultz)		5% iodine, Giemsa, Gimenez, Papanicolau	Leishman, Giemsa, Wright Giemsa, May-Grunwald Gieimsa, Papanicolau or silver impregnation
Immuno-fluorescent	Direct FITC labeled mouse monoclonal/ Reiter absorbed anti T. pallidum globulin Sensitivity : 73-100%	Direct Sensitivity: 70-90%	Indirect	Direct/Indirect	Indirect
Immunohisto-chemistry	Peroxidase/ alkaline phosphatase. Counterstain with haematoxylin	Horseradish peroxidase labelled			

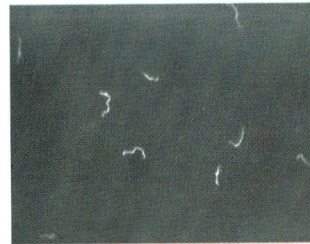
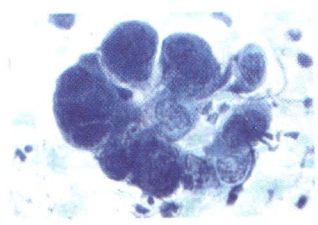
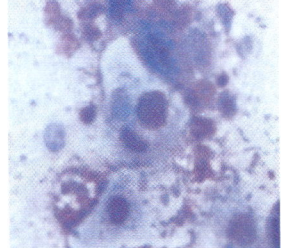
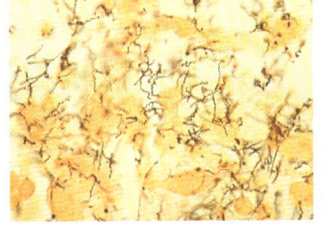

From L to R: *Treponema pallidum* seen in dark ground microscopy; Tzanck smear showing a multinucleated giant cell in HSV infection

From L to R: Giemsa stained smear showing intracellular basophilic Donovan bodies (x40); *Treponema pallidum* Levaditi's stain

Antigen detection

T. pallidum	Rabbit inoculation (intracerebral, intradermal, ocular, intravenous, scrotal) sensitivity 100% Local Lesions develop, infective for life, infection can be transferred to other animal by using minced lymph nodes or testes & serological test becomes reactive following inoculation
HSV	Classical ELISA; antigen capture ELISA Sensitivity: 70-95%; Specificity: 94-100% Rapid assays, less sensitive, specific,
H. ducreyi	Blot RIA – monoclonal antibody EIA – polyclonal antibody ELISA
C. trachomatis	Enzyme Immunoassay (EIA); Chlamydial lipopolysaccharide – Sensitivity: 71-97%; Specificity: 97-99%; Rapid assays (within 30 min.) on clinical specimens; Surecell chlamydia test; Clearview chlamydia test; Testpack chlamydia test – Sensitivity: 52-85%; Specificity: > 95%; Immunofluorescence (IF): MOMPFITC labelled antibodies - Sensitivity: > 90%; Specificity: > 95%

Culture and Identification

Herpes viruses

Parallel inoculation of 2 different cell lines:

- Human diploid fibroblast
- Vero cells
- Baby hamster kidney
- Rabbit kidney cells

Examine daily under stereoscopic microscope.

Cytopathic effects:

- Rounding of cells / ballooning
- Granular appearance
- Focal necrosis
- Multinucleated cells

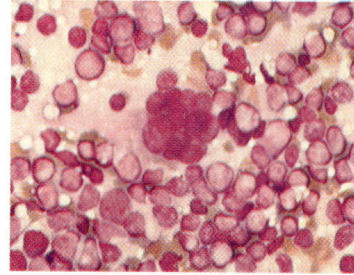

Cytopathic effects of Herpes viruses seen in Vero cell line

Hemophilus ducreyi

- Chocolate agar: incubate 3-4 days at 33-36°C with moisture and 5% CO_2

- Selective media: Gonococcal agar or Mueller Hinton based agar (5% horse chocolate agar or 1% haemoglobin, 1% isovitalex, & 5% fetal calf serum, 3mg/l vancomycin).

- After 24 hrs, colonies are small yellow-gray translucent or semi-opaque, often appear mixed, can be pushed intact across the agar surface.

- Can be grown on chorioallantoic membrane of chick embryo

- Strains may be preserved on moist slopes of Sheffield's medium for about 4 weeks at 4°C or in liquid nitrogen for upto 2 years.

- Catalase negative

- Oxidase positive

- Asaccharolytic

- ALA-porphyrin test negative (requires exogenous hemin)

- Positive nitrate reduction and alkaline phosphatase test.

- Susceptible to SPS

- Rapid kit systems: API-ZYM, Minitek system, RapDNH.

Chlamydia trachomatis

- Yolk sac of 6-8 day old chick embryo

- Cell cultures: Cycloheximide treated McCoy cells; HeLa 229 cell line of human carcinoma; Mouse fibroblast; BHK-21.

- Organ culture: Guineapig conjunctival epithelium

- Chylamydial inculsions detected by iodine staining of the cell carpet. Inclusions stain brown; host cells yellow colour; or stain with monoclonal antibodies and counterstain with rhodamine B: cells red, apple green inclusions. Cut off:10 EB

- In LGV, the entire cell carpet may contain inclusions.

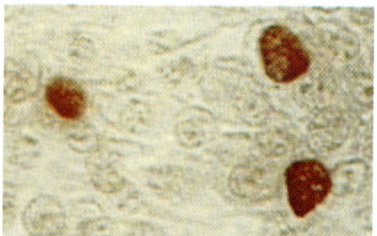

C. trachomatis inclusion bodies (brown) in a McCoy cell culture

Klebsiella granulomatis

- Fresh egg yolk containing media
- Modified Levinthal medium
- Yolk sac of embryonated hen's egg

- Human monocytes co-culture system (Hep-2 cell)
- Semisynthetic medium with lactalbumin hydrolysate.

Treponema pallidum
- Does not grow in artificial medium.
- Maintained by serial passage in rabbit testes.

Serology
Serodiagnosis of Syphilis
Non treponemal tests: Antibody to cardiolipin

1) Microscopic cardiolipin tests
- VDRL (Venereal Diseases Research Laboratory) test
It is a Slide flocculation test, rapid, simple and most widely used. Inactivated serum + freshly prepared antigen are used. Antigen: cardiolipin + lecithin + cholesterol. Can also be done on CSF.
- Unheated serum reagin test (USR)
Uses unheated serum and stabilized antigen. Antigen: VDRL antigen + choline chloride + EDTA. Plasma can also be used.

2) Macroscopic cardiolipin tests
- Rapid plasma reagin test (RPR) card test. Uses antigen suspension + serum qualitative and quantitative. Charcoal particles: better visualization of antigen antibody complexes.
- Reagin screen test. Modification of RPR. Uses a lipid soluble black dye.

- Toluidine red unheated serum test (TRUST). Uses toluidine red instead of charcoal.

3) Other Non treponemal Tests
- Wasserman CFT:
Inactivated serum + cardiolipin antigen + guinea pig complement
Sheep erythrocytes + antisheep erythrocyte antiserum (amboceptor)
Interpretation: Haemolysis – negative; No haemolysis – positive.
- Kahn test:
It is a tube flocculation test using dilutions of antigen + serum.
- Kahn verification test:
At 1°C for nonspecific antibody; At 37°C for specific antibody.
- Automated reagin test (ART)
- VDRL-ELISA: Visuwell Reagin Test
- Capture-S solid phase red blood-cell adherence assay: Serum/plasma can be used. Sensitivity: 94%; Specificity: 99.2%.

Disadvantages of non treponemal tests
- Prozone phenomenon
- Biological false postive reactions

Sensitivity and specificity of non-treponemal tests

Test	% Sensitivity				% Specificity
	Primary	Secondary	Latent	Late	Non-syphilis
VDRL	78	100	95	71	98
RPR	86	100	98	73	98
USR	80	100	95	-	99
TRUST	85	100	98	-	99

Detection of antibody to group specific antigen
- Reiter protein CFT (RPCF):
- Lipopolysaccharide-protein complex antigen from Reiter treponeme
- Lower sensitivity and specificity than those using *T. pallidum*.

Detection of specific antibody to *Treponema pallidum*
Microscopic tests
a) Fluorescent treponemal antibody absorption test (FTA-Abs): Indirect fluorescent antibody technique. Most sensitive. Technically most difficult.
Antigen: lyophilized suspension of *T. pallidum* (Nichol's strain)

Sorbent: Sonicate from Reiter treponeme (variant of *T. phagedenis*), which absorbs group antibody which reacts to non-pathogenic treponeme.
Conjugate: Fluorescein isothiocyanate labelled anti-human globulin
Modification: FTA-Abs Double Stain (DS)
Conjugate: Rhodamine isothiocyanate labelled anti-human globulin
Counterstain: FITC labeled anti *T. pallidum* conjugate

b) *Treponema pallidum* immobilization (TPI) test
Most specific. Serum and live *T. pallidum* and complement. Check for motility of *T. pallidum* under dark ground microscopy

c) *T. pallidum* agglutination test (TPA)

Antigen: formalin inactivated suspension of Nichol's strain

Serum and antigen incubated.

d) *T. pallidum* immune adherence test (TPIA):

T. pallidum suspension and serum and complement and heparinized blood. Immune adherence to erythrocytes in presence of antibodies.

Dark ground microscope - agglutination.

Haemagglutination methods

a) *T. pallidum* haemagglutination test (TPHA) - (Rathlev's procedure)

Indirect agglutination of sensitised erythrocytes; Antigen: formalinized tanned sheep erythrocytes sensitized with ultrasonicated material from *T. pallidum*

Non reactive: compact red button; Reactive: smooth mat of cells.

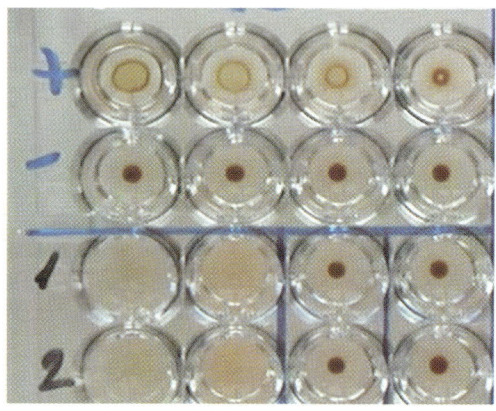

b) Microhaemagglutination assay (MHA-TP):

Microvolume haemagglutination - modification.

c) Other haemagglutination tests:

Use of sensitized turkey, sheep chicken cells and performed in microtitre plate.

d) Automated methods:

Pk-TP (Olympus); PK7100 or 7200

Twice as specific as RPR card test.

e) Particle agglutination test: (Serodia TP-PA)

f) Modified Microhaemagglutination test:

Gelatin particles used in place of erythrocytes. Thus eliminates the preabsorption test.

g) Latex agglutination test:

Cloned *T. pallidum* antigens bound to latex particles. It is easy, fast (< 30 min).

Sensitivity & specificity similar to FTA-Abs & TP-PA.

h) Western blot:

Detects IgG or IgM antibodies against *T. pallidum*. Test using IgG conjugate is as sensitive and specific as FTA- Abs. IgM Western blot may be the most sensitive test currently available for congenital syphilis.

i) Immunochromatographic test (ITC):

Variations of western blot. Uses one or more cloned antigens of *T. pallidum*. Has high

Specificity. It is rapid within 10-15 min. Does not require equipment or electricity.

j) Enzymeimmunoassay (EIA):

Two types: Sonicated *T. pallidum* as antigen; Syphilis-G test (Mercia Diagnostics)

Cloned antigens (47-,17-, & 15-kDa proteins).

Trep-chek - Limitations: Time & cost; Advantages: Capability to automate the test.

Large number of samples can be processed. Automated objective read out.

Detection of treponemal IgM

- FTA-Abs: IgM
- FTA-Abs: 19S IgM fraction
- IgM solid phase hemadsorption assay (SPHA)
- IgM capture ELISA
- Immunoblot: Western Blot.
  - For congenital syphilis, sensitivity > 80% and specificity > 90%.
  - In active neurosyphilis: IgM EIA (Captia Syphilis-M test).

Disadvantages of non treponemal tests:

· False positive reactions

· Varying rates of reactivity in early primary syphilis.

Sensitivity and specificity of treponemal tests

Test	% Sensitivity			% Specificity
	Primary	Secondary	Latent	Non-syphilis
FTA-Abs	84	100	100	97
TP-PA	88	100	100	98
FTA-Abs DS	86	100	100	98
EIA	90	100	100	99
Western blot	90	100	100	98

Serodiagnosis of HSV infection

- Complement fixation test
- Indirect immunoflurescence
- Neutralisation test
- Latex agglutination tests
- Haemagglutination tests
- EIA

  ELISA most widely used. It detects IgM or a rising IgG titre. These cannot differentiate between HSV1 & HSV 2. Extensive cross reactivity (common viral glycoprotein) Glycoprotein G ELISA. Differentiates between HSV1 & HSV 2.

Serodiagnosis of LGV

Antibodies to LPS & MOMP

- Complement fixation test: titres of > 256 indicates LGV; titres < 32 rules out LGV
- Microimmunofluorescence assay

  Antigen: formalin fixed elementary body of Chlamydia trachomatis titres of ≥ 512 indicates LGV

- Single antigen whole inclusion test:

  Antigen: inclusion of tissue culture infected with LGV- L_2 strain

- ELISA (chlamydiazyme) Low specificity, Cross reactions with *S. aureus* and gram negative bacteria
- Local antibody in genital secretion: not consistent.

Serodiagnosis of Klebsiella granulomatis

- Complement fixation test: low diagnostic value
- Indirect immunofluorescence

Molecular Diagnosis

Syphilis

- PCR
- 658 bp segment of gene for 47 kDa surface antigen
- Gene coding for 39 kDa basic membrane protein
- Gene sequence of tmpA which codes for 45 kDa membrane protein
- Genes for 4D antigen (oligomeric protein)
- Multiplex PCR (HSV, *H. ducreyi, T. pallidum*)

- RT PCR: T. pallidum 16 S RNA
- DNA probes :
- Dot blot assays : lack sensitivity

LGV

- Nucleic acid probes:
- Probe assay chemiluminescence enhanced (PACE) test (Gen Probe San Diego)
- Species specific
- Non-isotopic DNA probe
- For specific parts of r-RNA
- Sensitivity: 80-95%; specificity: 98-99%
- Nucleic acid amplification tests:
- PCR: Sensitivity: 90%; specificity: 99-100%
- LCR: Sensitivity: 90-97%; specificity: 99-100%
- Amplicor
- Strand displacement amplification
- Transcription mediated amplification (TMA) 16S-rRNA

H. ducreyi

- ^{32}P labelled DNA probes:
- Detects 10^4 CFU of H. ducreyi
- Used for confirming the identity
- Lack sensitivity to detect in clinical material.
- Multiplex PCR:
- Culture/clinical specimen (Taq polymerase inhibitors in DNA preparations of ulcer specimen)
- Less sensitive than those based on cultures.
- Hemi-nested PCR: *H. ducreyi* specific protein p27 (Sensitive and specific for clinical material)

Herpes simplex

- Samples: Pus, CSF
- DNA hybridization with radio labelled or biotin labelled probes, e.g. prof
- PCR

Klebsiella granulomatis

- PCR: Research tool

Treatment

Infective agent	Treatment	Resistance (R) / Ineffective (I)
HSV	Acyclovir, Famcyclovir, Valacyclovir, Foscarnet	Acyclovir (R) due to loss of functional gene TK or point mutations in TK or DNA polymerase gene
C. trachomatis	Doxycycline 200mg x 7 days (14), Erythromycin, Azithromycin, [Amoxycillin, Quinolones: 2nd line]	Erythromycin, Tetracycline (high MIC) Azithromycin, Quinolones.
H. ducreyi	Ciprofloxacin 500 mg PO as single dose or Ceftriaxone 250 mg IM as single dose or Azithromycin 1 g PO as single dose	Beta-lactamase, Tetracyclines, Co-trimoxazole (R)
T. pallidum	Benzathine penicillin 2.4 million units IM once Doxycycline or Tetracycline	Erythromycin, Azithromycin? (R); Insensitive to Quinolones, Rifampicin.
K. granulomatis	Tetracycline, Chloramphenicol, Erythromycin, Gentamicin, Co-trimoxazole, Norfloxacin	Penicillin, Ampicillin, Cephalosporins (I)

Syndromic approach to genital ulcer/sore

Patient complains of a genital sore or ulcer
↓

History & examine
↓

Only vesicles present → Sore/ulcer present → Educate & counsel
                    No   No   Promote & provide condoms
                                                      Offer HIV counselling & testing
Yes ↓   Yes ↓   if both facilities are available

Treat for syphilis if indicated[1]   Treat for syphilis
Treat for HSV-2[2]   & chancroid
                                      Treat for HSV-2[2]
↓   ↓

Educate & counsel on risk reduction.
Promote & provide condoms.
Offer HIV counselling & testing if both facilities are available.
Review in 7 days.
↓   ↓

Ulcer(s) healed? → Ulcer(s) improving → Refer to specialist
              No   No

Yes ↓   Yes ↓

Educate & counsel on risk reduction.   Continue treatment for a
Promote & provide condoms   further 7 days
Offer HIV counselling & testing
if both facilities are available.
Manage & treat partners

[1]Indications for syphilis treatment:
– RPR positive
– No recent syphilis treatment
[2]Treat for HSV-2 where prevalence
   is 30% or higher or adapt to local conditions

Prevention

General measures

- Health education and counseling
- Promote and provide condoms and other barrier methods
- Offer HIV counseling & testing if both facilities are available
- Contact tracing
- Sexual abuse needs to be considered when GUD is found in children beyond the neonatal period

Vaccine - Herpes simplex type 2

Type of vaccine	Pharmaceutical company or research group	Stage of development
gD2 subunit with AS04 adjuvant	GSK/NIH	Phase III (HSV seronegative women)
Live attenuated ICP8 DISC virus vaccine	Xenova	Phase I/II
Other live attenuated replication competent virus vaccines	Avant; AuRix	Phase I; Phase II

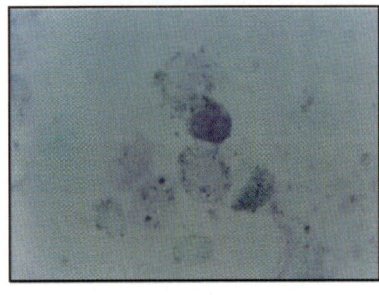

Donovan bodies produced by *Klebsiella granulomatis* from the lesion
stained by Wright's stain (100 X)

REFERENCES

1. Topley & Wilson's, Microbiology and Microbial Infections. Bacteriology, Vol. 1. 10th Ed. Eds. Borriello SP, Murray PR, Funkay G. John Wiley & Sons Ltd., West Sussex, UK. 2009.

2. Collee JG, Marion BP, Fraser AG, Simmons A. Mackie and McCartney Practical Medical Microbiology. 14th Ed. Churchill Livigstone, U.K. 1996.

3. Mims' Medical Microbiology. 5th Ed. Eds. Richard Goering, Hazel Dockrell, Mark Zuckerman, Ivan Roitt & Peter L. Chiodini. Saunders. 2013.

4. Harrison's principles of Internal Medicine. 17th Ed. Eds. Braunwald E., Fauci A. S., Kasper D. L., Hauser S. L., Longo D. L., Jameson J.L. McGraw-Hill, New Delhi. 2008.

5. Ananthanarayan & Paniker's Textbook of Microbiology. 9th Ed. Ed. Kapil A. Universities Press, Hyderabad. 2013.

6. Koneman EW, Allen SD, Janda WM, Schreckenberger P, Winn Jr. WC. In Color Atlas & Textbook of Diagnostic Microbiology. 6th Ed. Lippincott, Philadelphia. 2006.

7. Jawetz, Melnick & Adelberg's Medical Microbiology. 25th Ed. Eds. Brooks GF, Carroll KC, Butel JS, Morse SA, Mietzner TA. The McGraw Hill Companies, Lange, USA. 2010.

8. Sexually Transmitted Diseases Treatment Guidelines. U.S. Department of Health and Human Services, CDC. MMWR 2010; 59 (RR-12).

BACTERIOLOGY OF WATER, MILK AND AIR

BACTERIOLOGY OF WATER

Drinking water should be safe and pleasant to drink, i.e. cool, clear, colourless and devoid of disagreeable taste or smell. Hazards of water pollution is classified into :

A. **Chemical pollutants**: Detergents, solvents, cyanides, heavy metals, ammonia, sulphides, minerals, etc. derived from industrial and agricultural wastes.

B. **Biological hazards**: Water contaminated with sewage or other excreted material might cause intestinal or other systemic infections.

1. Organisms present in water without fecal contamination

   - Natural water bacteria – *Micrococcus, Pseudomonas, Serratia, Flavobacterium, Alkaligenes, Acinetobacters.*
   - Soil bacteria (washed into water) – *B. subtilis, B. megaterium, Klebsiella spp.*

2. Organisms present in water with fecal contamination

   - Intestinal bacteria (through sewage) – *E. coli, Klebsiella spp., Streptococcus fecalis, Clostridium perfringens.*
   - Sewage bacteria proper – *P. vulgaris, C. sporogenes,* filamentous bacteria (e.g. *Nocardia spp.*)

Bacteriological indicators of fecal contamination of water are based on organisms indicative of pollution of water by human/animal feces. They are

a) *Escherichia coli* and coliform as a whole
b) Fecal streptococci (*Streptococcus fecalis*) and
c) *Clostridium perfringens*

Bacteriological Diagnosis: Samples are collected in sterilised (autoclaved) glass bottles of 230 ml with ground glass stoppers, covered with kraft paper. Processing should be done within 1-3 hours.

Standard tests done are:

1. Presumptive coliform test by multiple tube method using double strength MacConkey's broth and membrane filter method and differential or confirmatory tests for *E. coli* **(Eijkman test)** and gas

production when grown in bile salt lactose peptone water at 44°C for *Streptococcus fecalis.*

2. Detection of *S. fecalis* by subculturing in glucose azide broth and incubated at 45°C and seen for azide production; and for *Clostridium perfringens* by inoculating litmus milk and incubating anaerobically at 37°C for 5 days and looked for stormy fermentation.

3. Colony count - Enumeration of viable bacteria (plate count), i.e. total number of viable bacteria per 1 ml of water. One ml test sample is mixed to 10 ml melted yeast agar at 45°-50°C and allowed to solidify. One plate is kept at 22°C and the other at 37°C for 18-24 hours.

Class	Presumptive coliform count/100ml	E. coli count/100ml
Class 1 Excellent	0	0
Class 2 Satisfactory	1-3	0
Class 3 Suspicious	4-10	0
Class 4 Unsatisfactory	> 10 (indicative of fecal contamination)	0, 1 or more

Water-borne diseases:

a) Viral – Viral hepatitis (A and E), Poliomyelitis, Rotavirus diarrhoea.

b) Bacterial – Cholera, Typhoid, Paratyphoid, Bacillary dysentery, *E. coli* diarrhoea, *C. jejuni* and *C. coli* diarrhoea, *Y. enterocolitica* diarrhoea.

c) Protozoal – Amoebiasis, Giardiasis, *Balantidium coli* diarrhoea.

d) Helminthic – Roundworm, Whipworm, Threadworm, Hydatid disease.

e) Due to aquatic host –
   - Cyclops - Guineaworm, Fish tapeworm
   - Snail - Schistosomiasis

BACTERIOLOGY OF MILK

Milk contain bacteria which are derived from udders of animals, use of unsterile milking equipments, improper cleaning of milker's hands, unclean udders and dust in milking shed, water used for adulteration, diseased animals (e.g. animals suffering fom mastitis, brucellosis, tuberculosis, etc.) and Carriers (of typhoid, paratyphoid, dysentery, food poisoning bacilli, coagulase positive staphylococci or hemolytic streptococci), who milk the animal.

Bacteria present in human milk:

- Prior to infant feeding - *S. epidermidis, S. mitis, Gaffkya tetragena* and *S. aureus.*

- After breast feeding - Organisms in large numbers, derived from maternal skin and infant's mouth.

Sources	Organisms
Utensils, udders	Lactic streptococci (e.g. *S. lactis*), *S. fecalis*, Lactobacilli, Staphylococci.
Environment	*Alkaligenes spp., Achromobacter*
Water, hands of milker	Coliform bacilli, *C. perfingens, C. butyricum*
Environment	*B. subtilis, B. cereus, P. vulgaris*

Milk-borne diseases:

1. Infections primarily of animals - Tuberculosis, brucellosis, streptococcal infections, enterotoxigenic staphylococci, salmonellosis, Q fever, cow pox, foot and mouth disease, anthrax, leptospirosis, milker's nodes, infection caused by *C. fetus, Y. enterocolitica.*

2. Contaminated milk by ticks, rats - Tick-borne encephalitis, *Streptobacillus moniliformis* (rat-borne).

3. Infections prirarily of man - Typhoid and paratyphoid fevers, cholera, shigellosis, enteropathogenic *E. coli*, staphylococcal food poisoning, streptococcal infections, tuberculosis, hepatitis.

Pasteurisation of milk. Eliminates risk of 90% infections, including more heat - resistant tubercle bacillus and Q fever organisms.

- Holder method 63°C for 30 minutes followed by rapid cooling to 5°C.

- Flash method 72°C for 15 seconds and then rapidly cooled to 4°C.

- Ultra-high temperature (UHT) method - rapidly heated in two stages between 125-132°C for a few seconds (2-3) only and then rapidly cooled.

Biological standards of milk as laid down under milk regulations act of 1965 and 1972 in England and Wales is as follows :

a) Untreated milk may contain 500 bacteria per ml

b) Pasteurised milk should not contain coliform in 0.1 ml milk

c) Sterilised milk must satisfy the turbidity test

d) Ultraheated milk must contain < 10 bacteria per 0.01 ml (i.e. 1000 bacteria/ml).

Bacteriological examination:

1. **Viable count** - Serial dilutions of milk (1:10, 1:100, 1:1,000) are done in sterile Ringer's solution, incorporated in yeast extract milk agar and incubated at 30-31°C for 72 hours. Colony count is given by number of colonies multiplied by dilution factor.

2. **Coliform test** is done in 3 tubes of MacConkey's fluid medium, incubated at 37°C for 48 hours and seen for production of acid and gas.

3. **Chemical tests:**

   a) Methylene blue should not be decolourised in 30 minutes by milk under standard conditions.

   b) Resazurin test

   c) Phosphatase test - concentration of p-nitrophenol should be < 10 µg/ml of milk

   d) Turbidity test - When milk is heated to 100°C for 5 minutes, soluble proteins in milk get denatured and this cannot be precipitated by ammonium sulphate.

4. **Detection of specific pathogens:**

   - Tubercle bacilli - 50 ml milk centrifuged at 300 r.p.m. for 30 minutes, deposit stained by AFB stain

   - Brucella – Cream inoculated on serum dextrose agar. Cream and centrifuged deposit of milk injected I/M in guineapigs.

   - In animals it is detected by demonstrating brucella antibodies by milk ring test and whey agglutination tests.

BACTERIOLOGY OF AIR

A man inhales about 500 cu. ft. of air in a day.

Sources of air polution :

1. Human sources – Bacteria spread in the environment in droplet during coughing and sneezing, e.g. tuberculosis, Q fever, psittacosis, coccidioidomycosis, etc.

2. Environmental sources – Soil and vegetations contain nonpathogenic organisms which become air borne, e.g. *Achromobacter, Sarcina, Micrococci*, coliform bacilli, *B. subtilis*, spores and fragments of moulds.

Hospital air contain droplet nuclei or skin scales carrying infective organisms.

Acceptable limits of air pollution:

   a) Factories, officers, homes – 50 per cubic feet

   b) Operation theatre – 10 per cu. ft.

   c) Dressing room, O.T. for neurosurgery – 1 per cu. ft.

Bacteriological examination:

1. **Settle Plate Technique (Qualitative, Sediment method)** – Petridish (blood agar plates) exposed for 30 minutes at various places in ward/O.T. It is than incubated at 37°C for 24 hours. Acceptable level of contamination in Ward/Dressing room is upto 10-15 colonies. Not a single colony of *S. aureus, Pseudomonas* or fungus should be present.

2. **Surface sampling (Scrape method)** – Uncovered plate is placed on fabric and the dish is sweeped back and forth (10 sweeps, 15 inches long).

   Sample floors and walls or done by using 10 cm^2 template. Swab is moistened in neutralizing buffer and than the floor is swabbed. Swab is replaced in buffer, shaped and plated out in duplicate in 1ml aliquots

   Desirable criteria for floor cleanliness:

   O.T. room – 0 to 5 organisms/cm^2

   Wards – 5 to 10 organisms/cm^2

3. Swab is rubbed in a path of 1/2″ width and 16″ long (to cover approx 8″). Five areas are swabbed for each group of similar utensils with 1 swab (total area 40″)

Rubbed slowly and firmly 3 times over the surfaces, reversing direction each time. First dipped in water, squeezed and then swabbed. Swab should be moist but not wet. Should be plated in blood agar within 2 hours of swabbing.

4. **Slit Sampler.** It is a large vaccum extractor with a nozzle, which is pointed towards the area to be sampled. A known volume of air (1 cu. ft. or 30-70 litres per minute) is directed onto a plate of blood agar for a given time (4-8 minutes), through a slit of 0.25 mm wide.

Plate is rotated mechanically for even spread of the organisms. Plates are incubated and colonies are counted per cubic metre of air, which gives the number of bacteria present in air.

Accepted levels are 100-500 BCP per cubic metre of air for conventional ventilated OT and 10 BCP per cubic meter for ultraclean ventilation. No more than are colony of *Clostridium perfringens* or *S. aureus* per plate is accepted.

Percentage of 1% of airborne organisms are pathogenic – *S. aureus* 0.1 to 10 organisms are present/cu. ft. of air; *S. pyogenes* 10/cu. ft. of air in rooms of patients having scarlet fever and streptococcal tonsillitis.

REFERENCES

1. Ananthanarayan & Paniker's Textbook of Microbiology. 9th Ed. Ed. Kapil A. Universities Press, Hyderabad. 2013.

2. Chakraborty PA. Textbook of Microbiology. 4th Ed. New Central Book Agency (P) Ltd., Kolkata. 2009.

3. Topley & Wilson's, Microbiology and Microbial Infections. Bacteriology, Vol. 1. 10th Ed. Eds. Borriello SP, Murray PR, Funkay G. John Wiley & Sons Ltd., West Sussex, UK. 2009.

VIROLOGY

24

Chapter

IMPORTANT DNA VIRUSES WITH THE DISEASES PRODUCED AND LABORATORY DIAGNOSIS

A. Pox virus

Configuration, Envelop, Symmetry: Double stranded (DS), nonenveloped, brick-shaped

Size: 300 nm (largest)

Diseases Produced:

1. Variola – small pox
2. Vaccinia used for small pox vaccination
3. Molluscum contagiosum
4. Cow pox

Laboratory Diagnosis:

1. Vesicle fluid inoculated in chorioallantoic membrane (CAM) of chick embryo – smaller 'pocks' by vaccinia virus
2. Ouchterlony procedure to find out antigen relatedness
3. Biopsy of skin – stained by H and E for Molluscum.

B. Herpes virus

Configuration, Envelop, Symmetry: DS, enveloped, icosahedral

Size : 100 nm

Diseases Produced:

1. Herpes simplex type 1 – herpes labialis, keratoconjunctivitis, dendritic keratitis, eczema herpeticum, encephalitis.
2. Herpes simplex type 2 – genital ulcers, cervical carcinoma, neonatal herpes.
3. Varicella zoster – chicken pox, herpes zoster.
4. Epstein-Barr virus – infectious mononucleosis, Burkitt's lymphoma, nasopharyngeal carcinoma.
5. Cytomegalovirus – cytomegalic inclusion disease in newborn, postnatal infections in adults, mainly hepatitis.

Laboratory Diagnosis:

Herpes simplex virus

a) Tzanck smear – multinucleated giant cells
b) Cowdry A type of inclusion bodies

c) In CAM – 'pock' formation
d) Diploid human embryo lung cell lines – focal degeneration
e) Serology – CFT, Neutralisation test
f) PCR – detection of HSV DNA directly from specimen.

Epstein-Barr virus

a) T.L.C. 10,000 - 90,000/cu.mm, abnormal lymphocytes
b) Paul - Bunnel test
c) Indirect IF test.

Cytomegalovirus

a) Large intranuclear "owl's eye" inclusion bodies in salivary glands, kidney, liver and spleen
b) Human embryonic fibroblast culture
c) ELISA for IgM and IgG antibodies
d) Rapid shell vial centrifugation assay – rabbit antimouse antibodies conjugated thiocyanate (FITC) is added and observed under fluorescent miscroscope
e) pp65 antigenemia assay – detects lower matrix protein pp65 of CMV under fluorescent iscroscope – bright apple green fluorescence
f) PCR
g) Hybrid capture DNA assay
h) bDNA assay.

C. Adeno virus

Configuration, Envelop, Symmetry: DS, non-enveloped, icosahedral

Size: 70-90 nm

Diseases Produced : The virus has many serotypes.

1. Acute febrile pharyngitis (types 1-7)
2. Pharyngoconjunctival fever (types 3, 7)
3. Acute respiratory disease (types 4, 7, 14, 21)
4. Follicular conjunctivitis (types 3, 4, 11)
5. Epidemic keratoconjunctivitis (types 8, 19, 37)
6. Hemorrhagic cystitis (types 11, 12)
7. Diarrhoea and vomiting (types 40, 41)
8. Disseminated infections in immunocompromised patients (types 5, 34, 35, 43-47).

Laboratory Diagnosis:

Samples from throat, conjunctiva, stool, etc.

1. In Hela or HEp2 cells – grape like clusters

2. Serology – CFT, Neutralisation test, Hemagglutination inhibition (HI) test

3. By hemagglutination with monkey and rat RBCs – differentiated into groups A-F.

D. Papova virus

Configuration, Envelop, Symmetry: DS, non-enveloped, icosahedral

Size : 45-55 nm

Diseases Produced:

1. HPV – Human papilloma or wart (types 6, 11)

2. Polyoma virus – papilloma in rabbits, mice and dogs

3. SV 40 – sarcoma in mice

4. BK and JC viruses – neuronal tumors in rodents.

E. Parvo virus

Configuration, Envelop, Symmetry: Single stranded (SS), nonenveloped, icosahedral

Size : 18-26 nm

Diseases Produced:

1. Disease in mice, rats

2. B19 – erythematous infectiosum.

F. Hepadna virus

Configuration, Envelop, Symmetry: DS, non-enveloped, icosahedral

Size: 42 nm Dane particle (Hepatitis B virus) 22 nm tubular and spherical particles (HBsAg)

Diseases Produced: Serum Hepatitis; 5-10% prevalence in India.

Laboratory Diagnosis:

1. ALT gradual rise over a longer period. Transaminase 500-2000 units (ALT > AST).

2. Serum bilirubin level is increased

3. Serological markers – HBsAg, HBeAg, Antibodies against HBc, HBe and HBs, viral DNA polymerase (in pre-icteric phase) detected by ELISA, RIA and HAT (hemagglutination test)

4. HBcAg in liver cells by immunofluorescence.

IMPORTANT RNA VIRUSES WITH THE DISEASES PRODUCED AND LABORATORY DIAGNOSIS

A. Picorna virus

Configuration, Envelop, Symmetry: SS, non-enveloped, icosahedral

Size: 20-30 nm

Diseases Produced:

1. Poliovirus Types 1, 2 and 3 – Poliomyelitis – inapparent infection, abortive or minor illness, non-paralytic and paralytic (spinal, bulbar and bulbospinal).

2. Coxsackie A virus Types 1-22, 24 – Aseptic meningitis, herpangina, hand foot and mouth disease.

3. Coxsackie B virus Types 1-6 – Bornholm disease, myocarditis, hepatitis, meningitis. Type B4 causes juvenile diabetes.

4. ECHO virus Types 1-9, 11-27, 29-34 – Aseptic meningitis, paralysis, respiratory disease, pericarditis, myocarditis, infantile diarrhoea.

5. Type 68 – Pneumonia, bronchitis; Type 69 – Not associated with disease; Type 70 – Acute hemorrhagic conjunctivitis; Type 71 – Meningitis; Type 72 – Hepatitis A virus (27 nm) causes Infectious Hepatitis.

6. Rhinovirus – 113 serotypes causes Common cold.

Laboratory Diagnosis:

Poliovirus: The virus is recovered from pharyngeal secretions in early stages and from stool upto 5 weeks.

a) Direct detection by electron microscopy

b) Human and monkey kidney cell lines – Neutralisation test

c) Serology – CFT.

Coxsackie A and B: Throat secretions or stool.

a) Intracerebrally into suckling mica.

b) Coxsackie B in monkey kidney cells.

ECHO virus : From throat secretions stools or CSF.

Growth in monkey kidney cells detected by neutralisation test.

Hepatitis A virus:

a) From stool by IF

b) Specific IgM antibody appears early; Anti HAV IgG is detected during convalescence

c) Serum bilirubin – 5-20mg/dl

d) Sharp rise of ALT within short duration.

Rhinovirus:

a) Nasopharyngeal secretions in monkey or human tissue cuttures at 33°C in roller drums

b) Intranasally in chimpanzees.

B. Orthomyxovirus

Configuration, Envelop, Symmetry: SS RNA in 8 pieces, enveloped, helical. Hemagglutinin (H1 to H3 – triangular) and Neuraminidase (N1 to N3 – mushroom shaped) are peplomers which are responsible for antigenic drift and shift.

Size: 80-120 nm

Diseases Produced: *Influenza* viruses types A, B, C – influenza, pneumonia, bronchitis.

Laboratory Diagnosis:

1. Nasal and throat washings in amniotic cavity of chick embryo – hemagglutination by guineapig RBCs for influenza A and B, fowl RBCs for influenza C
2. Kidney cell cultures – Hemadsorption
3. Antigens by IF, ELISA, RIA
4. Serology – Hemagglutination Inhibition (HI) test, CFT

C. Paramyxovirus

Configuration, Envelop, Symmetry: SS, enveloped, helical, H and N peplomers and hemolysin in all, except in RSV.

Size: 150-300 nm

Diseases Produced:

1. Parainfluenza viruses Types 1-4 – Acute respiratory illness, croup (laryngotracheobronchitis)
2. Measles virus – Measles
3. Mumps virus – Mumps
4. Respiratory Syncitial virus – Bronchiolitis in infants (60%), pneumonitis (30%)

Laboratory Diagnosis:

Parainfluenza virus

a) Mouth washings, throat swabs, nasopharyngeal swabs is inoculated in monkey kidney cells – growth is detected by hemadsorption
b) Antigen detection by IF or ELISA
c) Amniotic cavity of hen's egg – hemagglutination
d) Antibody detection – CFT, ELISA

Measles virus

a) Intranuclear cowdry A inclusion bodies
b) Multinucleated giant fibroblasts, monkey kidney cells
c) Specific IgM Antibody by HI, CFT and Neutralisation test

Mumps virus

a) Throat and salivary secretions – IF
b) Growth detected in monkey kidney and HEp-2 cells – by hemadsorption
c) CFT – rise in antibody titre

Respiratory Syncitial virus

a) Nasal and pharyngeal secretions – IF
b) In HeLa, HEp-2 or monkey cells and syncitia
c) CFT and Neutralisation test – rising antibody titre

D. Toga virus

Configuration, Envelop, Symmetry: SS, enveloped, icosahendral

Size : 50-70 nm

Diseases Produced:

1. Alpha viruses –
   a) Mosquito-borne – Western Equine Encephalitis (WEE), Eastern Equine Encephalitis (EEE), Venezuelan Equine Encephalitis (VZE)
   b) Chikungunya fever (hemorrhagic)
   c) Sindbis
2. Rubivirus – Rubella
   a) Postnatal (fever, macular rash, transient arthralgia)
   b) Congenital (50-70%, if mother is infected in first trimester). Triad of cataract, deafness and patent ductus arteriosus (PDA)

Laboratory Diagnosis of Rubella:

a) Postnatal rubella – From blood or throat swabs in Vero cell line or rabbit kidney – detected by interference
b) Specific IgG or IgM by ELISA
c) Congenital rubella – Rubell a specific IgM detection from pharyngeal secretions or urine

E. Flavi virus

Configuration, Envelop, Symmetry: SS, enveloped, icosahendral

Size: 50-70 nm

Diseases Produced:

1. Flavi virus (mosquito-borne) – Japanese B encephalitis, Dengue fever (hemorrhagic), Yellow fever (hemorrhagic)
2. Flavi virus (tick-borne) – Tick-borne encephalitis, Kyasanur forest disease (hemorrhagic).

Laboratory Diagnosis:

a) Specific IgG or IgM ELISA
b) Dot Blot ELISA

F. Buniya virus

Configuration, Envelop, Symmetry: SS, enveloped, helical

Size: 90-100 nm

Diseases Produced:

1. Mosquito-borne – Buniya virus (California encephalitis), Phlebo virus (Sandfly fever)
2. Tick-borne – Nairo virus (Nairobi sheep disease)

G. Arena virus

Configuration, Envelop, Symmetry : SS, enveloped, helical

Size: 50-300 nm

Diseases Produced:

1. Lymphocytic choriomeningitis (LCM) virus – Aseptic meningitis, pneumonia
2. Lassa virus – Hemorrhagic fever.

Laboratory Diagnosis:

Under electron microscope – grainy (sandy) appearance.

H. Filo virus

Configuration, Envelop, Symmetry: SS, enveloped, helical

Size: 800-1500 x 80 nm

Diseases Produced:

1. Marburg virus – Bronchiolitis
2. Ebola virus – Ebola fever.

I. Rhabdo virus

Configuration, Envelop, Symmetry: SS, enveloped, helical, bullet-shaped

Size: 75 nm

Diseases Produced:

1. Rabies virus – Rabies (hydrophobia); 25,000-30,000 deaths annually in India
2. Vesiculo virus – Vesicular stomatitis
3. Chandipura virus

Laboratory Diagnosis of Rabies:

a) Corneal smears or nuchal skin biopsies – antigen detection by Fluorescent Antibody Technique (FTA) after staining with specific rabies FITC conjugate

b) Immunoperoxidase staining technique

c) Virus isolation from saliva and CSF by electron microscopy

d) Rapid Rabies Enzyme Immunodiagnosis (RREID)

e) Postmortem diagnosis in humans:

   i) Intracerebral in suckling mice, demonstration of **negri bodies** (3-27 μm) in brain

   ii) IF of impression smears of brain FAT samples from hippocampus and brain

f) Infection in cell culture using Neuro 2a cell line

g) For demonstration of virus neutralising antibodies in CSF and serum (after 7-10 days) – Neutralisation test using mice (MNT) or incell culture – Rapid Fluorescent Focus Inhibition Test (RFFIT); ELISA (limited sensitivity); LAT; CIEP

h) Viral nucleic acid detection in saliva by nested PCR

i) RT-PCR of CSF and saliva.

J. Reo virus

Configuration, Envelop, Symmetry: DS, non-enveloped, icosahedral

Size: 60-80 nm

Diseases Produced:

1. Reovirus types 1, 2, 3 – No disease known
2. Rotavirus – Infantile diarrhoea
3. Orbivirus – Veterinary disease

Laboratory Diagnosis of Rotavirus:

a) Detection in stool by immune electron microscopy, ELISA or immunodiffusion

b) Serology – LAT, ELISA, CFT

K. Corona virus

Configuration, Envelop, Symmetry: SS, enveloped, helical, club-shaped peplomers (like crown)

Size: 80-160 nm

Diseases Produced:

1. a) Human Corona virus – Common cold

   b) Severe Acute Respiratory Syndrome (SARS) caused by SARS - CoV

2. Murine virus – Respiratory infection

3. Avian virus – Fowl leukemia

Laboratory Diagnosis:

Human Corona virus

a) From respiratory secretions – Viral antigen by ELISA

b) Culture in human embryonic trachea

c) Serology – CFT, ELISA, HAT

SARS - CoV

a) Detection of antibody to SARS - CoV by ELISA

b) Detection of SARS - CoV RNA by RT-PCR

c) Isolation of SARS - CoV in Vero cells

L. Calici virus

Configuration, Envelop, Symmetry: SS, non-enveloped, icosahedral

Size: 35-39 nm

Disease Produced: Causes diarrhoea in children

M. Retro virus

Configuration, Envelop, Symmetry: SS, enveloped, helical, reverse transcriptase (DNA polymerase).

VIRAL HEMORRHAGIC FEVERS

Virus	Virus group	Disease	Animal of origin	Lethality	Geographic distribution
Dengue virus	Flavivirus	Dengue hemorrhagic fever	Monkeys, man	±	India, Far East, Australia (Type 2 virus commonest)
LCM virus	Arenavirus	LCM (meningitis, pneumonia)	House mouse, hamster	-	Worldwide
Lassa fever virus	Arenavirus	Lassa fever	African bush rat	+	West Africa
Machupo virus	Arenavirus	Bolivian hemorrhagic fever	Field mouse	+	N.E. Bolivia
Junin virus	Arenavirus	Argentinian hemorrhagic fever	Field mouse, rodents	+	Argentina
Hantaan virus	Buniyavirus	Hemorrhagic fever, fever with renal syndrome (Korean hemorrhagic fever) Severe pulmonary syndrome Crimean Congo hemorrhagic fever (CCHF)	Mice, rates Cattle, sheep, goats, hares, birds	+	Far East, Scandivavia, E. Europe S.W. Asia Africa, W. Asia, Central Europe
Marburg virus	Filovirus	Marburg disease	Unknown	++	Africa (laboratory infection in Marburg)
Ebola virus	Filovirus	Ebola disease	Unknown	++	Africa (Sudan, Zaire)
KFD virus	Flavivirus	Kyasanur forest disease	Forest birds, animals	±	India (Karnataka)
Yellow fever	Flavivirus	Yellow fever	Wild monkies, man	±	Africa, Latin America
Chikungunya	Togavirus	Chikungunya fever	Not known, monkeys	±	India, Asia, Africa.

Signs and symptoms: Fever of insiduous onset, headache, significant myalgia and malaise, followed by shock and hemorrhage in severe cases. Fever may be as high as 41°C. Recovery is slow. Overall case fatality rate 2-88% in hospitalised paients.

Laboratory diagnosis:

1. Isolation of virus from blood, urine and throat washings
2. IgM antibody after first week of illness by ELISA
3. 4-fold rise in IgG antibody between acute and convalescent phase of the disease
4. Antigen detection by ELISA or PCR.
5. AST > 150 IU, proteinuria, etc.

Management:

1. Supportive treatment - fluid and electrolyte balance, treatment for shock and blood loss.
2. Mechanical ventilation, dialysis, neurological intensive care.
3. For Lassa fever - Ribavirin 30mg/kg I.V. loading dose, then 16mg/kg IV every 6 hours for 4 days, then 8 mg/kg IV every 8 hours for 6 days.

   Also against some Bunyaviridae in-vitro and in patients with CCHF.

Chemoprophylaxis:

1. In high-risk contacts of Lassa fever, ribavirin 500 mg orally every 6 hours for 7 days, also in CCHF contacts.

2. Isolation of the patient.
3. Notification to local State Health Departments.
4. Specimens handled with precautions as that followed for samples infected with HIV.
5. No vaccine is yet available.

Dengue Fever (DF) and Dengue Hemorrhagic Fever (DHF)

Mosquito borne RNA virus 40-50 nm, spherical, belonging to Flavivirus group.

Globally, 50-100 million cases of Dengue Fever (DF) and severeal thousand cases of Dengue Hemorrhagic Fever (DHF) per year.

2500 million people (2/5th of wold's population are at risk). Average case fatality rate of DHF is 5% and 5,00,000 cases reguire hospitalisation each year.

100-200 suspected cases are introduced into U.S. each ear by travellers.

Vector – *Aedes aegypti* and *Aedes albopictus* mosquitoes.

Mortality is 44% in DHF and Dengue Shock Syndrome (DSS)

No cross-immunity from each serotype.

A person can theoretically experience all four types of dengue infections.

Between 1780-1940, large epidemics occurred.

Following World War II, there was global pandemic.

There are 4 types or dengue viruses : Types 1, 2, 3, 4.

Major epidemics:

- 1953 in Philippines Dengue 2, 3, 4
- 1958 in Thailand Dengue 1
- 1963 in India at Kolkata
- 1996 in New Delhi – 1,20,000 to 1,60,000 cases, 4,000 hospital admissions and 20% developed DSS.

Positive tourniquet test – > 20 patches in a 2.5 cm² patch.

Signs and symptoms: Incubation period 3-4 days (average 4-7 days). Abrupt onset. High fever (40-41°C), 'saddleback', retrobulbar pain, muscle and joint pains, rash. DHF – a fatal complication of DF with high fever, haemorrhagic phenomera, hepatomegaly and circulatory failure.

Pathophysiology of Dengue Hemorrhagic Fever

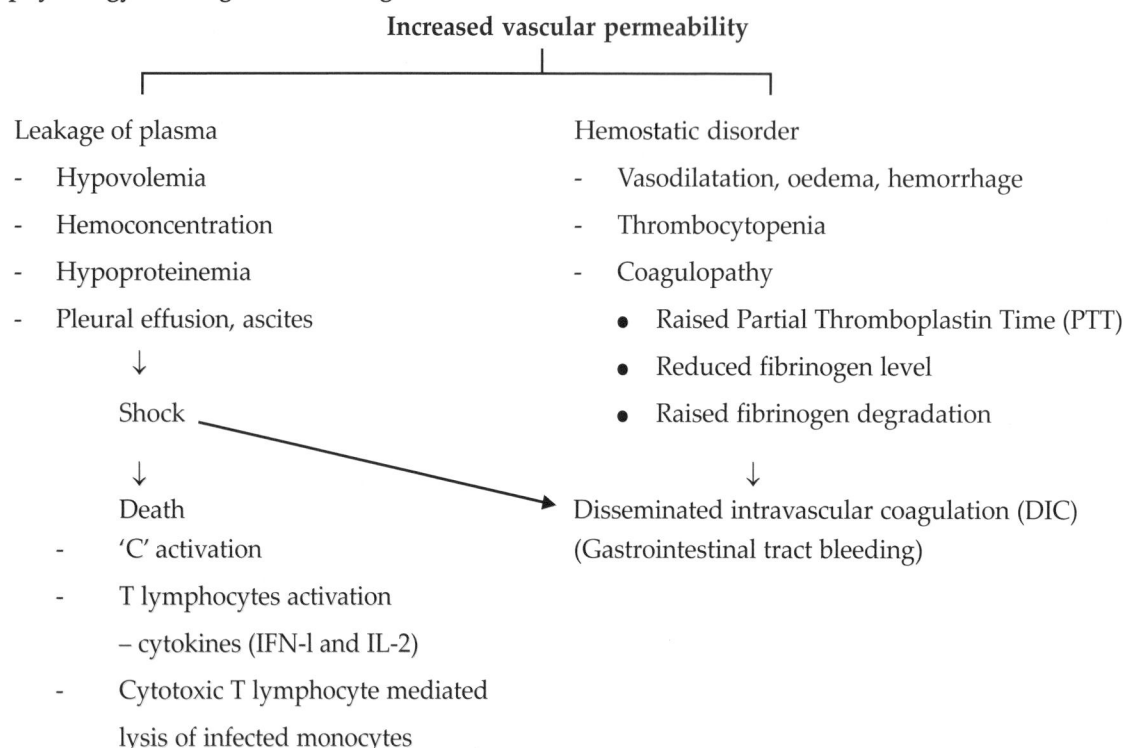

Increased vascular permeability

Leakage of plasma

- Hypovolemia
- Hemoconcentration
- Hypoproteinemia
- Pleural effusion, ascites

↓

Shock

↓

Death

- 'C' activation
- T lymphocytes activation
  – cytokines (IFN-l and IL-2)
- Cytotoxic T lymphocyte mediated
  lysis of infected monocytes

Hemostatic disorder

- Vasodilatation, oedema, hemorrhage
- Thrombocytopenia
- Coagulopathy
  - Raised Partial Thromboplastin Time (PTT)
  - Reduced fibrinogen level
  - Raised fibrinogen degradation

↓

Disseminated intravascular coagulation (DIC)
(Gastrointestinal tract bleeding)

Dengue Hemorrhagic Fever (DHF). A fatal complication of DF.

- Primarily disease of children < 15 years age. Develop around day 3-7 of illness.
- Acute febrile illness with facial flush, anorexia, headache, pains in muscles and joints (break-bone fever), hepatomegaly, generalised abdominal pain.
- Hemorrhagic manifestations (petechiae, epistaxis, bleeding gums, gastrointestinal hemorrhage, hematuria, menorrhagia), pleural effusion and ascites.

In severe DHF – *Circulatory failure:

- Cool and clammy skin, blotchy and congested
- Rapid and weak pulse
- Circumoral cyanosis.

* Profound shock with metabolic acidosis and severe bleeding. Death in 24-48 hours.

Laboratory diagnosis of Dengue

Caused by Dengue virus; Family: Flaviviridae; Genus: Flavivirus; Species: Dengue virus.

- Widely distributed in tropics and subtropics.

- Serotypes 1–4.

- Symptoms: Fever of sudden onset, retrobulbar pain, pain in back and limbs (break-bone fever), lymphadenopathy, maculopapular rash. Fever biphasic, lasts for 5-7 days.

- **Classic symptoms:** high fever, a petechial rash with thrombocytopenia and relative leucopenia and headache.

- **Complications: Dengue hemorrhagic fever (DHF)** and/or shock – probably a hypersensitive reaction to sequential viral infection in persons already sensitised by prior exposure to other serotypes.

WHO definition of Dengue Hemorrhagic Fever (DHF)

- Fever

- Haemorrhagic tendency [positive tourniquet test (> than 20 petechiae per square inch), spontaneous bruising, bleeding from mucosa, gingiva, injection sites, vomiting blood or bloody diarrhea].

- Thrombocytopenia (< 100,000 platelets per mm^3).

- Evidence of plasma leakage (rise in hematocrit level > than 20%).

- **Dengue shock syndrome (DSS)** is usually accompanied by hemorrhagic signs, is much more serious and results from increased vascular permeability leading to shock.

Virus isolation

- Intracerebral inoculation of 1-3 day old baby mice
- Mammalian cell cultures (LLC-MK2 cells)
- Intrathoracic inoculation of adult mosquito
- Mosquito cell cultures (most sensitive) - C6/36 clone of *A. albopictus* cells (most commonly used)
    - Identified by IFA with serotype specific monoclonal antibodies.
    - Can be performed with infected cell cultures, mosquito brain / tissue quashes or formalin fixed tissues.

Advantages

- Simple, reliable rapid
- Detects multiple viruses with more than one serotype

Disadvantages

- Time consuming
- BSL4 required as extremely dangerous

Electron Microscopy

Sample: Serum/buffy coat/tissue. Virus suspension is adsorbed on to a copper grid and an electron dense stain (phosphotungstic acid or uranyl acetate).

Advantage: Rapid

Disadvantages:

- Difficult for large number of samples;
- Requires expertise and training;
- Insensitive (10^3–10^6 particles/ml should be present)

Antigen detection test: NS1 antigen rapid test

Principle: Immunochromatographic test.

Samples: Human serum, plasma or whole blood.

- Sensitivity: 81.0 to 84.8%; Specificity: 100%

- Sensitivities of 80.5% and 48.0% for primary and secondary infections. DENV-1 was the most sensitive to be detected (97.5%), followed by DENV-3 (75%), DENV-4 (67%), and DENV-2 (56%)

Antibody detection: Rapid tests

Principle: Immunochromatography

- Panbio Dengue duo (IgM & IgG) - Sensitivity and Specificity > 90%
- SD Bioline - Sensitivity - 94%; Specificity - 89%
- Dengucheck (IgM & IgG) – Zephyr Biomedicals
- Dengue NS1 Ag + Ab Combo SD Bioline - Sensitivity: 92.8%; Specificity: 98.4%
- Bio-Rad Ag strip rapid test

Types of ELISA

- **IgM antibody capture ELISA** for acute phase sera – most widely used.
    - IgM antibodies persist for 2-3 months
    - Can identify serotype
    - 10% false negative
    - 1.7% false positive reactions

- **Indirect IgG ELISA**

  Advantage – Can differentiate primary and secondary stages.

  Disadvantage – Cross reactivity with other Flaviviruses.

- **IgG Sandwich ELISA and IFAT** using paired sera

  Disadvantage – Antigen cross reactivity with other flaviviruses.

Other serological tests

- Neutralisation tests
- Haemagglutination inhibition test
- Complement fixation test

Molecular methods

Real time PCR: Detects and differentiates all 4 serotypes.

Advantages: Easy; Non infective samples can be used; High sensitivity.

Disadvantages:

- False negative PCR due to PCR inhibiting products of tissue degradation
- Exceedingly low positive predictive value

CHIKUNGUNYA

- Chikungunya is a relatively rare form of mosquito-transmitted viral fever caused by an Alphavirus and usually associated with acute epidemic polyarthralgia.
- Spread by mosquito bites from *Aedes species.*
- The disease was first described by Marion Robinson and W.H.R. Lumsden in 1955, following an outbreak on the Makonde Plateau, along the border between Tanzaniya and Mozambique, in 1952.
- The name is derived from the Makonde word "kungunyala" meaning "that which bends up" (stooped posture developed as a result of the arthritic symptoms).
- Chikungunya is believed to have originated in Africa, maintained in 'sylvatic cycle' involving wild primates and forest dwelling mosquitoes such as *Aedes furcifer, A. luteocephalus, A. africanus or A. taylori.*
- It was subsequently introduced in Asia, where it is transmitted from human to human mainly by *Aedes aegypti* and, to a lesser extent by *Aedes albopictus* through an urban transmission cycle.

Classification

- Group: Group IV (+ ss RNA)
- Family: *Togaviridae*
- Genus: *Alphavirus*
- Species: *Chikungunya virus*

Epidemiology

- Chikungunya was first described in Tanzania, Africa in 1952.
- First Asian outbreak in 1958 in Bangkok, Thailand.
- The first outbreak in India was in 1963 in Kolkata.
- An outbreak of chikungunya was also discovered in Port Klang in Malaysia in 1999 affecting 27 people.
- In 2005-2007: 200 deaths on Réunion island, France. The European Network for Diagnostics of "Imported" Viral Diseases claims new phylogenetic variants of the virus been identified which are fatal.
- A widespread outbreak in India - primarily in Tamil Nadu, Karnataka, Kerala, and Andhra Pradesh. In August 2006 - thousands of cases detected in Rajasthan, Gujarat and Madhya Pradesh.

- In October 2006 more than a dozen cases reported in Pakistan.
- In December 2006, an outbreak of 3,500 confirmed cases in Maldives and over 60,000 cases in Sri Lanka, with over 80 deaths.
- A recent outbreak during June 2007 in districts of South Kerala, India claimed more than 50 lives. 7000 officially confirmed Chikungunya patients in these areas.
- In August/September 2007 some 160 people were infected in Italy's northern Ravenna region, resulting in one fatality.

WHO Reports of Chikungunya in India

- From February 2006 to 10 October 2006, the WHO Regional Office for South-East Asia has reported 151 districts in 8 states/provinces of India affected by chikungunya fever.
- More than 1.25 million suspected cases have been reported from the country, of which 752,245 were from Karnataka and 258,998 from Maharashtra provinces.
- In some areas reported attack rates have reached 45%.
- Most of the current strains of Chikungunya virus circulating in India are believed to be of East/Central African genotype. CHIKV isolates from earlier outbreaks had been of Asian genotype.
- The current Indian outbreak followed the explosive spread of the virus in La Réunion and other Indian Ocean islands. Phylogenic analysis of Chikungunya virus samples from La Réunion showed they were related but distinct from its East African ancestors.

Transmission

- The virus is transmitted by culicine mosquitoes, commonly *A. aegypti, A. albopictus* and *A. polynesiensis.*
- *Culex* and *Anopheles stephensi* have also been reported for the transmission in some cases.
- The common reservoirs for chikungunya virus are monkeys and other vertebrates.
- In the current outbreak suspected reservoirs were macaque monkeys, lemurs and bald mouse. In the epidemic period, men also act as reservoir.
- The role of cattles and rodents has also been reported in the transmission of the virus.
- Chikungunya virus usually shows a periodicity with occurrence of disease in the community with the silence interval of 3-4 years. The periodicity is probably due to their cycle in monkeys.
- Mother to child transmission has also been reported in recent studies in the Reunion Island.

Structure

- Lipid-enveloped; 50-70 nm in diameter.
- 11-12 kb genome; ++RNA.
- Structural proteins: Two glycoproteins – E1 and E2, inserted in lipid membrane.

  E1: attachment to cell surface, generates neutralizing antibody.

  E2: hemaglutinin activity.
- Non structural proteins: Protease and RNA – dependent RNA polymerase.

Phylogenetic Classification

- Phylogenetic analyses based on partial E1 sequences from African and Asian isolates revealed the existence of three distinct Chikungunya virus phylogroups:
  - First containing all isolates from West Africa,
  - Second containing isolates from Asia, and
  - Third corresponding to East, Central and South African isolates.
- Phylogenetically, CHIKV is more closely related to O' nyong-nyong virus.

Virus susceptibility

The virus is inactivated by acid pH, heat, lipid solvent, detergents, bleach, phenol, 70% alcohol and formaldehyde.

Vector (*Aedes species*)

- It is a diurnal vector with peak of activities at the end of the day and is a day biter.
- A body covered with scales decorated with white or silver plated spots is characteristic of the mosquito.
- *A. albopictus* is more active outdoors, while *A. aegypti* feeds and rests indoors.

Pathogenesis

- Detailed studies are not available.
- Once after inoculation, primary viral multiplication occurs in lymphoid and myeloid cells.
- The arthropod vectors acquire the virus by sucking blood during this period.
- The virus then spreads to the targeted organs and immune system starts functioning at this stage leading to the activation of both humoral and cellular immunity.

Symptoms

- Incubation period = 1 to 12 days (commonly 4 to 7 days).
- Fever upto 39°C, (102.2°F). Typically lasts for two days and then comes down abruptly. May reappear: "saddle back fever".
- A petechial or maculopapular rash usually involving limbs and trunk.

- Arthralgia or arthritis affecting multiple joints (debilitating).
- The symptoms could also include headache, conjunctival injection, and slight photophobia.
- Ocular inflammation from Chikungunya may present as iridocyclitis, and have retinal lesions as well.
- Other dermatological manifestations (e.g. erythema, freckle-like pigmentation, lichenoid eruption and hyperpigmentation in photo distributed areas, vesiculobullous lesions in infants, subungual hemorrhage, urticaria) observed in a recent outbreak of Chikungunya fever in Southern India, Western India (Surat) and Eastern India (Puri).
- Vomiting, haemetemesis and epistaxis
- Pedal oedema
- Recovery of patients depends on their age - Younger patients recover within 5 to 15 days; middle-agers within 1 to 2.5 months. Recovery is longer for the elderly. The severity of the disease as well as its duration is less in younger patients and pregnant women. No untoward effects of pregnancy are noticed following the infection.

Complications

- Infection can worsen pre-existing conditions.
- Infection aggravates conditions of patients who have heart, kidney, liver and respiratory problems as well as those with diabetes.
- Meningoencephalitis has also been seen in connection with chikungunya virus infection, according to physicians.

Laboratory Diagnosis

A) **Virus isolation**

- It is the most definitive test.
- Blood is collected during the first week of illness - transported on ice in heparinized tube.
- The virus produces cytopathic effects in BHK-21, HeLa and Vero cell lines. The cytopathic effects are confirmed by virus specific antiserum and results can take between 1-2 weeks.
- *A. albopictus* cell line has been used successfully for isolation of the virus.
- Virus isolation carried in BSL-3 laboratories to reduce the risk of viral transmission.

B) **Serological diagnosis**

- Blood specimen transport: Immediate at 4°C and not frozen. Delay in transport – blood specimen separated into sera and frozen. Tests are as follows:
  - Neutralization test
  - Hemagglutination/Hemagglutination inhibition test
  - Plaque reduction neutralization test
  - IgM capture ELISA

- Serologic diagnosis made by demonstration of four-fold rise in antibody titre in acute and convalescent sera (14 days apart) or demonstrating IgM antibodies specific for CHIK virus.

- Cross-reaction in NT/HAI test/ELISA occurs with other alphavirus antibodies such as O'nyong-nyong and Semliki Forest; however, the latter viruses are relatively rare in South East Asia but if further confirmation is required it can be done by plaque reduction neutralization test.

C) RT-PCR

- Recently, reverse transcriptase, RT-PCR technique for diagnosing chikungunya virus developed, using nested primer pairs amplifying specific components of three structural gene regions – Capsid (C), Envelope E2 and part of Envelope E1.

- Specimen for PCR - heparinized whole blood.

- PCR results for E1 and C genome either singly or together constitute a positive result for Chikungunya virus.

Existing laboratory network for diagnosing Chikungunya in South East Asia

- National Institute of Virology, Pune, India
- NIHRD, NAMRU-2, Indonesia
- Department of Medical Research, Indonesia
- NA, Myanmar
- Medical Research Institute, Sri Lanka
- NIH, Bangkok
- AFRIMS, Thailand

Treatment

- There is no specific treatment for Chikungunya.

- **Chloroquine** (250 mg/day): a possible treatment for the symptoms associated and as an antiviral agent to combat the Chikungunya virus.

- The fact sheet on Chikungunya advises against using Aspirin, Ibuprofen, Naproxen and other NSAIDs which are usually recommended for arthritic pain and fever.

- Supportive care with rest is preferred during the acute joint symptoms. Movement and mild exercise tend to improve stiffness and morning arthralgia, but heavy exercise may exacerbate rheumatic symptoms.

Preventive measures

- Protection against any contact with the disease-carrying mosquitoes: Insect repellent containing **NNDB, DEET** or **permethrin**, wearing long sleeves and trousers, and securing screens on windows and doors.

- The U.S. military vaccine "TSI-GSD-218", was developed using an Asian Chikungunya virus strain obtained from a Thai patient in 1962 by passaging in green monkey kidney(GMK) cells followed by formalin inactivation.

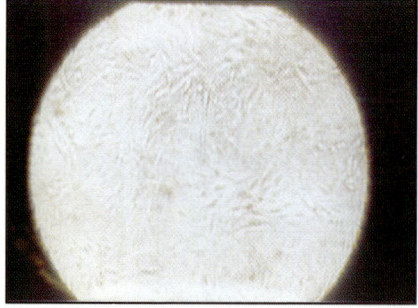 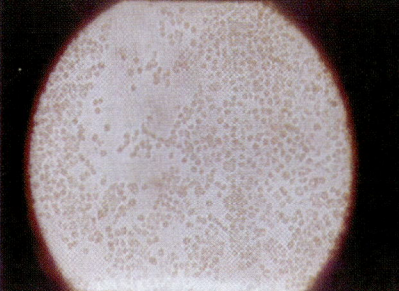 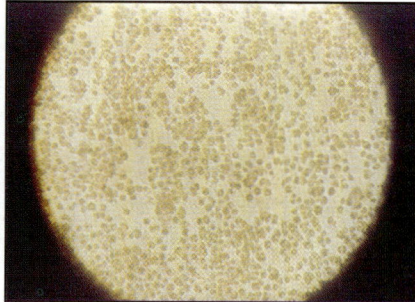

| Normal Vero cell line | Normal HeLa cell line (24 hours) | Cytotoxicity of HeLa cells after 72 hours showing rounding of cells |

REFERENCES

1. Topley & Wilson's, Microbiology and Microbial Infections. Virolgey, Vols. 1 & 2. 10th Ed. Eds. Mahy BWJ, Meulen VT. John Wiley & Sons Ltd., West Sussex, UK. 2009.

2. Ananthanarayan & Paniker's Textbook of Microbiology. 9th Ed. Ed. Kapil A. Universities Press, Hyderabad. 2013.

3. Jawetz E, Melnick JL, Adelberg EA. Review of Medical Microbiology. 11th Ed. Lange Medical Publications, Maruzen Asian Edition, Japan, 1974.

4. Gubler DJ. Dengue and Dengue Hemorrhagic Fever. ClinMicrobiol Rev. 1998; 11: 480-96.PMCID: PMC88892.

5. Dengue and severe Dengue. Updated May 2015. http://www.who.int/mediacentre/factsheets/fs117/en/

6. Karpe AS, Trivedi T, Padwal N, Yeolekar ME. Management of Dengue hemorrhagic fever and Dengue shock syndrome - an ICU experience. J Assoc Phys India 2005; 53: 371-2.

7. Chikungunya.www.wikipedia.com.

8. Lahariya C, Pradhan SK. Emergence of chikungunya in Indian subcontinent after 32 years: A review. J Vect Borne Dis 2006; 43: 151-160.

9. Laboratory diagnosis of chikungunya fevers. WHO. www.who.int.com.

10. Edelman R, Tacket CO, Wasserman SS, Bodison SA, Perry JG, Mangiafico JA. Phase II safety and immunogenecity study of live chikungunya virus vaccineTSI-GSD-218. Am Trop Med Hyg 2000; 21: 681-85.

HUMAN IMMUNODEFICIENCY VIRUS (HIV) INFECTION AND ACQUIRED IMMUNODEFICIENCY SYNDROME (AIDS)

25

Chapter

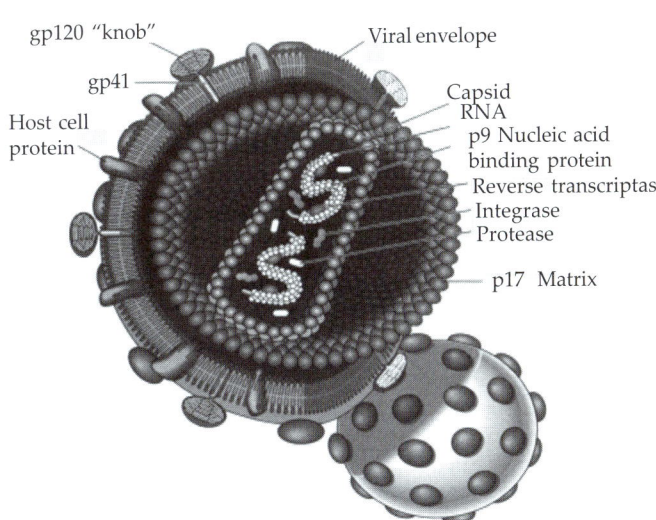

gp120 "knob"
gp41
Host cell protein
Viral envelope
Capsid
RNA
p9 Nucleic acid binding protein
Reverse transcriptase
Integrase
Protease
p17 Matrix

90-120 nm; Enveloped SS-RNA virus

HIV-1 genotypes

- There are 3 HIV-1 genotypes; M (Main), O (Outlayer), and N (New)
- M group comprises of a large number subtypes and recombinant forms
  - Subtypes - (A, A2, B, C, D, F1, F2, G, H, J and K)
  - Recombinant forms - AE, AG, AB, DF, BC, CD
- O and N group subtypes not clearly defined, especially since there are so few N group isolates.
- As yet, different HIV-1 genotypes are not associated with different courses of disease nor response to antiviral therapy.
- However, certain subgroups may be difficult to detect by certain commercial assays.

Subtypes

- 10 subtypes based on env and gag gene
- M:A – J subtypes
- Subtype A – Most prevalent worldwide; Subtype B – America & Europe; Subtype C – India & China.

	HIV-1	HIV-2
Spread	In western and Asian countries	In Africa, Asian and North American population
Infectivity	More	Less
Mode of transmission	Through bloodMother to child	Through sexual routeThrough blood
Incubation period	1-14 years (av. 6 years)	Longer than HIV-1
Prevalence	85%	4%

HIV Transmission

- Unprotected sexual contact with an infected partner
- Exposure of broken skin or wound to infected blood or body fluids
- Transfusion with HIV-infected blood
- Injection with contaminated objects
- Mother to child during pregnancy, birth or breastfeeding.

Type of exposure	% of total	Risk of transmission
Heterosexual intercourse with infected partner without condom use	70-80%	50%
Blood and blood products	3-5%	90%
Perinatal transmission	5-10%	30%
Needle sharing by drug users	5-10%	0.5-1%
Accidental needle-stick injury	< 0.001	0.4%
HIV positive semen, organ and cornea donation	–	Almost 100%

2014 Global Statistics

- 15 million people accessing antiretroviral therapy (March 2015)
- 36.9 million (34.3 million–41.4 million) people globally were living with HIV
- 2 million (1.9 million–2.2 million) people became newly infected with HIV
- 1.2 million (980,000–1.6 million) people died from AIDS-related illnesses

People living with HIV accessing antiretroviral therapy

- As of March 2015, 15 million people living with HIV were accessing antiretroviral therapy, up from 13.6 million in June 2014.
- 41% (38%-46%) of all adults living with HIV were accessing treatment in 2014, up from 23% (21%-24%) in 2010.
- 32% (30%-34%) of all children living with HIV were accessing treatment in 2014, up from 14% (13%-15%) in 2010.
- 73% (68%-79%) of pregnant women living with HIV had access to antiretroviral medicines to prevent transmission of HIV to their babies in 2014; new HIV infections among children were reduced by 58% from 2000 to 2014.

People living with HIV

- In 2014, there were 36.9 million (34.3 million-41.4 million) people living with HIV.
- Since 2000, around 38.1 million people have become infected with HIV and 25.3 million people have died of AIDS-related illnesses.

New HIV infections

- New HIV infections have fallen by 35% since 2000.
- Worldwide, 2 million (1.9 million-2.2 million) people became newly infected with HIV in 2014, down from 3.1 million (3.0 million-3.3 million) in 2000.
- New HIV infections among children have declined by 58% since 2000.
- Worldwide, 220,000 (190,000-260,000) children became newly infected with HIV in 2014, down from 520,000 (470,000-580,000) in 2000.

AIDS-related deaths

- AIDS-related deaths have fallen by 42% since the peak in 2004.
- In 2014, 1.2 million (980,000-1.6 million) people died from AIDS-related causes worldwide compared to 2 million (1.7 million-2.7 million) in 2005.

HIV/tuberculosis

- Tuberculosis-related deaths in people living with HIV have fallen by 33% since 2004.

- Tuberculosis remains the leading cause of death among people living with HIV, accounting for around one in five AIDS-related deaths.
- In 2013, the percentage of identified HIV-positive tuberculosis patients who started or continued on antiretroviral treatment reached 70% (up from 60% in 2012).

Regional Statistics of Asia and the Pacific

- In 2014, there were 5 million (4.5 million–5.6 million) people living with HIV in Asia and the Pacific.
- In 2014, there were an estimated 340,000 (240,000-480,000) new HIV infections in the region.
- New HIV infections declined by 31% between 2000 and 2014 - China, Indonesia and India account for 78% of new HIV infections in the region.
- In Asia and the Pacific, 240,000 (140,000-570,000) people died of AIDS-related causes in 2014.
- Between 2000 and 2014 the number of AIDS-related deaths in the region increased by 11%.
- Treatment coverage is 36% (32%–41%) of all people living with HIV in Asia and the Pacific.
- An estimated 3.2 million adults did not have access to antiretroviral therapy in Asia and the Pacific in 2014.
- Only two countries in Asia and the Pacific, Thailand and Cambodia, have more than 50% of all people living with HIV currently on antiretroviral treatment.
- There were 21,000 (16,000-27,000) new HIV infections among children in Asia and the Pacific in 2014.
- Since 2000, there has been a 27% decline in new HIV infections among children in the region.

Global targets for 2015

- Halve sexual transmission of HIV by 2015
- Reduce transmission of HIV among people who inject drugs
- Ensure that no children are born with HIV by 2015
- Increase access to antiretroviral therapy to get 15 million people on life saving treatment by 2015
- Reduce TB deaths in people living with HIV by 50% by 2015
- Reach a significant level of annual global expenditure (between $22 billion and $24 billion) in low and middle-income countries
- Eliminate gender inequalities and gender-based abuse and violence and increase the capacity of women and girls to protect themselves from HIV
- Eliminate stigma and discrimination against people living with and affected by HIV through promotion of laws and policies that ensure the full realization of all human rights and fundamental freedoms

Eliminate HIV-related restrictions on entry, stay and residence

Eliminate parallel systems for HIV-related services to strengthen integration of the AIDS response in global health and development efforts.

Fast track targets

	By 2020	By 2030
Treatment	90-90-90	95-95-95
New infections among adults	500,000	200,000
Discrimination	ZERO	ZERO

How are indicators different in 2015?

- **Key populations** (Indicators 1.7-1.14 and 2.1-2.5): provide disaggregation by site/administrative area
- **PMTCT** (Indicator 3.1) and ART (Indicator 4.1: since 2014 mid year reporting, countries are asked to report any available subnational data
- **AIDS Spending** (Indicator 6.1): has a refined conceptual framework of the National Funding Matrix, with revised classification of AIDS programmes and a new National Funding Matrix.
- **Intimate Partner Violence** (Indicator 7.1): additional comment box included for data on gender based violence towards women, men and key populations, including people living with HIV, that may be available for their country.
- **Discriminatory attitudes** (Indicator 8.1): When using data from DHS question 'Would you buy fresh vegetables from a shopkeeper or vendor if you knew that this person had the AIDS virus?' to respond to question 1, the numerator should only include "No" responses.
- **External economic support** to the poorest households (Indicator 10.2): has been updated with more information about the method of measurement.
- **Narrative report** is requested
- **National Commitments and Policy Instrument (NCPI)** is not requested.

Highlights of new things in 2013 and 2014 taken to 2015

- As in the last three reporting rounds, survey data not changed since the last reporting round do not need to be re entered (i.e. indicators 1.1, 1.2, 1.3, 1.4, 1.5, 1.6, 1.7, 1.8, 1.9, 1.10, 1.11, 1.12, 1.13, 1.14, 1.22, 2.2, 2.3, 2.4, 2.5, 7.1, 10.1, 10.2).
- Two indicators about male circumcision that were added in 2013 for the 16 countries with high HIV prevalence and low prevalence of male circumcision are still included and can be found in Appendix 5.
- Transgender as a possible disaggregation for sex workers (Indicators 1.7, 1.8, 1.9 and 1.10), introduced in the 2014 reporting round is still available.
- In the 2014 reporting round, the PMTCT indicator (Indicator 3.1) had updated language to clarify the disaggregations and the links to Spectrum.
- The indicator to measure coverage of PMTCT during breastfeeding was added directly after this indicator (labelled Indicator 3.1a)
- The indicator for ART coverage (Indicator 4.1) has the same denominator as in 2014, including all people living with HIV, not only those eligible for treatment. Further, the disaggregation of those newly initiated on ART (in the last 12 months) is still available as in 2014.
- As in the 2014 reporting, the 12 month ART retention indicator (Indicator 4.2) includes possible disaggregations for pregnancy status and breastfeeding status at initiation.
- The change in 2014 reporting remains on the indicator for co management of tuberculosis and HIV treatment (Indicator 5.1) where "adults" was changed to "adults and children" in the numerator and "advanced" deleted from "advanced HIV infection".
- The indicator Discriminatory attitudes towards people living with HIV (Indicator 8.1) is kept under target 8.
- Joint reporting of the Global AIDS Response Progress Reporting indicators and additional health sector indicators from WHO and UNICEF are included.

Countries	Adults & children living with HIV 2014	Children < 15 years estimated to be living with HIV in 2014	Adults & children newly infected with HIV 2014	Adults & children deaths due to AIDS 2014
Sub-Saharan Africa	25.8 m	2.3 m	1.4 m	790,000
Asia & Pacific	5.0 m	200,000	340,000	240,000
Latin America	1.7 m	33,000	87,000	41,000
E. Europe & Central Asia	1.5 m	17,000	140,000	62,000
Western & Central Europe & North America	2.4 m	3,300	85,000	26,000

Indian Scenario

People living with HIV/AIDS	2.39 million
Adults (15 years or above) HIV prevalence	0.31%
HIV prevalence in general population	0.36%
HIV prevalence in MSMs	5.69%
HIV prevalence in FSWs	5.38%
HIV prevalence in people attending STD clinics	3.6%
HIV prevalence among women attending antenatal clinics	0.35%

Risk of HIV transmission

- Blood is the single most important source of HIV, HBV and other blood borne pathogens in the occupational setting
- Other potentially infections body fluids (especially those containing visible blood)
  - Semen and vaginal secretions
  - Cerebrospinal fluid
  - Synovial fluid
  - Pleural fluid
  - Peritoneal fluid
  - Pericardial fluid
  - Amniotic fluid

The risk of HIV transmission is extremely low or negligible with other body fluids/secretions/excretions unless these contain visible blood. These include:

- Feces
- Nasal secretions
- Sputum
- Sweat
- Tears
- Urine
- Vomitus

Sterilization

- Autoclaving at 121°C, 15 lbs pressure for 20 minutes.
- Dry heat 170°C for 1 hr.
- Boiling for 20-30 minutes Chemical disinfection
- Sodium hypochlorite: 5gm/litre. (0.5 to 1% ordinarily, 5-10% for high organic matter content e.g. discarding tissues, etc.)
- Calcium hypochlorite: 1.4 gm/litre.
- Chloramine: 20gm/litre (Available chlorine 0.1%)
- Ethanol: 70%
- Formalin: 3-4%
- Glutaraldehyde: 2% for 30 minutes
- Povidone iodine (PVI)

HIV Replication

- Predominant receptor is the CD4 molecule present on T lymphocytes and macrophages.
- The first step of infection is the binding of gp120 to the CD4 receptor (CXCR4 and CCR5) of the cell, which is followed by penetration and uncoating.
- Once the gp41/36 of the virus fuses with the host cell membrane, the capsid is uncoated and a ribonucleoprotein complex capable of reverse transcription is formed.
- The RNA genome is then reverse transcribed into a DNA provirus which is integrated into the cell genome.
- This is followed by the synthesis and assembly of virus, followed by maturation of virus progeny by budding.
- gp120/140 binds to host cell receptors → Reverse transcription
- Proviral DNAsynthesis
- Integration with host cell DNA
- Viral proteins synthesis
- Synthesis of core-proteins
- Virus assembly
- Maturation by budding.

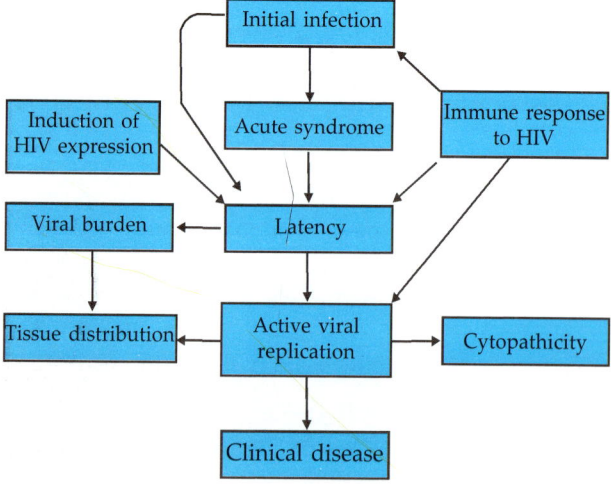

WHO HIV/AIDS Classification System

Stage I	Stage II
Asymptomatic	Minor Symptoms
Stage III	**Stage IV**
Moderate	AIDS

Symptoms

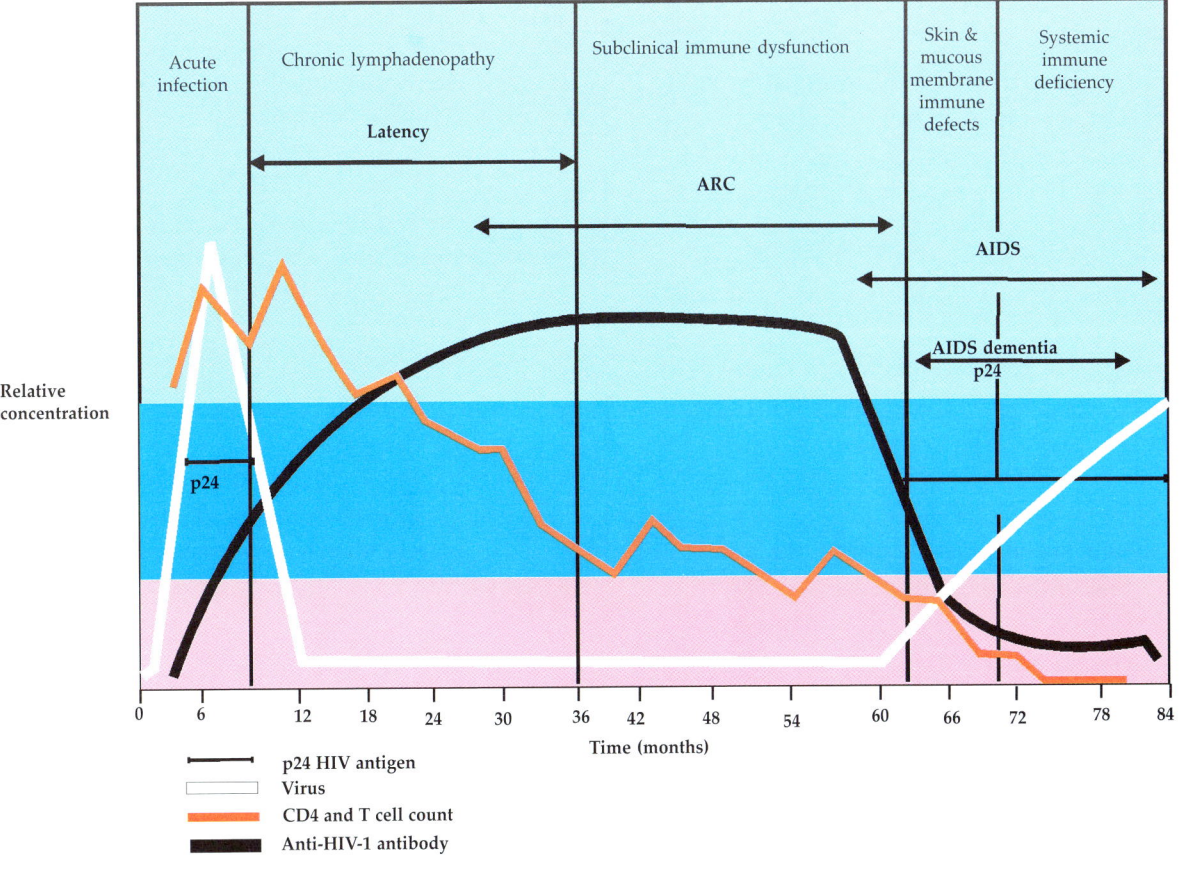

p24 HIV antigen
Virus
CD4 and T cell count
Anti-HIV-1 antibody

Serologic Profile of HIV-1 Infection

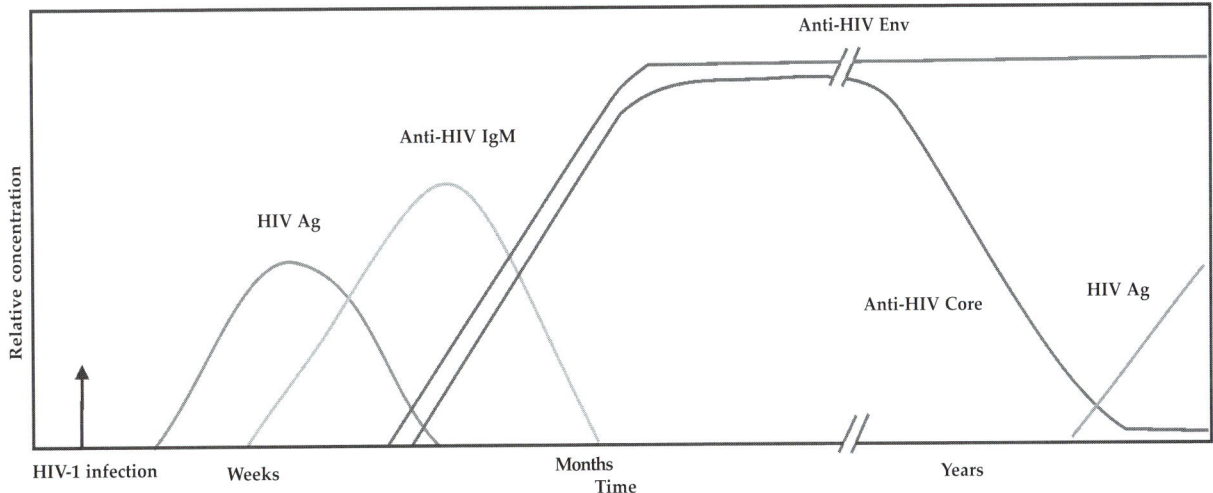

High risk behavior group

- HIV positive patients
- Commercial sex workers (CSWs)
- Professional blood donors
- Individuals sharing infected needles (IV drug users), razors, or through contaminated instruments (ear piercing, acupuncture, tattooing, etc.)

- STD patients
- Spouses of HIV infected persons
- Frequent recipients of blood and blood products
- Men having sex with CSWs
- Men having sex with males (MSM)
- Truck drivers

CDC Classification of Stages of HIV Disease

I. Acute Infection (Seroconversion)

II. Asymptomatic

III. Persistent Generalised Lymphadenopathy

IV. AIDS related complex (ARC) and AIDS

A. Constitutional disease, any one or more of:
  1. Fever persisting for > 14 days without infectious cause.
  2. Weight loss > 10 lbs. or 10% of body weight.
  3. Diarrhoea persisting > 30 days without definable cause.
B. Neurological disease as a direct consequence of HIV infection, any one or more of: Dementia, myelopathy or peripheral neuropathy
C. Secondary infectious disease.
  C1 - AIDS defining infection*
  C2 - Symptomatic or invasive disease due to any one:
    Clinically overt oral hairy leukoplakia
    Multidermal herpes zoster
    Recurrent salmonella bacteremia
    Nocardiosis
    Tuberculosis
    Persistent or recurrent oral candidiasis
D. One or more secondary cancers moderately indicative of a defect in cell mediated immunity.
E. Patients with other clinical findings or diseases not classified above which may be attributable to HIV infection or indicative of a defect in cell mediated immunity.

*AIDS - Defining Conditions

Infections

Pneumocystis jirovecii Pneumonia

Disseminated Atypical Mycobacterial Disease

Cerebral Toxoplasmosis

Oesophageal Candidiasis (Diarrhoea > 1 month)

Extrapulmonary Cryptococcosis

Cytomegalovirus (CMV) Retinitis

Herpes Simplex Virus (HSV)

Mucocutaneous Ulcer lasting > 1 month or Visceral Disease of any duration

Disseminated Coccidioidomycosis

Disseminated Histoplasmosis

Tuberculosis involving at least one Extrapulmonary site

Recurrent Non-typhoid Septicaemia

Extra Intestinal Strongyloidosis

Progressive Multifocal Leucoencephalopathy

Neoplasms

Kaposi's Sarcoma

Primary Lymphoma of the Brain

Other Non-hodgkins Lymphoma of B cell or Unknown Immunologic Phenotype

---

Opportumistic infections and tumors in AIDS

Viruses	Disseminated Cytomegalovirus (CMV) infection (including lungs, retina, brain)
	Herpes simplex virus (HSV) infection (lungs, gastrointestinal tract, CNS, skin)
	JC Papovavirus in brain - Progressive multifocal leucoencephalopathy (PML)
	Epstein-Barr virus (EBV) - hairy leukoplapia.
Bacteria	Mycobacteria (*M. tuberculosis, M. avium, M. kansasii* (disseminated extrapulmonary)
	Salmonella (recurrent, disseminated) septicemia
	Campylobacter
	Nocardia
	Legionella
Protozoa	*Toxoplasma gondii* (disseminated including CNS)
	Cryptosporidium (chronic diarrhoea)
	Isospora (diarrhoea persisting more than a month)
	Giardia lamblia
	Strongyloides stercoralis
Fungi	*Pneumocystis jirovecii* (pneumonia)
	Candida albicans (oesophagitis, lung infection)
	Cryptococcus neoformans (CNS)
	Histoplasma, Aspergillus, Coccidioides (disseminated, extrapulmonary)
Tumors	Kaposi's sarcoma (associated with Human Herpes Virus 8 (HHV8), 300 times as frequent in AIDS as in other immuno-deficiencies.
	Lymphomas - Hodgkin's and non-Hodgkin's types (B cell type in brain)
	(some are EBV related)
Others	Wasting disease (cause unknown)
	HIV encephalopathy (AIDS dementia complex)

Symptoms of AIDS

Major	Recent unexplained weight loss (> 10 kg/month)
	Prolonged fever and diarrhoea (> 1 month)
	Enlarged lymph nodes (> 1 month)
	Night sweats, weakness, fatigue
Minor	Generalised skin infections (e.g. herpes infection severe or recurrent)
	Persistent cough (> 1 month)
	Genital or anal ulcers (> 1 month)
	Fungal infections of mouth and throat (Candidial oesophagitis, oral thrush)
	Transient signs of meningitis, encephalitis or transient neuropathy

At least 2 major and 2 minor criteria should be present in order to label a case as AIDS.

CD4 + lymphocyte count in Opportunistic Infections

CD4 count	Opportunistic infections
400	Herpes zoster
300	Tuberculosis
	Oral candidiasis
200	*P. jirovecii*
	Oesophageal candidiasis
	Mucocutaneous herpes
100	Toxoplasmosis, Cryptococcosis, Coccidioidomycosis, *M. avium* complex, Cytomegalovirus infection
50	Cryptosporidiosis, PML

Treatment in opportunistic infections:

- In tuberculosis, CD4 count < 200 – INH and Rifampin
- In *P. jirovecii* pneumonia and toxoplasmosis, CD4 count < 200 – Trimethoprim - sulfamethoxazole (alternatively Dapsone/pyrimethamine)
- In fungal infections, CD4 count < 200 – Fluconazole
- In herpes simplex infection, CD4 < 200 – Acyclovir

Gastrointestinal manifestations in HIV infection and their causes

Clinical manifestations	Causes
Persistent diarrhoea (60% patients, recurrent episodes, significant weight loss, copious watery stool, blood/mucus may be present, nausea, vomiting, flatulence, abdominal pain may be present)	Cause not known (most common)
	Giardiasis
	Cryptosporidiosis
	Isospora belli infection } Common causes
	Strongyloidiasis
	Cyclospora infection, Microsporidiosis
	Salmonellosis, Shigellosis
	Campylobacteriosis
	M. avium-intracellulare complex
Dysphagia (20-30% patients, recurrent, refusal of feed, excessive salivation in children, in adults painful deglutition Perianal discomfort	Candidial oesophagitis (90% cases, oral thrush usually present)
	Herpesviral oesophagitis
	Cytomegaloviral oesophagitis
	Non-specific ulcerative lesions
	Herpesviral proctitis
	Herpesviral ulcerations
	Amoebic proctocolitis

Respiratory manifestation in HIV infection and their causes

Cough, dyspnoea, cyanosis (in 80% of all patients) and tachypnoea	*S. pneumoniae, Staphylococcus species,* *H. influenzae, M. tuberculosis,* *P. jirovecii,* Atypical mycobacteria, Atypical pneumonia organisms, Toxoplasma pneumonia, *Cytomegalovirus,* *C. albicans, C. neoformans, H. capsulatum,* Lymphoid Intesrstitial Pneumonitis (LIP)
Pleural effusion, hemoptysis	Tuberculosis
	Malignancy (Kaposi's sarcoma)

Neuropsychiatric manifestations and their causes	
Headache, altered personality, lethargy, dementia, ataxia, convulsions, alteration of sensorium, disturbed co-ordination, depression, irritability, photophobia	HIV encephalopathy, Tubercular and cryptococcal meningitis, Cerebral toxoplasmosis, Herpes virus and CMV infection, Cerebral lymphoma, Vascular lesions
Meningitis	HIV (Aseptic)
	Tubercular
	Cryptococcal
Neurologic signs, focal seizures	Pyogenic (bacterial, common in children) Abscess/SOL due to Toxoplasmosis Tuberculosis, Cryptococcosis, Lymphoma, Vascular lesions
Myelopathy (in 20% adults)	HIV/Other viruses
Peripheral neuropathy (in some)	Vasculitis
Visual impairment	CMV, Toxoplasmosis, Unknown causes

Cardiovascular system. Pericarditis, Myocardial infarction in a young person.

Hematologic. Thrombocytopenia, Anemia, Neutropenia.

Cutaneous. Macular rash, Alopecia.

Mucocutaneous. Ulcers (gingival, buccal, anal, penile), Conjunctivitis.

Tuberculosis and AIDS

In USA – *M. avium intracellular, scrofulaceum* (MAIS) common.

In India – *M. tuberculosis* common.

37% of HIV positive persons present with tuberculosis.

17% of them have neurotuberculosis (meningitis, tuberculoma or disseminated T.B.)

53% of AIDS deaths is caused by tuberculosis.

Mean duration from diagnosis of AIDS to death is 4½ months.

Pediatric HIV

Through vertical transmission – 12 to 40%

Through breast milk – 12%

Through needle stick injury – 0.3%

2% in developed countries

15% in Africa and Asia.

15,000 new HIV infections per day in world, of which 1,600-2,000 is in children below 15 years of age.

1-5 years age group is maximum affected.

Systemic manifestations

Common findings

Weight loss or failure to thrive

Persistent generalized lymphadenopathy

Hepatomegaly, Splenomegaly

Oral candidiasis

Anemia, Idiopathic thrombocytopenic purpura

Recurrent diarrhoea

Parotitis (persistent, repeated)

Lymphoid interstitial pneumonitis, *Pneumocystis jirovecii*

Mumps, Herpes zoster (involving more than one segment)

Chicken pox

Recurrent invasive bacterial infections

Cardiomyopathy, Hepatitis, Nephropathy

Microcephaly – Loss of milestones – HIV encephalopathy, CMV.

Less common findings

Chronic dermatitis

Acquired neurological manifestations (due to toxoplasmosis, cryptococcosis, etc.)

Kaposi's sarcoma

Malignancies such as lymphoma

For safety of blood and blood products

Licensed blood bank carry out the following tests : Hepatitis B and C screening, HIV antibody testing and VDRL test for syphilis.

All high risk staff should be vaccinated by Hepatitis B vaccine and booster given every 5 years.

HIV testing

1. **Unlinked anonymous testing:** Not directed at the individual level, but is important in public health surveillance of HIV infection - a method for measuring HIV prevalence in a selected population with the minimum of participation bias.

2. **Voluntary confidential testing:** Often done for diagnostic purposes. Counselling given, informed consent taken and confidentiality is maintained. This not only instils faith in the individual about the health care system in the community, but also encourages more and more people practising risk behavior to come forward for an HIV test.

3. **Mandatory testing:** When testing is done without the consent of the patient and the result could be linked to identify the person. It is recommended for ensuring blood safety and screening donors of semen, organs or tissues in order to prevent transmission of HIV to the recipient of the biological products.

Laboratory diagnosis of HIV

- **Tests that detect antibody**
- **Tests that detect antigen**
- **Tests that detect viral nucleic acid**
- **Tests that detect the virus**
- **Tests to estimate number/phenotype of T-lymphocytes**

Surrogate Markers of disease progression

- CD4 estimation: CD4/CD8 ratio
- Beta 2 microglobulin/Serum Neopterin
- p24 antigen/antibody
- Neutrophils/Lymphocytes/Platelets
- ESR/Hemoglobin concentration
- IL2 receptors

Antibody Detection

Most widely used and is the most effective way

Tests divided into two broad groups:

- Screening assays - designed to detect all infected samples/individuals and high sensitivity

- Supplemental assays - designed to identify samples or individuals with positive screening tests and high specificity

ELISA matrix may be

- Strips or wells of MWP
- Plastic beads
- Nitrocellulose paper

Conjugates are most often Ab (IgG/IgM/ Ig A) coupled with

- Enzymes (Alkaline phosphatase or Horse radish peroxidase)
- Fluorochromes
- Streptavidin

Generations of ELISA

- First generation kits use antigens derived from detergent disruption of viruses grown in human lymphocytes.

- Second generation kits use artificially derived recombinant antigens expressed from bacteria or fungi.

- Third generation kits use chemically synthesized oligopeptides of about 15-40 amino acids (synthetic peptides).

- Fourth generation kits use a combination of recombinant and synthetic peptides and can detect both HIV antigen (p24) and antibodies concurrently.

False positive result: Auto-immune diseases, Multiple pregnancies, Multiple transfusions, Antibody to Class II HLA Ag (HLA-DR4), Hyper gammaglobulinemia, Antipolystyrene antibodies, Chronic alcoholics, Patients with hepatitis, Hepatitis B immunisation, Technical error, etc.

False negative result: Infected but not yet seroconverted, Window period, Late stage disease (immune collapse), Technical error.

- ELISA/Rapid tests (E/R) used in strategies/ algorithms I, II, IIA, IIB & III
- Supplemental tests are E/R with different antigens and or different principle of test.
- Western Blot and Line Immunoassay are used only in problem cases e.g. in cases of indeterminate/ discordant result of E/R.

HIV Rapid Tests

- Detect HIV-1 and HIV-2
- Cannot differentiate
- Procedural control: anti-human IgG
- Whole blood or serum/plasma
- Widely available
- No additional reagents required
- Room temperature storage
- 15 minutes to result.

Different types of rapid tests are available. The various technologies on which rapid tests are based include:

- Immunoconcentration (flow through; dot blot assays)
- Immunochromatography (lateral flow assays)
- Particle agglutination (latex, gelatin, RBCs, etc.)

- Immunocomb (Dip stick/comb tests) (mostly ELISA based) Combinations as per principle
  1. Immunofiltration + Immunoassay + Agglutination
  2. Immunofiltration + Immunoassay + Lateral Flow/Immuno chromatography
  3. Immunofiltration + Agglutination + Lateral Flow/Immuno chromatography
  4. Immunoassay + Agglutination + Lateral Flow/Immuno chromatography

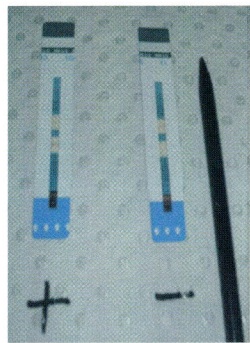

HIV Rapid Test

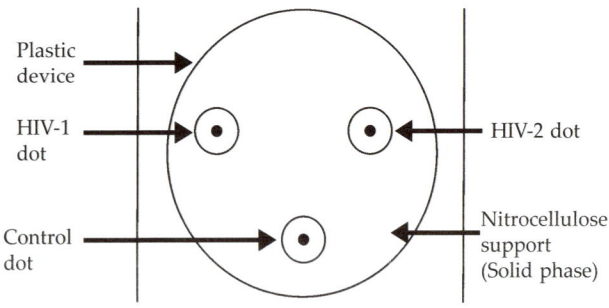

Dot Blot Test

WHO Recommended Strategies

- **Strategy I:** Test all samples with one EIA

- **Strategy II:** Strategy I with all reactives retested in a more specific test with different principle and/or antigen.

- **Strategy III:** Strategy II with reactives tested in a third test differing from the first two tests.

WHO Recommended Testing Strategies

Transfusion safety	Strategy I	
Surveillance	> 10%	I
	≤ 10%	II
Diagnosis	Strategy II	
Risk factors	> 10%	
No risk factors	≤ 10%	

Testing Algorithm

- An HIV Positive Status should be based upon the outcome of 2 or more tests

- When two test results disagree (one is reactive, the other non-reactive), the finding is called *"discordant"*. In this case, a third test must be performed.

- The combination and sequence of specific tests used in a given strategy

- An HIV Positive status is based upon the outcome of 2 or more tests

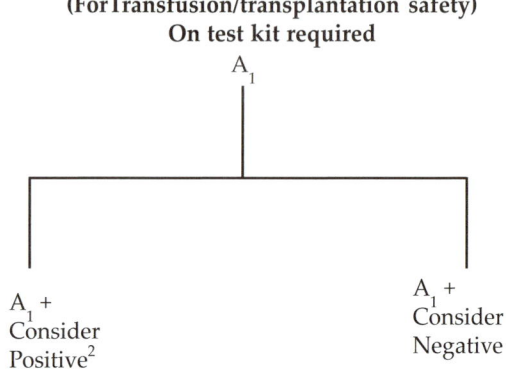

(For Transfusion/transplantation safety)
On test kit required

A_1

A_1 +
Consider
Positive[2]

A_1 +
Consider
Negative

(Destroy the unit of blood as per guidelines refer to ICTC for confirmation of status after consent

Strategy/Algorithm I

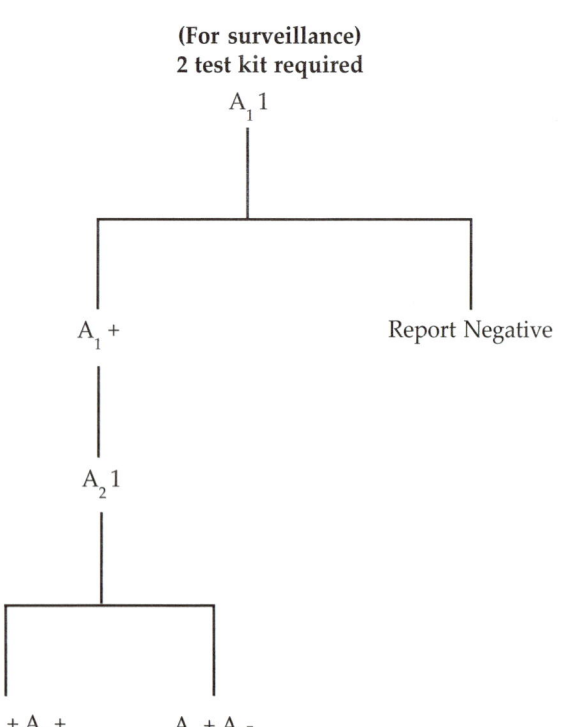

(For surveillance)
2 test kit required

A_1 1

A_1 +

Report Negative

A_2 1

$A_1 + A_2$ +
Report Positive

$A_1 + A_2$ -
Report Negative

Strategy / Algorithm II A

(Diagnosis of individual with AIDS indicator disease symptoms)
3 test kit required

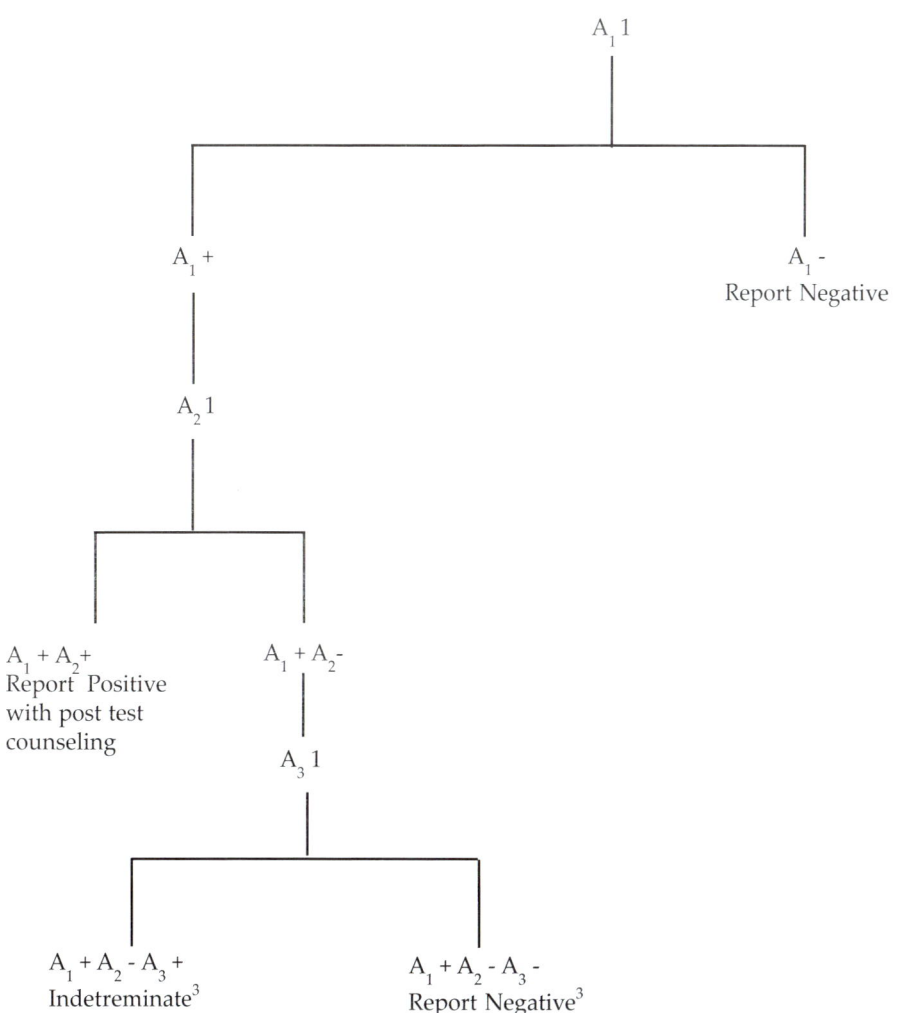

$A_1 1$

$A_1 +$

$A_1 -$
Report Negative

$A_2 1$

$A_1 + A_2 +$
Report Positive
with post test
counseling

$A_1 + A_2 -$

$A_3 1$

$A_1 + A_2 - A_3 +$
Indetreminate[3]

$A_1 + A_2 - A_3 -$
Report Negative[3]

Strategy/Algorithm II A

Strategy/Algorithm II B

- When two test results disagree (one is reactive, the other non-reactive), the finding is called "*indeterminate.*" In this case, a third test (tie-breaker) must be performed

Antibody testing limitations

- Difficulties in interpretation
- Limitations - 'window period'
- Antibodies appear within 3-4 weeks
  - Direct detection – HIV p24 antigen or DNA/RNA (NAT) – pre-antibody
  - Combo test – earlier detection
- Primary infection and therapy – delayed antibody response

Supplemental Tests

Recommended for validation of positive results on screening tests

- Western Blot assay (WB)
- Immunofluorescence assay (IFA)
- Line immunoassay (LIA)
- Radio-immunoprecipitation assay (RIPA)

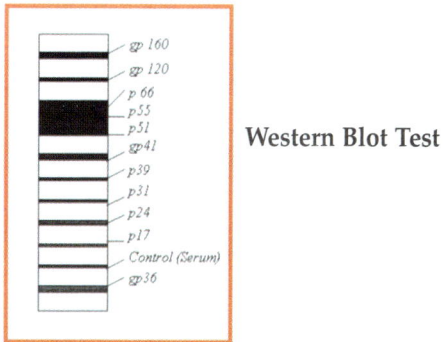

Western Blot Test

- Expensive – $80-100
  Technically more difficult
- Visual interpretation
- Lack standardisation
  - performance
  - interpretation
  - indeterminate reactions
- 'Gold Standard' for confirmation

Comparative costs :

- ELISA (Ab only) – $2 per test
- EIA (Ab/Ag combo) – $3.50
- Rapid test – $10-20 per test
- Western blot $80-100
- p24 antigen $30
- PCR - qualitative $80-100
- PCR - quantitative (viral load) $90-150
- DNA sequencing (resistance) $400–700
- Direct detection of HIV
  - To determine HIV status during window period
  - In health care workers following accidental exposure to contaminated blood etc.
  - Detection of HIV infection in babies born to HIV positive mothers
  - Indeterminate Western Blot results
  - Discordant results on antibody testing.
- **Detection of HIV-specific DNA by Polymerase Chain Reaction (PCR)**
- **Detection of p24 antigen by Conventional ELISA - Sensitivity 50-60 pg/ml**
- **Detection of HIV specific circulating antigen (p24 Ag)**

Indications	Limitations
Window period	30%
AIDS	50-60%
ARC	30-40%
Newborns/infants	20%
Asymptomatic	10%
HIV dementia/encepalopathy (CSF)	

Detection of Viral Nucleic acid

Template	Procedure
Viral RNA in plasma	RT - PCR
Viral RNA - cell associated	RT - PCR
Integrated proviral DNA	PCR

Virus culture

- Virus isolation is attempted from PBMC/plasma/ other body fluids
- Autologous (Direct)/heterologous PBMC (Co-culture method) activated with a mitogen (e.g. PHA) are cultured in-vitro at 37°C in 5% CO_2 upto 4 wks
- Presence of virus confirmed by p24 Ag assay/ detection of RT/demo of syncitium/IF.

Quantification of Viral Load

- Quantitation invaluable towards study of:
  - Viral replication dynamics
  - Central parameter in anti-retroviral therapy
  - Prognostic marker early in disease
- Three techniques used:
  - RT - PCR
  - NASBA
  - b DNA

Viral load assay

- Quantitation of HIV RNA in plasma is useful for determining free viral load, assessing the efficacy of antiviral therapy and predicting progression and clinical outcome.
- Baseline HIV viral load is predictive of survival at 10 years in patients with nearly identical CD4 counts.
- Assessment of baseline viral load prior to initiation of therapy is useful in patient management.

Lab diagnosis of HIV infection in "window-period"

- PCR
- p24 Ag assay (preferably ICD p24 assay; however has its limitations)
- Virus culture

Clinical AIDS/ARC

- Demonstration of HIV specific antibodies in serum or plasma
- Demonstration of p24 antigen
- Detection of viral RNA in plasma or cell associated viral RNA using RT- PCR

HIV infection (Infants and neonates)

- Culture isolation of HIV from PBMCs
- Demonstration of p24 antigen
- Detection of viral RNA in plasma or cell associated viral RNA using RT-PCR
- Detection of integrated proviral DNA in PBMC using PCR
- Detection of IgA and IgM anti-HIV antibody

Post Exposure Prophylaxis (PEP)

Various epidemiological and laboratory studies have shown that the risk of infection varies with type of exposure, such as:

- Type of needle (hollow bore vs. solid)
- Device visibly contaminated with patient's blood
- Depth of injury
- The amount of blood involved in the exposure
- The amount of virus in patient's blood at the time of exposure.
- Whether PEP was taken within the recommended time (< 2 hrs, up to 72 hrs)
- Ideally, prophylaxis should begin within 2 hours of exposure.

Risk for getting HIV after a needle stick, an injury with a sharp instrument or a splash

- About 1 in 300 or 0.3%
- Less than 0.1% if post-exposure medication taken
- The risk for infection from a bloody splash to mucous membranes or to open skin is very low - less than 1 in 1000

- An exposure to blood from a terminally-ill AIDS patient
- An exposure caused by a needle which was used in a blood vessel
- An exposure caused by a visibly-bloody device
- A deep puncture.

Donts:

- Do not panic
- Do not reflexively place pricked finger into mouth
- Do not squeeze blood from wound, this causes trauma and inflammation, increasing risk of transmission
- Do not use bleach, alcohol, betadine, or iodine, which may be caustic, also causing trauma

Dos:

- Remove gloves, if appropriate
- Wash site thoroughly with running water
- Irrigate thoroughly with water or saline if splashes have gone into the eye or mouth

Category	Definition and example
Certain work practices increase the risk of needlestick injury such as: - Transferring a body fluid between containers. - Poor healthcare waste management practices	- Recapping needles (Most important). - Failing to dispose of used needles properly in puncture-resistant sharps containers

Source HIV status	Definition of risk in source
HIV negative	Source is not HIV infected but consider HBV and HCV
Low risk	HIV positive and clinically asymptomatic
High risk	HIV positive and clinically symptomatic (see WHO clinical staging)
Unknown	Status of the patient is unknown, and neither the patient nor his/her blood is available for testing (e.g. injury during medical waste management the source patient might be unknown). The risk assessment will be based only upon the exposure (HIV prevalence in the locality can be considered)

Exposure	Status of source HIV positive & asymptomatic	Status of source HIV positive & clinically symptomatic	Status of source HIV status unknown
Mild	Consider 2-drug PEP	Start 2- drug PEP	Usually no PEP or consider 2-drug PEP
Moderate	Start 2-drug PEP	Start 3-drug PEP	Usually no PEP or consider 2-drug PEP
Severe	Start 3-drug PEP	Start 3-drug PEP	Usually no PEP or consider 2-drug PEP

- **Basic regimen** – Zidovudine 200 mg X 8 hourly in combination with Lamivudine 150 mg X 12 hourly for 4 weeks.
- **Extended regimen** – Zidovudine and Lamivudine along with Indinavir 800 mg X 12 hourly for 4 weeks.
- NNRTI – Efavirenz 200 mg X 8 hourly.

Prevention of needle stick injury

- Prevention is the best approach!
- Don't recap needles!

- Wash area with soap and water!
- Seek counseling and health care at once!

PPTCT (Prevention of Parent to Child Transmission) Program

- Risk of mother to child transmission:
    - Antepartum – 20-30%
    - Intrapartum – 50%
    - Postpartum (through breast milk) – 10-15%

- Four prolonged approach:
  - Primary prevention of HIV infection
  - Choice of termination of unwanted pregnancy in HIV positive women
  - Prevention of transmission from mother to child with ART
  - Provision of care and support for HIV positive mothers, their infants and their family.
- Treatment – Oral tablet of Nevirapine 200 mg given to HIV positive mothers during labour and Nevirapine syrup 2 mg/kg body weight given to neonates within 72 hours of birth.

Disease progression can be delayed by:

- Prevention and early treatment of opportunistic infections (OIs)
- Antiretroviral therapy
- Positive living
- Therapy should be started when:
  - CD4 count < 200/cu.mm.
  - Viral load > 30,000/ml
  - Symptomatic patients

Antiretroviral (ART) regimens

- 1 Protease inhibitor (PI) + 2 Nucleoside reverse transcriptase inhibitors (NRTIs)
- 1 Non-nucleoside reverse transcriptase inhibitor (NNRTI) + 2 NRTIs
- 2 PIs + 1 or 2 NRTIs
- 3 NRTIs

Side effects of treatment

- 66% had some side effects with AZT + 3TC
  - Headache
  - Nausea and vomiting
  - Fatigue
- 70% had side effects with PI's
  - Liver toxicity
  - Kidney stones
  - Diarrhea and abdominal distress.

Prevention

- A – Avoid pre-marital sex
- B – Be mutually faithful to each other
- C – Condom usage.
- D – Drugs should be avoided.

Standard precautions

- Barrier protection
- Hand washing
- Safe techniques
- Safe handling of sharp items

- Safe handling of specimens (blood, etc.)
- Safe handling of spill of blood/body fluids
- Use of disposable/sterile items
- Safe techniques including mechanical pipetting device
- Immunisation with Hepatitis B vaccine.

Action plans for AIDS Control in India

- Awareness, education and publicity
- Promotion of condom use
- Care of AIDS orphans
- Socio-medical approach to drug abuse
- Focus on Commercial sex workers (CSWs)
- Legislation for compulsory licensing of blood banks
- HIV testing facility for all STD ans surgical cases, antenatal cases, tuberculosis cases, periodic testind of medical and paramedical personnel
- Revitalizing RNTCP and National Mental Health Program
- Setting up of counselling centres at colleges (AAJ), universities, hospitals, medical colleges, health centres, for GPs and through voluntary organizations
- Role of voluntary organizations

HIV CONTROL PROGRAMS

The HIV epidemic in India is complex and heterogeneous, impacted by intricate and varied social structures. Evidence shows that the epidemic is moving outwards, from high risk groups to the general population and from urban centers to rural areas. Increasingly youth and women are getting infected. The epidemic can entail adverse consequences to our achievement of health and development goals, namely child mortality and poverty. AIDS related productivity losses can be substantial. Household surveys show a 9.24% decline in incomes and an increase of 10% in health spending.

India is estimated to have around 20.9 lakh persons living with HIV in 2011, at an estimated adult HIV prevalence of 0.27%. Adult HIV Prevalence has decreased from 0.41% in 2001 through 0.35% in 2006 to 0.27% in 2011. India has demonstrated an overall reduction of 57% in estimated annual new HIV infections (among adult population) from 2.74 lakhs in 2000 to 1.16 lakhs in 2011, reflecting the impact of scaled up prevention interventions. Declines in adult HIV prevalence and new HIV infections are sustained in most of the states including all the high prevalence states of South India and North East. However, rising trends have been noted in some other low prevalence states. Considerable declines in HIV prevalence have been recorded among Female Sex Workers at national level (5.06% in 2007 to 2.67% in 2011 and among Men who have sex with Men (7.41% in 2007 to 4.43% in 2011), though some pockets in the country show higher HIV prevalence among them with mixed trends.

In 1986, following the detection of the first AIDS case in the country (in Chennai), the National AIDS Committee was constituted in the Ministry of Health and Family Welfare. **National AIDS Control Organization (NACO)** was formed in 1992 – a division of the Ministry of Health and Family Welfare that provides leadership to HIV/AIDS control program in India through HIV/AIDS Prevention and Control Societies.

In **1992-1999**, a comprehensive **HIV/AIDS Control Project (NACP) Phase I** was launched during VIII Plan with the assistance of World Bank & WHO with an objective to slow the spread of HIV to reduce future morbidity, mortality and the impact of AIDS by initiating a major effort in the prevention of HIV transmission. A major expansion of infrastructure of blood banks, infrastructure for treatment of sexually transmitted diseases in district hospitals and medical colleges was created with the establishment of STD clinics. HIV sentinel surveillance system was initiated and the program led to capacity development at the state level with the creation of State AIDS Cells in the Directorate of Health Services in states and union territories.

Phase-II (1999-2006) was aimed at reducing spread of HIV infection in India and to strengthen India's capacity to respond to HIV epidemic on long term basis. Targeted Interventions (TIs) were started with a focus on High Risk Groups (HRGs), i.e. female sex workers (FSWs), men having sex with men (MSM), injectable drug users (IDUs) and bridge populations (truckers and migrants). Operationally the program also aimed to reduce blood borne transmission of HIV to less than 1%, attain awareness levels of not less than 90% amongst youth and others in the reproductive age group and increase condom use to not less than 90% amongst high risk categories. Package of services included Behavior Change Communication, Management of STDs and Condom promotion. The School AIDS Education Program was conceptualized. Voluntary counseling and testing centers (VCTC) were established in healthcare facilities to promote access to HIV counseling and testing. The interventions for prevention of parent to child transmission (PPTCT) were also started. Free antiretroviral (ARV) therapy was initiated in selected hospitals in the country. Increasing access to free ARV was one of the major achievements of NACP-II.

NACP III (2007-2012)

Objectives: To reduce the rate of incidence by 60% in the first year of the programme in high prevalence states to obtain the reversal of the epidemic and by 40% in the vulnerable states to stabilise the epidemic.

Goals: To halt and reverse the epidemic in India over the next five years by integrating programs for prevention, care, support and treatment.

This was achieved through **a four-pronged strategy** as follows:

A) **Prevent infections through saturation of coverage of high-risk groups with targeted interventions (TIs)** – Access to behavior change communication; Prevention services (condoms, STI services, needles and syringes); Treatment services (STI services, drug substitution for IDUs); & Creation of enabling environment at project sites;

   Scaled up interventions in the general population – STD control program; VCTC and PPTCT programs; Universal precautions and Post exposure prophylaxis; Blood safety; Improved access to quality condoms; Focused efforts on women, children and young people; Expanding HIV/AIDS response at workplace and Focused efforts on migrants, mobile populations and in cross border areas.

B) **Provide greater care, support and treatment to larger number of people living with HIV/AIDS (PLHA)** – Improved treatment access for opportunistic infections and continuation of care; Children affected and infected by HIV; Integration of prevention with care, support and treatment; Community care and support programs; Collaboration with PLHA network and Improving access to ART for PLHA and children.

C) **Strengthen the infrastructure, systems and human resources in prevention, care, support and treatment programmes at district, state and national levels –** State AIDS control societies; District AIDS prevention and control units; and Strengthening of the National AIDS control organization by capacity building and sustained technical training support to public and private agencies, mainstreaming HIV and partnership development, convergence with RCH, TB and MoHFW and coordination and partnership with donors.

D) Strengthen the nationwide **Strategic Information Management System (SIMS)** – Strengthening the computerized management system (CMIS) and making it more appropriate and user friendly; Developing community friendly information systems; Developing indicators for the state plans and institutional arrangement for collecting, analyzing and monitoring progress & Hardware and software procurements.

Targets achieved in NACP III

- Coverage of 2.34 million high-risk population annually across the country through a total of around 1,948 targeted interventions (TIs) for high risk groups (HRGs) and bridge population – FSWs (81%), IDUs (80%), MSMs (64%), Truckers (57%) and Migrants (40%).

- The overall condom distribution in the country has risen from 160 crores pieces in 2006-07 to 300.79 crores pieces in 2011-12

- Establishment of 4955 Integrated Counseling and Testing Centres (ICTC), which conducted 22 million tests in government sector and 12 million tests in private sector per year

- PPTCT program covered 75,600 HIV positive mothers with antiretroviral drug prophylaxis and Nevirapine prophylaxis to 11,981 infected mother-baby pairs at the time of delivery during 2011-12

- The coverage of STI services has been scaled up through STI clinics and 100.72 lakh STI/RTI patients managed during 2011-12

- During 2011-12, 93.2 lakh units blood were collected across the country. NACO supported Blood banks collected 55 lakh units; 84.3% of this was collected through voluntary blood donation

- Care, support and treatment services were provided through 355 ART centers, 725 Link ART Centers (LACs) and 253 Community Care Centers (CCC). About 5.16 lakh PLHIV were receiving ART by March 2012

- Coverage of 144,409 schools and colleges for AIDS awareness

- 380,000 personnel trained during NACP III

NACP Phase IV (2012-2017)

Goals and Objectives

- **Objective 1**: Reduce new infections by 50% (2007 Baseline of NACP III)

- **Objective 2**: Comprehensive care, support and treatment to all persons living with HIV/AIDS

Strategies of NACP IV

- Strategy 1: Intensifying and Consolidating Prevention services with a focus on HRG and vulnerable population

- Strategy 2: Expanding IEC services for (a) general population and (b) high risk groups with a focus on behavior change and demand generation

- Strategy 3: Comprehensive Care, Support and Treatment

- Strategy 4: Strengthening institutional capacities

- Strategy 5: Strategic Information Management Systems (SIMS).

New Operational Guidelines in 2013

This provides recommendations for strengthening the key aspects of the continuum of HIV care and improving linkages across the health system. It also addresses the implications of new clinical recommendations for laboratory services and supply systems for ARV drugs and other commodities. The 2013 guidelines represent an important step towards achieving universal access to ARV drugs for treating and preventing HIV, increasing the efficiency, impact and long-term sustainability of ARV programs & realizing the ultimate goal of ending the HIV epidemic.

When to start ART in people living with HIV

- ART should be initiated in all individuals with severe or advanced HIV clinical disease (WHO clinical stage 3 or 4) and individuals with CD4 count ≤ 350 cells/mm3.

- ART should be initiated in all individuals with HIV with CD4 count > 350 cells/mm^3 and ≤ 500 cells/mm3, regardless of WHO clinical stage.

- ART is recommended to be initiated regardless of WHO clinical stage or CD4 count for certain populations:

    - People with active tuberculosis (TB) disease who are living with HIV

    - People with both HIV and hepatitis B virus (HBV) infection with severe chronic liver disease

    - HIV-positive partners in sero discordant couples

    - Pregnant and breastfeeding women and children < 5 years of age.

First Line Regimen of ART for Adults

- First-line ART should consist of two nucleoside reverse transcriptase inhibitors (NRTIs) plus a non-nucleoside reverse-transcriptase inhibitor (NNRTI).

- TDF + 3TC (or FTC) + EFV as a fixed-dose combination is recommended as the preferred option to initiate ART (strong recommendation, moderate-quality evidence).

- If TDF + 3TC (or FTC) + EFV is contraindicated or not available, one of the following options is recommended:

    - AZT + 3TC + EFV

    - AZT + 3TC + NVP

    - TDF + 3TC (or FTC) + NVP (strong recommendation, moderate-quality evidence).

- Countries should discontinue d4T use in first-line regimens because of its well-recognized metabolic toxicities.

Principles for selecting the first-line regimen

- Choose 3TC (Lamivudine) in all regimens

- Choose one NRTI to combine with 3TC (Zidovudine - AZT/ZDV or Tenofovir Disoproxil Fumarate - TDF)

- Choose one NNRTI (Nevirapine - NVP or Efavirenz - EFV)

NACO Second Line Regimen

ARV drugs for 2nd line	Dosage	Dosing schedule
TDF + 3TC	Tenofovir 300 mg + Lamivudine 300 mg once daily in tablet form	Please advise the patients to consume all the pills simultaneously after meal (preferably dinner).
ATV/r	Tab. Atazanavir 300mg + Tab. Ritonavir 100mg Each tab to be taken once daily simultaneously	(Keep the drug interactions in mind).

ART recommendations for patients who develop TB within six months of starting a first-line or second-line ART regimen

First-line or second-line ART regimen	ART regimen at the time TB occurs	Management options
First-line ART	(AZT or TDF) + 3TC + EFV*	Continue with two NRTIs + EFV
	(AZT or TDF) + 3TC + NVP	Substitute NVP with EFV*
Second-line ART	Two NRTIs + PI	Substitute Rifampicin with Rifabutin in the ATT

Start ART as soon as TB treatment is tolerated (between 2 weeks and 2 months).

*EFV should not be used in the first trimester of pregnancy.

Prevention of Parent to Child Transmission (PPTCT)

- **Vision:** Women and children, alive and free from HIV
- **Goal:** To work towards elimination of pediatric HIV and improve maternal, newborn and child health and survival in the context of HIV infection.
- **Objectives:**
  1. To detect more than 80% HIV infected pregnant women in India
  2. To provide access to comprehensive PPTCT services to more than 90% of the detected pregnant women
  3. To provide access to early infant diagnosis to more than 90% HIV exposed infants
  4. To ensure access to anti-retroviral drugs (ARVs) prophylaxis or Anti-Retroviral Therapy (ART) to 100% HIV exposed infants
  5. To ensure more than 95% compliance with ART in HIV infected pregnant women and ARV/ART in exposed children.

WHO new guidelines (June 2013) recommend two options for pregnant and breast-feeding women living with HIV:

- Providing lifelong ART to all the pregnant and breast-feeding women living with HIV regardless of CD4 count or clinical stage OR
- Providing ART (ARV drugs) for pregnant and breast-feeding women with HIV during the mother to child transmission risk period and then continuing lifelong ART for those women eligible for treatment for their own health.

India has successfully launched (September 2012) the multi-drug PPTCT Option-B regimen in the three southern high prevalence States of Andhra Pradesh, Karnataka initially and subsequently in Tamil Nadu.

National Strategic Plan for Prevention of Parent to Child Transmission (PPTCT)

- All pregnant women living with HIV will be provided a triple-drug ART regimen regardless of CD4 count or clinical stage – Tenofovir (TDF) 300 mg + Lamivudine (3TC) 300 mg + Efaviranez (EFV) 600 mg.
- All HIV exposed infants – Syrup Nevirapine (NVP) first dose within 6 to 12 hours of delivery and continue daily for minimum 6 weeks. Infants of mothers who are receiving ART and are breastfeeding should receive six weeks of infant prophylaxis with daily NVP.
- If infants are receiving replacement feeding, they should be given 4 to 6 weeks of infant prophylaxis with daily NVP (or twice-daily AZT). Infant prophylaxis should begin at birth or when HIV exposure is recognized postpartum.
- Potential to reduce HIV transmission from mother to child to less than 5%.
- It will also help in increasing coverage of ART for those requiring anti retroviral treatment, provide early protection against mother to child transmission, reduce the risk of HIV transmission to HIV serodiscordant partners and improve mental health.

Key initiatives under Strategic Information Management during NACP-IV

- National Integrated Biological & Behavioural Surveillance (IBBS) among HRG & Bridge Groups
- National Data Analysis Plan
- National Research Plan
- Transforming SIMS into an integrated decision support system with advanced analytic and Geographic Information System (GIS) capabilities

- Institutionalizing Data Quality Monitoring System for routine program data collection
- Institutionalizing data use for decision making.

Program Targets of NACP IV

- By 2017, NACP- IV will cover 9 lakh FSWs, 4.40 lakh MSMs including Transgenders/ Hijras and 1.62 lakh IDUs through TIs. Over 16 lakh long distance truckers and 56 lakh high-risk migrants will be separately targeted as part of bridge population.
- Vulnerable sections of the population will be reached through ICTCs (280 lakh tests) and through expanded STI/RTI program, covering nearly 90 lakh people.

- 140 lakh pregnant women will be targeted, to prevent mother-to-child transmission in the community.
- Supply of 90 lakh units of safe-blood and enhanced use of blood products will be ensured.
- The program will provide first and second line ART to all who require it. It is estimated that there will be 10,05,000 people on ART (including 50,000 children who require first line ART and nearly 50,000 PLHIV who require second line drugs) by 2017.

REFERENCES

1. Ananthanarayan & Paniker's Textbook of Microbiology. 9th Ed. Ed. Kapil A. Universities Press, Hyderabad. 2013.

2. *naco.gov.in/NACO/Quick_Links/HIVData/* NACO, Department of AIDS Control, Ministry of Health & Family Welfare, Government of India. HIV Epidemic Fact Sheet, July 2014. Updated on 08 September, *2015.*

3. www.unaids.org/en/dataanalysis/.../epidemiologypublications.Fact sheet: 2014 statistics. Updated 24 August 2015.

4. www.unaids.org/sites/default/.../2015_GARPR_reporting_overview.pdf. Global AIDS Response Progress Reporting including the Additional UA Health Sector Indicators. Reporting overview for 2015. Updated February 2015.

5. Gupta V, Gupta S. Laboratory markers associated with prognosis of HIV infection. Indian J Med Microbiol 2004; 22: 7-15.

6. National AIDS Control Program Phase-IV (2012-2017). Strategy Document. Department of AIDS Control, Ministry of Health & Family Welfare, Government of India, New Delhi.

7. Antiretroviral Therapy Guidelines for HIV-infected Adults and Adolescents. May, 2013. Department of AIDS Control, National AIDS Control Organization, Ministry of Health & Family Welfare, Government of India, New Delhi.

8. Consolidated Guidelines on the Use of Antiretroviral Drugs for treating and preventing HIV infection. Recommendations for a Public Health Approach. June, 2013. World Health Organization, HIV/AIDS Program.

9. National framework for Joint HIV/TB Collaborative Activities. November, 2013. Central TB Division, Directorate General of Health Services & Basic Services Division, Department of AIDS Control, Ministry of Health & Family Welfare, Government of India, New Delhi.

10. National Strategic Plan. Multi-Drug ARV for Prevention of Parent to Child Transmission of HIV (PPTCT) under National AIDS Control Program in India. May, 2013 & Updated December, 2013. Basic Services Division, Department of AIDS Control, Ministry of Health & Family Welfare, Government of India, New Delhi.

PARASITOLOGY

IMPORTANT PROTOZOA WITH DISEASES PRODUCED AND THEIR LABORATORY DIAGNOSIS

I. *Entamoeba histolytica*

A. Location: Large intestine

Mode of Transmission: Ingestion of quadrinucleate cysts in food

Diseases produced:

1. Intestinal amoebiasis (acute and chronic amoebic dysentery)
2. Extraintestinal amoebiasis.

Laboratory Diagnosis:

1. In acute amoebic dysentery – Stool is offensive, mixed with blood and mucus, trophozoiles of *E. histolytica* are present, RBCs are in clumps, very few pus cells, Charcot Leyden crystals are present
2. In chronic amoebic dysentery – Stool shows cysts of *E. histolytica*. A purged stool shows trophozoites and cysts.
3. Culture – On Boeck and Drbohlav's or Philips medium
4. Serology – CIEP, LAT, ELISA, etc.
5. Detection of Isoenzyme pattern.

B. Location: Liver

Mode of Transmission: From ulcer in submucosa of intestine, trophozoites are carried to liver by radicles of portal vein

Disease produced: Amoebic liver abscess (postero-superior surface of right lobe is commonest site)

Laboratory Diagnosis:

1. Aspirated material – Anchovy sauce pus with trophozoites of *E. histolytica*
2. Stool – Cysts of *E. histolytica* in < 15% cases
3. Blood – Neutrophilic leucocytosis
4. Serology – CFT, FAT, precipitin test, ELISA
5. Skin test
6. Hepatic photoscan

Laboratory diagnosis of Amoebiasis:

Fifty million peple suffer worldwide from diarrhoea secondary to *E. histolytica* infection.

1. **Culture:**
   E. histolytica can multiply in vitro and in vivo. In vitro, it iis cultured in mono and diphasic media, in the presence of numerous and complex associated microbial flora (polyaxenic culture) single symbiont (monoaxenic culture mostly with *Trypanosoma cruzi)*. Significant progress has been achieved with axenic cultures from which antigens containing few foreign organisms can be prepared.

2. **Serological Tests:**
   - Complement fixation test (CFT)
   - Indirect hemagglutination test (IHA)
   - Latex agglutination test (LAT)
   - Indirect fluorescent antibody test (IFA)
   - Gel diffusion test (GDT)
   - Immunoelectrophoresis (IEP)
   - Counterimmunoelectrophoresis (CIEP)
   - Enzyme - linked immunosorbent assay (ELISA)
   - Skin test:
     1. Extract of *E.histolytica* culture in the presence of mycoplasma 81% positivity.
     2. Monoaxenic culture – delayed type reaction in 64% hepatic lesions.

   - **Study of Isoenzyme pattern in *E.histolytica*** Enzymes:
     1. Phosphoglucomutase
     2. Glucokinase
     3. Glucose phosphate isomerase
     4. Malate dehydrogenase (L-malate: $NADP^+$ oxidoreductase – oxaloacetate decarboxylating)
     5. NADP diaphorase

The analysis of isoenzyme pattern has demonstrated a difference in metabolic markers between pathogenic and non-pathogenic strain of *E. histolytica*.

A vertical polyacrylamide gradient gel (3-7%) was designed to facilitate the electrophoretic resolution (using thin layer starch gel electrophoresis) and classification of isoenzyme pattern. Biochemical markers have been important tools in the study and classification of parasite populations having similar structural features, but different functional properties, as in case of pathogenic and nonpathogenic strain of *E. histolytica*.

The asymptomatic carriers may harbour amoebas with pathogenic isoenzyme patterns as was expected from the fact that both invasive and luminal amoebiasis are transmitted through cysts.

Electrophoresis is carried out for 5 hours at 180V with 0.2M Tris NaOH, MgCl$_2$, EDTA, pH 7.4 buffer → gels were subsequently removed, transferred to a glass dish → stained → band developed in 30 minutes at 37°C.

Reeves and Bishcoff (1968) using cellulose acetate electrophoresis, classified 10 strains of *Entamoeba* into 2 groups using 5 enzymes.

Sargeant and Williams (1978) compared cultures of 14 stocks of *E. histolytica* and one only of *E. coli* by electrophoretic patterns of 3 enzymes (1, 3 and 4). Easily distiguished patterns divided *E. histolytica* in 3 groups, whilst a distinctly different pattern for *E. coli* was also seen. Ability to detect clearly identifiable groups of *E. histolytica* could be useful in answering epidemiological problems (WHO, 1969), although as yet any particular enzyme variant have not been associated with any behavioural character.

Vaccines for Amoebiasis

1. Parenterally administered recombinant antigen based vaccines :
   a) Serine – rich *E.histolytica* protein (SREHP) is a surface membrane protein, is highly immunogenic and when expressed in *E.coli*, was found to be safe and immunogenic in gerbils and monkeys.
   b) *E.histolytica* Gal/Gal NAc – specific lectin is a surface adhesin with specificity for galactose and N-acetyl galactosamine residues.
   c) 29 kDa cysteine – rich *E.histolytica* antigen is highly immunogenic.
   d) Amoebapore, a pore forming protein
   e) Neutral amoebic cysteine proteinase
2. Development of an oral vaccine :
   a) Recombinant *E. histolytica* antigen SREHP with cholera toxin or CT B subunit, a potent mucosal adjuvant tried in mice.

b) Attenuated *Salmonella, S.typhimurium* X 4550 used as a carrier for SREHP molecule. One problem is, there is lack of good animal models for intestinal disease.

II. Primary amoebic meninoencephalitis (PAME)

Caused by 1) *Naegleria fowleri* 2) *Acanthamoeba spp.* 3) *Hartmanella*.

Predisposing factors are diving or swimming underwater; diabetes, neoplasm, alcoholism, radiation therapy; patient on immunosuppressive, antimicrobial or chemotherapeutic agents.

Prognosis – usually fatal.

Laboratory diagnosis:

CSF – hazy to turbid with presence of red blood cells and increased leucocytes.

1. Centrifuged and examined under phase contrast microscopy:

   10-35 µm trophozoites with a single nucleus, motile by lobopodia – *Naegleria*.

   15-45 µm trophozoites with a single nucleus and pulsating vacuoles, motile by means of acanthopodia (fine tapering projection) – *Acanthamoeba*.

2. **Culture** in nonnutrient agar (tap water agar) for *Naegleria* incubated at 37°C for 7 days. Encystation occurs on 2nd or 3rd day. Thermophilic and grow at 43°-46°C.

   Smooth walled cysts present in soil and stagnant water (7-15 µm) – *Naegleria*.

   Double walled cysts (10-20 µm) – *Acanthamoeba*.

   PYG medium with 0.2% fetal calf serum for *Acanthamoeba*.

3. **Serology** – Complement fixation test
                     – Indirect immunofluorescence

Treatment – Parenteral, intrathecal and ventricular treatment with amphotericin B.

	Naegleria	*Acanthamoeba*
Disease	Primary amoebic meningoencephalitis (PAME)	Granulomatous amoebic meningoencephalitis (GAM)
Portal of entry	Nose	Upper respiratory tract
Predisposing cause	Swimming in contaminated water	Immune incompetence
Clinical course	Acute	Subacute or chronic
Pathological changes	Acute suppurative inflammation	Granulomatous inflammation
Tissue form	Trophozoite	Cyst and trophozoite
Culture	Usually positive	Usually negative
Trophozoite	10-35 µm, with single pseudopodium	15-45 µm, with thorn-like pseudopodia
Cyst	70-100 µm, with a smooth surface	9-27 µm, with a wrinkled surface
Flagellate form	Present	Absent
Leukocytes in cerebrospinal fluid	Predominantly neutrophils	Predominantly lymphocytes.

III. *Giardia lamblia*

Location: Small intestine

Mode of Transmission: Ingestion of cysts in food

Diseases produced: Giardiasis (steatorrhoea, acute or chronic enteritis)

Laboratory Diagnosis:

1. Trophozoites and cysts of *Giardia lamblia* in stool
2. Trophozoites in bile by duodenal intubation.

IV. *Cryptosporidium parvum*

Location: Small intestine

Mode of Transmission: Ingestion of cysts

Diseases produced: Cryptospridiosis (diarrhoea in immuno-competent and immunocompromised patients)

Laboratory Diagnosis:

1. Modified acid fast stain of stool - Demonstration of oocysts (round, double-walled)
2. Concentration of stool by Sheathar's sucrose floatation method
3. Serology - ELISA, LAT
4. PCR.

Characteristics	C. parvum	C. cayatanensis	Isospora belli
Size (µm)	4-6	8-10	20-30 x 10-19
No. of sporocyst	0	2	2
No. of sporozoites	4/oocyst/sporocyst	2	4
Sporulation time	Already sporulated after excretion	7-14 days	24-48 hours
Modified acid-fast stain	Red	Variable	Red

V. *Trichomonas vaginalis*

Location: Urogenital tract

Mode of Transmission: Sexual

Disease produced: Trichomoniasis

Laboratory Diagnosis:

1. Motile trophozoites of *T. vaginalis* in wet mounts from vaginal swabs and urine. Cystic phase is not seen in *T. vaginalis*.
2. Culture in Diamond's medium.

VI. *Trypanosoma brucei*

Location: Blood and tissues

Mode of Transmission: By bite of tsetse fly (*Glossina*) – Anterior station development

Disease produced: Sleeping sickness

Laboratory Diagnosis:

1. Trypomastigote forms of *T. brucei* in peripheral blood, bone marrow and C.S.F.
2. Serology – Agar gel precipitation (IgM level in CSF), CFT, Indirect FAT.

VII. *Trypanosoma cruzi*

Location: Blood and tissues

Mode of Transmission: By bite of reduviid bug followed by rubbing of fecal matter into the wound caused by the bite or contamination of conjunctiva with fingers – Posterior station development

Disease produced: Chagas' disease (acute and chronic)

Complications: Myocardial failure, neurological, dilatation of tubular organs.

Laboratory Diagnosis:

1. 'C' shaped trypomastigote form of *T. cruzi* in peripheral blood
2. Xenodiagnosis
3. Serology - CFT, Indirect FAT
4. Cruzin skin test
5. Biopsy of lymph node or muscle – Amastigote form of *T. cruzi*.

VIII. *Leishmania species*

Location: Blood and tissues

Mode of Transmission: By bite of sand fly (*Phlebotomus*)

Diseases produced:

1. *L. donovani* – Visceral leishmaniasis (Kala-azar) and post kala-azar dermal leishmaniasis (PKDL)
2. *L. tropica* – Cutaneous leishmaniasis (Oriental sore)
3. *L. brasiliensis* – Mucocutaneous leishmaniasis (Espundia).

Laboratory Diagnosis of Kala-azar:

1. Amastigote form of *L. donovani* in peripheral blood, bone marrow, lymph node and spenic pulp
2. Culture in Novy McNeal Nicolle (NNN) medium – Promastigote form
3. Blood – Progressive leucopenia

4. Brahmachari test – Formation of turbid ring at the point of contact between patient's serum and distilled water

5. Serology – Aldehyde test; Antimony test; CFT with Witebsky Klingenstein Kuhn (WKK) antigen; IFA test (Titre > 1:80 is positive, *Crithida* is used as antigen); Gel Diffusion Test (GDT) and CIEP using lyophilised antigen; IHA (Titre > 1:50); **ERAT** (Enzyme Revealed Antibody Test), where whole parasite is used as antigen and enzyme induced colour change is examined microscopically; **ELISA** – With sonicated or disrupted promastigotes (soluble antigen); Semiautomated ELISA is preferable.

6. Skin test – Delayed type of hypersensitivity (≥ 5 mm is positive).

7. For IHA and skin test, soluble antigen prepared from washed promastigotes are used.

For oriental sore and espundia:

1. Leishmanin skin test

2. Amastigote forms from skin and boipsy materials.

IX. *Plasmodium species*

Location: Blood and tissues

Mode of Transmission: By bite of female anopheles mosquito

Diseases produced:

1. *P. vivax* – Benign tertian malaria (relapses occur)

2. *P. falciparum* – Malignant tertian malaria (recrudescence is seen), Pernicious malaria, Blackwater fever

3. *P. malariae* – Quartan malaria

4. *P. ovale* – Ovale tertian malaria.

Laboratory Diagnosis of Malaria:

1. Peripheral blood at height of febrile parozysm shows trophozoites (ring form), schizonts and gametocytes of various plasmodia –
   i) Thick film for screening
   ii) Thin film for identifying species

2. Culture in RPMI 1640 medium

3. Blood - leucopenia with monocytosis (15-20%)

4. Serology (done in endemic areas and to detect latent infection) – Henry's melanin flocculation test, CFT, IF, PHA test.

X. *Toxoplasma gondii*

Location: Blood and tissues

Mode of Transmission: Ingestion of oocysts in raw meat; Contact with soil contaminated by cat's feces; Through placenta (abortion in females).

Diseases produced:

1. Congenital Toxoplasmosis – Cerebral, ocular, etc. (Pseudocyst in R. E. cells and tissue cyst in CNS and muscles)

2. Acquired Toxoplasmosis – Usually asymptomatic, occur as latent infection.

Laboratory Diagnosis:

1. Microscopy of stained smears of bone marrow, CSF or tissues removed by biopsy (lymph node or muscle)

2. Intraperitoneal inoculation in mice and guineapigs

3. Serology - IFA test; Goldman's test (fluorescent stain); Methylene blue dye test of Sabin and Feldman; Confirmatory test of Sabin and Ruchman; CFT of Sabin and Warren; Toxoplasmin skin test of Frenkel; IgG and IgM ELISA

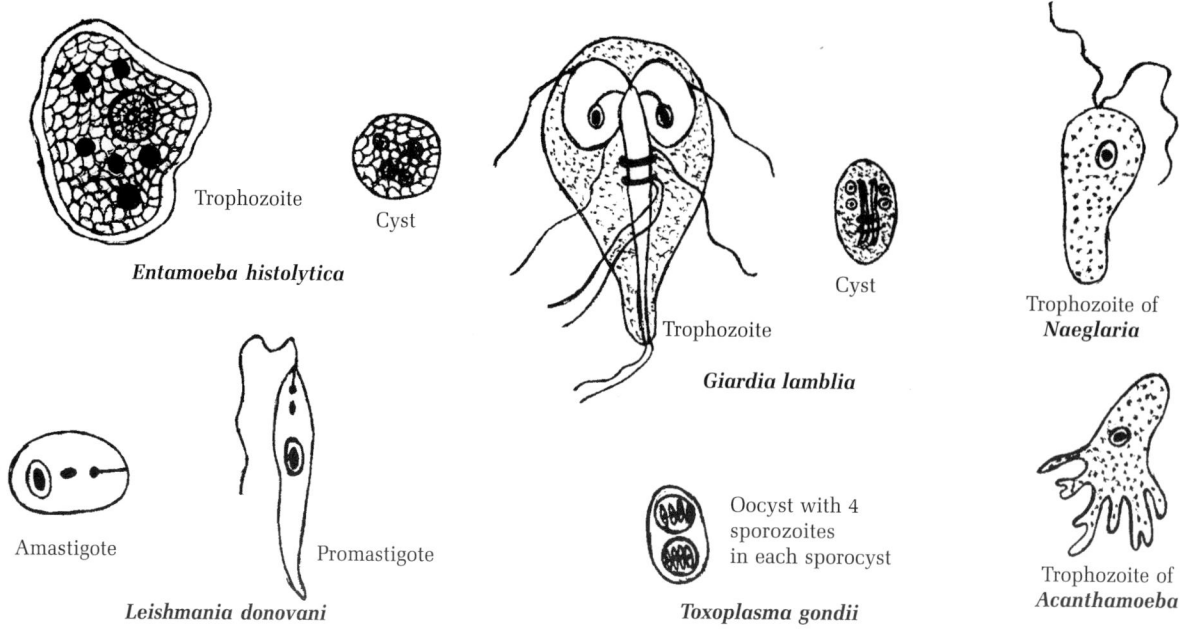

Trophozoite

Cyst

Entamoeba histolytica

Trophozoite

Giardia lamblia

Cyst

Trophozoite of *Naeglaria*

Amastigote

Promastigote

Leishmania donovani

Oocyst with 4 sporozoites in each sporocyst

Toxoplasma gondii

Trophozoite of *Acanthamoeba*

THE HELMINTHS

	Phylum Platyhelminthes		Phylum Nemathelminthes
	Class Cestoidea	Class Trematoda	Class Nematoda
Shape	Tape-like, Segmented	Leaf-like (flukes), Unsegmented	Elongated, Cylindrical, Unsegmented
Sexes	Not separate (Monoecious)	Monoecius, except Schistosomes (Diecious)	Separate (Diecious)
Alimentary canal	Absent	Present but incomplete (no anus)	Present and complete
Body cavity	Absent	Absent	Present
Head end	Suckers, often with hooks	Suckers, no hooks	No suckers, no hooks, well developed buccal capsule in some species

Classification

I. Cestodes

1. Pseudophyllidean
   Larval development in two intermediate hosts
   - First stage in cyclops (procercoid)
   - Second stage in fish (plerocercoid)
     e.g. *Diphyllobothrium latum.*

2. Cyclophyllidean
   Larval development in one intermediate host –

   a) Cysticercus proper – Larva transformed into bladder from which scolex of adult worm sprouts, e.g. *Taenia saginata, Taenia solium*

   b) Hydatid – Bladder multiply by budding and forms daughter and grand-daughter bladders with scolices inside each, e.g. *Echinococcus granulosus*

   c) Cysticercoid – Entire larva is solid, only proximal is vesicular with invaginated scolex, e.g. *Hymenolepis nana* (no intermediate host required, sometimes rat fleas act as intermediate host).

II. Trematodes

1. **Blood flukes** – *Schistosoma hematobium, S. mansoni, S. japonicum*

2. **Intestinal flukes** –
   a) Small intestine – *Fasciolopsis buski*
   b) Large intestine – *Gastrodiscoides hominis*

3. **Liver flukes** – *Fasciola hepatica, Clonorchis sinensis*

4. **Lung fluke** – *Paragonimus westermani*

Larval development:
- First intermediate host – Snail or mollusc
- First larval stage – Sporocyst
- Second larval stage – Second generation sporocyst (seen in schistosomes only)

One sporocyst liberate 1 lakh to 3 lakhs cerceriae.
- Third larval stage – First and second generation rediae (not seen in schistosomes)
- Final larval stage –

a) Cerceriae (phototactic) free in water penetrating directly into skin – *Schistosoma species*

b) Metacerceriae (encysted cerceriae)
   i) Encysted in water plants and vegetables – *F. hepatica, F. buski*
   ii) Encysted in second intermediate host (Freshwater fish) – *C. sinensis*
   iii) Encysted in second intermediate host (Crabs or crayfish) – *P. westermani*

III. Nematodes

1. **Intestinal Nematodes**

   a) In small intestine :
      i) *Ascaris lumbricoides* (Roundworm)
      ii) *Ancylostoma duodenale* (Old World Hookworm)
      iii) *Necator americanus* (New-World Hookworm)
      iv) *Strongyloides stercoralis*
      v) *Trichinella spiralis*

   b) In caecum and vermiform appendix :
      vi) *Trichuris trichiura* (Whipworm)
      vii) *Enterobius vermicularis* (Threadworm or Pinworm).

      i), ii), iii), vi) and vii) – **Oviparous;**

      iv) – **Ovoviviparous;** v) – **Viviparous.**

2. **Tissue Nematodes**
   Wuchereria bancrofti, Brugia malayi, Onchocerca volvulus, Loa loa, Dracanculus medinensis, etc. All are viviparous.

Important Cestodes

	T. saginata	T. solium	E. granulosus	H. nana
Life span Length Head	10 years 5-10m Large, quadrate, no rostellum or hooks	25 years 2-3m Small, globular with rostellum and hooks	6 months 3-6 mm Small with rostellar hooklets and 4 suckers	– 1-4 cm Globular, 4 suckers, hooklets
Proglottides	1000-2000 Expelled actively and singly	Below 1000 Expelled passively in chains of 5 or 6	Usually 3, occasionally 4 (terminal segment biggest – 2-3mm)	200
Eggs	31-43 µm, bile stained, oncosphere with 3 pairs of hooklets, embryophore brown, thick-walled and radially striated, sinks in saturated solution of common salt.	Same as T. saginata	32 - 36 µm x 25-32 µm, ovoid with 3 pairs of hooks (resembles Taenia)	30-45 µm, oval, nonbile - stained yolk granules, 2 knobs at both ends from where 4-8 polar filaments come out, 3 pairs of hooklets, floats in saturated solution of common salt.
Larval stage	Cysticercus bovis in musckles of cow or buffalo, 3-4 mm x 8-10 mm. Does not occur in man.	Cysticercus cellulosac in pig, 5mm x 8-10mm, opalescent, ellipsoidal, long axis lying parallel with muscle fibre.	Within the hydatid cyst, contains thousands of scolices.	Cysticercoid.
Mode of infection	Ingestion of improperly cooked infected beef.	Ingestion of improperly cooked infected pork.	Ingestion of eggs in dog's feces by intimate handling and fondling with dogs or taking uncooked vegetables contaminated with dog's feces.	Ingestion of food contaminated with eggs.
Pathogenicity	Vague abdominal discomfort, diarrhoea alternating with constipation, indigestion, etc.	Same as T.saginata Cysticerci localise in muscles, subcutaneous tissues, brain or eye, become calcified after 5-6 years.	Main site of cyst formation is liver. Other sites – lungs, spleen, kidney, long bones, also in brain, heart, orbit, etc. Rate of growth is very slow – 4cm diameter in 1 year.	Abdominal pain and diarrhoea with heavy infection.
Laboratory diagnosis	Demonstration of eggs in feces and segments of adult worm.	Same as T. saginata. For Cysticercosis – i) Biopsy of subcutaneous nodule ii) X-ray of skull and soft tissues for calcified nodules. iii) Eosinophilia. iv) Positive indirect hemagglutination test (HAT).	i) Casoni's test (Intradermal immediate hypersensitivity test) – wheal ≥ 5 cm in diameter with multiple pseudopodia. ii) Eosinophilia 20-25%. iii) Serological tests – Antigens used (a) thermolabile lipoprotein antigen S (b) thermostable lipoprotein antigen B. CFT, LAT, Bentonite flocculation test, Indirect HAT, IFA test, IEP, CIEP, Solid phase RIA, ELISA. iv) Skin test – Radio immuno sorbent test (RIST) for assay of total IgE and Radio allergo sorbent test (RAST) for specific IgE antibodies (30-90% positive). v) Cyst puncture – hazardous. vi) Hepatic photoscan – for location of site of cyst in liver.	Demonstration of eggs in stool.

N. B. Hydatid Cyst

1.　Two layers　　　–　a)　Outer cuticular (ectocyst) – thick (1mm), elastic, laminated hyaline membrane.

　　　　　　　　　　　b)　Inner germinal layer (endocyst) – thin (22-25 µm), vital layer which gives rise to brood capsules with scolices, forms outer layer, secrete hydatid fluid.

2.　Hydatid fluid　　–　Clear, colourless fluid (pale yellow), specific gravity 1.005-1.010, pH 6.7, slightly acidic, contains NaCl, Na_2SO_4, Na_3PO_4, Na and Ca salts of succinic acid, antigenic, highly toxic, granular deposit at bottom (hydatid sand – contains liberated brood capsules, free scolices and loose hooklets.

Important Trematodes I

	S. hematobium	*S. mansoni*	*S. japonicum*
Male	1-1.5 cm x 1 mm	1 cm x 1 mm	1.2-2 cm x 0.5 mm
Female	2 cm x 0.25 mm	1.4 cm x 0.25 mm	2.6 cm x 0.3 mm
Reunited intestine	Long, reuniting in the middle of body	Longest, reuniting in the anterior half of body	Short, reuniting in the posterior fourth of body
Eggs	150 x 50μm, has a terminal spine	150 x 60 μm, has a lateral spine	100 x 65 μm, has a lateral knob
Intermediate host (snail)	Bulinus, Planorbius	Biomphalaria, Australorbis	Oncomelania
Habitat	Vesical and prostatic venous plexuses	Mesenteric plexus of sigmoido-rectal area	Mesenteric plexus of ileo-cecal area
Cercerial dermatitis	+	+	+
Generalised anaphylactic reaction due to liberation of toxins	Rare	Intermediate	Common (Katayama fever in Japan)
Others	Painless terminal hematuria; Reversible granulomatous inflammatory reaction to eggs – Pseudotubercle or egg granuloma in bladder; followed by Irreversible fibrosis and calcification with degenerated eggs in centre.	Periportal cirrhosis; portal hypertension; cor pulmonale; dysenteric attack; transverse myelitis.	Clay pipe-stem cirrhosis; confluent progressive hepatic fibrosis; enlarged spleen; rupture of oesophageal varices (hematemesis); space-occupying lesion in brain; cor pulmonale.
Laboratory diagnosis	Demonstration of eggs in centrifuged deposit of urine; Cystoscopic biopsy of vesicle mucosa.	Demonstration of eggs by concentration of stool by formol ether or $ZnSO_4$; Rectal biopsy.	Same as *S. mansoni*. Ileal biopsy.
Eosinophilia (in early cases)	+	+	+
Serum aldehyde test	Often + (due to high globulin)	Often +	+

Serological Tests for Schistosomes :
CFT with cercerial antigen from snail's liver; Fairley's intradermal skin test; Indirect Fluorescent Antibody test (IFA); Circumoval Precipitin (COP) test; Miracidial immobilisation test.

Treatment :
Niridazole, Nilodin, Hycanthone, Tartar emetic, Fouadin, Anthiomaline, Antimony - dimercapto succinate.

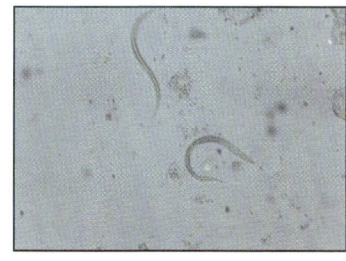

Larvae of *Strongyloides stercoralis* in saline mount of stool sample (10 X)

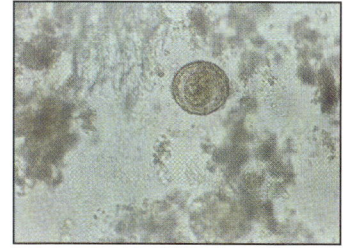

Fertilized ova of *Ascaris lumbricoides* in saline mount of stool sample (40 X)

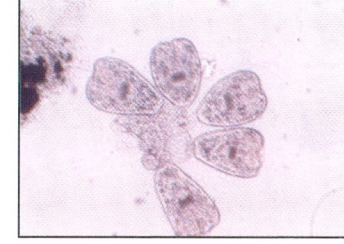

Aspirated fluid from liver showing hydatid cyst, scolices and daughter cysts (10 X)

Important Trematodes II

	F. hepatica	*F. buski*	*C. sinensis*	*P. westermani*
Adult	3 x 1.5 cm, leaf-shaped	2 x 7.5 cm x 8-20 mm, elongated, oval	10-25 mm x 2-3 mm, narrow, oblong	8-12 mm x 4-6 mm, thick, fleshy, egg-shaped.
Egg	140 x 80 μm, operculated, ovoid, bile-stained	140 x 80 μm, 25,000 eggs laid per day	35 x 20 μm, operculated, flask-shaped, terminal hook-like spine	80 x 55 μm, flattened opercula, oval, golden brown.
First intermediate host (snail)	Lymnaea	Segmentina	Bithynia	Melania
Second intermediate host	–	–	Ciprinoid fish	Crabs or crayfish
Definitive host	Man, sheep, goat, cattle	Man, pig	Man, pig, dog, cat	Man, tiger, leopard
Mode of infection	Ingestion of encysted cerceriae in blades of grass or water cress.	Ingestion of encysted cerceriae in water caltrop or water chestnut.	Eating raw or inadequately cooked fish containing encysted cerceriae in scale or flesh.	Eating raw or inadequately cooked crabs or crayfish containing encysted cerceriae in viscera, muscles and gills.
Geographical distribution	Worldwide	Worldwide	China, Japan, Korea, Vietnam.	China, Japan, Korea, India (Bengal, Assam, South India).
Habitat	In biliary tract	In small intestine	In biliary tract	In respiratory tract
Pathogenesis	Biliary colic, vomiting, persistent diarrhoea, tender hepatomegaly.	Chronic diarrhoea, asthenia, mild anemia.	Chronic diarrhoea, hepato-megaly, recurrent jaundice, stagnation of bile, cholangitis, intrahepatic calculi forma-tion, cholangiocarcinoma or anaplastic carcinoma.	Chronic cough with recurrent hemoptysis, enlarged liver, Jacksonian epilepsy, space-occupying lesion, generalised fever, lymphadenitis, cutaneous ulceration.
Sample/s for laboratory diagnosis	Feces or aspirated bile	Feces	Feces or aspirated bile	Sputum and feces
Eosinophilia	40-85%	–	40%	–
Treatment	Emetine injection, Bithionol	Tetrachloroethylene	Antimony compound, Chloroquine, Bithionol	Bithionol.

Important Nematodes I

	T. trichiura	*E. vermicularis*	*A. lumbricoides*
Male	3-4 cm	2-4 mm (dies after fertilising)	15-25 cm x 3-4 mm (has copulatory spicules)
Female	4-5 cm	8-12 mm (dies within 2-3 weeks)	25-40 cm x 5 mm (has vulvar waist)
Ova	50 x 25 μm, barrel-shaped, mucus plugs at both ends, bile-stained	50-60 μm x 30 μm, planoconvex, nonbile-stained	Fertilised – 60-75 μm x 40-50 μm, round, bile-stained; Unfertilised – 80 x 55 μm, elliptical, bile-stained, does not float in saturated solution of common salt.
Larva	200 μm	–	0.25 mm x 14 μm; 2 mm in lungs; 4 moults – 1 in soil, 2 in lungs (5th and 10th day), and 1 in intestine (25th - 29th day).
Mode of infection	Ingestion of embryonated eggs in food or drink containing rhabditiform larvae.	Ingestion of eggs containing larvae; Inhalation of dust containing eggs; Handling clothes and bed sheets of infected patients.	Ingestion of embryonated eggs with raw vegetables or drinking contaminated water; Inhalation of dust containing eggs.

Pathogenesis	Acute appendicitis; Heavy infection – mucus, diarrhoea with blood, abdominal pain, loss of weight; Massive trichiuriasis – prolapse of rectum.	Autoinfection; Retrograde infection; Pruritus periani et perinei, eczema around anus and perineum, salpingitis, nocturnal enuresis, inflamed appendix (in 2% cases).	1. Due to migrating larvae – Loeffler's pneumonia, rash, 20% eosinophilia. Can go to brain, spinal cord, heart or kidneys.
			2. Due to adult worms –
			  a) Spoliative action – PEM, night blindness.
			  b) Toxic action – fever, urticaria, oedema of face, conjunctivitis, irritation of upper respiratory tract.
			  c) Mechanical effects – intussusception, perforation through ulcer, intestinal obstruction.
			  d) Ectopic ascariasis – through mouth, nose, respiratory passage.
			  e) Wandering ascaris – cause appendicitis, obstructive jaundice, acute hemorrhagic pacreatitis, abscess in liver.

Laboratory diagnosis	Demonstration of ova in stool.	1. Demonstration of adult worms in stool. 2. Ova from perineal area by NIH swab.	1. Adult worms in stool. 2. Barium meal – string shadow. 3. Demonstration of ova in stool. 4. Scratch test – Intradermal test by ascaris antigen.
Treatment	Stilbazium iodide, Difetarsone, Thiabendazole, Mebendazole.	Piperazine hydrate, phosphate, citrate, adipate and tartarate. Pyrvinium and pyrantel pamoate, stilbazium iodide, thiabendazole, mebendazole.	Piperazine salts, tetramisole, pyrantel pamoate, thiabendazole, mebendazole.
Prophylaxis	Proper disposal of human feces; Not to consume uncooked vegetables and fruits grown in native gardens.	Prevent reinfection of individual; Prevent infection by contact; Washing of hands.	Proper disposal of human feces; Education of school children on sanitary laws and hygiene.

Important Nematodes II

	T. spiralis	S. stercoralis	A. duodenale
Male	1.4-1.6 mm (dies after 1 week)	1-2 mm	8-10 mm; Anterior end bends in the same direction as the body curvature (lives for 3-4 years), has copulatory bursa at posterior end with a single dorsal ray (Total rays - 13), 2 separate spicules.
Female	3-4cm (dies after 16 weeks)	2.5 mm	10-12.5 mm; (lives for 3-4 years), posterior end tapering with a spine, no copulatory bursa.
Ova	–	55 x 30 μm (immediately hatches into larva)	65 x 40 μm (oval, 4 blastomeres, nonbile-stained)
Larva	100 x 6μm, 1mm by 35 days, lives for average 6 months (10-31 years)	200-250 μm (rhabditiform); Longer and slender (filariform)	250 μm (rhabditiform); 500-600 μm (filariform); 4 moults – 2 in soil, 1 in lungs and 1 in intestine.
Habitat	Worldwide	Worldwide	Parallel 36°N to 30°S In India – Punjab and U.P.
Mode of infection	Eating raw or partially cooked pork containing encysted larvae	Filariform larva through skin penetration	Filariform larva through skin penetration by walking barefooted on fecally contaminated soil.
Pathogenesis	1. Stage of intestinal invasion (5-7 days). 2. Stage of larval migration (4-16 weeks). 3. Stage of encystation in striated muscles (diaphragm, intercostals, pectoralis major, deltoid, biceps, gastrocnemius).	Hyperinfection; 1. Skin lesions – rash at entry, wheal around anus. 2. Larva currens – moves in subcutaneous tissues 3-4 cm per hour for few weeks. 3. Pulmonary lesions – hemorrhages in lung alveoli, bronchopneumonia. 4. Intestinal – diarrhoea with blood and mucus.	1. Skin lesions – ancylostome dermatitis, creeping eruption. 2. Larva migrans – moves in subcutaneous tissues 1-2 cm per day for many months to 2 years (A. brasiliense, A. caninum, N. americanus). 3. Anemia – chronic blood loss (microcytic, macrocytic or dimorphic). Blood loss per worm – 0.2 ml/day. 4. N. americanus – 0.03 ml/day/ worm.
Laboratory diagnosis	1. Demonstration of adult worms in stool. 2. Biopsy of deltoid and gastrocnemius muscles for encysted larvae. 3. Eosinophilia – 15-50%. 4. Serological tests – CFT, Precipitation test, LAT, FAT. 5. X-ray. 6. Skin test with Bachman's antigen – erythematous patch within 15-20 minutes.	1. Demonstration of larvae in stool and duodenal washings. 2. Jejunal biopsy material, sputum, bronchoalveolar lavage, blood and urine for larvae. 3. Serological tests – Filarial CFT positive in 75% cases, IHA, IFA, ELISA.	1. Direct – Stool examination and duodenal intubation for ova. 2. Indirect – a) Eosinophilia. b) Occult blood in stool. c) Charcot Leyden crystals in stool. 3. Test tube filter paper method of Haroda and Mori (testing soil sample for presence of hookworm eggs). 1. Ferrous sulphate 200-400 mg thrice daily. Folic acid, Vitamin B12 – for anemia. 2. Bephenium hydroxynaphthoate (for Ancylostoma); Tetrachloroethylene (for Necator);

N.B. *N. americanus* –

Habitat – America, Australia, Far East, India (except in Punjab and U.P.), South Africa, Sri Lanka.

Anterior end bends in the opposite direction to the body curvature. Adult worms smaller and more slender.

Buccal capsule – 4 chitinous plates (2 on ventral and 2 on dorsal surface).
(*A. duodenale* has 6 teeth – 4 hook-like on ventral surface and 2 knob-like on dorsal surface).

Dorsal ray split from the base (Total rays-14); 2 spicules, fused at the tip.

Female has no spine at posterior end.

Vulval opening in front of middle of the body.
(*A. duodenale* vulval opening behind middle of the body).

Tissue Nematodes

	Geographical distribution	Adult Worm	Microfilaria	Vector
1. *Wuchereria bancrofti*	India, West Indies, China, Japan, Pacific, C. Africa, S. America.	In lymphatic system (inguino-scrotal lymph nodes), smooth cuticula, no appendage in head.	In blood, sheathed and periodic, tail-tip free from nuclei.	Culex pipiens fatigans (India and China); Anopheles (Melanesian islands) Aedes (Polynesian islands)
2. *Brugia malayi*	Indonesia, Thailand, Vietnam, Malaysia, Burma, Korea, China, Japan, India.	In lymphatic system, smooth cuticula, no appendage in head.	In blood, sheathed and periodic, nuclei upto tail-tip.	Mansonia; Anopheles
3. *Onchocerca volvulus*	C. Africa, America	In subcutaneous connective tissues, rough cuticula with annular and oblique thickenings.	In skin, unsheathed and non-periodic, tail-tip free from nuclei.	Simulium (black fly)
4. *Dipetalonema perstans*	Africa, S. America	In subcutaneous connective tissues, smooth cuticula with appendage in head.	In blood, unsheathed and non-periodic, nuclei upto tail-tip.	Culicoides
5. *Dipetalonema streptocerca*	W. Africa	In subcutaneous connective tissues in chimpanzees only.	In natives of W. Africa in skin, unsheathed and non-periodic, nuclei upto tail-tip, tail-end crook-like.	Culicoides
6. *Mansonella ozzardi*	W. Indies, America	Mesentery	In blood unsheathed and non-periodic, tail-tip free from nuclei.	Culicoides
7. *Loa loa*		In subcutaneous connective tissues and conjunctiva, rough cuticula with minute	In blood, sheathed and	

Characteristics	Classical Filariasis	Occult Filariasis
Developing forms	Developing adult worms	Microfilaria
Due to basic lesion	Epitheloid granuloma surrounding adult worm	Eosinophilic granuloma (allergic reaction)
Organs involved	Lymphatic system	Lymphatic system, lungs, liver and spleen
Microfilaria present	In blood	In affected tissues, but not in blood
Complement fixation test	Not sensitive	Highly sensitive
Response to diethylcarbamazine	No response	Responds

Pathogenicity of Classical Filariasis

1. Lymphadenitis – Inflammation of regional lymph nodes (in groin or axilla)

2. Lymphangitis due to:

   i) Mechanical irritation by adult worms

   ii) Secretion of toxic fluid by fertilised females and metabolites of growing larvae

   iii) Absorption of toxic products from dead worms

   iv) Secondary bacterial infection by *Streptococcus species*.

   Might cause epididymo – orchitis, funiculitis, retroperitoneal (abdominal) lymphangitis.

3. Lymphatic obstruction due to:

   i) Mechanical blocking of lumen by dead worms

   ii) Endothelial proliferation and thickening of walls of lymphatic vessels (obliterative endolymphangitis)

   iii) Excessive fibrosis of lymphatic vessels by recurrent and repeated attacks of lymphangitis

   iv) Fibrosis of afferent lymph nodes.

4. Effects of lymphatic obstruction:

   i) Lymph varix – varicosity of lymph vessels (dilatation)

   ii) Rupture of lymphangiovarix

   - Lymphorrhagia (lymph scrotum, lymphocele, lymphuria)

   - Chylorrhagia (chylocele, chyluria, chylous ascites, chylous diarrhoea, chylothorax)

   iii) Elephantiasis (solid oedema) – hypertrophy and hyperplasia of skin and connective tissue of affected parts.

Laboratory diagnosis of Filariasis

1. Direct peripheral blood smear – direct coverslip preparation to see microfilaria.

   To detect light infection :

   One ml laked blood and 10 ml of 2% formalin centrifuged – sediment examined for microfilaria. 1-5 ml heparinised blood passed through 5μ nucleopore filter and then filter stained on a slide.

2. Lymph node biopsy for adult worms.

3. Serological tests :

   When parasite load is very low or due to periodic fluctuation and quantitative variation in microfilaremia, parasite is not detected in peripheral blood smear.

Two types of antigens are used:

- Particulate antigens are prepared from adult worms by cryostat sectioning and are used for indirect immunofluorescence test. Microfilaria can also be used.

- Soluble antigens may be crude or purified by disrupting the worm or larvae. Prepared by freeze - drying, crushing and making lipid-free, followed by saline extraction procedures. Except *O. volvulus*, no other human filaria is available in sufficient quantity. *D. immitis* or *D. witei* is used as sources of antigen and also antigen is prepared from non-filarial parasites as *Ascaris suum*.

   i) Complement fixation test and skin test using *D. immitis* antigen (Swada antigen). The latter is less sensitive and specific purified extract of adult. Hence it is not used now. Wheal of > 7mm diameter, 15 seconds after intradermal injection of 0.2 ml antigen is considered as positive.

   ii) Gel precipitation and Immunoelectrophoresis. – Good specificity in loiasis, wuchereriasis and onchocerciasis. If performed using cellulose acetate membrane, speeds up the migration of serum proteins and parasitic antigens, hence detection is faster.

   iii) Indirect fluorescent antibody test

   - SAFA test - Soluble filarial antigen is impregnaned into disc of special paper. Suitable for objective assessment using fluorimeter and may be automated.

   - Test using microfilaria whole or fragmented by crushing or ultrasonication and performed in tubes. The test is highly sensitive.

   - Test using section of adult worm. Counterstained with Evans blue, non-specific fluorescence is not seen. 85-90% positivity, but cross-reactions are seen between various types of filariasis. Diagnosed by difference in intensity of fluorescence and titres.

   iv) Macrophage inhibition factor and rosette inhibition test done with *O. volvulus* antigen appear to be specific, but only used in research.

   v) CIEP by using high density microfilaria as antigen, plasma of patients are tested.

   vi) Indirect haemagglutination inhibition test

   vii) Radioimmuno polyethylene glycol assay

   viii) Immune radiometric haemagglutination assay

   ix) Enzyme-linked Immunosorbent Assay (ELISA).

Dracanculus medinensis (Guineaworm)

Habitat – Worldwide

Adult Female – 60 cm to 1m long, in subcutaneous tissues of legs, arms and back.

Larva – 650-150 μm, tadpole-like coiled.

Intermediate host – Cyclops.

Mode of infection – Drinking unboiled and unfiltered water containing infected cyclops.

Pathogenesis – Blister formation at the site where female worm comes out to discharge larvae.

Laboratory diagnosis:

1. Blister site bathed with water – female worm appears at surface of skin and discharges milky white fluid, which is seen under microscope for coiled embryos

2. Intradermal test with Dracanculus antigen – wheal within 24 hours

3. Eosinophilia in blood

4. X-ray to detect calcified worms.

Treatment:

Extraction of the female worm by tying head with a fine silk attached to match stick and rolled out inch by inch daily with gentle traction (total 15-20 days), until whole worm comes out.

Guineaworm Eradication Programme:

- Started in 1983-84.
- Total affected – 14 million in India in endemic areas.
- Seven states in India were affected – Rajasthan, Gujarat, M. P., Maharashtra, A. P., Karnataka and Tamil Nadu.
- Tempos (Abate) 1 ppm is added to all open water sources to destroy cyclops – at 4-6 weeks interval for 4 times, during the months of April to June.

LABORATORY DIAGNOSIS OF PARASITIC INFECTIONS

A. Microscopy:

1. Saline and iodine mounts for protozoal cysts and trophozoites, helminthic ova and larvae of *Strongyloides stercoralis*.

2. Perianal or perineal area swabbed with NIH swab and seen for ova of *E. vermicularis*.

3. Peripheral blood smears stained by Leishman's, Wright's or Giemsa stain for demonstration of malarial parasites, microfilaria, trypanosomes, etc.

4. Demonstration of parasites in urine (ova of *S. haematobium*, microfilaria), sputum (*P. westermanii*), CSF (free-living amoebae), hydatid fluid (scolices and hooklets of *E. granulosus*) and other body fluids by direct wet smear.

5. Modified acid fast staining with 10% H_2SO_4 for demonstration of oocysts of *Cryptosporidium*, *Isospora* and *Cyclospora species*.

6. Trichrome staining (Chromotrope 2R) for Microsporidia.

7. Concentration methods of stool – Floatation and sedimentation methods for helminths and Sheather's sucrose floatation method for *Cryptosporidium*.

8. Hemoconcentration for hemoflagellates and microfilaria

9. Immunofluorescence test – for *Toxoplasma*. Microscopy is relatively simple, economical and can be performed with case.

It is of limited use in larva migrans (developmental stages are not excreted in body fluids), in chronic infections (less number of parasites in body fluids) and in survey of parasitic infections of low incidence.

B. Culture:

1. Blood or bone narrow culture in Novy McNeal Nicolle medium for *L. donovani* and *T. cruzi*.

2. Blood culture in RPMI 1640 for *Plasmodium spp*.

3. Culture of vaginal discharge in Diamond's medium for *Trichomonas vaginalis*.

4. Stool culture in Boeck and Drbohlav's medium or Philips' medium for *E. histolytica*

5. Copro culture of stool for filariform larvae of hookworm by the method of Haroda and Mori.

C. Animal inoculation:
Isolation of *Toxoplasma gondii* by intraperitoneal inoculation in mice.

D. Xenodiagnosis:
For diagnosis of trypanosomiasis, etc.

E. Serological methods:
Useful in chronic cases or carriers where parasite load is very less or parasites are lodged in tissues of host. It also can diagnose asymptomatic or latent infections and some chronic infections. They are highly sensitive and specific, but are of high cost technology.

1. Detection of specific circulating antibodies - IFA, LAT, Bentonite flocculation test, IHA, CFT, double diffusion in gel, CIEP, Methylene blue Sabin and Feldman dye test (for *T. gondii*), direct agglutination test (leishmaniasis), Direct card agglutination test (CAT for African trypanosomiasis), India ink immunoassary (toxoplasmosis), ELISA, RIA etc.

   These tests are unable to differentiate between recent and past infection and asses degree of parasitic infection. They are also unreliable in patients on immuno-suppressive therapy, having underlying diseases and in diagnosis of congenital infetions (congenital toxoplasmosis).

2. Specific IgM antibodies can diagnose acute and congenital infections, e.g. IgM-IFA for congenital toxoplasmosis.

3. CIEP, reverse IHA and ELISA can detect circulating parasitic antigens in serum, CSF, urine and feces for diagnosing amoebiasis, toxoplasmosis, filariasis, hydatid disease, etc.

   a) Whole, particulate or crude extracts of parasites (*Leishmania, Trichinella,* etc.)

   b) Semipurified (Trypanosoma, Echinococcus, Schistosoma, etc.)

   c) Purified (Trypanosoma, Schistosoma, etc.)

Antigen detection helps in early detection of cases, diagnosis of acute and congenital infections, prognosis to assess parasitic infection in patients on chemotherapy, immunosuppressive therapy or underlying diseases with suppressed antibody responses.

F. **Skin test:** It is of limited valve as specifity is less, lack of standardised antigen, difficulty in standardisation and uniformity of the test.

1. Immediate hypersensitivity – In diagnosis of helminthic infections, e.g. Casoni's test for hydatid disease, Scratch test for ascariasis, etc.

2. Delayed hypersensitivity – In diagnosis of protozoal infections, e.g. amoebiasis, leishmaniasis, trypanosomiasis and toxoplasmosis.

G. Tests to detect enzymes:

1. p-LDH enzyme detection in *P. falciparum*

2. HRP in *P. vivax* and *P. falciparum* infections.

H. Molecular Methods:

1. Monoclonal antibodies against a variety of parasitic antigens are used for identification of antigens

2. DNA probes are used in diagnosis of falciparum malaria and other parasitic diseases

3. Restriction endonuclease digestion

4. PCR

5. Western blotting.

Ovum of
*Schistosoma
mansoni*

Ovum of
S. hematobium

Ovum of
S. japonicum

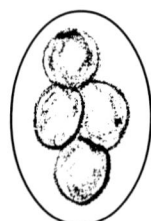

Ovum of
Hookworm

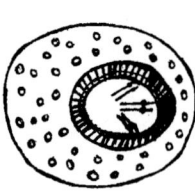

Ovum of *Taenia*

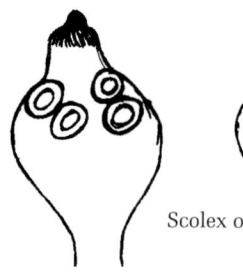
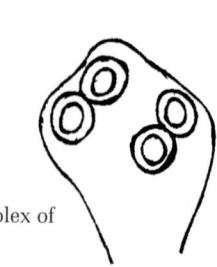

Scolex of

T. solium   *T. saginata*

Adult
Echinococcus granulosus

Ovum of
Fasciolopsis buski

Ovum of
Fasciola heptica

Ovum of
Paragonimus westermani

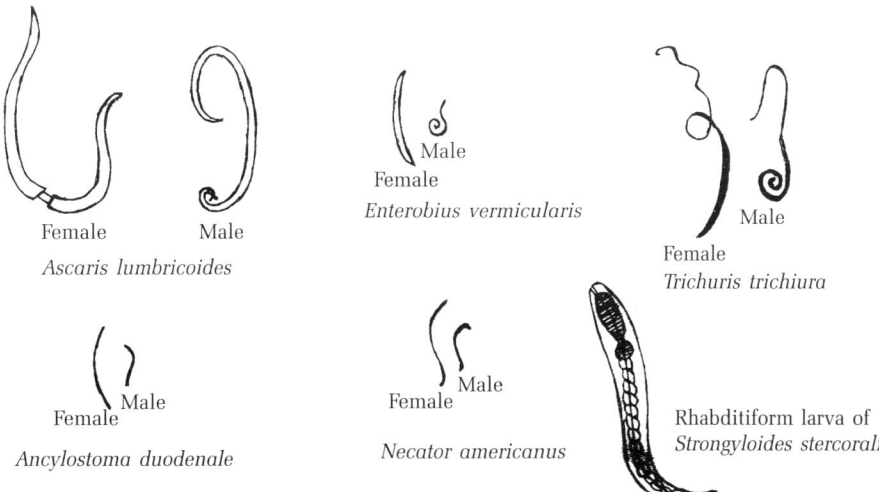

Female   Male

Ascaris lumbricoides

Male
Female
Enterobius vermicularis

Male
Female
Trichuris trichiura

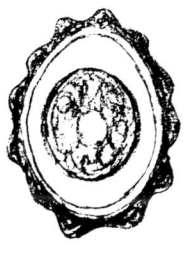

Male
Female

Ancylostoma duodenale

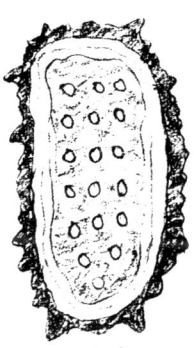

Female   Male

Necator americanus

Rhabditiform larva of
Strongyloides stercoralis

Fertilised ovum of
Ascaris
lumbricoides

Unfertilised ovum of
Ascaris
lumbricoides

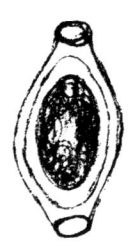

Ovum of
Trichuris
trichiura

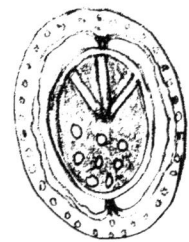

Ovum of *Enterobius*
vermicularis

Ovum of *Taenia*

Ovum of
Hymenolepis nana

PROTOZOAL INFECTIONS IN AIDS

Protozoal infection is one of the most common causes of morbidity and mortality in AIDS patients. Diarrhea is a major cause of morbidity and mortality in HIV-infected individuals. The number of immuno-suppressed (IS) individuals worldwide continues to increase each year as the human immunodeficiency virus (HIV) pandemic continues to spread unabated in many parts of the world, with an estimated 14,000 new infections occurring daily. Adults and children: 33.3 million (31.4-35.3 million); People newly infected with HIV: 2.6 million (2.3-2.8 million); Children newly infected with HIV: 370,000 (230,000-510,000); AIDS-related deaths worldwide: 1.8 million (1.6- 2.1 million).

All individuals affected by immunosuppression are at risk of infection by opportunistic enteric parasites (such as cryptosporidia, isospora, cyclospora, microsporidia), as well as those more commonly associated with gastrointestinal disease (*Giardia, Entamoeba*). Opportunistic enteric parasitic infections are encountered in 30-60% of HIV seropositive patients in developed countries and in 90% of patients in developing countries. The outcome of infection by enteric protozoan parasites is dependent on absolute CD4$^+$ cell counts, with lower counts being associated with more severe disease, more atypical disease and a greater risk of disseminated disease.

Classification

Group	Characteristics	Pathogen
Amebae	Simple amoeboid forms. Move by bulging and retracting their cytoplasm in any direction (pseudopodia).	*E. histolytica*
Ciliates	Move by rapid beating of cilia that cover the cell membrane.	*Balantidium* coli
Flagellates	Possess one or more flagella that give them a lashing motility.	*G. lamblia, T. vaginalis*, Trypanosomes
Apicomplexa	No special structures for locomotion. Reproductive cycle includes both immature and mature forms.	*T. gondii* and *Plasmodium species*
Coccidia	Represent a subphylum of the Apicomplexa.All stages of parasite development are intracellular.	*Cryptosporidium*, Isospora and *Cyclospora*
Microspora	Obligate, intracellular protozoa that produce spores. More than 160 genera and >1300 species, collectively called microsporidia.	*Enterocytozoon, Encephalitozoon, Nosema, Pleistophora*, etc.

Common protozoal infections in HIV/AIDS

- *Cryptosporidium parvum* – Cholera like watery diarrhoea, cholangiopathy, respiratory tract infections.
- *Isospora belli* – Chronic diarrhea and dehydration.
- *Cyclospora cayatenensis* – Chronic diarrhea.
- **Microsporidia** – Chronic diarrhea, Ocular, Renal and Respiratory tract infections.
- *Entamoeba histolytica* – Amoebic dysentery, ALA.
- *Giardia lamblia* – Diarrhea with malabsorption of fat.
- *Toxoplasma gondii* – Toxoplasmic encephalitis, Pneumonitis.
- *Leishmania donovani* – Visceral leishmaniasis.

Worldwide prevalence and Indian scenario of pathogenic enteric protozoa in HIV-infected persons

Enteric parasites	Worldwide Prevalence	Indian scenario
Cryptosporidium parvum	1.5%-37.3%	10.8%-11.5%
Microsporidia	2%-11%	1%-41%
Giardia lamblia	1.5%-17.7%	3.8%-8.3%
Isospora belli	0.8%-11%	2.5%-26.9%
E. histolytica/E. dispar	1.4%-10.3%	1.7%-7.7%
Cyclospora cayetanensis	1.9%-3.8%	-

Cryptosporidium parvum

Cryptosporidium parvum and *Cryptosporidium hominis* are the species most commonly associated with human cryptosporidiosis. In developed countries, incidence varies from 6-70% and in developing countries it varies from 8.7-48%.

Pathophysiologic features

Cryptosporidium parvum and microvillous membrane causes loss of microvilli and effacement resulting in malabsorption.

Second-signal pathways:

Nuclear factor-kB (NF-kB) – c-src system
↓
Production of cytokines – host-cell cytoskeletal
and chemokines – reorganization
↓
IL-8 – dysfunction of tight junctions
↓

Trigger an inflammatory reaction and stimulates anti-apoptotic survival signals in directly infected cells.

C. parvum induces secretion of 5-hydroxytryptamine (5-HT) and prostaglandin E2 (PGE$_2$) into the lumen and "Enterotoxin" activity which produces chloride secretion in vitro. *C. parvum* induces apoptosis in epithelial cells, resulting in damage to the epithelial barrier. HIV type 1 infection amplify damage by the action of soluble factors (tat protein). It produces varying degrees of villous atrophy by an unknown mechanism, resulting in malabsorption. Both humoral and cell-mediated immunity are involved in the resolution of cryptosporidiosis and resistance to infection. Inflammatory chemokines cause early development of the inflammation. Prostaglandin E_2 stimulates mucin production. The CD4+ T-cell has a protective role. In acute infection, specific IgG and IgM antibodies occur in serum. There is no protective immunity against reinfection.

Clinical features

- Asymptomatic infection: No change in bowel habits and 3 stools/day

- Transient infection: Diarrhea lasts for < 2 months, then a complete remission of symptoms and loss of cryptosporidium from fecal specimens.

- Chronic diarrhea: Lasting 2 months/more, with persistence of the parasites in stool or in biopsy specimens.

- Fulminant infection: In patients pass at least 2 lit. of watery stool daily (CD4+ T-cell count < 50 per mm).

- The diarrhea in patients with AIDS is usually watery, the stool frequency 10 per day, 10% drop in body weight and severe malabsorption.

- Low grade fever, nausea, vomiting, crampy abdominal pain.

- Extra intestinal cryptosporidiosis include cholangiopathy in biliary tract and respiratory tract infections in lungs.

Life Cycle

- Host - Man
- Habitat - The microvillus border of the small intestinal epithelial cells.
- Infective stage - oocysts.
- Transmission:
  i) Human to human: feco-oral route or indirectly by fomites.
  ii) Through the environment: By water.
  iii) Autoinfection: Thin walled cyst.

Laboratory diagnosis

- Laboratory diagnosis of cryptosporidiosis traditionally relies on special staining techniques, such as modified acid-fast, Kinyoun's safranin methylene blue and Giemsa stains, where 4-6 μm diameter, spherical oocysts can be detected. These may be confused with yeast cells which are pear-shaped and show evidence of budding.

- Concentration of stool can be done by Sheather's sucrose floatation technique. Direct immuno-fluorescent microscopy is used to detect oocysts in stool smears.

- Fluorescent stains used are Auramine O, Auramine-rhodamine, Auramine-carbol fuchsin, Acridine orange or 4, 6-diamidino-2-phenylindole (DAPI).

- Immunofluorescence with polyclonal and monoclonal antibodies which adds to the specificity in diagnosis. Several commercial companies have developed rapid diagnostic tests that are simple to perform e.g. lateral-flow immunoassays, immuno-chromatograhic assays and direct fluorescent-antibody tests.

- ELISA for antigen detection (by microwell ELISA) using polyclonal or monoclonal antibodies coated on to the wells of the microtiter plate. Sensitivity 94%; Specificity 100%.

- Latex agglutination assay where antisera are coated with latex beads.

- Antibody detection is done to study the prevalence of infection in a community. This cannot diagnose acute infection due to persistence of antibodies. Sensitivity is 94-97% and specificity is 98-100%. Commercial kits are available - ProSpecT, IDEIA, Color VUE. The use of enzyme-linked immunosorbent assay (ELISA) for the detection of *Cryptosporidium* antigen in stools has sensitivity of 94% and specificity of 100%.

- PCR → amplifies a 400 base sequence of the cryptosporidial DNA with no apparent cross-reactivity with human DNA. A TaqMan PCR assay targeting the 18S ribosomal DNA (rDNA) has allowed sensitive detection of *Cryptosporidium* species in stools. While being more expensive and time-consuming, PCR and ELISA have shown superior sensitivity for the detection of *Cryptosporidium* species compared to conventional staining and microscopy.

- Flow cytometry.

Treatment: Nitazoxanide 500 mg twice daily for 3 days.

Isospora belli

It is a coccidian parasite of humans that causes isosporiasis in patients with AIDS.

Isospora was first described by Virchow in 1860. The first case of human infection by *I. belli* was described among military personnel in First World War in 1915. First case from India was in 2001.

Pathophysiology

Invasion of epithelial cells of the proximal part of small intestine with cell damage. Electron microscopy shows mucosal alterations, including shortened villi, and eosinophilic infiltration of the lamina propria in the proximal part of the small intestine. Villus architecture may return to normal with treatment.

Life Cycle

- Mode of infection - By water and food contaminated with mature oocysts.
- Host - Man
- Infective stage - Oocyst
- Habitat - Small intestine

Clinical features

- Diarrhea without blood and inflammatory cells which is watery, profuse, foul smelling.
- Fever and abdominal pain.
- Acute self-limiting diarrhea resolves spontaneously in a normal host.
- In severely immunocompromised patients, severe chronic diarrhoea leading to a wasting syndrome and sometimes death can occur in AIDS patients.

Laboratory diagnosis

- Diagnosis is by direct visualization of elongated 20-30 μm by 10-20 μm pink oocyst with tapering ends, in feces by Modified Z-N stain.
- Concentration of stool by formol ether sedimentation can be done in asymptomatic or partially treated oocyst passers.
- A mucosal biopsy may be required sometimes for a definitive diagnosis.
- Oocysts have the property to autofluoresce under UV rays. The fluorescent oocysts are demonstrated in epifluorescence illumination using a 450-490 excitation filter.
- Enterotest demonstrate oocyst in duodenal samples.
- Detection of *I. belli* by PCR has been used as an additional diagnostic tool in clinical laboratories. Conventional PCR using primers based on 18S rDNA sequences shows excellent sensitivity and specificity.
- Recently, a real-time PCR targeting the internal transcribed spacer 2 region of the rRNA gene for the detection of *Isospora belli* DNA in fecal samples has been developed with 100% specificity and sensitivity.

Treatment

Trimethoprim-sulfamethoxazole i.e. Co-trimoxazole (80/600 mg) is the drug of choice. Initially 2 tablets QID for 10 days, then 2 tablets BD for 3 weeks.

Cyclospora cayatenensis

It was first described in the human faeces in 1979. There are 17 *Cyclospora species*. *C. cayetanensis* is the only species causing disease in humans. In 1996 a large outbreak of cyclosporiasis occurred with 1465 cases in 20 states. It was associated with contaminated raspberries from Guatemala.

Pathophysiology

Infection of the small intestine produces inflammation with leukocyte infiltration of the lamina propria. On histological examination, villous atrophy and crypt hyperplasia. The exact pathogenic mechanism is not well understood, believed to be mediated by endotoxin like substances.

Life cycle

- Mode of infection - By water and food contaminated with sporulating oocysts.
- Host - Man
- Infective stage - Sporulating oocyst.
- Habitat - Small intestine.

Clinical features

- Diarrhea, abdominal cramping, weight loss, and nausea.
- Diarrhoea watery and profuse.
- Symptoms typically wax and wane for several weeks and may persist for several months.
- *Cyclospora* associated with prolonged diarrhea in immunocompetent as well as immuno-compromised populations.
- Progressive immune suppression (CD4$^+$ T cell counts of < 200 cells/μl in HIV-infected individuals) results in frequent severe relapses which may last from 4 to 7 weeks, resulting in severe malnutrition and significant morbidity and mortality.

Laboratory diagnosis

- Diagnosis of *Cyclospora* oocysts may be problematic, as most laboratories fail to recognize them in direct fecal smears. They appear as 8-12 μm, pink, round oocysts with a distinct well defined wall in Modified Z-N stain and in safranin methylene blue stain, oocysts are reddish orange in colour.
- Special stains such as modified acid-fast auramine or modified iron-hematoxylin are usually required for definitive diagnosis.
- *Cyclospora* have the property of being autofluorescence under UV epifluorescence at a wave length of 320-380 nm.
- DNA amplification by PCR can also be performed using primers CYCF1E/CYCR2B.
- Histopathological examination of jejunal biopsy specimens from infected individuals shows mild to moderate acute inflammation of the lamina propria and surface epithelial disarray.
- Sporulation assay confirms the diagnosis of cyclosporiasis, oocysts should be examined over a 2-3 week period for evidence of sporulation. Stool sample and 2-3 volumes of potassium dichromate are taken. An air space is left over the specimen to promote aeration. At room temperature, oocysts sporulate in 5-14 days. If the number of oocysts is

small, it is centrifuged and the pellet is examined. Prior to examination, the potassium dichromate solution should be diluted with equal volume of distilled/deionized water. Wet mount is examined, **epifluorescence** (to identify the oocysts) followed by DIC, phase contrast microscopy, or conventional bright-field microscopy (to identify the sporocysts/ sporozoites). Sporulated mature oocysts contain two sporocysts, each of which contains two sporozoites.

Treatment

- In immunocompetent patients - Trimethoprim-sulfamethoxazole, 160 mg and 800 mg orally QID for 7 days.
- In patients with AIDS - Trimethoprim-sulfamethoxazole, 160 mg and 800 mg orally QID for 10 days, followed by prophylaxis TDS for a week to prevent relapse.

Microsporidia

These are a group of obligate intracellular protozoan parasites, first recognized as cause of pe´brine (pepper) disease pathogens in silkworm industry in France and Italy in Mid-17th century. The first human case was reported in a Japanese boy in 1959. Since 1985, it has been identified as a cause of opportunistic infection in persons with AIDS. The phylum Microsporidia consists of nearly 160 genera with > 1300 species, but only 7 genera *Enterocytozoon, Encephalitozoon, Pleistophora, Trachipleistophora, Vittaforma, Brachiola* and *Nosema* have been described as pathogens in humans. Most common species are *Enterocytozoon bieneusi, Encephalitozoon hellem, E. intestinalis* and *E. cuniculi*. The highest rates are found in homosexual men in Sweden.

Cell-mediated immunity appears to be critical for protection against the microsporidia. An effective Th1 cytokine response is important in the immune response to infection. The role of humoral immune responses is not known.

Life Cycle

- Mode of infection - Ingestion or inhalation.
- Host - Man.
- Infective stage - Spores
- Habitat - Small intestine.

Clinical Features

- Diarrhea is the most common clinical manifestation.
- In AIDS patients (< 100 CD4+ T cells/ml), it causes persistent diarrhea, fever, loss of appetite and weight loss.
- *Enterocytozoon bieneusi*/HIV co-infections are can occur in the hepatobiliary system (cholangitis), in the maxillary sinus (invasive sinusitis) and in the respiratory system (pulmonary infections).

- *Encephalitozoon species* can cause encephalitis, sinusitis, hepatitis, myositis, keratitis, pneumonia or peritonitis.

Laboratory diagnosis

- Traditionally, light microscopy based on modified trichrome stains (e.g. Weber Green and Ryan Blue) has been used by most laboratories for the diagnosis of enteric microsporidiosis. Spores appear as 1-2 μm diameters, bright pink, ovoid or pyriform in shape, with a pinkish belt-like stripe inside.

Modified Trichrome stain

- Chromotrope 2R stain - 0.8-1 μ, oval, bright pink in colour. Inside is transparent or has a pinkish belt-like stripe, mostly in clusters.
- Quick-Hot Gram Chromotrope stain - Spores stain dark or deep violet, belt-like stripe is enhanced.
- Duodenal aspirate - fixed with 10% formalin, centrifuged at 1500 rpm for 20 minutes. Smears made from sediment, air-dried and then stained by any of the above methods.

- Immunofluorescence tests using the chitin binding fluorochromes Uvitex 2B, Fungifluor, calcofluor white, and Fungiqual A are also available and offer a sensitive and rapid method for detecting microsporidial spores in stool, intestinal fluid, biopsy imprint, and tissue specimens.

- An immunofluorescent-antibody test (IFAT) for the detection of *E. bieneusi* and *E. intestinalis* is also available which is cheaper and more rapid.

- Staining of tissue sections by the Warthin-Starry staining method is an effective diagnostic tool for the microscopic detection of microsporidia and has better diagnostic capabilities than the hematoxylin and eosin stain.

- Molecular based PCR assays are also available for the diagnosis of *E. bieneusi* and *E. intestinalis* infection, but PCR inhibitors interfere with the test.

- Fluorescent in-situ hybridization (FISH) technology utilizes a fluorescence-labeled probe that binds to complementary nucleic acid in the specimen. It has been used in humans with probes against the SS rRNA or intergenic regions of rRNA to detect *E. bieneusi* and *E. hellem*. These methods detect either more microsporidia-positive samples or more infected cells within samples, than traditional histochemical staining. Disadvantages are that this is laborious and less sensitive than PCR.

- Microarray technology commonly employs an array of target-complementary oligonucleotides printed on a "chip". The array is able to detect all four species (*E. bieneusi, E. cuniculi, E. hellem* and *E. intestinalis*) of microsporidia at a sensitivity of 10^2 spores/100 μl of fecal sample.

Treatment: Albendazole, 400 mg twice daily for 2-4 weeks.

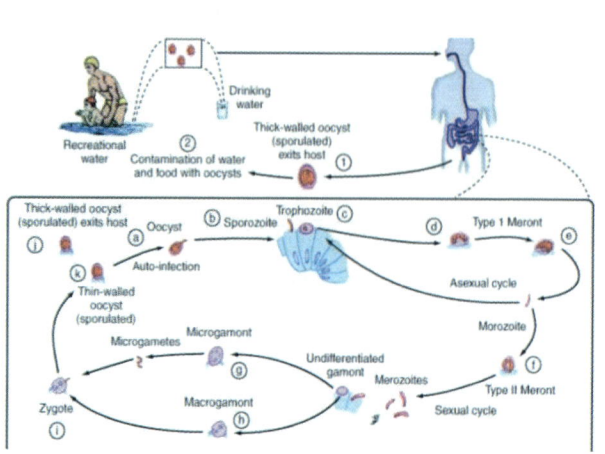

Life Cycle of Cryptosporidium species

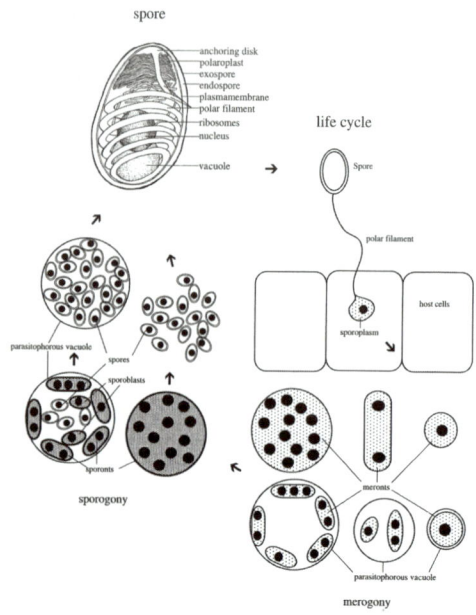

Life Cycle of Microsporidia

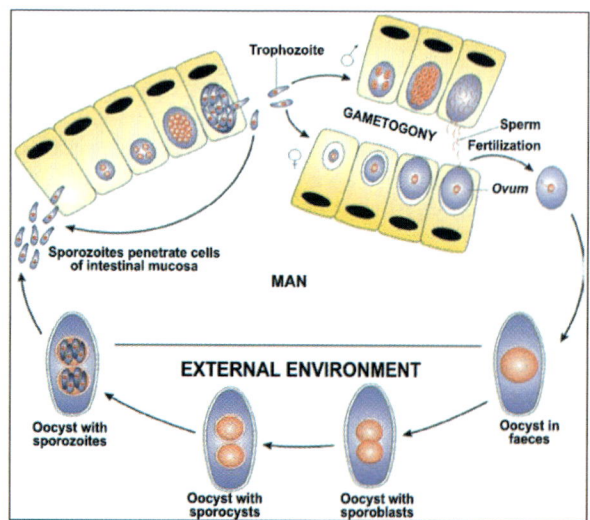

Life Cycle of Isospora belli

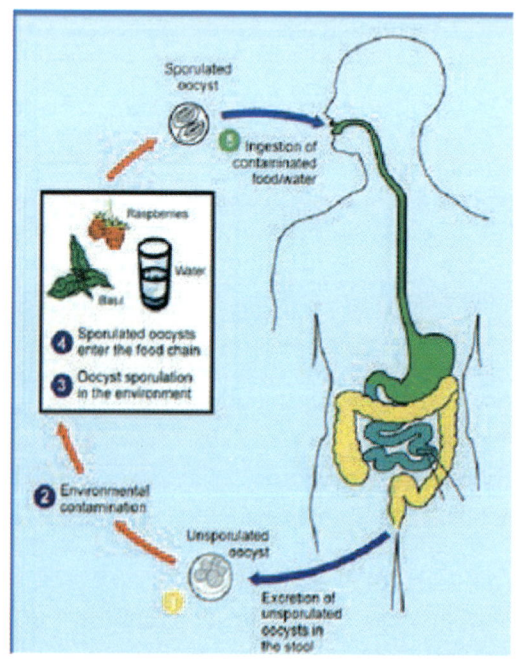

Life Cycle of Cyclospora species

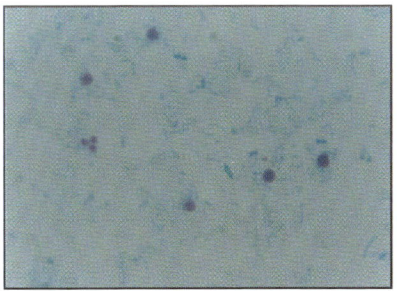 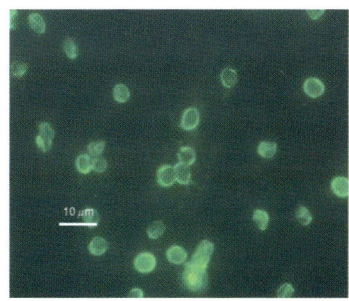

Modified Z-N stain (L) and Fluorescent Microscopy (R) showing oocysts of *Cryptosporidium species*

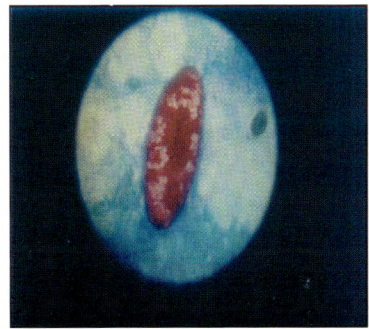 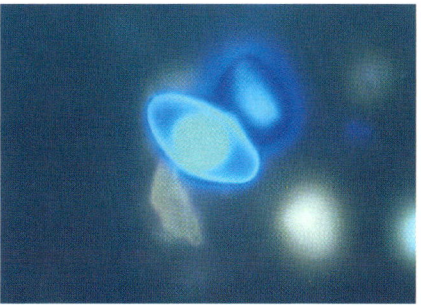

Modified Z-N stain (L) and Autofluorescence (R) showing oocysts of *Isospora belli*

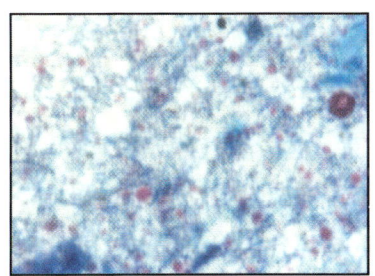 

Modified Z-N stain (L) and Autofluorescence (R) showing oocyst of *Cyclospora species*

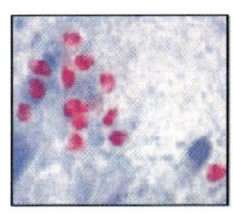 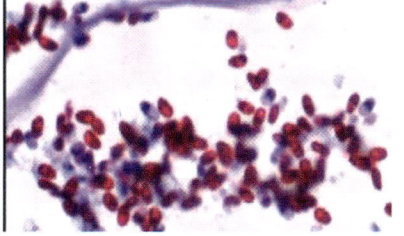

Modified Trichrome stain (L) and Gram Chromotrope stain (R) showing spores of microsporidia

Entamoeba histolytica

Entamoeba histolytica is a nonflagellated amoeboid protozoan parasite. The genus *Entamoeba* includes six species (*E. histolytica, Entamoeba dispar, Entamoeba moshkovskii, Entamoeba polecki, Entamoeba coli,* and *Entamoeba hartmanni*) that are capable of infecting the intestinal lumen of humans. Recent studies have reported *E. moshkovskii* as an enteropathogen in patients presenting with gastrointestinal symptoms. There have also been no adequate studies examining the pathogenic potential of this organism in IS groups and thus further study is needed to assess the true pathogenicity of this organism. It infects approximately 50 million people worldwide and accounts for > 100,000 deaths annually. It is more common in HIV-infected patients with a CD4+ cell count of < 200 cells/il. It is common in homosexual men. Transmission in these men occur by oral-genital or oral-rectal sex.

Host immune response to intestinal amebiasis

Stomach acid is a first line of defense against enteropathogens, but amebic cysts are highly resistant and excyst in the lumen of the intestine. Mucin is the main constituent of the protective mucus layer. Injured IECs release potent chemokines to migrate immune cells to the site of invasion. Activated macrophages release TNF-α, stimulating PMNs and macrophages to release reactive oxygen species (ROS) and NO, which kill the parasite. ROS and NO may also contribute to tissue destruction. IFN-γ released by lymphocytes activates macrophages and PMNs.

Mechanisms of action

Trophozoites attach to the host tissue surface via Gal/GalNAc lectin. Amebae secrete cysteine proteases, which disrupt the mucus layer and facilitate tissue invasion. E. histolytica trophozoite produces arginase, converts L-arginine to ornithine, this depletes the L-arginine supply that macrophages use to produce NO. COX in ameba or ameba-exposed macrophages produces the immunoregulatory molecule PGE2. PGE2 suppresses macrophage, elevating cAMP levels, which inhibits NO and TNF-α production. Amebic surface protein, confers resistance to ROS.

Life Cycle

- Host - Human.
- Infective stage - Quadrinucleate cyst.
- Invasive stage - Trophozoite.
- Habitat - Large intestine
- Mode of infection - Ingestion of food and water contaminated with quadrinucleate cyst.

Clinical manifestations

- Diarrhea with 10 stools/day, abdominal pain and tenderness
- Invasive or extraintestinal disease is seen in 0.1 to 1% of patients.
- Liver being the most common site involved and the lungs are the second most common site of invasive amoebiasis. Invasive amoebiasis is commonly seen in HIV infection. Amoebic liver abscess and amoebic colitis is more commonly seen. Invasive amoebiasis involving the heart, brain, and genitourinary tract are also seen.

Laboratory Diagnosis

- Stool microscopy, antigen detection in stool and stool culture in Boeck and Drbohlav's medium/NIH medium can be employed for diagnosis.
- Antigen-based ELISA kits that are specific for E. histolytica use monoclonal antibodies against the Gal/GalNAc-specific lectin, against serine-rich antigen and against a lectin-rich surface antigen.
- Detection of antibodies can be helpful in the case of ALA where patients do not have detectable parasites in feces. ELISA is performed in patients with ALA in intestinal and extraintestinal infections. Serum IgG antibodies persist for year, whereas serum IgM antibodies are short lived and can be detected during the present or current infection.
- Commercial antigen capture and antibody-detecting ELISA kits are also available along with rapid lateral-flow cartridge tests. Detection of serum anti-LC3 IgG can be used for diagnosis of ALA and acute amebic colitis.
- PCR is more useful than antigen capture ELISA for detection of E. histolytica in stools due to the higher sensitivities observed in PCR and the reduced chance of cross-reactivity with other Entamoeba species. A number of PCR assays are available for detection and/or differentiation of Entamoeba species.
- Serological tests like latex agglutination assays and indirect immunofluorescence assays have also been developed. However, serological testing in patients with intestinal disease is normally not recommended. The sensitivity of serological tests is 82% in those with invasive intestinal disease.
- Entamoeba dispar, Entamoeba moshkovskii and E. histolytica are morphologically identical, though genetic differences have confirmed the separation of these three as independent species. Therefore, staining of fixed fecal smears with iron-hematoxylin or Ziehl-Neelsen stain can determine the presence of the Entamoeba histolytica/E. dispar/E. moshkovskii complex within a stool specimen, while other techniques such as PCR or ELISA must be employed for differentiation.
- Amoebic liver abscess can be diagnosed by microscopy of anchovy sauce pus, where motile trophozoites will be seen and culture of pus in NIH medium.

Treatment

- For invasive disease - Metronidazole (750–800 mg TDS for 6–10 days) or Tinidazole (2 g once daily for 10 days) followed by Paromomycin (500 mg three times a day for 7 days)
- For intestinal disease - Paromomycin (500 mg three times a day for 7 days).

Giardia intestinalis

Giardia intestinalis (Giardia lamblia) is a common and ubiquitous flagellated protozoan parasite with a worldwide distribution. Humans become infected by ingestion of cysts, which develop into trophozoites after excystation. Giardiasis is not considered a major cause of enteritis in HIV-infected patients. With progressive immunosuppression following reduced CD4+ counts, the risk of symptomatic *Giardia* infections is increased. There has been an increase in the prevalence of giardiasis in the male homosexual population. Antigiardial host defenses are B-cell dependent. T cells also appear to be important in the intestinal elimination of the organism.

Life Cycle

- Host - Man
- Infective stage - Cyst
- Habitat - Small intestine
- Mode of infection - Ingestion

Clinical features

In acute infection, sudden onset watery, foul-smelling diarrhea and nausea are seen for days to weeks. In chronic symptoms, diarrhea, abdominal discomfort with progressive immunosuppression following reduced CD4 counts for years.

Laboratory diagnosis

- Giardia infection can be easily diagnosed by stool microscopy by identification of cysts and trophozoites in stained or unstained fecal smears.
- Immunofluorescence assay using anti-Giardia fluorescein-labelled monoclonal antibodies are used to detect Giardia specific antigens.

- Stool antigen detection uses a monoclonal antibody-based antigen capture ELISA used to detect Giardia specific antigens in stool specimens. Sensitivity 88-98% and a specificity of 87-100%.
- Enzyme immunoassays and immunochromatographic assays for detection of *G. intestinalis* in stool are available in the form of commercial kits.
- A number of PCR assays are also available for the detection of *Giardia* in stool specimen.
- Duodenal contents can be examined by string test and duodenal biopsies or aspirates are obtained by oesophagogastro-duodenoscopy. The biopsy specimen is prepared by touching a glass slide to the fresh biopsy specimen and staining for trophozoites by Giemsa and Trichrome stain.

Compared to microscopy, the coproantigen assays are less time-consuming and easier to perform. However, conflicting data are there in regard to the performance of these rapid assays. Some researchers have reported excellent sensitivity and specificity, while others have reported the rapid tests to be generally less sensitive than conventional microscopic methods.

Treatment

Metronidazole, 2 g daily for 3 days or Tinidazole 2 g single dose.

Blastocystis hominis

Organisms round, 6-40 μm, usually characterised by a large central body (resembles a large vacuole).

More amoebic form seen in diarrhoeal fluid.

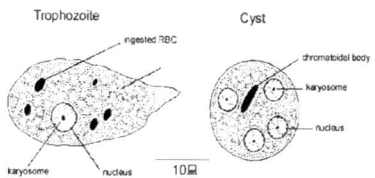

Trophozoite and cyst of *Entamoeba histolytica*

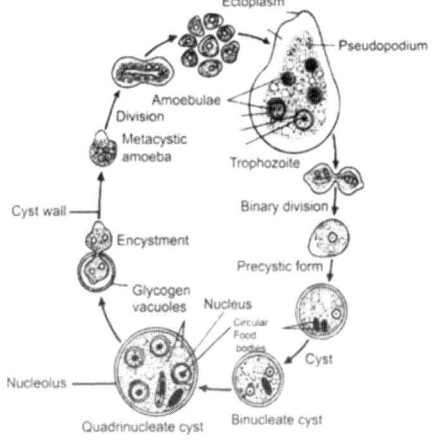

Life Cycle of *Entamoeba histolytica*

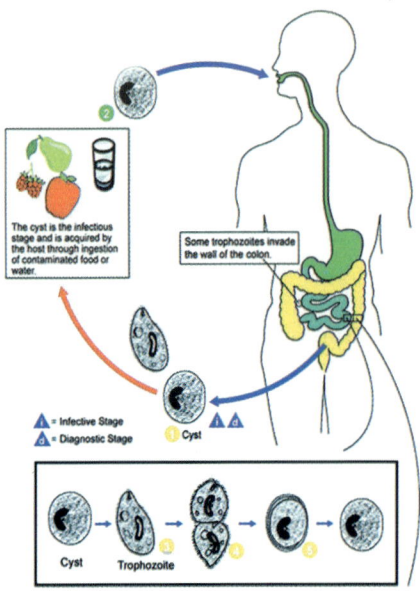

Life Cycle of *Giardia lamblia*

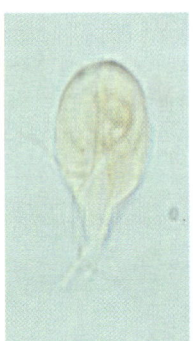

Trophozoite of *G. lamblia*

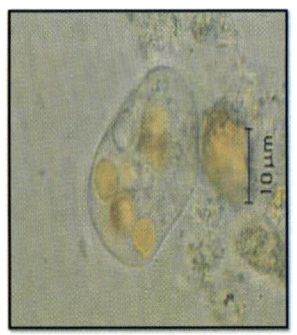

Trophozoites of *E. histolytica*

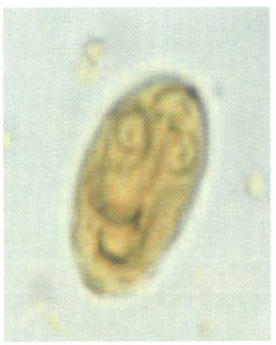

Cyst of *G. lamblia*

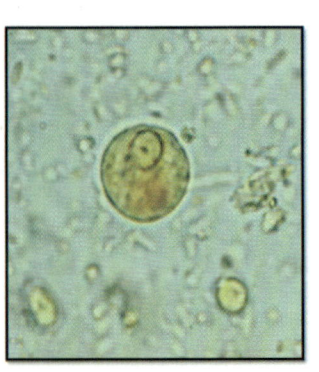

Cyst of *E. histolytica*

Toxoplasma gondii

Toxoplasma gondii is an obligate intracellular parasite. The parasite was discovered by Nicolle and Manceaux in blood, spleen, and liver of a North American rodent *Ctenodactylus gondii* in 1908 and was named *Toxoplasma gondii* in 1909. In 1937 it was found in a congenitally infected child by Wolf, Cowan, and Paige and in 1970, the life cycle of this parasite was fully described.

Toxoplasmosis is a common opportunistic infection among immunocompromised patients, usually due to reactivation of latent infection but can result from acute infection as well. Toxoplasmosis in these persons may lead to lethal meningoencephalitis, focal lesions of CNS and less commonly myocarditis and pneumonitis. Determination of IgM antibodies against Toxoplasma encourages rapid diagnosis and treatment. The global seroprevalence is 46.1%.

Seroprevalence in Indian population is 16.3%-30.8%. Seroprevalence in pregnant women (worldwide) varies from 7-51.3%.

Pathophysiology

In acute infection, tachyzoites proliferate in the gastrointestinal and in extraintestinal sites cause disruption and death of cells, producing necrotic foci, which are surrounded by mononuclear cells. Acute necrotizing encephalitis, pneumonitis and myocarditis are also seen in immunodeficient and normal host. IgM antibody appear in first week of the infection and then declines in the next few months. IgA antibody is detectable in acute infection (since the titre can last for >1 year). IgG antibody appears late, attains peak within 6-8 weeks following infection and then declines over the next two years. CMI plays a role in resistance to reinfection. In patients with AIDS, inducer T-lymphocytes is markedly decreased. This may contribute to the severe manifestations of toxoplasmosis.

Life Cycle

- Tachyzoites, tissue cysts and oocysts are important stages of parasite. All these three stages are infectious to humans.
- Definitive host - Cat

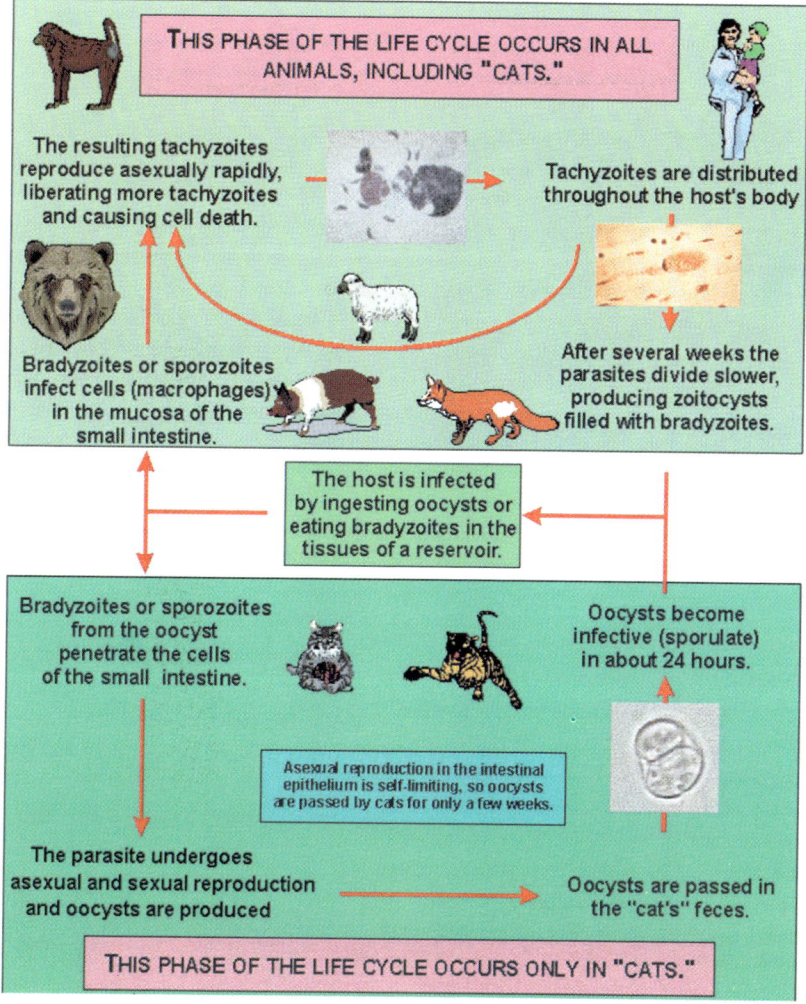

Life Cycle of Toxoplasma gondii

- Intermediate host - Man and other mammals
- Infective form - Sporulating oocyst, tissue cysts, tachyzoites.
- Habitat - Reticuloenndothelial cells and other nucleated cells.

Clinical Features

- Severe encephalitis is seen in 40% of AIDS patients.
- Lymphadenopathy is classical sign. Deep cervical LN, supraclavicular, suboccipital, axillary and inguinal lymph nodes are also involved.
- In brain - altered mental status, seizures, sensory defect, cerebellar signs, and neuropsychiatric manifestations.
- In lung - pneumonitis with prolonged febrile illness, cough and dyspnoea (AIDS with CD4 count < 50 cells/mm^3).
- In other immunocompromised patients - CNS is mainly affected and manifests as encephalitis/meningoencephalitis/SOL.

Laboratory diagnosis

- Microscopy detects tachyzoites or cysts. Specimens are bone marrow aspiration and brain biopsy. Staining methods used are Giemsa stain, Periodic acid Schiff stain, Gomori methenamine silver stain and Haematoxylin & eosin stain.
- Radiodiagnosis: Cerebral toxoplasmosis usually causes unifocal & multifocal lesions and less likely diffuse encephalitis. Computed tomography (CT scan) – Typical findings of hypodense lesions with ring enhancing and perilesional edema in 80% of cerebral toxoplasmosis cases. Magnetic resonance imaging (MRI) – Focal lesions and diffuse cerebral encephalitis without visible focal lesions, are shown in ~20% of these cases.

    A CT scan or MRI is useful for assessment of patients who had responded to initial empirical treatment. An evidence of either clinical or radiographic improvement within 3 weeks of initial therapy provides a confirmed diagnosis of patients with cerebral toxoplasmosis.

- Serodiagnosis is done by Sabin and Feldman dye test and ELISA for IgG/IgM antibodies. *T. gondii* excretory/secretory antigens (ESAs) are an excellent serological marker for the diagnosis of cerebral toxoplasmosis in AIDS patients. ESAs stimulates humoral and cellular immunities in order to control Toxoplasma infection. Anti-ESA IgG antibodies are also present in CSF sample of AIDS patients with cerebral toxoplasmosis which can be determined by ESA-ELISA and immunoblot techniques.

- Molecular diagnosis by PCR techniques are suitable for patients with AIDS. Samples used are CSF, amniotic fluid or blood. Clinical samples should be collected before or up until the first 3 days of anti-Toxoplasma therapy because the diagnostic sensitivity will be reduced after the first week of this specific treatment. In CSF samples, sensitivity varies from 11.5%-100% and specificity is 96-100%. In blood samples sensitivity varies from 16%-86%.

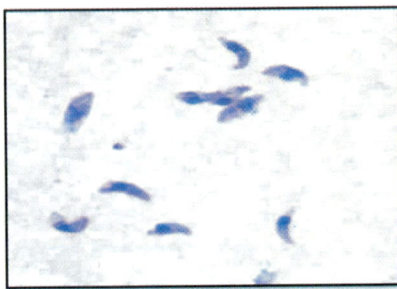

Toxoplasma gondii **tachyzoites (comma shaped), stained with Giemsa**

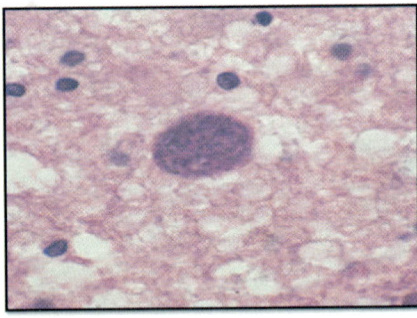

Toxoplasma gondii **cyst in brain tissue stained with hematoxylin and eosin**

Treatment

- In AIDS patients: Pyrimethamine, 200 mg orally, followed by 50-70 mg/day; Folinic acid, 10 mg/day orally; Sulphadiazine, 4-8 mg/day orally for 6 weeks.
- Suppressive therapy include: Pyrimethamine, 50 mg/day orally; Sulphadiazine,1-1.5 g/day orally; Folinic acid, 10 mg/day.

Leishmania donovanii

Both Leishman and Donovan reported the parasite simultaneously in 1903. Leishman detected it in spleen smear of soldier in England, who died of fever contracted at Dum Dum in Kolkata. Donovan found it in spleen smear of a patient suffering from Kala-azar in India. More than 12 million people are affected worldwide. New cases reported annually are 2 million. HIV/AIDS increases the risk (100 to 1,000 times) of developing Visceral Leishmaniasis (VL) in endemic areas. Leishmaniasis in HIV-infected patients is mainly seen as VL.

HIV-Leishmania coinfection

The clinical triad of fever, splenomegaly, and hepatomegaly is found in patients with low CD4 counts (< 50 CD4 cells/mm^3). In these patients, leishmaniasis can present with gastrointestinal involvement, ascites, pleural or pericardial effusion, involvement of lungs, tonsils, and skin. The parasite load is heavy in these patients, the presence of leishmania amastigotes in the bone marrow can often be demonstrated.

Morphology

Amastigote (Leishman-Donovan body):

- Human phase – reside in macrophage, elliptical, no free flagellum.
- Nucleus: located at one side
- Has cytoplasm
- Kinetoplast: slender, rod shaped
- Axoneme: arises from kinetoplast to margin of the body.

Promastigote:

- Vector phase – long, slender, 15-25 um in length and 1.5-3.5 um in breadth
- Reside in the gut of sandfly
- Spindle shaped with 1 free flagellum
- Have nucleus, cytoplasm, basal body, axoneme

Life Cycle

- Host - man and sandfly
- No sexual development in the host
- Residing site - macrophage
- Infective stage - promastigote
- Infective route - inoculation of sandfly *Phlebotomous argentipes*
- Reservoir host - dog
- Infection could also be via transfusion

Clinical features

- Pyrexia
- Splenic enlargement
- Liver enlarged
- Haematological abnormalities – anemia and leucopenia (1000/mm^3)
- Hypergammaglobulinemia
- Other features – Hyperpigmentation of skin is seen in Indian patients so called

  Kala azar; Lymphadenopathy in African and Chinese forms, rare in Indian form.

Laboratory diagnosis

- **Microscopy**:
  - Splenic smear: LD bodies
  - Bone marrow aspiration: Sternum/iliac crest – LD bodies positive in 54-86% of cases.
  - Other tissue: Liver tissue can be used
  - Peripheral Blood: Amastigote form

- **Indirect evidences:**
  - Aldehyde test
  - Antimony test
  - Leishmanin skin test (Montenegro test)
  - Complement fixation test – This test depends upon the presence of certain immune bodies in blood sera of kala-azar patients. Antigen is prepared from human tubercle bacillus by Witebsky, Klingenstein and Kuhn, so called as W. K. K. antigen. Used for early diagnosis of disease (within 3 weeks of infection)

Treatment: Antimony compounds

Sodium antimony gluconate (SAG), 600 mg daily for 6-10 days by IV route

Pentamidine isethionate, 4 mg/kg/day IM for 10 days

REFERENCES

1. Chatterjee KD. Parasitology (Protozoology and Helminthology in relation to clinical medicine). 12th Ed. Chatterjee Medical Publishers, Calcutta. 1980.

2. Parija SC. Textbook of Medical Parasitology: Protozoology and Helminthology. 4th Ed. Paperback. 2013.

3. Parija SC. Review of Parasitic Zoonoses. 1st Ed. A.I.T.B.S. Publishers Distributors, Delhi. 1990.

4. Xian Ming C hen, Keithly JS, Paya CV. Cryptosporidiosis. N Engl J Med 2002; 346: www.nejm.org

5. Franzena C, Müllerb A. Microsporidiosis: human diseases and diagnosis. Microbes and Infect 2001; 3: 389-400.

6. Stark D, Barratt JLN, Hal S van, Marriott D, Harkness J, Ellis JT. Clinical Significance of Enteric Protozoa in the Immunosuppressed Human Population. Clinical Microbiology Reviews, 2009; 22: 634-50.

7. Pollok RC, Farthing MJ. Enteric viruses in HIV-related diarrhoea. Mol Med Today 2000; 6: 483-87.

8. Mathers CD, Loncar D. Projections of global mortality and burden of disease from 2002 to 2030. PLoS Med 2006; 3: e442.

9. Mohandas K, Sehgal R, Sud A, Malla N. Prevalence of intestinal parasitic pathogens in HIV seropositive individuals in Northern India. Jpn J Infect Dis 2002; 55:83-4.

10. De A, Patil K, Mathur M. Detection of enteric parasites in HIV positive patients with diarrhea. Indian J Sex Transm Dis 2009; 30:55-6.

11. Garcia LS, Shimizu RY, Bernard CN. Detection of Giardia lamblia, Entamoeba histolytica/Entamoebadispar, and Cryptosporidium parvum antigens in human fecal specimens using the triage parasite panel enzyme immunoassay. J ClinMicrobiol 2000; 38:3337-40.

12. Jayalakshmi JB, Appalaraju, Mahadevan K. Evaluation of an enzyme-linked immunoassay for the detection of Cryptosporidium antigen in fecal specimens of HIV/AIDS patients. Indian J PatholMicrobiol 2008; 51:137-8.

13. ten Hove RJ, Lieshout L van, Brienen EA, Perez MA, Verweij. Real-time polymerase chain reaction for detection of Isospora belli in stool samples. DiagnMicrobiol Infect Dis 2008; 61:280-3.

14. Lalonde LF, Gajadhar AA. Highly sensitive and specific PCR assay for reliable detection of Cyclosporacayetanensis oocysts. Appl Environ Microbiol 2008; 74:4354-8.

15. Añé MS, Fernández FAN, Avila JP, Bringuez MB, Viamontes BV. Emergence of a new pathogen: Cyclosporacayetanensis in patients infected with human immunodeficiency virus. Rev Cubana Med Trop 2000; 52:66-9.

16. Endeshaw T, Kebede A, Verweij JJ, Zewide A, Tsige K, Abraham Y, et al. Intestinal microsporidiosis in diarrheal patients infected with human immunodeficiency virus-1 in Addis Ababa, Ethiopia. Jpn J Infect Dis 2006; 59:306-10.

17. Didier ESJ, Orenstein M, Aldras A, Bertucci D, Rogers LB, Janney FA. Comparison of three staining methods for detecting microsporidia in fluids. J Clin Microbiol 1995; 33:138-45.

18. Patil K, De A, Mathur M. Comparison of Weber green and Ryan blue modified trichrome staining for the diagnosis of microsporidial spores from stool samples of HIV-positive patients with diarrhoea. Indian J Med Microbiol 2008; 26: 407.

19. Cisse A, Ouattara OA, Thellier M, Accoceberry I, Biligui S, Minta D, et al. Evaluation of an immunofluorescent-antibody test using monoclonal antibodies directed against Enterocytozoonbieneusi and Encephalitozoon intestinalis for diagnosis of intestinal microsporidiosis in Bamako (Mali). J ClinMicrobiol 2002; 40:715-8.

20. Dowd SE, Gerba CP, Enriquez FJ, Pepper IL. PCR amplification and species determination of microsporidia in formalin-fixed feces after immunomagnetic separation. Appl Environ Microbiol 1998; 64:333-6.

21. Fotedar RD, Stark N, Beebe, Marriott D, Ellis J, Harkness J. Laboratory diagnostic techniques for Entamoeba species. ClinMicrobiol Rev 2007; 20:511-32.

22. Fotedar RD, StarkN, Beebe, Marriott D, Ellis J, Harkness J. PCR detection of Entamoeba histolytica, Entamoeba dispar, and Entamoeba moshkovskii in stool samples from Sydney, Australia. J ClinMicrobiol 2007; 45:1035-7.

23. Leo M, Haque R, Kabir M, Roy S, Lahlou RM. Tests for the Rapid Diagnosis of Amebiasis. J Clin Microbiol 2006; 44:4569-71.

24. Amar CF, Dear PH, McLauchlin J. Detection and genotyping by real-time PCR/RFLP analyses of Giardia duodenalis from human faeces. J Med Microbiol 2003; 52:681-3.

25. Oster NH, Gehrig-Feistel, Jung H, Kammer J, McLean JE, Lanzer M. Evaluation of the immunochromatographic CORIS Giardia-Strip test for rapid diagnosis of Giardia lamblia. Eur J ClinMicrobiol Infect. Dis 2006; 25:112-5.

26. http://www.dpd.cdc.gov/dpdx.

MALARIA

WHO's division of Control of Tropical Disease (CTD) classifes the world into 4 malaria zones:

- There was never malaria or was eradicated without specific measures – 27% of world's population.
- Eradication compaign of 1960s was successful and has since been maintained – 31%.
- Malaria has resurged after earlier reductions – 33%
- Endemic malaria has been never substantially changed – 9%.

Malaria is a common tropical disease caused by *Plasmodium species* through the bite of female anopheles mosquito.

- Malaria occurs between 60°N and 40°S
- It is endemic in 91 countries, with 40% of world's population at risk
- New cases occurring annually – 300 to 500 million, 90% of them are in Africa.
- Deaths – 1.5 to 2.7 millions (1 million children in developing countries).

In India

- Number of malaria positive cases – 2.8 million
- *Plasmodium falciparum* – 1.09 million
- Malaria kills one child every 30 seconds. It is the major factor contributing to maternal deaths during pregnancy:
- Blood slide examination (BSE) – 308.2 lakhs.
- Positive cases – 0.3 lakh
- Slide positive rate (SPR) – 1.06%
- *P. falciparum* infections – 0.01 lakh
- Slide falciparum rate (SFR) – 0.45%

Malaria are of four types:

- Malignant tertian malaria caused by *P. falciparum* (25-30% incidence)
- Benign tertian malaria caused by *P. vivax* (70% incidence)
- Ovale tertian malaria caused by *P. ovale* (< 1% incidence)
- Quartan malaria caused by *P. malariae* (< 1% incidence)

- Mixed infection with 2 or more species 4-8% 12 parasites/cu. mm. of blood is required to cause disease.
- Breeding season of anopheles mosquito is July to November.
- Optimum temperature 20-30°C
- Relative humidity > 60%.
- They are not found above 2000 to 2500 m height.
- Those with sickle cell trait have a milder form of malaria.
- Duffy negative persons are resistant to vivax infection.

Symptoms and signs:

- Fever with chills and rigors coming down with profuse sweating. Paroxysms of fever every 72 hours with *P. malariae* and every 48 hours with other types.
- Headache and myalgia are also present at height of fever (40-41°C)
- Liver moderately enlarged
- Splenomegaly (mild to massive) is always present when liver is enlarged
- Anemia – normocytic normochromic anemia.

Complications:

- In *P. falciparum* infection :
  - Pernicious malaria (Cerebral malaria, Algid malaria, Septicemic malaria)
  - Black water fever
- Other complications :
  - Disseminated intravascular coagulation (DIC)
  - Renal failure
  - Pulmonary oedema
  - Lactic acidosis
  - Nausea, severe vomiting and diarrhoea with or without blood and pus
  - Hypoglycemia, metabolic acidosis, hypokalemia.

Mortality:

- In *P. vivax* – due to rupture of spleen, immunocompromised state, repeated attacks in malnourished children.

- In *P. falciparum* – Cerebral malaria; Renal, hepatic and pulmonary involvement (80% mortality); Severe anemia and shock, hyperthermia and leucocytosis; Signs of gram negative sepsis; DIC > 3% parasitemia.

Characteristics of Plasmodium species

Characteristic	*P. falciparum*	*P. vivax*	*P. ovale*	*P. malariae*
Pre-erythrocytic cycle (days)	5-7	7-8	9	14-16
Erythrocytic cycle (hrs)	48	48	50	72
Red cell involvement	Usually invade younger cells, No change in cell size, Maurer's dots inside	Reticulocytes, Cell size enlarged, Schuffner's dots inside	Reticulocytes, Cell size enlarged Oval, Fimbriated with James's dots	Normal RBC, Ziemann's dots inside infected RBCs
Morphology				
a) Trophozoites	Only ring forms, Multiple ring in a single RBC, accolé formation	Irregularly shaped large rings, single ring in a RBC	Single size in a RBC	Band or rectangular forms of tropho-zoites, slightly amoeboid
b) Gametocytes	Sickle-shaped or crescentic, larger than RBC, host cell hardly recognized	Spherical or globular, larger than RBC	Oval, same size as RBC, host cell slightly enlarged enlarged	Round or oval, same size of RBC, host cell not
Pigment color	Black	Yellow-brown	Dark brown	Brown-black
Relapses (due to exo-erythrocytic cycle)	No relapses, Recrudescence is seen (persistence of blood infection)	Yes	Yes	Yes

Young ring form in RBC   Adult ring in enlarged RBC with Schuffner's dots   Merozoites   Female gametocyte   Male gametocyte

Accolé formation

Ring forms in RBC   Mature ring form with Maurer's dots, RBC not enlarged   Merozoites   Female gametocyte   Male gametocyte

Different forms of *Plasmodium falciparum* in erythrocytes

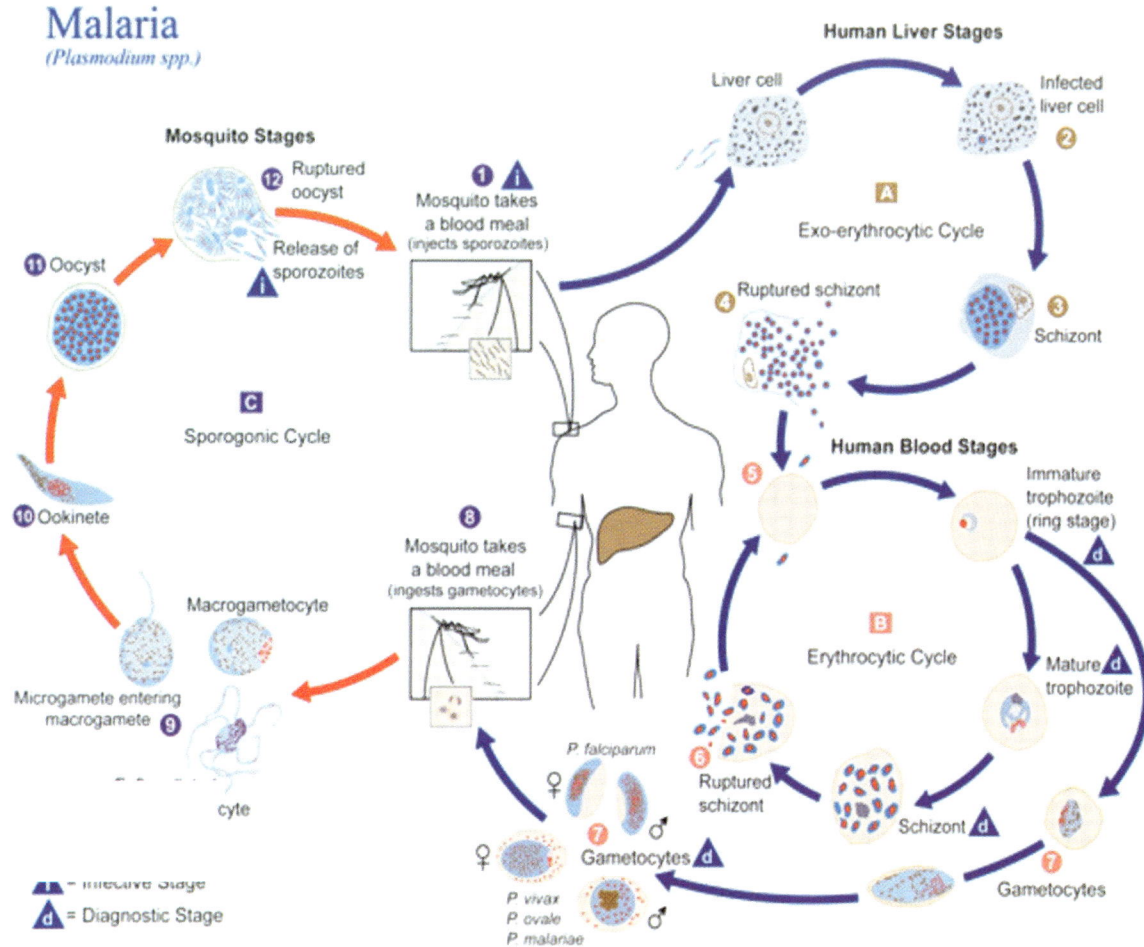

Life Cycle of *Plasmodium species*

Laboratory Diagnosis of Malaria

Malaria remains the world's most devastating human parasitic infection. Malaria affects over 40% of the world's population. WHO estimates that there are 350-500 million cases of malaria worldwide, of which 270-400 million are falciparum malaria, the most severe form of the disease.

- **Direct demonstration of Plasmodium**
  - Light microscopy
  - Peripheral blood smears
  - Staining techniques and quantification of the parasite.

- **Fluorescent microscopy**
  - Direct staining
  - Quantitative buffy coat examination.

- **Antigen detection**
  - Detection of Pf HRP 2
  - Detection of pLDH

- **Molecular techniques**
  - DNA probes
  - RNA probes
  - PCR

Light Microscopy

No other diagnostic means has superseded light microscopy for the routine diagnosis of malaria.

- Blood collected from finger prick or ear lobe is ideal.
- Best time to collect blood – few hours after the febrile paroxysm reaches its peak.
  - Thick films – 10 times more sensitive than thin films; Can detect 50 parasites/ml of blood.
  - Thin films – Specific; Less sensitive; Can detect 200 parasites/ml of blood.

The various Romanowsky's modifications currently available are Leishman's, Wright's, Giemsa and Jenner's. While Leishman's gives the most exquisite results in the thin film, Giemsa has proved to be the best all round stain.

Comparison of thin and thick films

Characteristics	Thin Film	Thick Film
Area on slide	250-450 mm^2	50-90 mm^2
Blood volume	1 microlitre	3-5 microlitres
Mean thickness	0.0025 mm	0.06-0.09 mm
Mean difference in concentration	1	30
Volume in 100 microscopic fields	0.005-0.007 microlitre	0.1-0.25 microlitre
Time for examination	200-300 field/20-25 minutes	100 fields/5 minutes
RBCs	Fixed	Hemolysed
Parasite morphology	Intact	Distorted

Quantification of parasites: Parasite density is directly proportional to the severity, duration and outcome of the infection. It also grades the response to therapy and monitors resistance to antimalarials. Quantification is done by:

- The percentage of infected RBCs
- The number of parasites/white blood cell
- The number of parasites/microscopic field

All the above can be transformed into cell index of parasites/ml of blood.

Estimating parasite numbers/ml of blood from thick film: The average number of parasites/high power field X 500. Ten to 50 fields should be examined. The no. of parasites/ml of blood = WBC count X parasites counted against 100 WBC/100.

Fluorescent Microscopy

- Direct staining of blood film by Acridine Orange (AO) stain: Simple and rapid.
- Fluorescent dye conjugates with gamma globulin of serum to be tested. If such conjugated gamma globulin contains malarial antibody, it will adhere to the relevant malarial parasite. Glistening particle under fluorescent microscope.
- Excited at 490 nm and exhibits apple green fluorescence. Stains nucleic acid from all cell types (Howell-Jolly bodies).
- Sensitivity: < 100 parasites/ml 41-93%; Specificity: 93%.

Quantitative Buffy Coat (QBC) – Becton Dickinson USA

A plastic cylinder positioned in the capillary tube expands the length of the buffy coat. The tubes are examined under a fluorescent microscope. The infected RBCs are most abundant at the granulocyte - erythrocyte interface.

Procedure

- EDTA venous blood is taken in glass test tube.

- A bored capillary tube coated with AO is charged with anticoagulated blood and spun in a microcentrifuge.
- Supernatant is discarded.
- The buffy coat and red cells below it are immediately transferred below to a slide.
- Tube is examined under fluorescent microscope.

Advantages

- Red cells are not destroyed.
- Parasites are stained well and appear more clearly defined in thick films.
- Malarial pigment in white cells is more easily detected.

Antigen Detection

The new generation antigen capture tests are capable of detecting fewer parasites and producing a rapid result.

- **_P. falciparum_ Histidine Rich Protein2 (Pf HRP2) detection:** This antigen is synthesized only by asexual forms of P. falciparum. The synthesis begins with immature trophozoites and is found in the infected RBC cytoplasm and cell membrane. Only 5% is found free in human plasma and even lower in the CSF. Therefore whole blood is the preferred sample.
- **Plasmodium Lactate Dehydrogenase (pLDH) detection:** The plasmodium LDH does not cross-react with human LDH. It is produced by all the four species of plasmodia in the asexual and sexual stages. The test is a direct sandwich ELISA using dipstick. It can differentiate between P. vivax and P. falciparum infection.
- **Aldolase.**

Rapid Tests with HRP2:

- Paracheck Pf (Orchid Biomedical Systems)
  100% sensitive; 93% specific
  Field-based blinded single microscopy in India since 2003

- ParaSight F (Becton Dickinson, USA)
- Malaquick (ICT, Australia).
- ParaHIT f (Span Diagnostic Ltd.)

  Sensitivity 77-98.1%; Specificity 91.2% with > 100 parasites/ml

- ICT Malaria Pf (ICT Diagnostics)

  Sensitivity: 100%; specificity: 84.5%.

Optimal (Flow Inc., USA)

Uses a panel of monoclonal antibodies that can bind to active pLDH:

- One specific for *P. falciparum*
- The other is a pan-specific, which reacts with all 4 species of Plasmodium infecting humans
- No cross reaction with human LDH or Rheumatoid factor or any nonspecific reactions with heterophile antibodies.
- Clearance of the parasite in blood stream during therapy is parallel with decrease pLDH level. Hence useful in:
  - Monitoring drug response.
  - Detecting drug resistant malaria.

Cultivation of Plasmodium

- Trager & Jensen (1976): RPMI 1640 medium is used.
- Malaria parasite is maintained in continuous culture in human erythrocytes , incubated at 38°C, in RPMI 1640 medium with human serum, with 7% CO_2 & O_2 (1 or 5%).
- Parasite is derived from an infected *Aotus trivirgatus* (Owl monkeys).

Molecular Diagnosis

- Ribosomal RNA has been detected by the radioimmunoassay technique.
- PCR amplification of the DNA, RNA and an enzyme Qâ replicase of *P. falciparum* have been developed primarily to screen blood donors in hyperendemic areas.
- Nested or multiplex PCR: can identify species.

  Sensitive & specific (almost 100%): Can detect 5 parasites or less/ml of blood.

- The advantage of molecular technique is that blood spots on filter paper can be used as a sample. The disadvantage is that they are money, labour and time consuming.

Comparison of diagnostic tests of malaria

Methods	Asexual parasites / µl	Sensitivity %	Specificity %	Cost (Rs.)	Total time
Giemsa stain	4	-	-	0.30	80 mins
AO stain (400X)	100	85-95	85-95	0.10	9 mins 10 sec
QBC	84	70-99	70-96	100	10 mins
Pf HRP2	30	86-100	84-99	98	16 mins
pLDH	60	83	99	150	16 mins
DNA	5	-	-	-	3-4 hrs
rRNA	10	-	-	-	3-4 hrs
PCR	0.3	100	100	-	3-4 hrs

Antimalarial Resistance

Drug-resistant *P. vivax*

Chloroquine-resistant *P. vivax* malaria was first identified in 1989 among Australians living in or travelling to Papua New Guinea. *P. vivax* resistance to chloroquine has also now been identified in Southeast Asia, on the Indian subcontinent, and in South America. *P. vivax* from Oceania, show greater resistance to primaquine than *P. vivax* isolates from other regions of the world.

Drug-resistant *P. falciparum*

Chloroquine-resistant *P. falciparum* first developed independently in three to four areas in Southeast Asia, Oceania, and South America in the late 1950s and early 1960s. Since then, chloroquine resistance has spread to nearly all areas of the world where falciparum malaria is transmitted. *P. falciparum* has also developed resistance to nearly all of the other currently available antimalarial drugs, such as sulfadoxine/ pyrimethamine, mefloquine, halofantrine, and quinine. Although resistance to these drugs tends to be much less widespread geographically, in some areas of the world, the impact of multi-drug resistant malaria can be extensive. Most recently, a low-grade resistance to artemisinin-based drugs has emerged in parts of Southeast Asia.

GPARC (Global Plan for Artemisinin Resistance Containment) – Goals and recommendations

To contain or eliminate artemisinin resistance where it already exists and to prevent artemisinin resistance where it has not yet appeared by: Stopping the spread of resistant parasites; Increasing monitoring and surveillance to evaluate the artemisinin resistance threat; Improving access to diagnostics and rational treatment with ACTs; Investing in artemisinin resistance-related research; Motivating action and mobilizing resources.

Tier I – Areas for which there is credible evidence of artemisinin resistance – In areas for which there is credible evidence of artemisinin resistance an immediate, multifaceted response should be launched. The goal is to contain and, if feasible, eliminate the resistant parasites. Intensify and accelerate malaria control to reach universal coverage of at-risk populations as soon as possible, including:

- Parasitological diagnosis for all patients with suspected malaria
- A full course of quality-assured ACTs plus primaquine for confirmed cases, in compliance with current.
- Vector control, as locally appropriate, to lower transmission and minimize the spread of resistant parasites.
- Programmes to reach mobile and migrant populations with adequate prevention, diagnosis and treatment.
- Trends in the numbers of confirmed malaria cases and deaths at health facilities and at community level.
- Malaria positivity rates with diagnostic tests, trends in treatment failure rates and the results of any special surveys conducted in the affected area.

Tier II – In areas with significant inflows of mobile or migrant populations from tier I areas, including those immediately bordering tier I –

- Intensified monitoring of therapeutic efficacy to track the spread of artemisinin resistance and ensure that the recommended first-line treatment remains effective
- Education and enforcement to eliminate the use of oral artemisinin-based monotherapies and substandard and counterfeit antimalarial medicines.

Tier III – Areas with no evidence of artemisinin resistance and limited contact with tier I areas. Implementation and scaling-up of effective control measures should be continued, including increasing access to parasitological diagnosis for all patients suspected of having malaria and increasing coverage with vector control to limit malaria transmission.

In addition, tier III areas should undertake two other components of good control:

- Monitor the therapeutic efficacy of first- and second-line treatments every 24 months
- In areas in which there is extensive use of oral artemisinin-based monotherapies or poor quality drugs, introduce or enforce actions to eliminate their use.

Management of Malaria

	Chloroquine	Quinine hydrochloride	Mefloquine	Halofantrine	QingHaosu
Treatment	600 mg (base), followed by 300 mg at 6, 24, 48 and 72 hours	650 mg twice daily for 7 days + Doxycycline 100-200 mg per day	1.5 mg single dose orally	For *P. falciparum* 200 mg every hour for 3 doses	From herbs by Chinese for *P. falciparum* (Artemisinin)
Prophylaxis	300mg (base) once weekly for 1 week before exposure, during exposure and for 4 weeks after exposure	200 mg daily for 1-2 days before exposure, during exposure and for 4 weeks after exposure	250 mg (base) once weekly for 1 week before exposure, during exposure and for 4 weeks after exposure (for chloroquine-resistant cases).	25mg-500 mg/ tablet, 3 tablets to carry to endemic area when medical treatment is not immediately available.	
Treatment of relapse	**Primaquine** 15 mg twice a day for 15 days				

Preventive measures:

- Proper, sewage drainage, prevention of water stagnation.
- Larvicides are spread over water to kill mosquitoes and stop breeding
  - DDT 1 gm/sq.m. for 2 rounds.
  - HCH 0.2 gm/sq.m. for 3 rounds.
  - Malathion 2 gm/sq.m. for 3 rounds at intervals of 6 weeks.
- Use of mosquito nets and anti-mosquito repellant creams.

REFERENCES

1. Chatterjee KD. Parasitology (Protozoology and Helminthology in relation to clinical medicine). 12th Ed. Chatterjee Medical Publishers, Calcutta. 1980.

2. Parija SC. Textbook of Medical Parasitology: Protozoology and Helminthology. 4th Ed. Paperback. 2013.

3. www.cdc.gov/malaria/diagnosis_treatment/diagnosis.html

4. Tangpukdee N, Duangdee C, Wilairatana P, Krudsood S. Malaria Diagnosis: A Brief Review. Korean J Parasitol. 2009; 47: 93-102.

5. Hawkes M, Kain KC. Advances in malaria diagnosis. Expert Rev Anti Infect Ther. 2007;5: 485-95.

6. Lee SW, Jeon K, Jeon BR, Park I. Rapid diagnosis of vivax malaria by the SD Bioline Malaria Antigen test when thrombocytopenia is present. J ClinMicrobiol. 2008;46:939-42.

7. Tagbor H, Bruce J, Browne E, Greenwood B, Chandramohan D. Performance of the OptiMAL dipstick in the diagnosis of malaria infection in pregnancy. Ther Clin Risk Manag. 2008;4:631-6.

8. Bhandari PL, Raghuveer CV, Rajeev A, Bhandari PD. Comparative study of peripheral blood smear, quantitative buffy coat and modified centrifuged blood smear in malaria diagnosis. Indian J Pathol Microbiol. 2008;51: 108-12.

9. Reyburn H, Mbakilwa H, Mwangi R, Mwerinde O, Olomi R, Drakeley C, Whitty CJ. Rapid diagnostic tests compared with malaria microscopy for guiding outpatient treatment of febrile illness in Tanzania: randomised trial. BMJ. 2007;334:403.

10. www.cdc.gov/malaria/malaria_worldwide/.../drug_resistance.html?. Accessedon 20.03.2014.

11. Global plan for Artemisinin resistance containment (GPARC). January 2011. http://www.who.int/malaria/publications/atoz/9789241500838/en/

MYCOLOGY

28

Chapter

IMPORTANT FUNGI

Type	Site	Disease	*Causative fungi*	Growth form	Cultural characteristics on SDA	Microscopy
A. Superficial Mycoses **1. Cutaneous**	a) Stratum corneum of skin	i) Tinea versicolor	*Malassezia furfur*	Y (Yeast)	SDA overlaid with a film of olive oil – creamy colonies in 5-7 days	KOH mount of skin scales – round yeast cells and short curved non-branched hyphae.
		ii) Tinea nigra	*Exophiala werneckii*	F (Filamentous)	Light to dark grey in colour	KOH mount of skin scrapings – brown branched septate hyphae and budding cells.
	b) Hair shaft of axilla	White piedra	*Trichosporon beigelii*	F	Blastospores, mycelium and arthrospores	KOH mount of hair – branched hyphae and arthrospores within and outside hair.
	c) Hair shaft of beard and scalp	Black piedra	*Piedraia hortae*	F	Asci and ascospores	KOH mount of hair – club-shaped asci, each containing 8 fusiform ascospores.
	d) Epidermis, hair and nails	Dermatophytosis (Ringworm) i) Tinea corporis, cruris, pedis, unguinum	*Trichophyton rubrum*	F	White cottony, occasionally pink, powdery, reverse wine red.	10% KOH mount for skin and hair, 20% KOH mount for nails – septate hyphae. **LPCB mount** – few thin-walled pencil-shaped macroconidia, abundant tear-drop microconidia.
			T. mentagrophytes	F	White-pinkish granular, powdery, cottony. Reverse– buff.	Thin-walled club-shaped macroconidia. Plenty round globose microconidia in grape-like clusters.
		ii) Favus	*T. schoenleinii*	F	Glabrous white-cream colony, radial grooves. Reverse – white brown.	Branching antler-hyphae with swollen tips, spiral hyphae, chlamydospores.
		iii) Endothrix hair infection	*T. violaceum*	F	Waxy, light to deep violet	Hyphae distorted, thickened ends, contain cytoplasmic granules, conidia rare.
	e) Epidermis and hair	i) Tinea capitis and corporis, Ectothrix hair infection	*Microsporum audouinii*	F	Velvety, salmon pink. Reverse – tan to pink.	Comb-like hyphae with terminal chlamydospores, microconidia rare. Thick-walled spindle-shaped multiseptate macroconidia with 4-6 septa.
		ii) Tinea capitis	*M. gypseum*	F	Powdery, cinnamon coloured. Reverse – light tan.	Thick-walled spindle-shaped macroconidia (4-6 septa), few microconidia.
	f) Epidermis and nails	Tinea capitis, cruris, pedis and unguinum	*Epidermophyton floccosum*	F	Powdery, greenish brown (khaki coloured). Reverse – yellow brown.	Large club-shaped multiseptate macroconidia in clusters of 2 or 3, **no microconidia.**

Important Fungi Cont...

Type	Site	Disease	Causative fungi	Growth form	Cultural characteristics on SDA	Microscopy
2. **Subcutaneous**	a) Subcutaneous tissue (usually affects upper limb)	Sporotrichosis (Severe ulcerating skin lesions)	*Sporothrix schenckii*	D (Dimorphic)	Conidia and hyphae in culture at 25°C. Blastospores in culture at 37°C.	Septate hyphae and cluster of oval spores. In tissues and room tempereature – cigar shaped yeast cells without mycelia (**Asteroid bodies**).
	b) Subcutaneous (S/C) tissues (usually affects foot)	Mycetoma (Madura foot) Eumycotic	*Madurella mycetomi, M. grisea, Pseudoallescheria boydii, Phialophora jeanselmei*	F F	Thick filaments > 1µm diameter, with septae and chlamydospores	Black grains Black grains White grains White grains
		Actinomycotic (bacterial)	*Nocardia brasiliensis Streptomyces somaliensis, Actinomadura pelletieri*	F	Thin filaments < 1 µm diameter, no septae or chlamydospores	White grains Yellow grains Red grains
	c) S/C tissues (usually nasal mucosa)	Rhinosporidiosis	*Rhinosporidium seeberi*	Y	Does not grow on SDA	Tissue section stained with H and E – Sporangium containing numerous endospores.
B. Deep Mycoses **1. Systemic**	d) S/C tissues	Chromoblastomycosis (Demetiaceous soil fungi)	*Exophiala, Fonseceae, Cladosporium (C. bantianum)*	D	Greenish grey to black after 4-6 weeks incubation at 25°C.	Brown pigmented hyphae. In tissue – dark brown yeast like bodies with septae (**Sclerotic bodies**).
	a) Lungs mainly, also disseminated when RE system is involved. In cases of leukemia and kidney transplants.	Histoplasmosis	*Histoplasma capsulatum;* *H. duboisii* (skin infections)	D	SDA with cycloheximide and chloramphenicol, incubated at 25°C – white fluffy colony, septate mycelium and large round tuberculate macroconidia.	1. Wright or Giemsa stain – small, oval, budding yeast cells (not broadbased), packed within macrophages. 2. Intraperitoneal inoculation in mice. 3. Histoplasmin skin test. 4. Serological tests – CFT, LAT, EIA.
	b) Lungs mainly, also skin and bone	Blastomycosis	*Blastomyces dermatitidis*	D	SDA with chloramphenicol only septate mycelium with conidia borne on lateral hyphal branches.	1. 10% KOH – thick walled yeast cells with broadbased buds. 2. Skin test and serology cross react with *H. capsulatum* and *C. immitis.*
	c) Lungs. Dissemination in 1% cases.	Coccidioidomycosis	*Coccidiodes immitis*	D	On SDA at 25°C for 3 weeks – little red pigment. Thickwalled arthroconidia in chains alternate with empty space and septate hyphae. Alternate arthroconidia, barrelshaped.	1. Large spherules containing numerous endospores. 2. Serological tests – LAT, CFT. 3. Coccidioidin skin test.
	d) Skin, mucosa, lymph node and lungs.	Paracoccidioidomycoses	*Paracoccidioides brasiliensis*	D	On SDA with yeast extract at 25°C for 2-4 weeks – mycelial phase with chlamydospores.	1. Large number of chains of yeast cells usually with multipolar budding. 2. Serological tests – Precipitin test, CFT.

Important Fungi Cont...

Type	Site	Disease	*Causative fungi*	Growth form	Cultural charac-teristics on SDA	Microscopy
2. Opportu-nistic (Predisposing factors – chronic diseases, diabetes, immuno-suppression)	a) Lungs, meninges. Sometimes skin, lymph node and bones.	Cryptococcosis Meningitis; Pulmonary – coin lesions, cavities, pleural effusion, skin infection.	*Cryptococcus neoformans var neoformans* in pigeon and chicken droppings. *C. neoformans var gatti* in euca-lyptus trees.	Y	Smooth, mucoid, cream-coloured colonies on SDA. On niger/bird seed agar – brown to black colonies within 7 days due to production of phenoloxidase.	1. Budding yeast cells with prominent polysaccha-ride capsule, by India ink. 2. Intraperitoned inoculation in mice. 3. Assimilate inositol. 4. Rapid hydrolysis of urea. 5. Positive L-DOPA ferric citrate test–black pigment. 6. Serological tests – Preci-pitation, CIEP, LAT, EIA, Dot Blot, Coagglutination. 7. Evaluation of bD glucan in blood, CSF, etc. 8. Detection of D-mannitol.
	b) Skin, mucosa, sometimes endo-cardium, meninges, gastrointe-stinal tract.	Candidosis	*Candida albicans*	Y with pse-udohyphae	Smooth, dry, creamy white colonies.	1. Gram stain – gram positive budding yeast cells with pseudohyphae. 2. Chlamydospore formation on corn meal – Tween 80 agar at 20°C. 3. Germ tube formation within 3 hours on incubation in human serum at 37°C. 4. Skin test with candida extract. 5. EIA. 6. Detection of bD glucan in blood and detection of D–arabinitol.
	c) Lungs mainly, also dissemi-nated.	Aspergillosis	*Aspergillus flavus*	F	Velvety yellow to green. Reverse – yellow to brownish.	Rough spiny conidiophore. Septate hyphae. On LPCB mount – conidia arranged in chains on elongated cells called sterigmata, borne on expanded ends (vesicles) of conidio-phores. Uniseriate and biseriate phialides covering entire vesicle.
	d) Ears – Oto-mycosis; Eyes – Oculo-mycosis.		*A. niger*	F	Wooly, white turning black. Reverse white to yellow.	Phialides uniseriate on upper two-thirds of vesicle, short smooth conidiophore.
			A. fumigatus		Gray – green colony, powdery. Reverse – white to tan.	Biseriate phialides covering entire vesicle, long smooth conidiophore. Serology – EIA. Detectio of D-mannitol.
	e) From nose to brain, orbit, sinuses and lungs also.	Mucormycosis	*i) Mucor, Absidia*		White fluffy wooly growth.	Broad nonseptate hyphae. On LPCB – branched hyaline sporangiophores termi-nating in globose sporangia containing numerous sporangiospores – no rhizoids.
		Mucormycosis	*ii) Rhizopus*		White fluffy wooly growth.	Unbranched brown sporangiophores arising opposite rhizoids.

Penicilliosis caused by *Penicilliosis marneffei.* First isolated in Manipur in India in an HIV positive patient.
- **Symptoms** – Fever, weight loss, cough, hepatosplenomegaly, diarrhoea, lymphadenopathy, disseminated typical skin-lesion like *Molluscum contagiosum.*
- Yeast like form at 37°C, grows on brain heart infusion agar.
- It reproduces not by budding, but they develop cross wall (median septa) and then splits into two.
- At room temperature – diffusible red pigment, typical long chains of *Penicillium.*

Urease positive fungi :
Cryptococcus neoformans, Candida krusei, Candida tropicalis, Trichosporon beigelii, Rhodotorula sp.

A. Serological Methods for Diagnosis of Fungal Infections

These are rapid methods of diagnosis.

Ideal diagnostic test should have a
- Good sensitivity and specificity
- High positive and negative predictive value
- Useful for monitoring
- Simple, rapid and inexpensive.

There is a need for improved standardisation.

1. *Aspergillus*
   - Aspergillus precipitin test for diagnosis of fungus ball.
   - *A. fumigatus* precipitin test for diagnosis of allergic bronchopulmonary aspergillosis.
   - Latex agglutination for *Aspergillus* species.

2. *Candida*
   - ELISA
   - BALISA (Biotin Avidin linked enzyme assay).

   In valve replacement, there is recurrence of candida endocarditis, which can be diagnosed by serology.

   Serology is also useful in candida endophthalmitis.

3. *Cryptococcus*
   - Antigen detection methods are available. In cryptococcal meningitis, antigen testing is done in CSF. Antigen persists for long time because of persistence of polysaccharides even when fungal cells disappear.
   - Latex agglutination test is 90% sensitive. False positivity is seen in rheumatoid arthritis, *T. beigelli* infection, *Capnocytophaga* infection.
   - CIEP
   - ELISA; Dot Blot
   - Detection of mannan antigens and fungal metabolites (D-arabinitol) by EIA.

4. *Histoplasma* – Antigen detection (available in USA).

5. *Coccidioides* – Exoantigen test detects 3 precipitin bands (F, HL and HS bands)
   - TP band is seen in atypical isolates.

B. Rapid Identification Tests for Yeasts

Plasma-β-D-glucan assay or G-test – β-D-glucan, a component of cell wall of fungi is detected in plasma of patients with deep mycosis using factor G, a horse-shoe crab coagulation factor. Sensitivity is 90% and specificity is 100% in patients with disseminated fungal infections. It helps in early diagnosis, but is unable to determine the species of fungi causing infection. This test needs further evaluation.

C. Routine antifungal susceptibility testing is done in casitone medium with amphotericin B and imidazoles. Colonies are inoculated in yeast nitrogen base and glucose medium. 106 yeast/ml is the standard inoculum.

Standard strains   – *C. albicans* ATCC 90028
                   – *C. tropicalis* ATCC 750.

D. Molecular methods for diagnosis of fungal infections

Conventional methods are of low sensitivity and may require days to weeks or even months for identification of fungi. Therefore several PCR - based assays have developed.

Polymerase chain reaction (PCR):

1. Utilises a highly conserved region of 18S rRNA, primers are designed to amplify all medically important fungi except zygomycetes.

2. PCR can amplify and detect DNA from as few as 5 *Candida* colony-forming units after 3 hours incubation of blood cultures.

3. PCR assays incorporating an internal plasmid to monitor sample inhibition, is used for diagnosis of aspergillosis and candidosis.

   Species specific DNA probes are absorbed to the solid phase, which can then hybridise with biotinylated or digoxigenin-labelled PCR amplicons produced from *Candida species* or *Aspergillus* species DNA.

4. Level of circulating *Aspergillus* DNA can be measured by PCR-EIA in invasive aspergillosis.

5. Detection of *Pneumocystis* DNA in BAL fluid, biopsy specimens and oropharyngeal secretions by PCR.

PNEUMOCYSTIS JIROVECII

Delanoe and Delanoe (1912) named the organism after Antonio Carinii.

In 1960s and 1970s interstitial pneumonia was recognised in children.

Sequence analysis of 16 S ribosomal RNA gene as well as protein encoding genes revealed it to be member of kingdom fungi.

- *P. jirovecii* causes Pneumocystosis, which is primarily a disease of the lungs.

- Mode of infection – by inhalation.

Life Cycle (proposed)

- Asexual – Thin-walled cysts, trophic forms multiply by binary fission.

- Sexual – Haploid cells by conjugation becomes a trophozoite – diploid (precyst – by meiosis to early cyst – matures into cyst – undergoes excystment – cyst attaches to **Type I pneumocyte** and multiply – thick- walled cysts (5-8 µ) with with 8 nuclei – spills in alveolar lumen – blocks gas exchange – respiratory distress.

Diseases produced:

1. **Pneumonia** – 80% AIDS patients get *P. jirovecii* pneumonia (PCP) with low CD4 count < 200/cu.mm. Symptoms – dyspnoea, dry cough, night sweats, cyanosis, few rales. Patients on aerosolised pentamidine have greater chance of pneumatoceles leading to bronchopleural fistula.

2. **Extrapulmonary** – seen in 0.5% to 3% patients with PCP. Sites affected – mastoid antrum, choroid, skin, small bowel, ascites with small nodules in stomach and small intestine, liver, spleen and hilar lymph nodes.

Laboratory diagnosis:

Samples – Open lung biopsy (best sample), has 100% sensitivity and specificity

– Bronchoalveolar lavage fluid has 96-98% sensitivity and specificity

– Tracheal aspirate or endotracheal secretions

– Induced sputum has 60% sensitivity

– Naturally expectorated sputum has 20% sensitivity

Induced sputum is concentrated by N-acetyl L-cysteine – NaOH and bovine serum albumin.

A. **Staining techniques**

1. **Cyst wall stains:**
   a) Gomori methenamine silver (GMS) stain has highest specificity. False negativity 10%.
   b) Toluidine blue O – cysts dark blue against a pale blue background (screening test).
   c) Calcofluor white has 100% sensitivity and 95% specificity.
   d) Cresyl ethyl violet.

2. **Non Cyst wall stains:**
   a) Wright Giemsa
   b) Diff Quick – stains trophozoites and nuclei of cysts, false negativity 10-15%.
   c) Direct and Indirect Immunofluorescence with monoclonal antibodies has sensitivity > 90%. Few false positives are seen.

B. **X-ray chest** – Diffuse, asymmetric interstitial infiltrates, intercostal retraction. In children, prominent plasma cell infiltration is seen.

C. **Histopathology** of lung biopsy – alveolar thickening, frothy cosinophilic, honey combed exudate.

autopsy – lungs have a rubbery feeling.

D. **Culture** – Human lung cell line A549
– Human embryonic lung cell
– Dulbecco's Minimum Essential Medium(MEM) 5 fold increase
– RPMI 1640 (maintained for 7 days).

E. **Serology** – Rise in antibody titre.

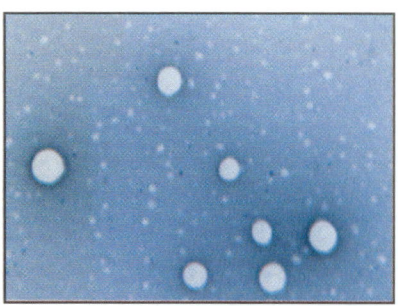

Negative staining by India ink showing capsules of *Cryptococcus neoformans*

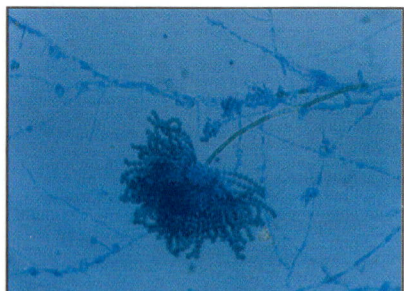

Lactophenol cotton blue (LPCB) mount showing *Aspergillus species* (40 X)

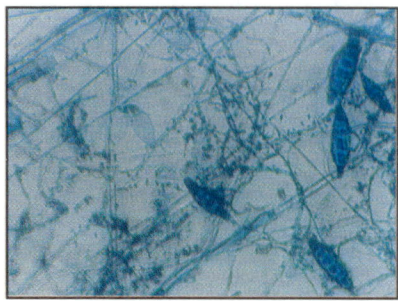

LPCB mount showing spindle-shaped macroconidia of *Microsporum species* (40 X)

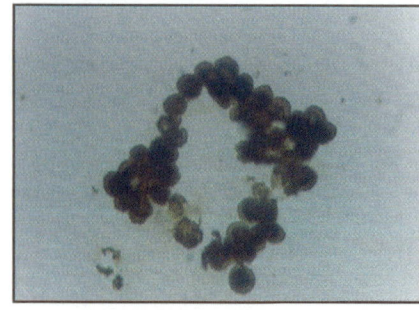

Sputum sample stained by Gomori Methenamine Silver (GMS) stain showing *Pneumocystis jirovecii* (100 X)

Septate hyphae

Aseptate hyphae

Yeast cells with budding

Chlamydospores

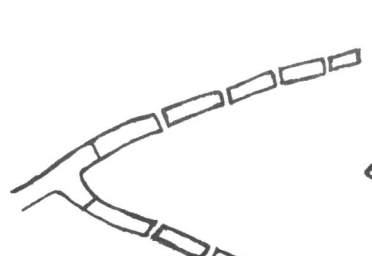

Arthrospores

Yeast cells attached
to pseudohyphae

Sporangium of
Mucor containing
sporangiospores

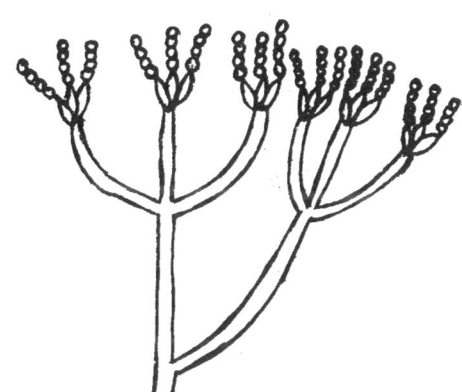

Penicillium species
with unicellular
conidiospores

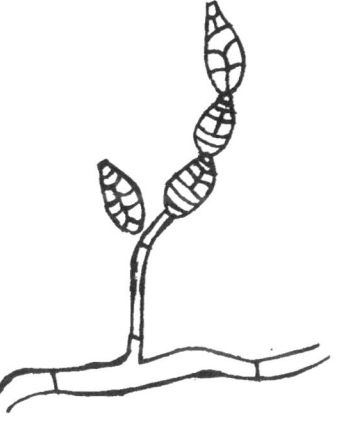

Multicellular
conidia of
Alternaria species

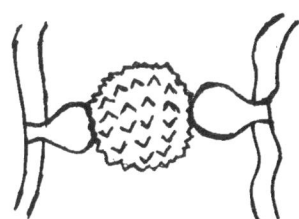

Zygospore

Ascospore
in Ascus

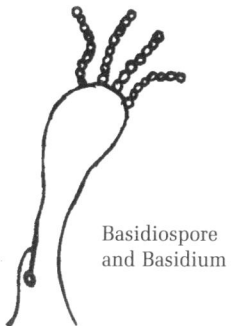

Basidiospore
and Basidium

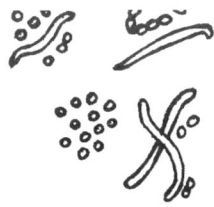

Malassezia furfur with
short hyphae and clusters of
yeast cells

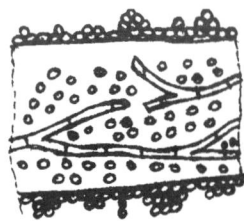

Ectothrix
infection

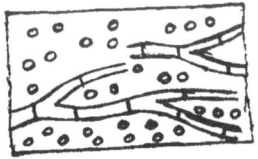

Endothrix infection

Trichophyton

Microsporum

Epidermophyton

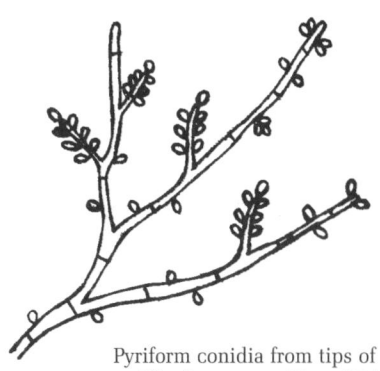

Pyriform conidia from tips of
conidiophores at 25ºC on SDA
Sporothrix schenckii

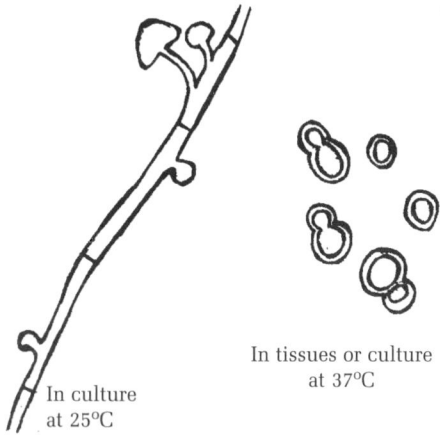

In culture
at 25ºC

In tissues or culture
at 37ºC

Blastomyces dermatitidis

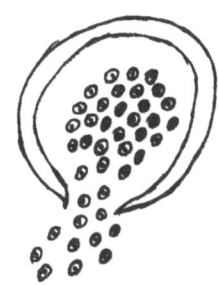

Sporangium of *Rhinosporidium
seeberi* with sporangiospores

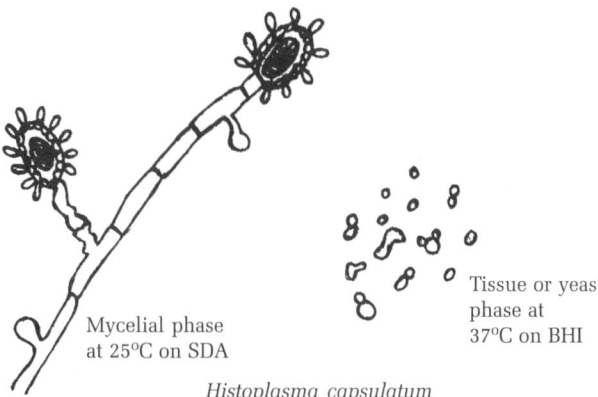

Mycelial phase
at 25ºC on SDA

Tissue or yeast
phase at
37ºC on BHI

Histoplasma capsulatum

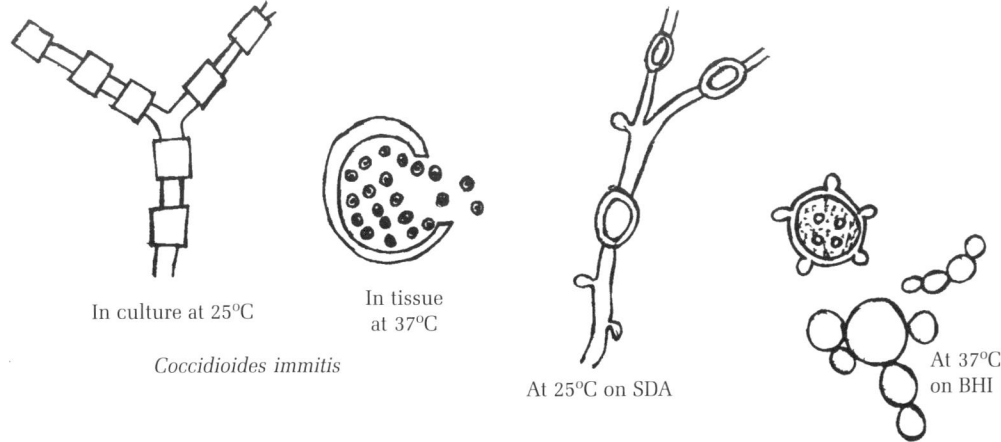

In culture at 25ºC   In tissue at 37ºC

Coccidioides immitis

At 25ºC on SDA

At 37ºC on BHI

Paracoccidioides brasiliensis

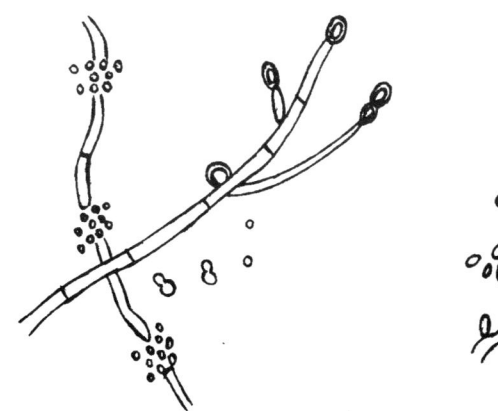

Candida albicans

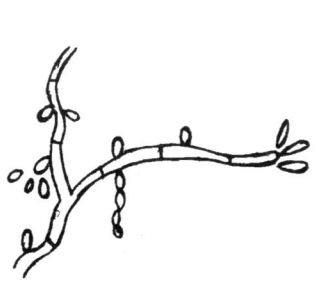

Candida tropicalis

Candida krusei

Candida glabrata

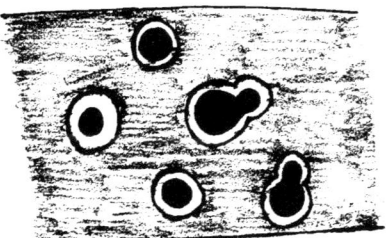

Cryptococcus neoformans
with capsule

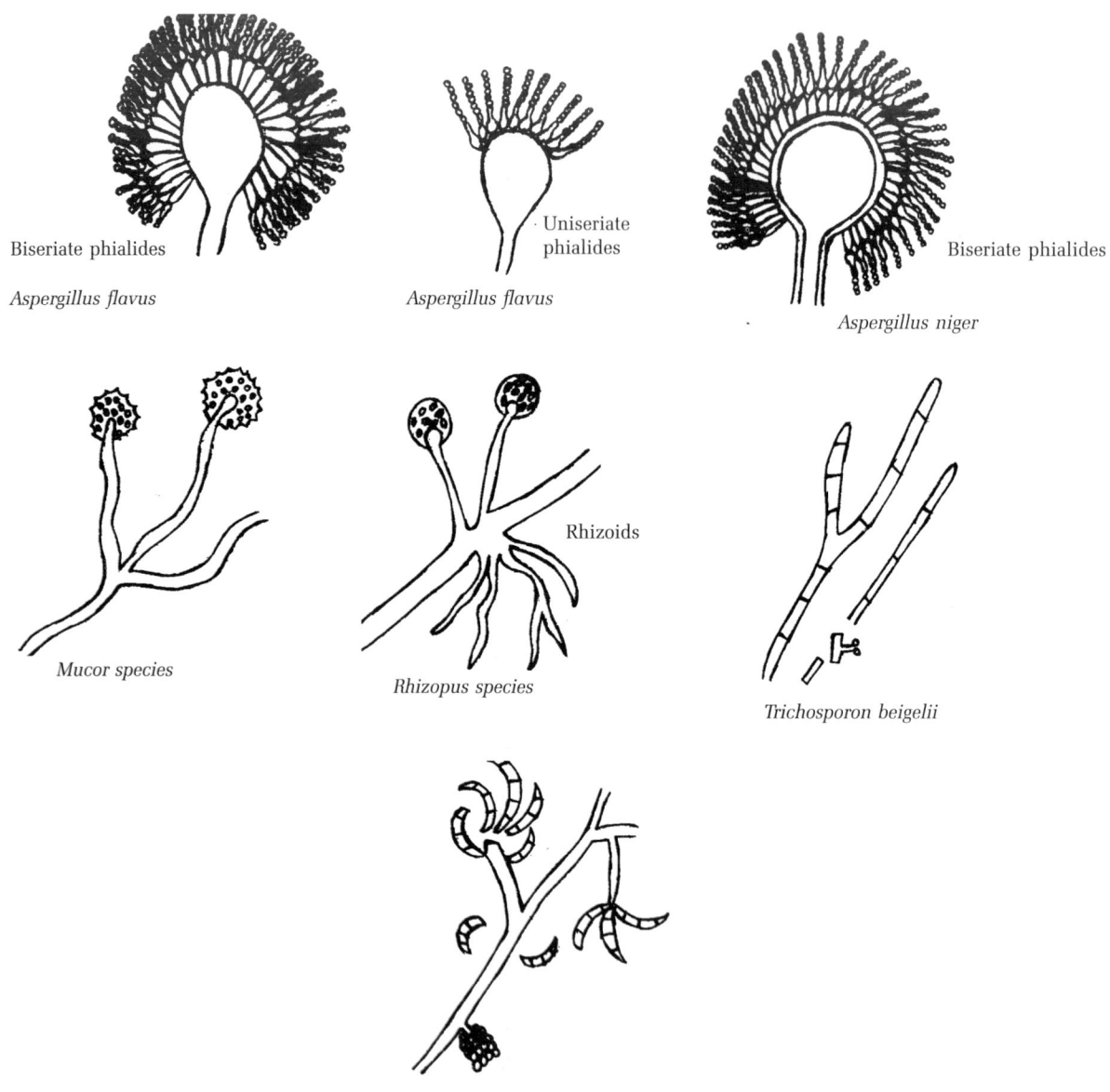

Biseriate phialides

Aspergillus flavus

Uniseriate phialides

Aspergillus flavus

Biseriate phialides

Aspergillus niger

Mucor species

Rhizoids

Rhizopus species

Trichosporon beigelii

Fusarium species

REFERENCES

1. Ananthanarayan & Paniker's Textbook of Microbiology. 9th Ed. Ed. Kapil A. Universities Press, Hyderabad. 2013.

2. Chander J. Textbook of Medical Mycology Hardcover. 3rd Ed. Mehta Publishers, New Delhi. 2009.

3. Odds, FC. Candida infections: an overview. Crit Rev Microbiol 1987; 15:1-5. doi:10.3109/10408418709104444. PMID 3319417.

4. Abraham N. Candida: Diagnostic and Therapeutic Approaches. Positive Health Online. http://www.positivehealth.com/article/colon-health/candida-diagnostic-and-therapeutic-approaches. 2001; 62.

5. Vincent JL, Anaissie E, Bruining H, Demajo W, EI-Ebiary M, Haber J, et al. Epidemiology, diagnosis and treatment of systemic Candida infection in surgical patients under intensive care. Intensive Care Med 1998; 24: 206-16.

6. Pappas PG. Invasive candidiasis. Infect Dis Clin North Amer 2006; 20: 485-506. doi:10.1016/j.idc.2006.07.004. PMID 16984866.

7. Lalla RV, Patton LL, Dongari-Bagtzoglou A. Oral candidiasis: pathogenesis, clinical presentation, diagnosis and treatment strategies. J California Dental Assoc 2013; 41: 263-8. PMID 23705242.

CLINICAL PATHOLOGY

Chapter

PROCESSING OF RESPIRATORY SPECIMENS

Upper respiratory tract consists of:

- Nose
- Throat
- Oropharynx
- Nasopharynx
- Larynx

Lower respiratory tract consists of:

- Trachea
- Bronchi and Bronchioles
- Alveolar air sacs

Clinical	Bacteria	Virues	Fungi
Common cold (rhinitis, coryza)	Rare	Rhinovirues Coronavirus Parainfluenza virus Adenoviruses Respiratory syncytial virus	Rare
Pharyngitis and tonsiliits (tonsillopharyngitis)	Group A β-hemolytic streptococci *Corynebacterium diphtheriae* *Neisseria gonorrhoeae* *Mycoplasma pneumonia* *Mycoplasma hominis* (type 1) Mixed anaerobes	Adenoviruses Coxsackievirues A Influenza virues Rhinovirus Coronavirus Parainflueza viruses Epslein-Barr viruses, cytomegatovirus Herpes simplex virus	*Candida albicans*
Epiglotitis and laryngotracheitis (group)	*Haemophilus influenzae* type b *Corynebacterium diphtheriae* *Haemophilus influenzae*	Respiratory syncytial virus Parainfluenza virues Parainfluenza virues	Rare
Bronchitis and	*Mycoplasma pneumoniae*	Respiratory syncytical virus Adenoviruses Herpes simplex virus	
Pneumonia	*Streptococcus pneumoniae* *Staphyiococcus aureus* *Streptococcus pyoigenes* *Haemophilus inluenzae* *Klebsiella pneumoniae* *Esherichia coli* *Pseudomonas aeruginosa* *Mycoplasma pneumoniae* *Legionella species* Anaerobic bacteria *Mycobacterium tuberculosis* and other *Mycoplasma species* *Coxiella burnelli* *Chlamydia psittaci* *Chlamydia trachomatis* *Chlamydia pneumoniae*	Adenoviruses Parainfluenza viruses Respiratory syncylial virus Influenza viruses Varicella-zoster virus Measles virus Cytomegalovirus Herpes simplex virus Hanlavirus (Muerto Canyon)	*Histoplasma capsulatum* *Blastomyces dermitidis* *Paracoccidiodes brasiliensis* *Coccidioides immitis* *Candida albicans* *Filobasidiella (Cryplococcus) neolomans* *Aspergillus funigatus* and other *Aspergillus species* *Pneumocystis jirovecii*

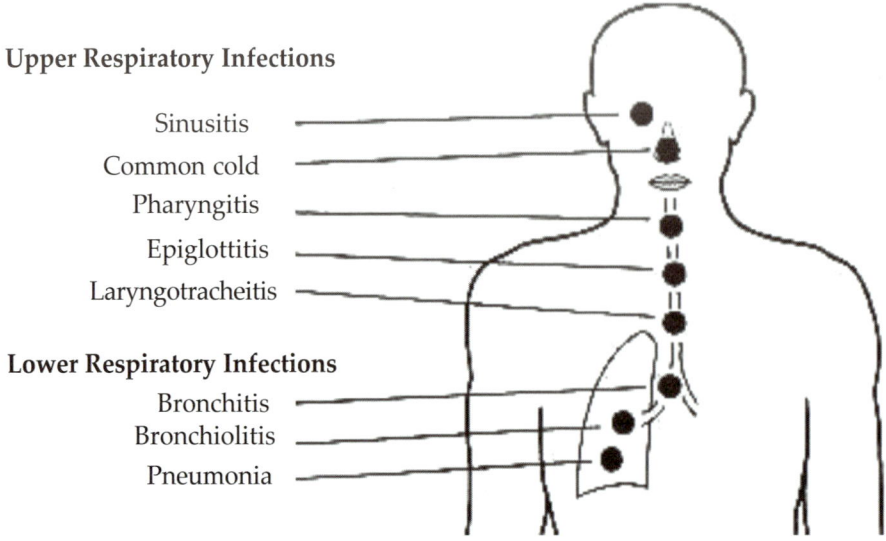

Upper Respiratory Infections

Sinusitis
Common cold
Pharyngitis
Epiglottitis
Laryngotracheitis

Lower Respiratory Infections

Bronchitis
Bronchiolitis
Pneumonia

Types of respiratory samples

Nose and sinuses

- Nasopharyngeal swab
- Nasopharyngeal aspirate
- Sinus washings

Throat, pharynx

- Posterior pharyngeal swabs
- Nasopharyngeal swab
- Tonsillar swabs

Lower respiratory tract

- Sputum
- Tracheal/Endotracheal aspirate
- Bronchial secretion/bronchoalveolar lavage
- Lung biopsy

Nasopharyngeal swabs

- Used to detect healthy carriers
- Rarely used in diagnosis of disease

Types of swabs

Swab tip	Shaft	Use	Disadvantages
Cotton	Wood	Non-fastidious bacteria genital, respiratory, gut, fistulae, wounds. Tip may be too bulky for some orifices	Fatty acids toxic to some fastidious bacteria and chlamydia. Serum tipped < toxicity
Cotton	Plastic	Can be bulky	Not easily broken into transport media
Cotton	Wire (rigid)	Fine tips for NPA, EAM, urethra, cervix	Require sterile wire cutters or another tube for transport
Cotton	Wire (woven/flexible)	As above	As above
Dacron™ & rayon	Wood, plastic, wire (rigid), wire (woven)	Increased isolation of *S. pyogenes* because of inert and absorptive capacity. Used for detection of viruses. Small tip useful for genital swabs, fungi	-
Calcium alginate	-	Can be useful to detect *Chlamydia spp.*	Toxic to ureaplasma, lipid-enveloped viruses, cell cultures, N gonorrhoea
Flocked nylon swabs i.e. medical grade polyamide	Plastic with moulded break point	Respiratory viruses, PCR of Bacteria and viruses	Cost higher than cotton-tipped swabs but surface area collects more sample which is easily eluted

Nasopharyngeal aspirate

- Ideal specimen for microbiological processing in cases of URTI.
- Procedure is same as of nasopharyngeal swab collection only wall suction is attached to catheter and approx. 1 ml of saline is instilled into nostril and then aspirated.
- Method of choice for viral cultures.

Sinus washings

- Pus collected or aspirated from sinus or saline wash should be subjected to microbiological examination.
- Nasal washings are obtained by rubber suction bulb by installing and withdrawing 5 ml of sterile saline.

Throat swabs

Swab is rubbed over the tonsillar areas and the posterior pharynx, specifically targeting any inflamed areas.

Sputum

- Ideal specimen should be representative of LRTI
- Collection of sample:
  - Early morning sample should be collected because of overnight pooling of secretion leading to higher microbial load.
  - In case of TB, two samples are collected.
  - In case where sputum production is scanty, it should be induced with nebulization with saline.
  - Patient is instructed to brush and gargle with water before specimen collection to reduce contamination with oral commensal flora.

Endotracheal (ET) aspirates

- Lower respiratory tract may be sampled either by introducing catheter through larynx into trachea, or by aspirating through the ET tube which is in situ.
- It should be processed by quantitative or semi-quantitative method.

Procedure of ET aspirate

- Two types of suctioning are - open type and closed type
- Two methods of suctioning are - deep suctioning and shallow suctioning
- In general, the external diameter of the suction catheter should not be more than ½ the internal diameter of the artificial airway. The ratio of this internal to external diameter should be 0.5 in adult and 0.5 to 0.66 in infants and smaller children.
- Negative pressure should be 80 to 100 mmHg in neonate and less than 150 mmHg in adult.
- Suctioning event should not exceed more than 15 seconds.

Transtracheal aspiration

- Invasive procedure which gives tracheal secretions for culture without contamination.
- A needle is inserted through crico-thyroid ligament into trachea and passed to level of tracheal bifurcation, some saline introduced and then withdrawn.
- Generally not recommended for sample collection.

Bronchial secretions

These are obtained with the use of bronchoscope.

Bronchoscopes are of two types:

Rigid (capable of sampling only central airways)

Flexible

- Contains fiber-optic system that transmits image from tip of instrument to eyepiece
- It can be maneuvered to small bronchioles
- In addition, it has interior channels through which suctioning secretion or obtaining tissue samples or delivering oxygen is possible.

Bronchial wash

- Bronchial washing is performed to check the lower respiratory tract for irregularities and take to tissue samples wherever indicated.
- The physician injects saline through the bronchoscope into the lung and then suctions it back out
- By checking the wash return fluid, we can diagnose bleeding, pneumonia.

Bronchoalveolar lavage (BAL)

- It involves injection of 30 to 50 ml of physiologic saline through flexible bronchoscope that has been inserted into the peripheral branches of the tracheal tree.
- Saline is then aspirated and subjected to quantitative and qualitative cultures.

Bronchial brushings

Protected brushing may be preferable to BAL in patients with more severe pulmonary dysfunction.

- Advance bronchoscope to the orifice of the area of concern
- Advance the PSB catheter 3 cm from the scope
- Eject the distal carbon wax plug
- Advance brush into sub-segment and rotate brush within secretions
- Retract brush into catheter sleeve and remove entire catheter from bronchoscope
- Wipe distal portion of catheter with 70% alcohol, then advance brush portion and cut bush with sterile scissors and place in 1 ml of sterile saline
- Sent for quantitative bacterial cultures.

Lung puncture and biopsy

- Per-cutaneous aspiration or needle biopsy may be performed under fluoroscopic guidance if localized lesion is present in the lung.
- Open lung biopsy is most invasive procedure and is reserved for extreme situations.

Procedure:

- Bronchoscopic biopsy is used for collecting lung tissue near bronchi.
- Needle biopsy procedure- used when lesion is close to chest wall
- Site is marked cleaned and local anesthetic is injected.
- Small superficial incision is made on site and needle is inserted under ultrasound or fluoroscopic guidance and the sample is collected.

Transportation of samples

- For microbiological processing, a maximum 2 hrs time limit between collection and delivery to lab is recommended.

- Sterile test tubes or transport medium can be used e.g. Amies or modified Stuart's transport medium.
- If delay in processing is anticipated, the samples should be refrigerated at 4-8°C for upto 24 hours.
- A freezer at -20°C can be used if periods of storage are more than 24 hours.
- Prolonged storage of samples ma be done at -70°C.
- Samples for viral culture must not be frozen.

Bacteriological processing

- Gram stains of all respiratory samples are made.
- Special stains used are:
  - Albert's stain for *C. diphtheriae*
  - India ink preparation for capsule demonstration e.g. in *Streptococcus pneumoniae* and *Hemophilus influenzae*
  - Ziehl-Neelsen stain for *Mycobacterium tuberculosis*, Atypical mycobacteria, *Nocardia spp.*
  - Giemsa stain for *Chlamydia pneumoniae*, *Mycoplasma pneumoniae*, *Neisseria spp.*

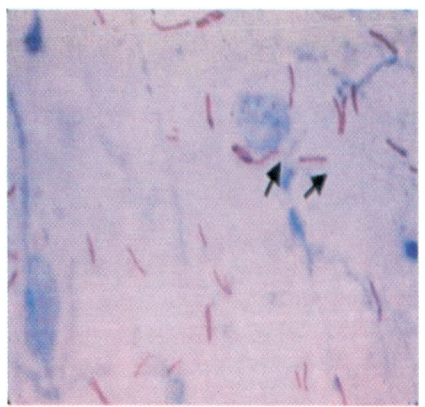

Albert's stain showing *C. diphtheriae*

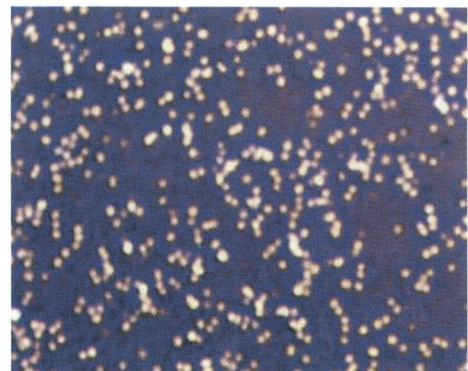

Negative stain showing capsule

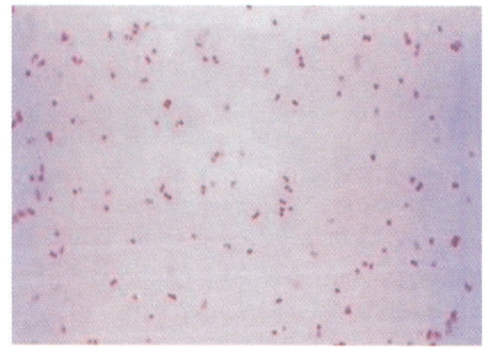

Ziehl-Neelsen stain showing acid fast bacilli

Giemsa stain showing *N. gonorrhoeae*

From Gram stain of sputum, assessment of quality of sputum is done.

Bartlett's devised grading system

No. of neutrophils per 10X	Grade
< 10	0
10-25	+1
> 25	+2
Presence of mucus	+1

No. of epithelial cells per 10X	Grade
10-25	-1
> 25	-2

- Average no. of cells in 20 to 30 fields is taken
- Final score of 0 or less shows lack of active inflammation.

Murray & Washington's Grading System

Group	Epithelial cells per 10X	Leucocytes per 10X
1	25	10
2	25	10-25
3	25	25
4	10-25	25
5	< 10	25

Processing

- All samples are plated culture media.
- Most commonly used media are MacConkey, Chocolate and Blood agar.
- Specific media can be used for specific organism, if requested.

Gram positive cocci

Streptococcus pyogenes – Sheep blood agar or Pike's medium.

Streptococcus pneumoniae – Blood agar and Chocolate agar.

Gram positive bacilli

Corynebacterium diphtheriae – Loffler's serum slope, Potassium tellurite medium.

Gram negative cocci

Neisseria meningitidis – Blood, Chocolate agar, Muller Hinton-starch caesin hydrolysate agar are ommonly used; Modified Thayer Martin agar.

Neisseria gonorrheae – Chocolate agar, Muller Hinton agar, Thayer Martin agar.

Gram negative bacilli

Bordetella pertussis – Bordet and Gengou agar, Regan-Lowe medium.

Hemophilus influenzae – Fildes agar, Levinthals medium.

Legionella pneumophila – Buffered charcoal yeast extract (BYCE) supplemented with L-cysteine.

Chlamydia pneumoniae – Inoculation into embryonated eggs, experimental animals, and tissue culture.

Mycoplasma pneumoniae – Mycoplasma agar, media enriched with 20% horse serum and yeast extract and with penicillin and thallium acetate.

- For *Mycobacterium tuberculosis*, concentration of sputum is done by Petroff's method or NALC method and it is subcultured on Lowenstein-Jensen medium or Middlebrook medium and incubated at 37°C.
- All bronchoscopic specimens should be subjected to quantitative or semi quantitative culture.

Quantitative	Semiquantitative
Pour plate method	Standard loop method
Surface Viable count by spreading method	Filter paper method
Surface Viable count by Miles & Misra method	Dip slide method

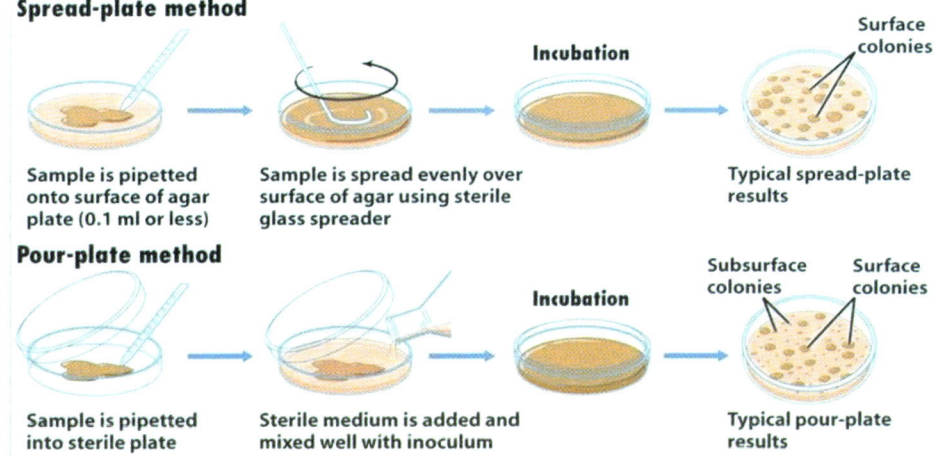

Spread-plate method

Sample is pipetted onto surface of agar plate (0.1 ml or less)

Sample is spread evenly over surface of agar using sterile glass spreader

Incubation

Surface colonies

Typical spread-plate results

Pour-plate method

Sample is pipetted into sterile plate

Sterile medium is added and mixed well with inoculum

Incubation

Subsurface colonies   Surface colonies

Typical pour-plate results

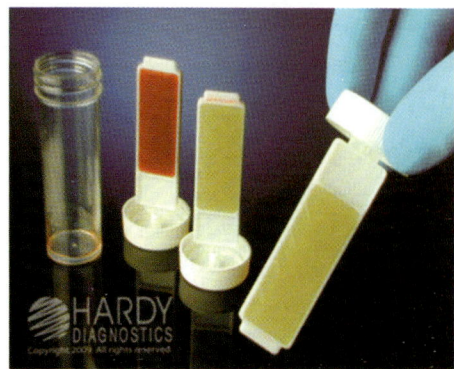

For Fungi

- Wet mount - demonstrate morphology of fungi
- KOH mount - Due to dissolution of tissue debris, fungal elements are better appreciated.
- Along with Gram stain specific stains can be used like LPCB, PAS stain.
- GMS stain is used for *Pneumocystis jirovecii*.
- Media used are:
  - Non specific media with addition of antibacterial e.g. Brain heart infusion agar and
    Kelly's medium to demonstrate dimorphism.
  - Specific media e.g. Sabouraud dextrose agar, Corn meal agar.

For Viruses

- Specimens should be transported in Viral Transport Medium.
- VTM consists of 10gm veal infusion broth + 2 gm bovine albumin fraction in sterile distilled water, along with 0.8 ml Gentamicin Sulphate and 3.2 ml Amphotericin B.
  Commonly used balanced salt solution is Hank's BSS, Phosphate Buffer Saline (PBS).
- Viruses can be demonstrated by Phase contrast or Electron microscopy or Fluorescent microscopy.
- For cultivation of viruses, three methods include:
  - Animal inoculation
  - Inoculation into embryonated eggs
  - Tissue culture

For Parasites

- Saline mount – to demonstrate eggs, larvae, trophozoites
- Iodine mount – to stain nuclei of eggs.
- Special staining techniques e.g. Iron-hematoxylin stain.

URINE ROUTINE EXAMINATION

Composition of normal urine

Sr. No.	Parameters	Values
1	Volume	600-2000 ml
2	Specific gravity	1.003-1.030
3	Osmolality	300-900 mOsm/kg
4	pH	4.6-8.0
5	Glucose	< 0.5 gm
6	Proteins	< 150 mg
7	Urobilinogen	0.5-4.0 mg
8	Porphobilinogen	0-2 mg
9	Creatinine	14-26 mg/kg (men), 11-20 mg/kg (women)
10	Urea nitrogen	12-20 gm
11	Uric acid	250-750 mg
12	Sodium	40-220 mEq
13	Calcium	50-150 mg
14	Formiminoglutamic acid (FIGlu)	< 3 mg
15	Red cells, epithelial cells and white blood cells	< 1-2 / HPF

Indications for urinalysis

- Suspected renal diseases like Glomerulonepritis, Nephrotic syndrome, Pyelonephritis and renal failure.
- Detection of urinary tract infection.
- Detection and management of metabolic disorders like diabetes mellitus.
- Differential diagnosis of jaundice.

Collection of urine

- For routine examination of urine, a wide mouthed glass bottle of 20-30ml capacity, which is dry, chemically clean, leak-proof and with a tight fitting stopper is used.
- About 15ml of midstream sample is cleanly collected.
- Preservatives should be avoided.

Preservatives

- Hydrochloric acid for 24 hrs urine for adrenaline, noradrenaline, vanillyl mandelic acid and steroids.
- Boric acid is a general preservative.

Methods of Collection

a) **Midstream specimen**:
  - After voiding initial half of urine into the toilet, a part of urine is collected in the bottle.
  - First half serves to flush out contaminated cells and microbes from urethra and perineum. Subsequent stream is from bladder.

b) **Clean catch specimen**:
- For bacteriological culture.
- It avoids contamination with vaginal flora.

c) **Catheter specimen**: In bedridden ill patients or in patients with obstruction of urinary tract.

d) **Infants**: A clean plastic bag is attached around the baby's genitalia and left in place for sometime. Also suprapubic aspiration is done.

Physical examination

Parameters:

- Volume
- Colour
- Appearance
- Odour
- Specific gravity
- pH

Volume

- Average 24 hrs urinary output in adults is 600-2000ml. Volume varies according to fluid intake, diet and climate.
- Abnormalities:
  - **Polyuria**: urinary output > 2000 ml/24 hrs seen in diabetes mellitus, diabetes insipidus, chronic renal failure or diuretic therapy.
  - **Oliguria**: urinary output < 400 ml/24 hrs Seen in febrile states, acute glomerulonephritis, congestive cardiac failure or dehydration.
  - **Anuria**: urinary output < 100 ml/24 hrs or complete cessation of urine output. Seen in acute tubular necrosis complete urinary tract obstruction

Colour

- Normal urine colour in a fresh state is pale yellow or amber and is due to presence of various pigments collectively called urochrome.

Different colours of urine

Sr. No.	Colours	Conditions
1	Colourless	Dilute urine (DM, DI, overhydration)
2	Red	Hematuria, hemoglobinuria, porphyria, myoglobinuria,
3	Dark brown or black	Alkaptouria, chloroquine and primaquine, metronidazole
4	Brown	Hemoglobinuria, hepatocellular jaundice
5	Yellow	Concentrated urine
6	Yellow green or green	Biliverdin, obstructive jaundice, promethazine, propofol
7	Deep yellow with yellow foam	Bilirubin
8	Orange or orange brown	Urobilinogen, Rifampicin
9	Milky white	Chyluria
10	Red or orange fluorescence with UV light	Porphyria

Appearance

Normal: Clear

Appearance	Diagnosis	Cause
White or cloudy on standing in alkaline urine	Disappear on addition of a drop of acetic acid	Amorphous phosphates
Pink and cloudy in acid urine	Dissolve on warming	Amorphous urate
Varying grades of turbidity	Microscopy	Pus cells
Uniformly cloudy, do not settle at bottom after centrifugation	Microscopy, Nitrate test	Bacteria

Odour

- Freshly voided urine has a typical aromatic odour. After standing, urine develops ammoniacal odour (formation of ammonia when urea is decomposed by bacteria).
- Abnormal odours:
  - Fruity: Ketoacidosis, starvation
  - Mousy or musty: Phenylketonuria
  - Fishy: UTI with *Proteus*, Tyrosinaemia.
  - Ammoniacal: UTI with *Escherichia coli*, old standing urine
  - Sulfurous: Cystinuria

Specific gravity (SG)

- Specific gravity is a measure of concentrating ability of kidneys and is determined to get information about tubular function.

- Affected by proteinuria and glycosuria
- Normal SG of urine is 1.003 to 1.030
- Causes of increase in SG of urine are:
  - Diabetes mellitus (glycosuria)
  - Nephrotic syndrome(proteinuria)
  - Fever and Dehydration.
- Causes of decrease in SG of urine are:
  - Diabetes insipidus
  - Chronic renal failure
  - Methods for measuring specific gravity inometer method
  - Refractometer method
  - Reagent strip method

Reaction and pH

- If pH is < 7.0 - acid , > 7.0 – alkaline , 7- Neutral
- Normal pH range - 4.6 to 8.0
- Methods for determination of reaction:
  - **Litmus paper test:** Blue litmus turns red to acid; red litmus turns blue to alkaline
  - **pH indicator paper:** Reagent area (impregnated with bromothymol blue and methyl red) of indicator paper strip is dipped in urine and colour change is compared with colour guide provided

pH meter: An electrode of pH meter is dipped in urine sample & pH is read directly from digital display.

- **Reagent strip test:** the test area contains polyionic polymer bound to H+, on reaction with cations in urine, H+ is released causing change in colour of pH sensitive dye.

Chemical examination

Chemical examination of urine is carried out for the following substances:

i. Proteins
ii. Glucose
iii. Ketones
iv. Bilirubin
v. Bile salts
vi. Urobilinogen
vii. Blood
viii. Hemoglobin
ix. Myoglobin
x. Nitrate or leukocyte esterase

i) Proteins

- Normally, kidney excrete 150 mg/24 hrs.
- Normal proteins include proteins from plasma (albumin) and proteins derived from urinary tract (Tamm-Horsfall protein, secretory IgA, and proteins from tubular epithelial cells, leucocytes and other desqaumated cells).
- Abnormal proteins are Bence Jones proteins.

Causes of proteinuria

- Glomerular or tubular urinary diseases
- HIV associated renal disease and treatment with nephrotoxic antiretroviral drugs like cidofovir, tenofovir, adefovir.
- Pyogenic or tuberculous pyelonephritis
- Nephrotic syndrome
- Eclampsia
- Urinary schistosomiasis usually accompanied by both proteinuria and haematuria
- Occasionally in diabetes (diabetic nephropathy).

Tests for detection of proteinuria

- **Heat and acetic acid test** (boiling test) – based on the principle that proteins get precipitated when boiled in an acidic solution. It is not recommended.
- **Reagent strip test:**
  - Urine protein strip test detect mainly albumin.
  - The test area is impregnated with the indicator tetrabromophenol blue (bayer) or tetrabromophenolphthalein ethyl aster (Roche) and buffered to an acid pH. In presence of protein there is a change in the colour of the indicator from light yellow to green blue depending on amount of protein present.

Commercially available protein test strips

- **Albustix** (Bayer) which measures albumin semiquantitatively:

  Scale: Negative, trace, +0.3g/dl, ++1g/dl, +++3g/dl, ++++20g/dl or more.

- **Unistix** (Bayer) which measures proteins semiquantitatively and also detects glucose.

- **Combur 3 Test E** (Roche) measures protein semi quatitatively and also detects glucose and blood

  Sensitivity: 5-10 mg/dl

  Cost: 3000 INR for 100 strips.

False strip test reactions

- Contaminated with
  - Disinfectants which contain quaternary ammonium compounds or chlorhexidine.
  - Vaginal or urethral secretions.
- Strongly alkaline urine

Sulphosalicyclic acid test

- Based on the precipitation of protein, mainly albumin by sulphosalicyclic acid.
- Reagent: Sulphosalicyclic acid 200g/l
- **Method:**
  - Take two tubes and label one 'C' for comparison and other 'T' for test.
  - Pour 2 ml of clear urine into each tube

- Using pH papers, test the reaction of the urine. If neutral or alkaline, add a drop of glacial acetic acid to each tube and mix.
- Add 2-3 drops of sulphosalicylic acid reagent to the tube 'T'
- Holding both tubes against a dark background, examine for cloudiness in tube 'T' compared with tube 'C'.

Results: It ranges from no cloudiness to cloudiness with precipitate.

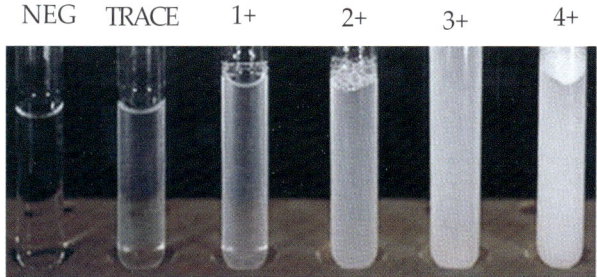

| NEG | TRACE | 1+ | 2+ | 3+ | 4+ |

Sensitivity: 20 mg/dl

False sulphosalicylic acid test reactions:

- Patient is receiving
  - Tolbutamide
  - Intensive therapy with penicillin,
  - Para aminosalicylic acid or sulphonamides.
- High concentration of urates in the urine.

Bence-Jones proteins in urine

- Abnormal low molecular weight globulin consisting of monoclonal free light chains of immunoglobulins (either kappa or lambda).
- Seen in multiple myeloma which is malignant disease of the plasma cells.
- Urine protein electrophoresis is required to detect the paraprotein 'M' (monoclonal immunoglobulin).

ii) Glucose

- Normally < 500 mg/24 hrs or < 15mg/dl of glucose is excreted in urine. Presence of detectable amount of glucose in urine is called as glycosuria or glucosuria.

Causes of glycosuria:

1) Glycosuria with hyperglycemia
   a. Endocrine diseases like Diabetes Mellitus, Acromegaly, Cushing's syndrome, Hyperthyroidism.
   b. Drugs like corticosteroids or thiazides.
   c. Alimentary glycosuria (lag storage glycosuria)
      - After a meal there is rapid intestinal absorption of glucose leading to trasient elevation of blood glucose above renal threshold. Occurs in patients with gastrectomy or gastro-jejunostomy and in hyperthyroidism.

2) Glycosuria without hyperglycemia:
   a. Fanconi's syndrome
   b. Toxic renal tubular damage

Tests for detection of glucose in urine

1) Benedict's test (copper reduction method):
   cupric ions (blue) + sugar alkali cuprous oxide (red) + cuprous hydroxide (yellow) heat

   Benedict's test can be performed using:
   a) A solution reagent
   b) Dry reagent

The dry reagent uses chemicals to generate heat necessary for boiling the mixture whereas in solution reagent method heat is applied to boil the mixture.

Solution reagent method

- Benedict's reagent
- Glucose control solution

Method:

Three or more heat resistant glass tubes are taken and labelled as positive control, negative control and tests

Add 2.5ml of Benedict's reagent into each test tube

Add to each tube as follows:

Positive - 0.2 ml glucose control solution

Negative - 0.2 ml of distilled or boiled filter water

Test - 0.2 ml of fresh patient's urine sample

↓ Mix

Place tubes in a heat block set at 100°C or in container of boiling water for 5 mins

Remove tubes and examine for precipitate and change in colour.

Results: Appearance of solution ranges from Blue-neg and Green to Red with scale of sugar concentration from trace to 2 g% respectively.

Sensitivity: 200 mg reducing substance per dl of urine.

Appearance	Sugar concentration
Blue, clear and cloudy	Nil
Green, no precipitate	Trace
Green, with preciptate	About 0.5 g%
Brown and cloudy	About 1 g%
Orange and cloudy	About 1.5 g%
Red and cloudy	2.0 g% or more

Dry reagent method

Benedict's dry reagent is prepared by grinding and mixing together

Copper 11 sulphate, 5-hydrate	10g
Citric acid	150g
Sodium hydroxide or potassium hydroxide	

Method:

In this Benedict's reagent is added to each tube to a depth of 10mm & one pellet of sodium hydroxide is added to each tube and gently shaken, mixture begins to boil. Continue shaking until boiling stops and observe change of colour and precipitate

Results: Range from Blue – nil, Orange to Green brown precipitate with sugar concentration of trace to 2 g% respectively.

Appearance	Sugar concentration
Blue clear or cloudy	Nil or less than 0.2 g%
Slight orange precipitate with blue supernatent	About 0.2 g%
Persistant orange precipitate	About 0.5 g%
Orange precipitate turning brown during boiling	About 1 g%
Green brown precipitate, rapidly turning dark brown	2.0 g% or more

Reagent strip method

This test is specific for glucose and preferred over Benedict's test. It is based on glucose oxidase-peroxidase reaction.

Stage I:

Glucose + oxygen $\xrightarrow{\text{Glucose oxidase}}$ Glucolactonate + hydrogen peroxide

Stage II:

Hydrogen peroxide + Chromogen $\xrightarrow{\text{Peroxide}}$ Oxidised Chromogen (blue) + H_2O

Strips:

- *Diastix* (Bayer) – detects 5.5 mmol/l of glucose and measures it semiquantitatively up to 111 mmol/l

- *Diabur-test 5000* (Roche) – detects 5.5 mmol/l of glucose and measures it semiquantitatively up to 280 mmol/l

- Sensitivity: 100mg glucose/dl
  Cost: 275 INR for 50 strips.

iii) Ketones

Excretion of ketone bodies (acetoacetic acid, beta hydroxybutyric acid and acetone) in urine is called as ketonuria (> 5mg/dl).

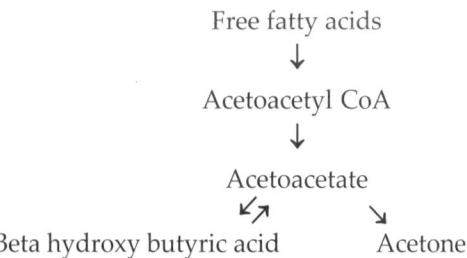

Free fatty acids
↓
Acetoacetyl CoA
↓
Acetoacetate
↙↗ ↘
Beta hydroxy butyric acid   Acetone

Causes of ketonuria

- Untreated diabetes

- Starvation when fat metabolism is increased

- Diet low in carbohydrates

- Severe dehydration following prolonged vomiting or diarrhoea.

Methods to detect ketonuria:

1) Rothera's test (classic nitroprusside reaction)

2) Acetest tablet test

3) Ferric chloride (Gerhardt's test)

4) Ketone reagent strips

Rothera's test

Acetoacetate + sodium nitroprusside $\xrightarrow{\text{alkaline pH}}$ Purple colour

Reagent strip test

Acetoacetate + sodium nitroprruside + Glycine $\xrightarrow{\text{alkaline pH}}$ Purple colour

Sensitive to 1-5mg/dl of acetoacetate and 10-25 mg/dl of acetone.

Acetest test

- Rothera's test in form of tablet.

- It is more sensitive than reagent strip test.

Gerhardt's test

- Not specific since drugs like salicylate give similar reaction.

- Sensitivity is 25-50 mg/dl for acetoacetate.

Ketone reagent strips

- **Ketostix** – which detect 0.5-1.0mmol/l of acetoacetate. Less sensitive to acetone

  Scale: negative, + (small), ++ (moderate), +++ (large)

- **Kito-diabur test** – measure urine glucose from 5.5mmol/l to 280mmol/l

  Sensitivity-5-10mg/dl of acetoacetate

  Cost: 299 INR for 50 strips.

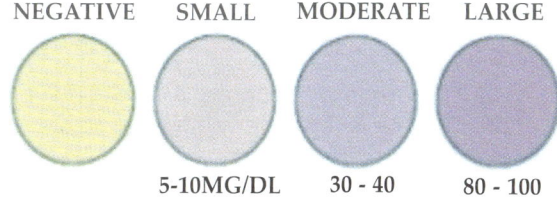

NEGATIVE   SMALL   MODERATE   LARGE

5-10MG/DL   30 - 40   80 - 100

iv) Bilirubin

- Bilirubin (breakdown product of hemoglobin) is undetectable in urine of normal persons. Presence of bilirubin in urine is bilirubinuria.
- Bilirubinuria indicates conjugated hyper-bilirubimemia (obstructive or hepatocellular jaundice) because only conjugated bilirubin is water soluble.
- > 8.4 μmol/l (0.5 mg%) is bilirubinuria.
- Bilirubin in urine is absent in hemolytic jaundice, because unconjugated bilirubin is water insoluble.

- **Increases in:**
  - i. Haemolytic disease
  - ii. Paralytic ileus or enterocolitis
  - iii. Hepatocellular damage
  - iv. Cirrhosis of liver
  - v. Early stage of viral hepatitis

- **Decreases in:**
  - i. Obstruction of bile ducts
  - ii. Reduction of intestinal bacterial flora.

Tests for detection of bilirubin

- Fouchet's test – Barium chloride is used to precipitate the sulphates in urine. Any bilirubin if present gets attached to precipitated barium sulphate. When fouchet's reagent is added to the precipitate, the iron (ferric) chloride oxidizes the bilirubin to green blue biliverdin.
- Bilirubin strip test – based on coupling of bilirubin with 2, 6-benzene-diazonium fluroborate in an aicd medium to give reddish violet azo dye.

  Strip test have sensitivity of 7-14 mol/lbilirubin.

Test	Procedure	Colour	Picture
Foam test	5 ml of urine in test tube is shaken	Yellowish foam	Urine
Gmelin's test	3 ml conc nitric acid + 3 ml urine	Variable (yellow, red violet, blue green)	
Lugol iodine test	4 ml lugols iodine + 4 drops urine	Green colour	
Fouchet's test	5 ml urine +2.5 ml of 10% barium chloride = precipitate Add fouchet's reagent on precipitate	Blue green	
Reagent strip test	Bilirubin + diazo reagent	Shades of pink	XFGATN + ++ +++

v) Urobilinogen

- Unconjugated bilirubin excreted into duodenum through bile is converted by bacterial action to urobilinogen in intestine. Major part is eliminated in faeces. A portion of it is absorbed in blood, which undergoes enterohepatic circulation, a small amount is not taken by liver is excreted in urine
- Normally about 0.5-4 mg of urobilinogen is excreted in urine in 24 hrs

Tests for detection of Urobilinogen

- **Ehrlich's test:** Urobilinogen reacts with *p*-dimethylaminobenzaldehyde to form a red condensation product. Intensity corresponds to concentration of urobilinogen present
- **Urobilinogen strip test:** Based on Ehrlich's reaction. For roche strips, a pink-red azo dye is formed by 4-methoxy-benzene-diazonium fluroborate when it combines with urobilinogen.

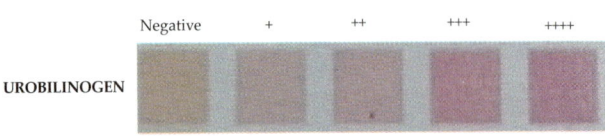

vi) Blood

Presence of intact RBC in urine - hematuria

Causes:

a) In urinary tract: Glomerular infection, calculus, TB, pyelonephritis, trauma, hydronephrosis, BHP, carcinoma prostate

b) Hematological: Sickle cell disease

Tests for detection:

1) Microscopic examination
2) Chemical tests
    i) Benzidine test
    ii) Orthotoludine test
    iii) Reagent strip test

vii) Hemoglobinuria

Presence of free hemoglobin

Causes:

a) Severe falciparum malaria
b) Trauma
c) Extensive burns
d) Mismatched blood transfusion
e) Paroxysmal nocturnal hemoglobinuria

Tests for detection: i) Benzidine test
                            ii) Orthotoludine test
                            iii) Reagent strip test

Chemical tests for significant bacteriuria

- **Catalase test**
- **Griess Nitrite test:** Gram negative bacteria reduce nitrates to nitrites by enzyme nitrate reductase which are detected by reagent strip test.
  Escherichia coli is the commonest organsim.
- **Leucocyte esterase test:** Detects esterase enzyme released in urine from granules of leucocytes. Positive in pyuria.
- **Glucose Oxidase test**
- **Triphenyl Tetrazolium Chloride (TTC) Reduction test**

Microscopic examination

- Also known as "Liquid biopsy".
- Urine sample is centrifuged for 5 minutes at 1500 rpm.
- Supernatant is poured off.
- Tube is tapped at bottom to resuspend the sediment.
- Drop of this is placed on glass slide with a cover slip on it.
- Presence of pus cells > 5/h.p.f. is suggestive of UTI.
- Urinary sediment
- Cells
    - Red blood cells
    - White blood cells
    - Epithelial cells
    - Oval fat bodies
- Casts
    - Cellular
    - Red blood cell
    - White blood cell
    - Renal epithelial cells
- Non cellular
    - Hyaline
    - Granular
    - Waxy
    - Fatty
- Crystals
    - Uric acid, calcium oxalate
    - Amorphous urate
    - Calcium carbonate
    - Phosphates
    - Cysteine, cholestrol
    - Bilirubin
    - Leucine, tyrosine, sulfonamide
- Organisms
    - Bacteria
    - Yeast cells

- *T. vaginalis*
- *S. hematobium*
- Other
  - Sperm

Cells

Red blood cells - Anucleate biconcave disks about 7 μm in diameter (isomorphic).

- They may appear :
  Swollen (hypotonic urine)
  Crenated (hypertonic urine)
  Dysmorphic (glomerular pathology).

- Quantity of red cells are reported as number of cells per high power field.

White blood cells - Granular with nuclei.

- Presence of white cells in urine is called as Pyuria.
- Sterile pyuria
- Normally 0-2 cells/hpf are seen.
- Significant pyuria:
  One WBC per 7HPF in uncentrifuged urine.

Cells	Apperance/Site	Presence	Picture
Squamous epithelial cells	Lower urethra and vagina	Contamination of urine with vaginal fluid	
Transitional epithelial cells	Renal pelvis, ureters, urinary bladder, upper urethra	Transitional cell carcinoma or after catheterization	
Renal tubular epithelial cells	Polyhedral, columnar and have granular cytoplasm	Acute tubular necrosis, allograft rejection, salicylate or heavy metal poisoning	
Oval fat bodies	Degenerated renal tubular epithelial cells filled with highly refractile lipid droplets	Nephrotic syndrome	

Organisms

Bacteria – Detected by Microscopy. Reagent strip test for significant bacteriuria (nitrate, leucocyte esterase)

Yeast cells – Round or oval structures same as RBC, but in contrast show budding and are not soluble in 2 % acetic acid. Seen in immunocompromised states and diabetes mellitus.

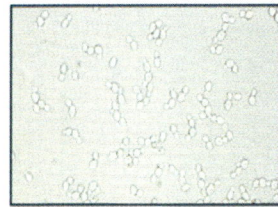
Yeast cells

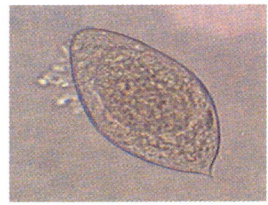
Egg of *S. hematobium*

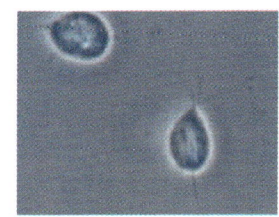
Trichomonas vaginalis

Casts

Precipitate of a protein that is secreted by tubules (Tamm-Horsfall protein)

- Indicative of disease of renal parenchyma.
- Urinary casts are cylindrical, cigar shaped that are formed in distal renal tubules and collecting duct, and take the shape and diameter of lumina of the renal tubules.
- Types: a) Noncellular; b) Cellular

Genesis of casts in urine

Stasis, low pH and high salt concentration of filtrate in renal tubules

↓

Denaturation and precipitation of Tamm Horsfall protein

↓

Hyaline casts

↓

Entrapment of cellular elements in hyaline matrix to form cellular casts

↓

Degeneration of cells within casts to form coarse granular casts

↓

Prolonged stay of casts in tubules with further degeneration of cells to form waxy casts

↓

Broad casts

Non cellular casts

Type	Formation	Seen in	Picture
Hyaline cast	Tamm Horsfall proteins	Transiently after muscle exercise fever	
Granular casts	Degenerated cellular debris	Strenous muscle exercise Glomerulonephritis Pyelonephritis	
Waxy casts	Hyaline casts remain in renal tubules for long time	End stage renal disease	
Fatty casts	Tamm Horsfall proteins filled with fat globules	Nephrotic syndrome	
Broad casts	Form in dialated distal tubules	Chronic renal failure Severe renal tubular obstruction	

Cellular casts

Type	Formation	Seen in	Picture
Red cell casts	Red cells in Tamm Horsfall protein matrix	Glomerular pathology	
White cell casts	White cells in Tamm horsfall protein	Pyelonepritis	
Renal tubular epithelial cell casts	Sloughed off renal tubular epithelial cells	Acute tubular necrosis Heavy metal poisoning Allograft rejection	

Crystals

- Crystals are refractile structures with definite geometric shape due to 3 dimensional arrangement of its atoms and molecules
- Crystals are divided into two main types:
    - Normal (seen in normal urinary sediment)
    - Abnormal (seen in diseased states)

Crystals in acid urine

Type	Shape	Seen in	Picture
Uric acid crystals	Diamond /rosette/plates	Gout, Leukemia	
Calcium oxalate	Envelope shaped	Ethylene glycol poisoning	
Amorphous urates	Compact masses	Gout, Uric acid stone	

Crystals in alkaline urine

Type	Shape/formation	Picture
Calcium carbonate crystals	Grouped in pairs/ dumbell shaped	
Phosphates crystals	Triple phosphates (ammonium, magnesium, phosphate) - coffin lids/fern like, calcium hydrogen phosphate	
Ammonium urate crystals	Covered with spines	

Abnormal crystals

Type	Shape	Seen in	Picture
Cysteine crystals	Hexagonal	Cysteinuria	
Cholesterol crystals	Flat rectangular with notched corners (stair case pattern)	Nehprotic syndrome Cholesterolemia	
Bilirubin crystals	Bead like or fine needle	Obstructive liver disease	
Leucine crystals	Spherical with radial striations	Cirrhosis	
Tyrosine crystals	Fine, needle like	Tyrosinemia	
Sulfonamide crystals	Sheaves of needles	Sulfonamide therapy	

STOOL ROUTINE EXAMINATION

Collection

- Universal precautions should be followed
- Stool sample is collected in a clean, dry, wide mouth, leak proof, screw-capped container, free from disinfectants and antiseptics
- Should be uncontaminated with urine or other body secretions, such as menstrual blood
- Collected with a clean tongue blade or similar object
- Rectal swab can be collected from infents and children
- It should not be contaminated with urine
- Should not be collected from bed pan
- Deliver immediately after collection
- Warm stools are best for detecting ova or parasites
- Do not refrigerate specimen for ova or parasites
- If the stool should be collect in 10% formalin or PVA fixative, storage temperature is not critical
- Because of the cyclic life cycle of parasites, three separate random stool specimens are recommended
- Some *coliform* bacilli produce antibiotic substances that
- destroy enteric pathogen. Refrigerate specimen immediately
- A diarrheal stool will usually give accurate results
- A freshly passed stool is the specimen of choice
- Stool specimen should be collected before antibiotic therapy, or as early in the course of the disease
- If blood or mucous is present, it should be included in the specimen

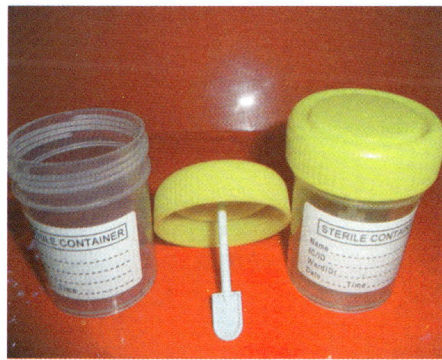

Interfering factors

- Patients receiving tetracyclines, anti-diarrheal drugs, barium, bismuth, oil, iron, or magnesium may not yield accurate results
- Bismuth found in toilet tissue interferes with the results
- Stool should not be collected from the toilet bowl. A clean, dry bedpan is the best
- Lifestyle, personal habbits, environments may interfere with proper sample procurement

Macroscopic Examination	Normal values
Amount	100-200 g/day
Colour	Brown
Odour	Varies with pH of stool and depend on bacterial fermentation
Consistency	Plastic, not unusual to see fiber, vegetable skins
Size and shape	Formed
Gross blood, Mucous, Pus, Parasites	None
Adult parasites and segments of tapeworm	None

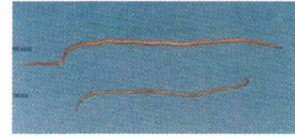

Adult Roundworm   **Adult Tapeworm**

Microscopic Examination	Normal values
Fat	Colorless, neutral fat (18%) and fatty acid crystals and soaps
Undigested food	None to small amount
Meat fibers, Starch, Trypsin	None
Eggs and segments of parasites	None
Yeasts	None
Leukocytes	None

Chemical Examination	Normal values
Water	Up to 75%
pH	Neutral to acid or alkaline – 6.5-7.5
Occult blood	Negative
Urobilinogen	50-300 mg/24hr
Porphyrins	Coprophyrins: 400-1200 mg/4 hrs Uroporphyrins:10-40 mg/24 hrs
Nitrogen	< 2.5 g/24 hrs
Bile	Negative in adults; positive in children
Trypsin	20-950 units/g (positive in small amounts in adults; present in greater amounts in normal children
Osmolarity	Used 200-250 mOsm with serum osmolarity to calculate osmotic gap
Sodium	5.8-9.8 mEq / 24hr
Chloride	2.5-3.9 mEq / 24 hr
Potassium	15.7-20.7 mEq /24 hr
Lipids (fatty acid)	0-6 g / 24 hr

Clinical implications

Fecal consistency may be altered in various disease states:

- Diarrhea mixed with mucous and red blood cells is associated with Typhus, Typhoid, Cholera, Amoebiasis, Large bowel cancer
- Diarrhea mixed with mucus and white blood cells is associated with Ulcerative colitis, Regional enteritis, Shigellosis, Salmonellosis, Intestinal tuberculosis
- "Pasty" stool is associated with a high fat content in the stool
    - A significant increase of fat is usually detected on gross examination
    - With common bile duct obstruction, the fat gives the stool a putty- like appearance.
    - In cystic fibrosis, the increase of neutral fat gives a greasy, "butter stool" appearance.

- **Stool Odour**

    Normal value varies with pH of stool and diet. Indole and sketole are the substances that produce normal odour formed by intestinal bacteria putrefaction and fermentation.

Clinical implications:

- A foul odor is caused by degradation of undigested protein.
- A foul odor is produced by excessive carbohydrate ingestion.
- A sickly sweet odor is produced by volatile fatty acids and undigested lactose

- **Stool pH**

    Increased pH (alkaline) – protein break down, villous adenoma, colitis, antibiotic use

    Decreased pH (acid) – carbohydrate malabsorption, fat malabsorption, disaccharidase deficiency

- **Stool color**

    Yellow to yellow-green: severe diarrhea

    Green: severe diarrhea, bile

    Black: resulting from bleeding into the upper gastrointestinal tract (>100 ml blood)

    Tan or Clay colored: blockage of the common bile duct

    Pale greasy acholic (no bile secretion) stool: pancreatic insufficiency.

    Maroon-to-red-to-pink: possible result of bleeding from the lower gastrointestinal tract (e.g. tumors, hemorrhoids, fissures, inflammatory process)

    Blood streak on the outer surface of usually indicates hemorrhoids or anal abnormalities.

Blood in stool can arise from abnormalities higher in the colon

In some case where the transit time is rapid, blood from stomach or duodenum can appear as bright or dark red or maroon in stool.

- **Blood in Stool**

    Dark red to tarry black indicates a loss of 0.50 to 0.75 ml of blood from the upper GI tract

    Positive for occult blood may be caused by carcinoma of colon, ulcerative colitis, adenoma, diaphramatic hernia, gastric carcinoma, diverticulitis, ulcers.

- **Mucous in Stool**

    Translucent gelatinous mucous clinging to the surface of formed stool occurs in spastic constipation, mucous colitis, emotionally disturbed patients, excessive straining at stool

    Bloody mucous clinging to the surface suggests neoplasm, inflammation of the rectal canal

    Mucous with pus and blood is associated with ulcerative colitis, bacilliary dysentery, ulcerating cancer of colon, acute diverticulitis, intestinal tuberculosis.

- **Fat in Stool**

    Increased fat or fatty acids is associated with the malabsorption syndromes – nontropical sprue, Crohn's disease, Whipple's disease, Cystic fibrosis, enteritis and pancreatic diseases

- **Urobilinogen in Stool**

    Normal values: 125-400 Ehrlich units/24 hrs; 75-350 Ehrlich units/100 g

Clinical Implications:

- Increased values are associated with Hemolytic anemias
- Decreased values are associated with complete biliary obstruction, severe liver disease, infectious hepatitis, oral antibiotic therapy that alters intestinal bacteria flora, infants are negative up to 6 months of age.

- **Bile in Stool**

    Bile may be present in diarrheal stools

    Increased bile levels occur in Hemolytic anemia

- **Trypsin in Stool**

    Decreased amounts occur in pancreatic deficiency, malabsorption syndromes, screen for cystic fibrosis

- **Leukocytes in Stool**

    Large amounts of leukocytes in chronic ulcerative colitis, chronic bacilliary dysentery, localized abscess, fistulas of sigmoid rectum or anus

Mononuclear leukocytes appear in typhoid

Polymorphonuclear leukocytes appear in shigellosis, salmonellosis, Yersinia, invasive *Escherichia coli* diarrhoea, ulcerative colitis

Absence of leukocytes is associated with cholera, non specific diarrhea, viral diarrhea, amebic colitis, noninvasive *E.coli* diarrhoea, toxigenic bacteria *Staphylococcus spp.* and *Clostidium spp.*, Parasite – *Giardia lamblia.*

- **Porphyrins in Stool**

  Increased fecal coproporphyrin is associated with coproporphyria (hereditary), porphyria variegate, protoporphyria, hemolytic anemia

  Increased fecal protoporphyrin is associated with porphyria veriegata, protoporphyria, acquired liver disease

- **Stool Electrolytes**

  Normal values: Sodium 5.8-9.8 mEq/24 hrs

  Chloride 2.5-3.9 mEq/24 hrs

  Potassium 15.7-20.7 mEq/24 hrs

Clinical Implications:

- Idiopathic proctocolitis – Increased Sodium and Chloride; Normal Potassium

- Cholera – Increased Sodium and Chloride.

Microscopic examination

- Saline mount with 0.9% saline
- Iodine mount with Lugol's iodine
- Trichrome stain with Chromotrope 2R
- Modified Z-N stain using 10% sulphuric acid and Kinyoun's cold method
- Buffered methylene blue mount

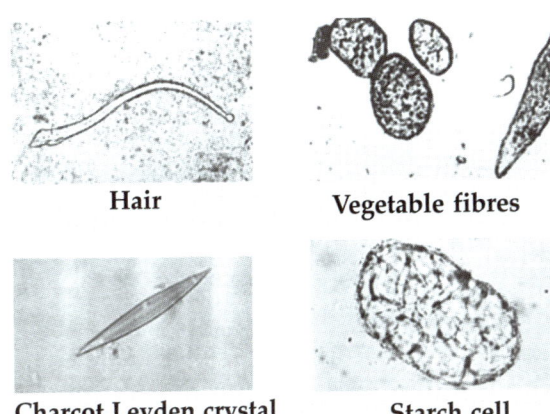

Hair   **Vegetable fibres**

Charcot Leyden crystal   **Starch cell**

Structures in feces that require differentiation from parasites

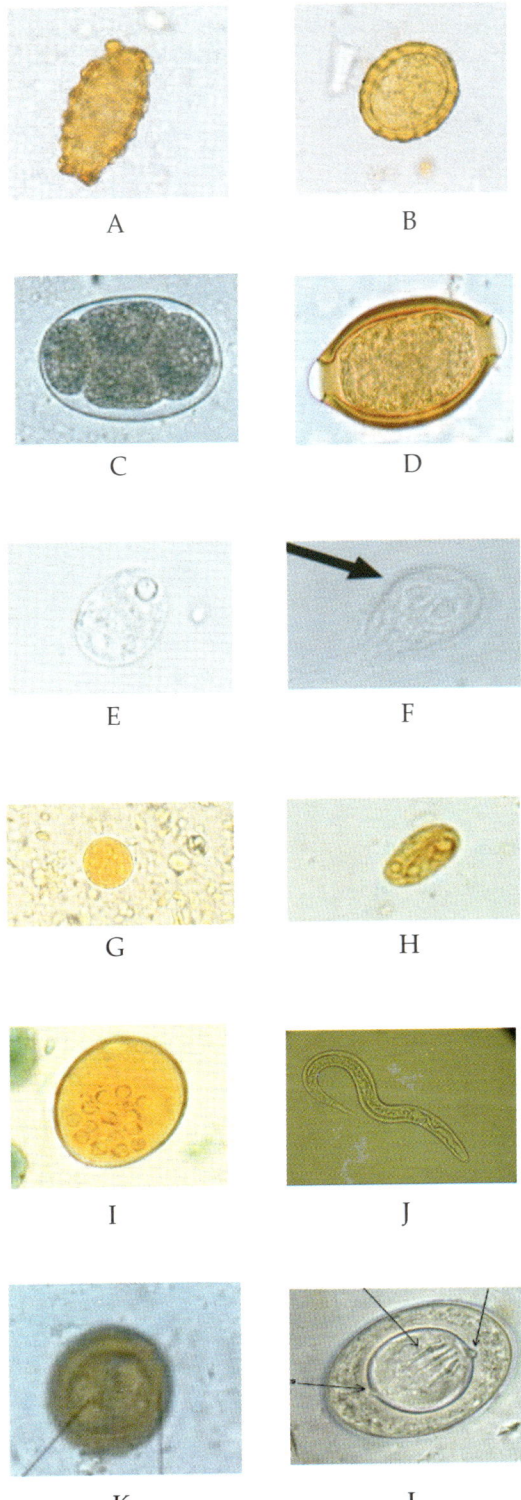

Ova seen in saline mount – A, Unfertilised and B, Fertilised egg of *Ascaris*; C, Egg of Hookworm; D, Egg of *Trichuris trichiura*; E, Trophozoite of *E. histolytica*; F, Trophozoite of *G. lamblia*; G, Cyst of *E. histolytica*; H, Cyst of *G. lamblia*; I, Cyst of *E. coli*; J, Larva of *S. stercoralis*; K, Egg of Tenia; L, Egg of *H. nana.*

Bile stained eggs	Non bile stained eggs
Trichuris trichiura	Ancylostoma duodenale
Ascaris lumbricoides	Necator americanus
Tenia saginata	Enterobius vermicularis
Tenia solium	Hymenolepis nana

Buffered methylene blue mount (BMB)

- Also called as Quensel's stain or Nair stain.
- BMB stains amoebic trophozoites.
- BMB stain is appropriate only for fresh unpreserved specimens.
- The nucleus will stain dark blue while the cytoplasm will stain light blue.

Stoll's technique for counting helminth eggs

- Weigh 3 grams of feces in a screw cap container.
- Add 42 ml of water to give (1/15) dilution of the feces.
- If the feces are formed specimen, use sodium hydroxide 0.1 mol/l solution instead of water.
- Using a rod, break up the feces and mix it with the water.
- Cap the container and shake hard to complete in mixing.
- Using a pipette, pick 0.15 ml of the suspension and transfer this slide.
- Cover the slide with coverslip.
- Examine systematically the entire preparation, using 10x objective. Include in the count any eggs laying outside the edges of the coverslip because these are contained in 0.15 ml sample.
- Multiply the number of eggs counted by 100 to give the number of eggs per gram of feces.
- If the specimen isn't formed, the following additional calculation is necessary to give the number of eggs per gram.
- Fluid specimen -- X5
- Unformed watery specimen ----------------------- X4
- Unformed soft specimen ----------------------------- X3
- Semiformed specimen ---------------------------------- X2
- Calculate the number of eggs per day, by multiplying the number of eggs per gram by the total weight of 24 hrs fecal specimen.
- Calculate the number of burden worm by dividing the number of eggs per gram on number of eggs the parasite laying per day.

Concentration procedures

The number of parasitic forms in fecal specimens is often too low to be observed microscopically in direct wet mounts or in stained smear preparations. Concentration procedures must therefore be used for their detection.

Techniques used are:
- Sedimentation techniques
- Floatation techniques

Formol Ether Sedimentation method

Faeces are emulsified in formal water, the suspension is strained to remove large faecal particles, ethyl acetate is added and the mixed suspension is centrifuged. Cysts, oocysts, eggs and larvae are fixed and sedimented and the faecal debris is separated in a layer between the ether and formalin. Faecal fat is dissolved in the ether.

Reagents used are:
- Ethyl acetate
- 10% Formol water
- 0.9% NaCl

Procedure

Mix stool with saline
↓
Filter the emulsion through fine mesh gauze
↓
Centrifuge the supernatant at 2000 rpm for 10 minutes
↓
Decant the supernatant and wash the sedim clear
↓
Add 10 ml of 10% formalin to the sediment
↓
Mix and let stand for 5 minutes to effect fixation
↓
Add 1 to 2 ml of ethyl acetate, stopper the tube and shake vigorously Centrifuge at about
1500 rpm for 10 minutes
↓
Four layers should result:
a) a top layer of ethyl acetate
b) plug of debris
c) layer of formalin
d) sediment

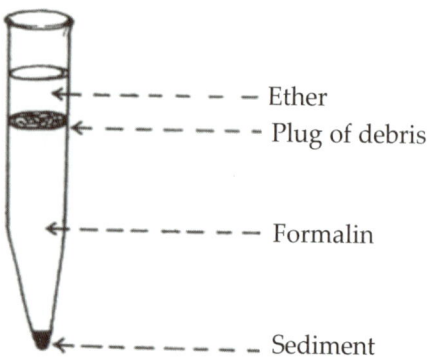

— Ether
— Plug of debris
— Formalin
— Sediment

Formal ether sedimentation technique

Floatation techniques

- Saturated sodium chloride floatation technique
- Zinc sulphate floatation technique
- Sheather's floatation technique

Saturated sodium chloride floatation technique

Stir sodium chloride into hot distilled water until no more can be dissolved

↓

Add few more grams of salt so that a layer of the undissolved salt remains in the bottom of the container

↓

Mix well and leave the undissolved salt to sediment

↓ cool, filter

Check specific gravity 1.2

↓

Emulsify 1 ml of feces with satuarated salt solution

↓

Add more salt solution to the container till it is full and convex meniscus is formed

↓

A glass slide is laid on top of the container for 20-30 min

↓

Slide is lifted, turned over and cover slip placed on top

↓

Examine under 10X and 40 X objectives

Eggs floating on saturated salt solution	Eggs that sink in saturated salt solution
Fertilised egg of *Ascaris lumbricoides*	Unfertilised eggs of *Ascaris lumbricoides*
Enterobius vermicularis	Eggs of *Taenia saginata* and *T. solium*
Ancylostoma duodenale	Eggs of all trematodes
Trichuris trichiura	
Hymenolepis nana	

Zinc sulphate floatation technique

Principle

A zinc sulphate solution is used which has a specific gravity of 1.180-1.200. Faeces are emulsified in the solution and the suspension is left undisturbed for the eggs and cysts to float to the surface. They are collected on a cover glass.

It is recommended for concentrating the cysts of *G. lamblia, E. histolytica/E. dispar,* eggs of *T. trichiura and Opisthorchis species.*

Procedure

Emulsify 1 g of faeces with zinc sulphate solution

↓

Strain to remove larger faecal particles

↓

In test tube again add zinc sulphate solution to the brim

↓

Place a clean, grease free cover slip on top of the tube

↓

Leave undisturbed for 30-45 minutes

↓

Lift the cover slip, place it facing downwards on a slide.

Sheather's floatation technique

- Used for recovery of Cryptosporidium oocysts in fecal specimens.
- Sheather's sucrose solution
    - Sucrose   500 g
    - Distilled water   320 ml
    - Phenol   615 g

Procedure

Heavy suspension of feces in saline

↓

Strained through gauze

↓

Mix with equal volume of Sheather's solution

↓

Prepare mount

↓

Observe under phase contrast microscopy.

Trichrome staining for fecal smears

- Wheatley Trichrome technique a modification of the original Gomori technique
- It contains chromotrope 2R and light green SF as the primary staining agents
- Easy to perform and rapid
- The refractive index of protozoan cysts and some helminth eggs is near that of water
- Staining procedures are required for detailed study of their internal structures
- Permanent stains enhance the identification of *Entamoeba histolytica* and the detection of *Giardia lamblia.*
- Reagents:
    - Chromotrope 2R   0.6 g
    - Light green SF   0.3 g
    - Phosphotunstic acid   0.7 g
    - Acetic acid   1.0 ml
    - Distilled water   100 ml

Procedure:

Prepare smear on a slide
↓
Immerse smear in Schauddin's fixative for 30 mins
↓
Dried, fixed smear
↓
Place slides in 70% ethyl alcohol for 3 mins
↓
Alcohol-iodine working solution for 10 mins
↓
Wash with 70% alcohol one for 5 mins and one for 2-5 mins
↓
Trichrome staining solution for 10 mins
↓
Acidified 90% ethyl alcohol for 5 seconds
↓
Dip slide in 100% ethanol
↓
Remove alcohol with xylene
↓
Examine under oil immersion 100X

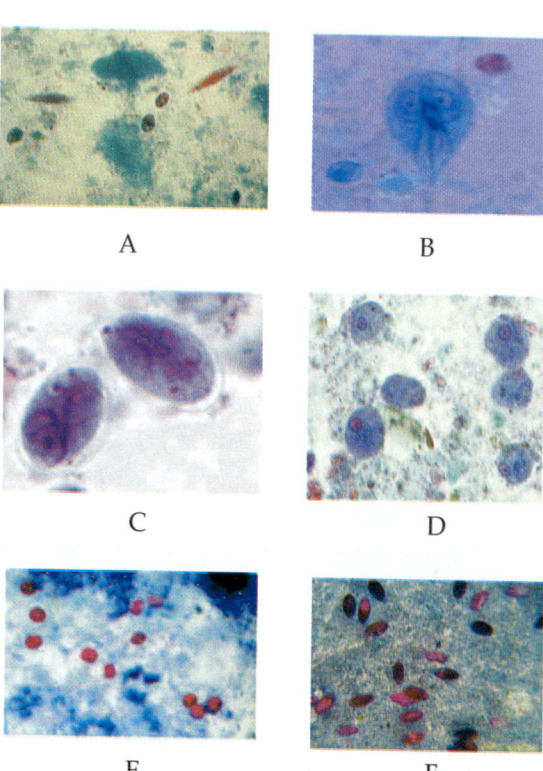

Trichrome stain showing – A, Charcot Leyden crystals seen in amoebic dysentery; B, Trophozoite of *G. lamblia*; C, Cyst of *G. lamblia*; D, Cyst of *E. histolytica*; E, Oocysts of *Cryptosporidium* species; F, Oocysts of *Isospora belli*.

Quality control for Trichrome staining

- Smears containing parasitic organisms of known staining properties be stained

- Properly fixed smears produce a blue-green cytoplasm of protozoan trophozoites with a slight tinge of purple
- Nuclei and inclusions are red, tinged with purple

Transport of specimen for stool culture

- In case of delay in processing for more than 1 hour, part of specimen is transferred to cotton swab and inserted into a container of Cary-Blair transport medium
  — For rectal swabs, moisten 2 swabs in an appropriate transport medium (e.g. Cary-Blair, Stuart, Amies, buffered glycerol-saline, etc.).
  — In case of infants, insert swab 1-1.5 inches into rectum and gently rotate. Place both swabs into the tube deep enough that medium covers the cotton tips. Break off top portion of sticks and discard.

Stool Preservatives

- 10% Formalin
- Methiolate-Iodine-Formalin (MIF)
- Low viscosity polyvinyl alcohol (LV-PVA)
- Sodium acetate-Acetic acid-Formalin (SAF)
- Schauddin's fixative

10% Formalin

- 10% Formalin is used for helminths eggs, larvae and for protozoan cyst
- Long shelf life
- Formal-ether concentration
- Disadvantage: Permanent staining procedures can't be performed from formalin preserved stool samples

Methiolate-Iodine-Formalin

- The preserved material permits concentration techniques
- The major disadvantages are the short shelf life (due to iodine) and permanent stained smears can't be prepared from MIF preserved material

LV-PVA

Polyvinyl alcohol	10 g
95% ethyl alcohol	62.5 ml
Mercuric chloride	125 ml
Glycerine	3 ml

- This fixative is recommended for the preservation of the trophozoite and cyst stages of the intestinal protozoa, and also suitable for helminthes eggs and larvae
- The PVA is a plastic resin that serves as adhesive for the stool material

- Has a long shelf life (months to years)
- Concentration methods can't be performed from the specimen preserved by PVA
- Advantage - permanent stain can be prepared from stool specimen preserved by PVA
- Disadvantage-Specimen preserved by PVA can't be used with immunoassay kits
- Toxic, because contain mercury compound
- Good routine fixative for protozoan cyst and trophozoites, helminthes eggs, and larvae
- Has long shelf life (months to years)

- The preserved stool samples permits concentration techniques, monoclonal detection kits, and permanent staining
- Disadvantage - Poor adhesive properties when SAF preserved samples are used to prepare permanent stained smears

Schauddin's fixative

- Saturated solution of mercuric chloride in distilled water	200 ml
- 95% or absolute alcohol	100 ml
- Glacial acetic acid	5 ml

Characteristics	Formalin 10%	PVA	SAF	MIF
Toxicity	+/-	+++ (duo to Hg)	+/-	+/-
Shelf life	Long(months)	Long(months/years)	Long(months/years)	Limited
Preparation	Easy	Difficult	Easy	Easy
Quality of fixation	Egg: ++	Egg: ++	Egg: ++	Egg: ++
Formol ether concentration	Cyst: ++	Cyst: +++	Cyst: ++	Cyst: ++
Permanent stained smear	Troph's: +/-	Troph's: +++	Troph's: +++	Troph's: +/-

Test	Use
Direct saline mount	Trophozoites, cysts, ova, larvae
Direct Iodine mount	Protozoal cysts
Faecal concentration	Recovery of protozoal cysts, helminth, ova and larvae
Trichrome staining	Protozoal cysts and trophozoites
Modified Z-N Staining	Oocysts of cryptosporidium, cyclospora, isospora

REFERENCES

1. Harsh Mohan. Pathology Practical Book Paperback. 3rd Ed. Jaypee Brothers Pvt. Ltd., New Delhi. 2012

2. Bronchoscopy/John Hopkins Medicine Health Library. http://www.hopkinsmedicine.org/healthlibrary/test_procedures/pulmonary/bronchoscopy_92,p07743/

3. www.livestrong.com/article/161434

4. Journals.cambridgepublishing.com.au

5. AARC Clinical Practice guided Endotracheal Suctioning of Mechanically Ventilated patients with Artificial airways, 2010

6. Cheesebrough M. District Laboratory Practice in Tropical Countries. Part 1. 2nd Ed. 2012

7. Kawthalkar SM. Essentials of Clinical Pathology. Jaypee Brothers Pvt. Ltd., New Delhi.2010; pp 3-27

8. Chatterjee K. D. Parasitology (Protozoology and Helminthology in relation to Clinical Medicine). 12th Ed. Chatterjee Medical Publishers, Calcutta. 1980

9. Karyakarte R, Damle A. Medical Parasitology Paperback. 2nd Ed. Books & Allied (p) Ltd., Kolkata. 2009

ARTHOPODS OF MEDICAL IMPORTANCE

Chapter

I. CLASS INSECTA

- Body divided into 3 segments – head, thorax and abdomen
- Have a pair of antenna and 3 pairs of legs
- Winged or wingless
- Life cycle is complete with egg, larva, pupa and adult stages.

Vector	Characteristics	Disease/s transmitted	Causative agent/s of disease/s	Mode of transmission
1. Mosquitoes	• Single pair of wings • Only female suck blood • Prominent proboscis			
a. Anopheles	• Spotted wings • Sits at an angle to surface • Palpi as long as proboscis • Plain abdomen • Breeds sunlit water & bites during night • Dull brown in colour	Malaria	*Plasmodium species*	By bite of female mosquitoes
b. Culex	• Plain wings • Rests sitting parallel to surface • Palpi shorter than proboscis • Spotted abdomen • Breeds in dirty waters & bites during night	Filariasis Japanese encephalitis (JE)	*Wuchereria bancrofti* JE virus	By bite of female mosquitoes By bite of female mosquitoes
c. Aedes	• Black & white in colour • Plain wings, legs spotted • Rests sitting parallel to surface • Palpi shorter than proboscis • Thorax has silvery markings & abdomen has 11 band ring • Breeds in small water collections & bites during day & night	Dengue fever Yellow fever Chikungunya	Dengue virus Yellow fever virus Chikungunya virus	By bite of female mosquitoes
d. Mansonoides	• Yellow or brown in colour • Spotted wings, legs, abdomen • Rests sitting parallel to surface • Palpi shorter than proboscis • Big scales with light & dark shades of same colour • Associated with air filled roots of water plants and bite at night	Filariasis	*Brugia malayi*	By bite of female mosquitoes

Vector	Characteristics	Disease/s transmitted	Causative agent/s of disease/s	Mode of transmission
2. Flies	• One pair of wings • Female flies suck blood usually			
a. **Sandfly** (*Phlebotomus*)	• Very small & hairy • Wings are erect at rest • Hops, do not fly • In damp & dusty places away from daylight	Kala-azar Oriental sore Sandfly fever	*Leishmania donovani* *L. tropica* Sandfly fever virus	By bite of female flies
b. **Black fly** (*Simulium*)	• Bigger than sandfly, black body • Snub nose & humped thorax • Transparent wings • In forests or places of dense vegetation & bites on sunny dys	Onchocerciasis	*Onchocerca volvulus*	By bite of female flies
c. **Deer fly** (*Chrysops*)	• Big fly with banded wings • Active on sunny days	Loiasis	*Loa loa*	By bite of female flies
d. **Tsetse fly** (*Glossina*)	• Medium sized fly • Mouth parts project out while resting • Only bite during day time when it is very hot & damp	Sleeping sickness	*Trypanosoma brucei*	By bite of both female & male flies
e. **Housefly** (*Musca domestica*) (non-biting)	• Medium sized blackish grey fly • Hairy body & legs • Thorax has 2-4 longitudinal stripes	Cholera Typhoid Amoebiasis Hepatitis A & E Poliomyelitis	*V. cholerae* *S. typhi* *E. histolytica* HAV & HEV Polio virus	By mechanical transmission
3. Fleas (combless)	• Body laterally compressed • Wingless • Powerful third leg which helps in jumping • Both male & female suck blood • Ectoparasites of animals & birds			
a. **Rat flea** (*Xenopsylla*)	—	Plague Endemic typhus	*Y. pestis* *R. mooseri*	By bite of infected fleas
b. **Human flea** (*Pulex irritans*)	—	Cestode infection	*H. diminuta*	By ingestion of infected fleas
4. Bugs	• Body flattened dorso-ventrally • Both sexes have sucking mouth parts • Both adult & nymph stages can bite			
a. **Bed bug** (*Cimex*)	• Wings - absent	—	—	Found in dark crevices of bed
b. **Reduviid bug**	• Wings - present • Rests outdoor in hilly parts	Chagas' disease	*T. cruzi*	Defecates while biting, so (*Triatoma*) contaminates the wound

Vector	Characteristics	Disease/s transmitted	Causative agent/s of disease/s	Mode of transmission
5. Lice	• Body compressed dorso-ventrally • Wings absent • Adults & nymphs suck blood • Permanent ectoparasites in man			
a. **Head louse** (*Pediculus capitis*)	• Body heavily pigmented • Eggs attached to hair • Found on heads of persons	Epidemic typhus	*R. prowazeki*	By bite followed by crushing
b. **Body louse** (*Pediculus corporis*)	• No black pigment in body • Eggs attached to clothing • Found on body & clothing of persons	Epidemic typhus Trench fever Relapsing fever	*R. prowazeki* *Ro. quintana* *B. recurrentis*	By bite followed by crushing

II. CLASS ARACHNIDA

- Wings and antenna absent
- Body is sac-like
- Adults have 4 pairs of legs.

Vector	Characteristics	Disease/s transmitted	Causative agent/s of disease/s	Mode of transmission
1. **Ticks**	• Ectoparasites of mammals • All stages take blood meal			
a. **Hard tick** (*Hemaphysalis, Ixodes*)	• Body covered with tough chitinous shield called scutum, which in males covers the back completely and in females covers only upper half • Mouth parts visible from dorsal aspect • Take only a single blood meal in their life time	Rocky Mountain Spotted fever Indian tick typhus Q fever Tularaemia Lyme disease Kyasanur forest disease	*R. rickettsii* *R. conori* *Coxiella burnetii* *F. tularensis* *B. burgdorferi* KFD virus	By bite of tick By bite of tick By contact with coral fluid By bite of tick By tick of tick By bite followed by crushing
b. **Soft tick** (*Ornithodorus*)	• Scutum absent • Mouth parts not visible from dorsal aspect • Take several blood meals in their life time	Relapsing fever	*B. duttoni*	By bite of tick
2. **Mites**	• More hairy than tick • Phytophagus mostly • Few eat blood meals • Ectoparasites of rodents			

Vector	Characteristics	Disease/s transmitted	Causative agent/s of disease/s	Mode of transmission
a. **Scrub mite** (*Leptotrombidium*)	• Larvae are orange in colour & has 3 pairs of legs • Body transparent when unfed, require blood meals • Adults & nymphs are harmless	Scrub typhus	*R. tsutsugamushi*	By bite of infected larva
b. **Mouse mite** (*Allodermanyssus*)	• Elongated body • 4 pairs of legs at the anterior end	Rickettsial pox	*R. akari*	By bite of infected adult mite
c. **Itch mite** (*Sarcoptes scabei*)	• Globular body • 4 pairs of short stumpy legs • Ectoparasites of human skin	Scabies	–	By burrowing of female mite into the skin
d. **House dust mite** (*Dermatophagoides*)	• Body globular • Wavy pattern on the skin	Dust allergy	–	By inhalation of mite allergens in dust

III. CLASS CRUSTACEA

-   Body has 2 segments – cephalothorax and abdomen
-   2 pairs of antenna
-   Found in fresh waters

Vector	Characteristics	Disease/s transmitted	Causative agent/s of disease/s	Mode of transmission
1. **Small crustaceans** (water fleas)				
a. **Cyclops**	• Body elongated • Female has egg sacs on either side of body • Found in tanks, canals, step wells • A single median eye	Dracanculosis	*D. medinensis*	By ingestion of infected cyclops
b. **Diaptomus**	• One antenna longer than body • Eggs in a single sac	Diphyllobothriasis	*D. latum*	By ingestion of infected diaptomus
2. **Big crustaceans**				
a. **Crabs**	• Body dorso-ventrally compressed • Eyes borne on a movable stalk	Paragonimiasis	*P. westermani*	By ingestion of infected crabs
b. **Crayfish** (*Lobster*)	• Body laterally compressed • Eyes borne on movable stalk • Abdomen elongated & held in extended position	Paragonimiasis	*P. westermani*	By ingestion of infected crayfish

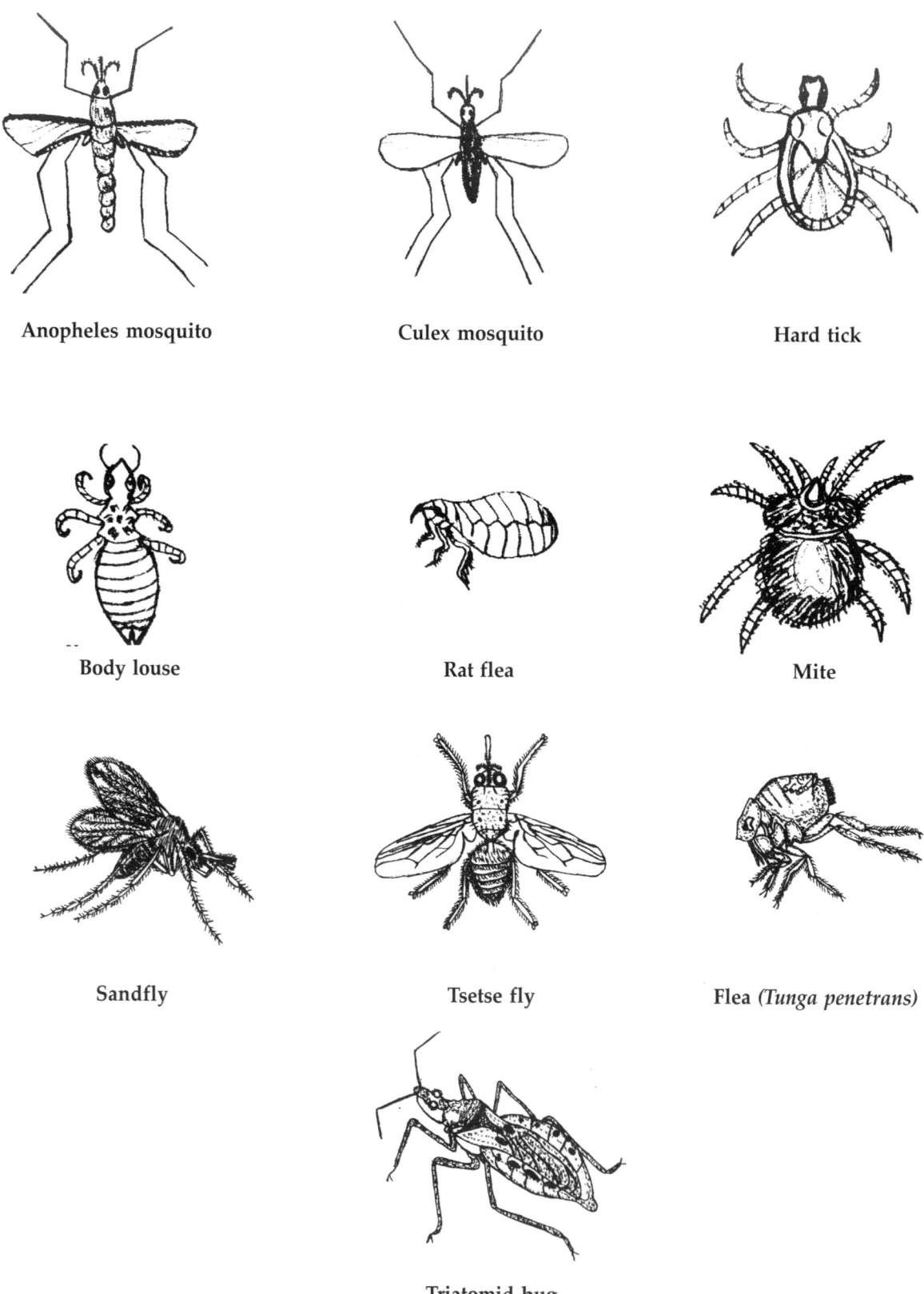

Anopheles mosquito   Culex mosquito   Hard tick

Body louse   Rat flea   Mite

Sandfly   Tsetse fly   Flea (*Tunga penetrans*)

Triatomid bug

REFERENCES

1. De A. Practical and Applied Microbiology. 5th Ed. The National Book Depot, Mumbai. 2014.

2. Gulati AK, Pal D. Diagnostic and Practical Microbiology. 1st Ed. Medical Allied Agency, Calcutta. 1985.

COMMON LABORATORY ANIMALS USED IN MICROBIOLOGY AND INSTITUTIONAL ANIMAL ETHICS COMMITTEE

31

Chapter

		Rabbit	Guineapig	Mice
1.	Body temperature (^{O}F)	101.6 - 102.4	99.6 - 102	99.3
2.	Respiration (per minute)	55	80	–
3.	Pulse (per minute)	135	150	120
4.	Gestation period (days)	28-30	60 - 70	19 - 21
5.	Mating age	6 - 9 months	12 - 20 weeks	6 - 8 weeks
6.	Weaning age	6 - 8 weeks	14 - 21 days	19 - 21 days
7.	Litter size (per year)	4	3	8 - 12
8.	Weight	0.9 - 6.75kg	200 - 1000g	25 - 28g; Weaning age – 7g
9.	Cage size	2 x 2 x 1½'	14 x 9 x 8"	6 x 12 x 6" for 6 mice 30 x 18 x 6' for 50 mice (a colony)
10.	Number kept in cage	1 female (doe) and 1 male (buck)	5 - 10 female (sows) and 1 male (boar)	4 female and 1 male
11.	Humidity	45%	45%	60%

12.	Common diseases caused	a) Coccidiosis
		b) Pseudotuberculosis
		c) Snuffle (by *P. septica*)
		d) Pneumonia
		e) Intestinal infection
		   (by *S. typhimurium*)
		f) Syphilis (by *T. cuniculi*)
		g) Ear canker (by mites)
		h) *Taenia pisiformis*
		   (dog tapeworm)
			a) Pseudotuberculosis
			b) Abscess in lymph nodes
			   (by Gr. C streptococci)
			c) *S. typhimurium*
			d) Coccidiosis
			e) Toxoplasmosis
			f) Viral (paralysis and
			   pneumonia)
				a) Mouse typhoid (*S. typhimurium* and
				   S. enteritidis)
				b) Mouse pox (acute or chronic
				   electromelia)
				c) Streptobacillus moniliformis
				   (acute and chronic)
				d) Pneumonia virus of mice
				e) Lymphocytic choriomeningitis virus
				f) *Taenia taeniaeformis* (cat tapeworm)
				g) *Giardia muris*

13.	Sites of blood collection	a) From ear vein (20 - 30ml)
		b) By cardiac puncture (50ml/kg)
			a) From marginal ear vein
			b) By cardiac puncture
			   (15ml)
				a) From great vessels of axilla and
				   (0.5ml)
				b) Retroorbital plexus (1ml)

14.	Different routes of inoculation (amount inoculated)	a) Intratesticular (0.2 - 0.4ml)
		b) Intracutaneous (0.5ml)
		c) Intracerebral (0.45ml)
		d) Conjunctiva (few drops
		   instilled)
			a) Intraperitoneal (5ml)
			b) Subcutaneous (5ml)
			c) Intracutaneous (0.2ml)
			d) Conjunctiva (few drops
			   instilled)
				a) Intraperitoneal (2ml)
				b) Subcutaneous (1ml)
				c) Intravenous (0.5ml into tail vein)
				d) Intracerebral (0.03ml)
				e) Intranasal (0.1ml)
				
				Suckling mice
				(< 48 hours old)
				a) Intraperitoneal (0.05ml)
				b) Subcutaneous (0.03ml)
				c) Intracerebral (0.03ml)
				(Beyond 48 hours)
				d) Intragastric (0.1ml) with
				   2% Evans blue/ml

| 15. Uses in Microbiology | a) Intratesticular for *T. pallidum*
b) Conjunctiva for *Listeria, Shigella*, Enteroinvasive *E. coli* (EIEC)
c) Intracutaneous for *C. diphtheriae* (virulence test)
d) Intracerebral for rabies
e) Differentiates human and bovine strains of *Mycobacterium tuberculosis*
f) Preparation of immune sera | a) Intraperitoneal for many viruses, *Clostridia, F. tularensis, Leptospira, M. tuberculosis, Nocardia, Spirillum minus, Y. pestis, Rickettsia*.
b) Subcutaneous for *F. tularensis, Y. pestis*.
c) Intracutaneous for *C. diphtheriae* (virulence test)
d) Conjunctiva for *Shigella*, EIEC
e) Source of complement required for CFT and other tests | a) Intraperitoneal for *S. pneumoniae, Borrelia, Clostridia, C. neoformans, H. capsulatum, Nocardia, S. minus, T. gondii, C. immitis, B. dermatitidis, Rickettsia*
b) Subcutaneous for *B. anthracis*
c) Intracerebral for *C. neoformans*, Rabies virus
d) Foot pad for *M. leprae*
Suckling mice
a) Intraperitoneal for *Coxsackie* A and B
b) Intragastric for *Enterotoxigenic E. coli* (ETEC) for detection of ST toxin
c) Intracerebral for herpes viruses, arboviruses. |

Uses of other animals and birds

1. Rat
   a) Intraperitoneal for *Borrelia, T. gondii*
   b) Subcutaneous for *Y. pestis*

2. Hamster
   a) Intraperitoneal for *L. donovani, Clostridia, Leptospira*
   b) Intradermal for *L. tropica, L. brasiliensis*

3. Armadillo – Intraperitoneal for *M. leprae*

4. Monkeys – Intraperitoneal for HAV, HBV, Polio virus

5. Chimpanzees–   Intraperitoneal for HAV
                  Intranasal for Rhinoviruses

6. Embryonated hen's egg
   a) Yolk sac for Rabies virus, *Rickettsia, Chlamydia*
   b) Amniotic cavity for influenza virus and parainfluenza virus (for isolation)
   c) Allantoic cavity for influenza virus (for vaccine production)
   d) Chorioallantoic membrane for variola, vaccinia, herpes simplex viruses 1 and 2

Care of the cages

1. Once a week all laboratory animals should be transferred to clean cages.

2. Cages should be scrubbed thoroughly with soap and water, then placed in autoclave or hot air oven.
   OR, Cages are boiled in soapy water and immersed overnight in 3% lysol (exvept for rabbits and mice, because smell of lysol causes cannibalism among them).

3. For litter – ½ to 1" thick bed made of soft wood, saw dust or sugarcane pith is used.

Anaesthetizing animals

1. For short duration – with ether (0.5ml)

2. For long duration – with Pentothal sodium (28ml/kg intraperitoneal or intravenous)

Killing animals

1. Physical ways
   a) Breaking spinal cord in cervical region or damaging brain
   b) Head suddenly thumped against edge of a sink (in mice and guineapigs)

2. Chemical ways
   a) Volatile substances – chloroform (toxic for mice), coal gas, nitrogen gas, ether (rapid action)
   b) Non-volatile substances – Pentothal sodium I/V 3-4 times the anaesthetic dose, saturated magnesium sulphate I/V, 5-10ml of air I/V.

Signs of death in animals

- Body cold, still and rigor mortis set
- Decapitated
- Complete necropsy done
- Heart removed.

Dead animals are first covered with 10% lysol and then destroyed in furnace or incinerator.

INSTITUTIONAL ANIMAL ETHICS COMMITTEE (IAEC)

Objectives of IAEC

- Ensure quality and consistency in review of research proposals.
- Prevent infliction of unnecessary pain & sufferings on animals prior, during & after experiments.
- Follow CPCSEA (Committee for the Purpose of Control and Supervision of Experiments on Animals) guidelines under the provision of section 15 of PCA act 1960 (Ministry of Environment &

Forests, GOI-2008) & the Gazette of India 1998 for experiments on animals.

IAEC functions

- Ensure that animals in course so injured that their recovery involves serious sufferings are euthanized as per specific norms.
- Experiments on animals are avoided whenever possible to do so & propagate principles of 3 R's (Reduce, Refine & Replace use of animals in experiment).
- Ensure that experiments on larger animals are avoided when it is possible to achieve the same results by experiments on small laboratory animals like guinea pigs, rats, mouse, frogs and pottery.
- Ensure that required records are maintained with respect to experiments performed on animals.
- See that experiments are **NOT** performed merely for the purpose of acquiring manual skills.
- Also note that animals in the experiment are properly looked after both before and after the experiments.

Contract research animal experiments

Amendment of rule-12, in the breeding and experiments on animals rule 1998, this rule amended to allow to establish, to undertake contract research as per provision of PCA-Act 1980 & the rules made their under.

Record keeping and archiving

All the following documents must be stored for a period of five years:

- Curriculum Vitae (CV)/Biodata of all members of IAEC.
- Minutes of the meetings duly signed by the Chairperson and CPCSEA nominee.
- Copy of all correspondences with members, researchers and other regulatory bodies.
- Copy of existing relevant national and international guidelines on research ethics & laws along with amendments.
- All study documents/study projects should be archived for minimum of 5 years period after completion of the study.
- A copy of filled proforma related to the projects shall remain with the Principal Investigator (PI) for minimum 5 years period.

Follow up procedures

- Deviations from project are **NOT** allowed without the permission from IAEC.
- Any new information related to the study should be communicated to IAEC.
- Premature study termination to be notified with reasons along with summary of the data obtained so far.

- Unused animals to be surrendered to animal house.
- Change of investigators should be done with approval of IAEC.

Communicating the decisions

- Will be communicated to PI by the member-secretary in writing.
- Suggestions for modifications & reasons for rejection shall be communicated to the PI.

IAEC members updating

- All relevant information on animal ethics will be brought to the attention of members of IAEC by the member-secretary.
- Institute members will be encouraged to attend national and international training programs/ conferences /seminars in the field of research related to animal ethics in order to help improve the quality of research projects/animal ethics committee submissions & review.

Element of review

- Scientific design and conduct of the study.
- Approval of scientific review committee and regulatory agencies.
- Assessment of predictable risks/harms to the animals.
- Protocol and proforma of the study.
- Plans for data analysis and reporting.
- Adherence to all regulatory requirements and applicable guidelines.
- Competence of investigators, researchers and supporting staff.
- Facilities and infrastructure in the animal house.

Membership duration and responsibilities

- Duration is of 3 years.
- One third ($1/3^{rd}$) should be fresh members otherwise members can serve more than one term.
- Member can be replaced in the event of long term non-availability (3 consecutive meetings)
- Authority of replacement to be with the Director, confidentiality of all decisions/discussions during the meeting to be maintained and members to sign a confidentiality form at start of their term.
- Each member to submit the declaration to maintain confidentiality of documents submitted to them during their membership period.
- No IAEC meeting can be held in absence of the nominee. Even no decisions pertaining to animal allotment & projects or approval of projects can take place in absence of IAEC nominee.

Documentation of research proposals

- Title of the project, names of PI & Co-investigators with designations.
- Name of other institute/hospital/field area where research is to be conducted.
- Endorsement of the HOD.
- Protocol of the proposed research.
- Ethical issues in the study & plans to address those
- Proposal to be submitted with all relevant annexures like proforma, CV of outside members, undertaking, etc. to be used in the study.
- Any other information regarding the study.
- Agreement to submit 6 monthly progress report & final report at the end of study.
- PI to give details of other ongoing research projects related to animal studies (title, date of starting, study duration, source and amount of funding)

Application procedure

- All proposals to be submitted in prescribed application form, copies of which will be available with the Member-Secretary.
- All relevant documents should be enclosed with application.
- Required number of copies of the proposal along with application & documents in prescribed format duly signed by the PI & Co-investigator/ collaborators should be forwarded by the HOD.
- Member-secretary to acknowledge the receipt & indicate any lacunae.
- Missing information to be supplied within 2 weeks.
- Date of meeting will be intimated to the PI who should be available to give clarification.
- The decision of IAEC will be communicated in writing. Revised copies/documents in proper number to be submitted within a stipulated period of time as specified in the communication.

Independent consultants

- IAEC may call upon subject experts as consultants for review of selected research protocols.
- These experts may be specialists in ethical /legal aspects, specific subjects or methodologies, or representative of CPCSEA. They will not take part in the decision making process.

Review procedure

- IAEC meetings to be held on scheduled intervals as prescribed (every 6 monthly, for which the month will be decided at the end of previous meeting).
- Additional meetings to be held as & when required.
- It is mandatory to call IAEC meetings every 6 months even if there are no projects, to discuss matters related to maintenance of animals in the animal house.

- The Animal House Incharge (I/C) must be present with all available records at every meeting.
- Proposals will be sent to members at least 2 weeks in advance.
- PI/Co-investigators should be available during the meeting & may be invited to offer clarifications.
- Independent experts/consultants may be invited to offer their opinion on specific research proposals.

Decision making

- IAEC decision would normally be taken by consensus. If divergent views are expressed by the members, these may be recorded in the minutes and a broad consensus be recorded as per understanding of the chair.
- Decision may be to approve, reject or revise the proposals. Specific suggestions for modifications and reasons for rejection should be given.
- Only members will make the decision. The decision shall be taken in absence of investigators and consultants.
- A member shall withdraw from the meeting during the decision procedure concerning an application where a conflict of interest arises. This shall be indicated to the chairperson prior to the review of application and recorded in the minutes.

Laboratory animal ethics

- All scientists working with laboratory animals must have a deep ethical consideration for the animals.
- Ethical point of view, such considerations are taken care at the individual level, at the institutional level and finally at the national level.

Quorum requirements

A minimum of 5 members, including 2 outside members is required for the quorum. Presence of CPCSEA member or nominee is mandatory for every meeting. The proposal of emergent need may be considered through circulation that will be decided by the Dean.

Offices/conduct of the meetings

- Chairperson to conduct all meetings of IAEC. If reasons beyond control in case of non-availability, the alternate chairperson will be elected by the members present from amongst themselves.
- The member-secretary will be responsible for organizing the meetings, maintaining the records & communicating with all concerned. Also will prepare the minutes of the meeting & get them approved by the chairperson & nominee of CPCSEA prior to communicating it to PI. No meeting to be held without the presence of CPCSEA member/nominee.
- Prior to finalizing the project, the IAEC chairperson to call for a preliminary meeting with

the project leaders/guides and ask them to detail their projects.

- After finalizing the proposals, the IAEC protocols dealing with small animals should be posted to the nominee one month in advance to the IAEC meeting and 16 copies with check list of each protocol dealing with large animals to be sent to CPCSEA, New Delhi office for scrutiny by the sub-committee members.

- Clearance of large animal project will be subject to the approval received from the sub-committee for large animal projects.

- The CPCSEA may authorize an additional representative to attend the IAEC meeting. Such authorized personnel should be permitted to attend the meeting and take part in the proceedings.

IAEC Members

Should have the following members in the committee (usually an odd figure):

- Chairperson
- Secretary
- Co-Secretary (optional)
- Nominee of CPCSEA
- Scientist
- Biological Scientist
- Veterinarian (Sr. V. O.)

Transport of laboratory animals

- The transport of animals from one place to another is very important and must be undertaken with care.

- The main considerations for transport of animals are:
  - The mode of transport
  - The containers
  - The animal density in cages, food and water during transit

- Protection from transit infections, injuries and stress.

Mode of transport of animals

- Depends on the distance, seasonal and climatic conditions and the species of animals.

- Animals can be transported by road, rail or air taking into consideration of above factors.

- The transport stress should be avoided.

The transport containers are cages or crates.

Requirements

- They should be of an appropriate size.

- Permissible number of animals should only be accommodated in each container to avoid overcrowding and infighting

- They should enable these animals to have a comfortable, free movement and protection from possible injuries.

- The food and water should be provided during transit.

Breeding and Genetics

- For initiating a colony, the breeding stock must be procured from CPCSEA registered breeders or suppliers

- Genetic makeup and health status of animal must be known.

- In case of an inbred strain:
  - The characters of the strain
  - Their gene distribution
  - The number of inbred generation must be known for further propagation.

- The health status should indicate their origin, e.g. conventional, specific pathogen free or transgenic gnotobiotic or knockout stock.

Criteria for considering transport:

Requirements for transport of laboratory animals by Road, Rail and Air

	Mouse	Rat	Hamster	Guineapig	Rabbit	Cat	Dog	Monkey
Maximum no. of animals per cage	25	25	25	12	2	1 or 2	1 or 2	1
Material used in transport box	Metal Cardboard Synthetic material	Metal Cardboard Synthetic material	Metal Cardboard Synthetic material	Metal Cardboard Synthetic material	Metal Cardboard Synthetic material	Metal	Metal	Bamboo/ wood/ metal
Space/animal (sq.cm.)	20-25	80-10	80-100	160-180	1000-1200	1400-1500	3000	2000-4000
Minimum height of box (cm)	12	14	12	15	30	40	50	48

Sanitation and Cleanliness of Animal House

- Animal house rooms, corridors, storage spaces and other areas should be cleaned with appropriate detergents and disinfectants (Lysol)
- Mops and brooms, cleaning utensils should not be transported between animal rooms.
- Removal of animal waste to be done at least twice a day.
- Cages to be sanitized before placing the animals.
- Racks to be washed/sanitized at least once a month, Wire bottom rodent cages washed at least once every two weeks. Watering & feeding devices cleaned at least twice/week.
- Cages can be disinfected by rinsing at the temperature of 180°F/82.2°C or higher for period long enough to ensure destruction of vegetative pathogenic organisms.
- Make available the extra cages at all times.
- Maintain systematic cage-washing schedule
- Disinfection be done by chemicals (Lysol) and equipments to be rinsed free of chemicals.
- Organic compounds of the urine of animals (like rabbits, guineapigs) that adheres to cages or surfaces necessitates removal by acid solutions.
- Water bottles/watering equipments to be washed and then sanitized by rinsing with water of at least 82.2°C or appropriate chemical agent like hyperchlorate to destroy pathogenic organisms.
- Deodorizers or chemical agents other than germicides not be used to mask animal smell.
- Waste disposal to be done regularly and frequently. Most preferred method is incineration.

Animal Food

- Food to be palatable, non-contaminated & nutritionally adequate unless the expt. protocol requires otherwise.
- Food to be sufficient amounts to ensure normal growth in immature animals & maintain normal body weight, reproduction & lactation in adult animals.
- Lab animal diet should not contain additives like hormones, rodenticides, metals antibiotics, toxicants, insecticides.
- Areas where diets are processed/stored to be kept clean & enclosed to prevent entry of insects, other creatures.
- Meats, fruits, vegetables, perishable items to be refrigerated sos/stored since they are source of biological & chemical contamination.
- Animals to have continuous access to fresh & uncontaminated drinking water. Periodic monitoring of microbial contamination of water is necessary.

Animal bedding

- Bedding to be absorbent, free of toxic chemicals/other substances that should injure animals or personnel. Should not be of type that is readily eaten away by animals.
- Should be used in amounts sufficient to keep animals dry between cage changes.
- Bedding to be replaced/removed by fresh materials at least twice per week. This will keep animals clean and dry.

Desirable criteria:

- Sterilizable
- No formation of deleterious products
- Readily available and easily stored
- Uncontaminated, unlikely to be chewed/eaten by animals
- Non-toxic, non-malodorous
- Disposable by incineration, chemically stable, non-hazardous, non-palatable, non-injurious to animals and personnel.

Animal housing

Cages and No. (as per Annexure 3 of CPCSEA)
Annexure 3A

Animal	Weight (gm)	Floor area (cm²)	Cage height (cm²)
Mice	< 10	38.70	12
	≤ 15	51.60	12
	≤ 25	77.40	12
	> 25	96.70	12
Guineapig	< 350	387.00	18
	> 350	≥ 651.4	18
Rat	< 100	109.6	14
	upto 200	148.3	14
	upto 300	187.0	14
	upto 400	258.0	14
	upto 500	387.0	14
	> 500	451.5	14
Rabbit	< 2000	1.5	14
	upto 4000	3.0	14
	upto 5400	4.0	14
	> 5400	5.0	14
	Female/mother with kids	4.5	14

Record keeping

Following records to be maintained in the animal house:
- Animal house plan, typical floor plan, all fixtures, etc.
- Health record of staff
- Animal house staff record (tech/non-tech)
- SOPs regards animals

- Breeding, stock purchase/sales record
- Minutes of the IAEC meetings
- Death record of animals
- Experiments conducted/no. of animals used
- Clinical record of sick animals
- Training record of staff involved in animal activities
- Water analysis report (once/3 to 4 months).

Annexure 3B for calculating no. of mice/cage

Recommended floor area/animal (cm^2)	38.7	51.6	77.4	96.7
Wt. of animal (gms)	< 10	= 15	= 25	> 25
Maximum no. of animals*	08	07	04	03

* For a cage size 24 X 14 cm == floor area 336 cm^2

Calculating no. of animals/cage

- Floor area of cage
- Recommended floor area/animal
- Weight of animal
- In c/o breeding pairs, 3 adults (1M:3F) along with pups from delivery up to weaning stage are allowed.

Annexure 3C for calculating no. of rats/cage

Recommended floor area/ animal (cm^2)	109.60	148.30	187.00
Wt. of animal (gms)	< 100	= 200	= 300
Max. no. of animals*	06	05	04

* For a cage size 32.5 cm X 21 cm = 682 cm^2

Biomedical waste

Responsibility of every individual who generates waste.

Requirements

- Colour coded bags (black, red, yellow depends on type of waste generated)
- Labels (all bags to carry labels mentioning hospital name, date, type of waste and area of generation)
- Bins of appropiate sizes to hold the bags.
- Sharps disposals cans/empty 5 litre cans (punture/leak proof with screw capped lid and handle)
- Needle burner
- Scissor to cut empty I/V infusion bottles and used gloves

- Sodium hypochlorite solution
- Log books
- Gloves and mask for the person transporting waste bags

Daily animal house rounds

- Animal house rounds will be taken between 9.30 am and 11.00 am every day.
- Rounds taken along with Veterinary doctor and resident posted in animal house
- Staff taking the rounds to ensure the following:
  - Cleanliness of animal cages and trays
  - Diet and water provided for the animals for the day
  - Animal counts taken and noted on the board and register
  - Presence of injured animals. If yes, then proper veterinary care
  - Cleanliness of animal rooms including passages and dusting of furniture and ceiling
  - Cleanliness of Experiment rooms and room of doctor I/C, residents room and servants room
  - Substitute/badlee arrangement in case of absenteeism of a particular staff member
  - Management of problems with respect to manpower, rodent infestation and civil needs.
- The animal house I/C will also keep an account of items in stores vis a vis usage.

Cages: Parameters/Dimensions required

Rat cage area = 42.1 cm x 29 cm = 1221 sq. cm;
Height: 19 cm (required ht = 14 cm)
Can keep 4-5 rats/cage if each animal is of 400 gm Wt.

Small Rat Cage Area = 35 cm x 20 cm = 700 sq.cm;
Height: 15.5 cm
Can keep 3-4 animals/cage

Mice cage area = 30 cm x 16 cm = 480 sq.cm; Height: 16 cm
Can harbour 2 animals/cage.

Rabbit cage area = 50 cm x 65 cm = 3250 sq.cm;
Height: 17.70 cm
Can keep 2 animals/cage (of 1.8 kg wt.) or only one rabbit (of 4 kg wt.)

Guineapig cage area = 90cm x 150 cm = 13,500 sq.cm.
Can keep 20 animals/cage (of wt. 350 gm).

REFERENCES

1. De A. Practical and Applied Microbiology. 5th Ed. The National Book Depot, Mumbai. 2014.
2. Collee JG, Marion BP, Fraser AG, Simmons A. Mackie and McCartney Practical Medical Microbiology. 14th Ed. Churchill Livigstone, U.K. 1996.
3. www.envfor.nic.in/mef/guidelines_IEAC.pdf.
4. cpcsea.nic.in/Content/109_1_ReconstitutionofIAEC.aspx.

AUTOMATION IN MICROBIOLOGY

32

Chapter

"The mechanization of man will lead to a situation when man will offer flowers to big machines and say, 'Please make me a nut and bolt in your system'. We are losing our identity".

- Bertrand Russell

- Webster's Dictionary defines 'Automation' as the "Automatically controlled operation of an apparatus, process or system by mechanical or electronic devices that take the place of human organs of observation, effort and decision".

- In 1973, the AutoMicrobic System (AMS) (McDonnell Douglas Corp., St. Louis, Mo.) was born. It incorporated a disposable miniaturized plastic specimen-handling system, solid-state optics for microbial detection, and a minicomputer for control and processing and today is recognized as the first generation of the Vitek instruments.

- In 1976, Williams commented at the Symposium on Rapid Methods and Automation in Microbiology that "... medical bacteriologists have tended to impose their own circadian rhythm on their bacterial cultures and to work in units of 18 to 24 hours".

- Automation is expensive as regards equipments, chemicals and reagents. It may be of value when the volume of laboratory load is very high and medical practitioners are in need of the investigations made by automated machines.

Uses of Automation

- For species identification due to the increasing volumes of clinical specimens processed by clinical laboratories, perceived cost-effectiveness, and convenient interfaces with laboratory and hospital information systems.

- Able to decrease the in-laboratory turn-around time compared to that required for standardized methods.

- Assist the microbiologist in sample preparation, plating, streaking, pipeting, inoculation, colony count and staining.

- Supply physicians with susceptibility profiles within 4-6 hours to help them guide antimicrobial therapy.

- Measurement of bacterial ATP

- ELISA techniques.

- Assist in detection and identification of bacteria by the following methods:
  - Particle counting
  - Microcalorimetry
  - Light scatter techniques (Autobac I)
  - Chromatography
  - Electrical conductivity
  - Radiometric measurement (BACTEC)
  - Line Probe Assay (LiPA)

Advantages

- Provides more accurate and reproducible results.
- Saves time and manpower.
- Assists in bacterial identification, biochemical characterization, media preparation, serological investigations and susceptibility testing.
- Helpful in busy laboratories.

Drawbacks

- Expensive, specially for developing countries.
- Remove the microbiologist from the fundamental basics of microbiology.
- Students may be misdirected from understanding the classical techniques of bench microbiology and depend only on electronic calculations, computerized print-outs and biotype numbers.
- Microbiologists may forget the cellular, cultural and biochemical properties of microorganisms.

Automated systems

- Specimen collection - Air samplers
- Sample preparation - Dilumats, Blenders, Pulsifiers
- Reagents preparation - Masterclaves
- Inoculation and pouring of samples - Pourer stackers, Peristaltic pumps
- Analysing samples, recording and validation of results
- Automated blood culture systems
- Automated bacterial identification systems

- Automated antimicrobial susceptibility testing
- Automated ELISA system
- Molecular methods.

Microbial Air sampler - Sampling is an essential step in the airborne contamination monitoring. The new Sampl'air generation offers superior performance and complete traceability.

Automated pourer stacker - Provide hands-free and automated operation. Has single carousel with a capacity of 320 plates of 55 or 90 mm. Pouring rate up to 750 petri dishes per hour. Has serial plate cooling device. It is compact and easy-to-clean without any tool.

Dilumat - Saves time and achieves total traceability by performing all 3 operations at once (weighing, dilution, storing). Total traceability: Sample, diluting agent and operator for quality control purposes. Self-regulating automatic calibration system is present. It is automatic, convenient articulated arm, particularly useful for laminar flow bench.

Automated colony counter - Allows to count all types of microorganisms in water, milk or air. Two counting modes: Semi-automatic count via mouse clicks upon colonies and fully automatic count (up to 90% time saving, one second counting). Professional reports either printed or saved as an Excel file or data management worksheet. Automated colony counter for both pour plates and spiral plates are present.

Media preparation station/Masterclave media preparators - State-of-the-art technology, lets automatically and rapidly prepare 1 to 60 liters of the highest-quality culture media and automatic distribution in different plates, tubes or bottles.

Automatic dispenser - Used to distribute fluids, media, reagents and serum in constant quantities Into different tubes, plates and bottles. Automatic slide stainer PREVI_Color_Gram (BioMeriéux). Used to stain slides and smears for microscopy, e.g. Gram stain and Leishman stain.

PREVI™ Isola system (BioMeriéux) - Automates and standardises tedious task of sample processing and inoculating plated media.

BD-Innova™ - A versatile, state-of-the-art Pre-analytical Automated Microbiology Specimen Processor. Improves clinical laboratory efficiency, thus taking care of shortage of laboratory professionals and an increase in workload. Reduce laboratory errors associated with specimen handling. Full range of specimen containers, streak patterns, protocols and workflows are available.

Automated Blood Culture Systems

Represents an important advance in clinical microbiology practice. It alerts the microbiologist that a culture is positive, after which the relevant bottle can be removed for Gram stain and subculture. Media selected for subculture can be chosen based on the Gram reaction and morphology of the microbes. If organisms are not visualized, a blind subculture should be performed and the bottle returned to the instrument for continued incubation. Multiple studies have demonstrated that bottles need to be incubated only for five days with the continuous monitoring system.

BacT/Alert System (Organon Teknika, Durham NC)

- Contains activated charcoal for effective neutralization of antimicrobials.
- Measures carbon dioxide derived pH changes by a colorimetric sensor in the bottom of each bottle.
- The sensor is separated from the broth medium by a semi-permeable membrane.
- Can detect organisms from blood and sterile body fluids.
- Can also detect mycobacteria and fungi.
- Color change – blue to light green to yellow as pH reduces.

BACTEC 9050/9240/9120 (BD Diagnostic systems)

- A radiometric devise used for early detection of 14C-labeled CO_2 in blood culture bottles.
- Measures the production of carbon dioxide by metabolizing organisms by using fluorescent light.
- Contains resin.
- Allows the laboratory to detect the presence of bacteria within one hour and therefore report of blood cultures given in 48 hours, instead of 10-14 days.
- Mainly used for aerobic and anaerobic bacterial blood cultures. Also used for T.B. culture and fungal blood cultures.
  - BACTEC™ Lytic/10 Anaerobic/F Medium
  - BACTEC™ Myco/F Lytic Medium
  - BACTEC™ PEDS PLUS™/F Medium
  - BACTEC™ Plus Aerobic/F Medium
  - BACTEC™ Plus Anaerobic/F Medium
  - BACTEC™ Standard Anaerobic/F Medium
  - BACTEC™ Standard/10 Aerobic/F Medium
  - BACTEC TB-460
  - BACTEC MGIT960
- **BACTEC TB-460** contains enriched Middlebrook 7H9 liquid medium with 14C-labeled palmitic acid as the sole carbon source. It is a radiometric method and is considered to be the 'gold standard' for drug

susceptibility testing to first line anti-tuberculous drugs. Can also be done for second line drugs. Has the capacity to detect drug resistance faster than with the solid media-based methods. Major disadvantage is the need for disposal of the radioactive waste.

- **BACTEC MGIT960** is based on the principle of oxygen consumption and a fluorescence signal. The tubes are incubated and controlled inside the MGIT960 apparatus. Approved by US FDA for detection of drug resistance to first line anti-tuberculous drugs. All readings are performed inside the machine and results are printed as susceptible or resistant.

- **TREK ESP culture system (TREK Diagnostic System, Cleveland, OH)**

  Differs from the BacTAlert and the BACTEC 9240/9120 system in the following ways:

  - Production of carbon dioxide is done manometrically by monitoring changes in the headspace pressure by sensitive detector attached to the bottle.

  - Both gas consumption and production are monitored.

Comparison of Three Automated Blood Culture Systems

Characteristic	BacTAlert 3D	BACTEC™FX	Versa TREK
Manufacturer	BioMeriéux	BD Diagnostics	Trek Diagnostic System
Principle	Uses an internal colorimetric sensor for the detection of CO_2 produced by microbial metabolism. dioxide by metabolizing organisms by using fluorescent light.	A radiometric devise used for early detection of 14-labeled CO_2 in blood culture bottles. Measures the production of carbon the bottle with an external pressure sensor.	Production of carbon dioxide is done manometrically by monitoring changes in the headspace pressure by sensitive detector attached to
Presence of charcoal	Contains activated charcoal	Charcoal not present	Charcoal not present
Presence of resin	Does not contain resin more detection.	Contains resin Plus Aer/Plus Anaer – 20%	Does not contain resin
Detects organisms in	Blood, sterile body fluids, mycobacteria, fungi	Blood, mycobacteria, fungi	Blood, sterile body fluids, mycobacteria, fungi
Delayed vial entry	+	+	Only one delayed vial entry
Capacity	120, 240, etc. bottles	100, 200, etc. bottles	Only 240 and 528 bottles
Advantages	1. Time to detection of mycobacteria is shortest in BacT Alert than in BACTEC.	1. The BACTEC system detects positives significantly faster. 2. BACTEC better than BacTAlert in recovering gram-positive and gram-negative bacterial pathogens in the presence of beta-lactam antibiotics, gentamicin/penicillin, and vancomycin. 3. The difference between the two systems in terms of microbial recovery was attributable largely to differences in the ability to grow staphylococci better by BACTEC than by BacTAlert. 4. Superior method for rapid detection of bacterial contaminants in cord blood 5. Gram stain smear can be seen clearly as there is no precipitate.	1. Significantly more streptococci and enterococci as a group were detected by VTI than by BacTAlert. Moreover, significantly more microorganisms were detected by VTI for patients receiving antimicrobial therapy.

Drawbacks	1. Activated charcoal appears as a black precipitate on the smear, which may be difficult to interpret 2. If large no. of WBCs are present in the sample, it may be false positive 3. Minimum 4 ml sample for pediatric and 10 ml for aduts	1. Not FDA approved for sterile body fluids. 2. For pediaric – 0.5 to 3 ml to be inoculated and 5-10 ml for adults	1. High false positivity rate. 2. No resins present. 3. Only one delayed vial entry. 4. Not good for recovery of *Candida spp.* as compared with BACTEC.

Bacterial identification systems

Combine a series of differential media or substrate in a single package, selected to aid in identifying members of a group of bacteria.

Advantages:

- Compact construction
- Easily visible chemical reactions
- Standardized quality control
- Long shelf life
- Aerobic incubation and short incubation time.
- Direct inoculum from a plate and easy handling of kits.
- Minimal storage space.
- Expanded computer services and updated data banks.

Disadvantages:

- Considerable training time is required for familiarity with test reactions.
- Interpretation of color reactions is sometimes difficult because of weak reactions.
- Data base includes only common clinical isolates from human sources.
- Misidentifications are possible for organisms not listed in data base.
- Additional testing, including GLC is sometimes required.

API Systems

- Test performed in cupules.
- Detects Gram-negative bacilli, Gram-positive cocci and Gram-positive bacilli.
- Substrate utilization time – Growth in 2 hours and no growth in 72 hours.
- Detected by colorimetry, fluorescence or turbidity.

Vitek 2 System (BioMeriéux)

- Identification as well as antimicrobial susceptibility testing.

- Integrated modular system: Filling – sealer unit, reader, incubator, computer control module, data terminal and a multicopy printer.
- Detects bacterial growth and metabolic changes in microwells of thin plastic cards using a fluorescence based technology – Enterobacteriaceae, GNB, GPC.
- ID-GNB Card: 64 well plastic card – 41 fluorescent biochemical tests, 18 enzymatic tests, 18 fermentation tests, 2 decarboxylase tests and 3 miscellaneous tests.
- Result interpretation after a 3 hours incubation period.

VITEK®MS (Automated microbial identification within minutes)

Mass spectrometry (MS) is an analytical technique for determining the elemental composition of a sample. The MS principle consists of ionizing chemical compounds to generate charged molecules and to measure their mass-to-charge ratio. Such molecular "signatures" can be used for rapid bacterial identification (ID) from isolated colonies. The use of mass spectrometry for bacterial ID is especially suitable and cost-efficient for laboratories with high volumes of samples. Mass spectrometry is not adapted for antibiotic susceptibility testing (AST).

- **MALDI-TOF technology** (Matrix Assisted Laser Desorption Ionization Time-of-Flight) used by VITEK MS examines the patterns of proteins detected directly from intact bacteria. The sample to be analyzed is mixed with another compound, called a matrix. The mixture is applied to a metal plate and irradiated with a laser. The matrix absorbs the laser light and vaporizes, along with the sample, in the process gaining an electrical charge (ionization). Electric fields then guide the ions into the time of flight mass spectrometer, which separates them according to their mass to charge (*m/2*) ratio, and ultimately the quantity of each ion is measured. Detection is achieved at the end of the flight tube.

The Phoenix System (BD Microbiology Systems)

- Newly developed, fully automated system for identification and ABS testing.
- Disposable panels and an instrument that performs automatic reading at 20 minutes interval during incubation.
- Results available in 2-12hours.
- Gram Negative identification segment: 45 biochemical, 16 fluorogenic, 14 fermentation, 8 carbon source, 5 chromogenic and 2 miscellaneous substrates.

Microscan Walkway (Dade Microscan

- Identification as well as antimicrobial susceptibility testing.
- Fully automated instrument.
- Incubates any combination upto 96 conventional or rapid Microscan panels simultaneously, automatically adds reagents to conventional panels, reads and interprets results, all without operator intervention.
- Rapid panels use fluorescent labeled compounds and require only 2 hours incubation.

Sensititre Autoidentification system (TREK Diagnostic Systems)

- Fully automated : Simultaneously incubates, reads, interprets plates
- Continuously process the sample, but allows access any time during sample run
- Unique feature: User defined incubation temperature to obtain maximum growth for identification
- First reading at 4 hours, thereafter after 16 hours upto 24 hours.

Omnilog ID System (Biolog)

- Fully automated: Simultaneously incubates, reads, interprets plates
- Continuously process the sample, but allows access any time during sample run
- Unique feature: User defined incubation temperature to obtain maximum growth for identification
- First reading at 4 hours, thereafter after 16 hours upto 24 hours.

Other bacterial identification systems

Product	Organisms identified	Performed in	Substrate utilization time	Detection method
BBL Crystal	Enterobacteriaceae, GNB, GPC, GPB, *Neisseria spp.*	Microplate	Growth in 3 hrs and no growth	Color/Fluorescence
BBL Minitek	Enterobacteriaceae, *Neisseria spp.*	Microplate with substrate impregnated discs	Growth in 4 hrs and no growth in 48 hrs	Color
Micro-ID	Enterobacteriaeceae	Substrate chambers in plastic strip	No growth in 4 hrs	Color
Rapid ID	Enterobacteriaeceae, GNB, GPC, *Neisseria, Hemophilus*	Self inoculated tray with substrate reaction wells	No growth in 2-4 hrs	Color

Autobac I system

- Does semi-quantitative MIC susceptibility testing and gives results within 4-6 hours.
- Determination of the biochemical profiles of Gram-negative bacilli and thus their identification.
- Screening of urine samples for the presence of bacteria within 3 hours and thus negative urine samples may not be cultured.

MS-2

- Used for qualitative MIC testing within 3-6 hours.
- Identification of enterobacteria within 5-6 hours.

Automicrobic system

- Used for rapid qualitative MIC suseptibility testing.
- Identification of Enterobacteriaceae.
- Identification of medically important yeasts.

- Direct analysis of urine samples for the identification and count of bacteria.
- Wells made of plastic.
- Growth detected after 6-8 hours by increase in turbidity.

Flow cytometry is an analytical method that allows the rapid measurement of light scattered and fluorescence emission produced by suitably illuminated cells. The cells or particles are suspended in liquid and produce signals when they pass individually through a beam of light. Detects bacteria in body fluids, blood and urine.

Applications

- Rapidly detect organisms responsible for the infection and identify the type of microorganism on the basis of its cytometric characteristics.

- Susceptibility testing of slow-growing microorganisms such as mycobacteria and fungi.
- Results are obtained rapidly in less than 4 hours. Offers susceptibility results even before the microorganism has been identified.
- The most outstanding contribution is the detection of mixed populations, which may respond to antimicrobial agents in different ways.
- To study the immune response in patients, detect specific antibodies and monitor clinical status after antimicrobial treatments.

Electrical impedance

- Measure changes in the flow of electric current passing through the medium where the specimen is incubated.
- With continuous monitoring, the first sign of impedance change is indicative of microbial growth.

Bioluminescence (ATP assay)

- Dephosphorylation of ATP caused by bacteria.
- Luciferin converted to oxiluciferin and light with the help of Luciferase enzyme.
- Measured by Luminometer.

Nephelometry

- Photometers are placed at angles to the suspension and scattered light generated by laser is measured.
- Dependent on the number and size of the particles in suspension.

Limulus Amoebocyte Lysate Assay

- Measured amount of fluid is to be tested.
- Lysates of amoebocytes (horse shoe crab) in the presence of minute amounts of endotoxin (< 0.1 ng/ml) from the cell wall of Gram-negative bacteria.
- Incubated at 37°C for ½ to 4 hours.
- Gel formation is taken as positive.

Automated electrophoresis machine

- CIEP machine is employed for detection of viral or bacterial antigens as well as antibodies in patient's serum.
- Common specimens examined by this system are:
- CSF for diagnosis of *Hemophilus influenzae*, *Neisseria meningitidis*, *Listeria* and *Streptococcus pneumoniae*.
- Sputum for diagnosis of *Streptococcus pneumoniae*.
- Serum for diagnosis of *Candida*, *Aspergillus* and HBsAg.

Gas liquid chromatography (GLC)

- For genus level identification of most obligate anaerobes based on types of fatty acids.

- Volatile fatty acids – isobutyric, butyric, etc.
- Non-volatile – succinate, fumarate, etc.
- A form of partition chromatography which uses a stationary liquid phase and a carrier gas phase.
- Volatile fatty acids are separated according to their partition coefficient between the gas and the liquid phase.
- Volatile products from the column are detected by a detector and recorded.
- Results within 1 hour of sample collection.

High performance liquid chromatography (HPLC)

- Have special resins bearing number of ion-exchange groups on their surfaces and the sample is retained in the column based on their affinity for the charged ionic groups.
- Also known as ion-exchange chromatography.
- Decreased analysis time and improved resolution than GLC.
- Determines levels of antimicrobials in body fluids, other therapeutic drug monitoring, detection of products of microbial metabolism or cellular constituents.

Automated ELISA

- Used for detection of antibodies against many viral diseases, bacterial infections, parasitic infestations and fungal diseases.
- Automated machine is easy to operate.
- Gives rapid, specific and reliable results.

Nucleic acid based automated methods

- Amplification – Real Time PCR, LCR, etc.
- Strand Displacement Amplification (SDA).
- Hybridization – Dot blot; In-situ hybridization with fluorescently labeled oligonucleotides (FISH); Oligonucleotides immobilisd on a solid support (biochips/DNA chips); Solid-phase hybridization (LiPA).
- Sequencing and enzymatic digestion of nucleic acid.
- Combined methods – Restriction fragment length polymorphism (RFLP) with DNA probe analysis; PCR-RFLP; PFGE; Ribotyping.

Real-Time PCR

- Speed of the test and a lower risk of contamination.
- Requires skilled technical personnel.
- Can be done directly from clinical samples.
- Rapid detection of drug resistance in tuberculosis by using TaqMan probe, fluorescence resonance energy transfer probe, molecular beacons and biprobes.

The Line Probe Assay (LiPA)

1. Solid-phase hybridization technique for rapid detection of drug resistance in tuberculosis – INNO-LiPA Rif TB Assay.

2. Based on reverse hybridization of amplified DNA from cultures or sputum samples with probes of the *rpo*B gene of *M. tuberculosis*, immobilized on a nitrocellulose strip.

3. Geno Type MTBDR assay for simultaneous detection of resistance to INH and RIF.

REFERENCES

1. Koneman EW, Allen SD, Janda WM, Schreckenberger P, Winn Jr. WC. In Color Atlas & Textbook of Diagnostic Microbiology. 6th Ed. Lippincott, Philadelphia. 2006.

2. Gradwohl R. B. H. Gradwohl's Clinical Laboratory Methods and Diagnosis; A textbook on laboratory procedures and their interpretation. 7th Ed. Mosby - Hardcover. 1970.

3. Stager CE, Davis JR. Automated systems for identification of bacteria.Clin Microbiol Rev 1992; 5: 302-27. PMCID: PMC358246.

4. Automated Microbial Identification and Quantitation - Technologies for the 2000s. Ed. Olson W. P. CRC Press LLC, London. 1996.

5. O'Hara CM. Manual and Automated Instrumentation for Identification of Enterobacteriaceae and other Aerobic Gram Negative Bacilli. Clin Microbiol Rev 2005; 18: 147-62.

NEWER IMMUNOLOGICAL AND MOLECULAR DIAGNOSTIC METHODS IN MICROBIOLOGY

33

Chapter

I. IMMUNOLOGICAL METHODS

A. **Labelled Antibody Assays** – Antibodies (Abs) are labeled with specific markers for detection of Antigens (Ags) and haptens.

1. **Radioimmunoassay (RIA)** – Ab labeled with ^{125}I or ^{131}I.

   - Berson and Yallow introduced this in 1961.
   - Successfully utilised for quantitative assay of all hormones, drugs and HbsAg
   - Measures Ags or Haptens upto picogram quantities.
   - It is based on competition under test conditions for a fixed amount of specific Ab between a known fixed level of radioactive labeled Ag and the unknown level of unlabeled (test) Ag. After Ag-Ab reaction, Ag is separated into free and bound fractions and their radioactivity is determined. The concentration of unknown (unlabelled) Ag is calculated from the ratio of bound and total labels by comparison with reference curve.

2. **Immunofluorescence (IF)** – Ab marker, fluorescein isothiocyanate.

   Fluorescence is a property of certain dyes, which absorb light in ultraviolet region (200-400 nm) and emit a characteristic wavelength of light (500-600 nm) in the visible region.

   - Fluorescein dyes – fluorescein isothiocyanate (blue green fluorescence), lissamine rhodamine (orange red fluorescence).
   - Dyes are conjugated to Abs, locates and identifies Ags in tissues– it was first introduced by Coons and Kaplan (1950).
   - Fluorescent group + Ab serves as label which enables Ab to fluoresce in situ when viewed under a fluorescent microscope.
   - More sensitive than precipitation or CFT. Detect as little as 1 μg protein per ml of body fluid.

Direct fluorescent antibody (DFA) test – for antigen detection e.g. *C. trachomatis, G. lamblia, C. parvum, P. carinii,* etc.

Indirect fluorescent antibody (IFA) test – to detect both antigens or antibodies. Antibodies against CMV, measles, etc.

3. **Enzyme-Linked Immunosorbent Asaay (ELISA)** (enzymes like peroxidase or alkaline phosphatase are used coupled to Ab).

   - Extremely sensitive – detect picogram (10^{-20g}) quantities of protein.
   - Methods – Indirect and Direct.

   a) **Indirect method** – 2 steps:

   (i) Step I – Ag and Ab react to form Ag-Ab complexes.

   (ii) Step II – A second Ab directed against the immunoglobulin (Ig) component of the immune complex is labeled by one of the markers and mixed with Ag-Ab complex formed in Step I. The labeled second Ab reacts with Ig components of immune complex and localises the Ag-Ab reaction. Second Ab is usually directed against Ig isotype characteristic of first Ab, e.g. if first Ab is of human origin, the second is an antihuman Ig (AHG).

   b) **Direct method** – Single step.

   Specific label is coupled directly to Ab and this labeled Ab is employed to detect its specific Ag.

   a) has advantages over b):

      i) One reagent is used to detect a multitude of antigens

      ii) More sensitive than direct test.

Direct method is useful to screen large number of serum for Ab directed against a single antigenic determinant.

It is simple and sensitive, require only μ litre quantities of test reagents. Used widely for detecting a variety of Abs and Ags such as hormones, toxins and viruses.

Principle – Test done on solid phase.

- Direct or indirect depending on whether the enzyme is conjugated to primary or secondary Ab.
- Enzyme (horse radish peroxidase, alkaline phosphatase) gives rise to a colour change on adding specific substrate (O-phenyl-diamine-dihydrochloride for peroxidase, p-nitrophenyl phosphate for alkaline phosphatase), only when Ag and Ab react specifically and this is measured by spectrophotometry .
- Detects either Ag or Ab. Ab is conjugated with enzyme by adding glutaraldehyde.

Techniques:

a) Double Ab technique for detection of Ag (Sandwich ELISA)

Performed in polystyrene tubes (Macro - ELISA) or in polyvinyl microtitre plates (Micro - ELISA).

i) For assaying Ag, Ab is coated on a plastic surface.

ii) Test solution is placed on Ab coated well or tube surface and incubated for 2 hours at 37°C. Ab specifically reacts with and immobilise the Ag molecules present in the test solution. Washed with buffer to remove untreated material.

iii) Enzyme (alkaline phosphatase) linked second Ab is added and incubated at 37°C for 1 hour. Ag is thus 'sandwiched' between the first Ab and enzyme-linked Ab (second Ab).

iv) Thorough washing with buffer solution is done, which removes excess amount of enzyme-linked second Ab which is not bound to Ag molecules.

v) A specific substrate is added to detect this specifically bound enzyme and left at room temperature till positive control shows development of specific colour.

vi) Phosphatase enzyme splits substrate to a yellow compound – colour is read visually or estimated colorimetrically.

It is useful in detection of hepatitis B Ag, hormones, toxins and Human chorionic gonadotropin (H.C.G.).

b) Indirect Method for detection of Ab

Serum to be tested is applied to the known Ag-coated well in microtitre plate. Plate is incubated and washed. AHG labeled with enzyme is added and incubated. Then the procedure is same as a).

Example: Detecting HIV Ab in patient's serum:

i) Wells are coated with purified HIV virus, which is a solid phase Ag.

ii) Patient's serum and conjugate (human anti-HIV labeled with enzyme horse radish peroxidase) are added.

iii) Substrate (urea peroxide and OPD solution) is added.

iv) Stop solution (1 mol/L) H_2SO_4 is then added to stop the reaction.

v) Absorbance of solution in the wells is read in an ELISA reader at 492 nm.

Cut-off value = 0.5 x (mean absorbance of negative controls + mean absorbance of positive controls)

If S < cut-off value, test is positive.

If S > cut-off value, test is negative.

Applications:

Bacterial – Capsular antigens of *S.pneumoniae, H. influenzae* type b, *N. meningitidis* and cell wall antigen of group A streptococci, *Chlamydia trachomatis*, Toxin A/B of *C. difficile*.

Parasites – *Giardia lamblia, E.histolytica, Cryptosporidium*.

Fungal – *Coccidioides immitis*.

Viral – Rotavirus in stool, RSV in respiratory tract infection, HIV-1 p24 antigen and HBs Ag in serum.

B. Blotting Techniques

Indications:

1. Those organisms which are difficult or costly to culture.

2. Slow growing organisms.

3. When organisms are present in low concentration in specimens.

4. Those organisms which are hazardous to propagate in laboratories.

Principle:

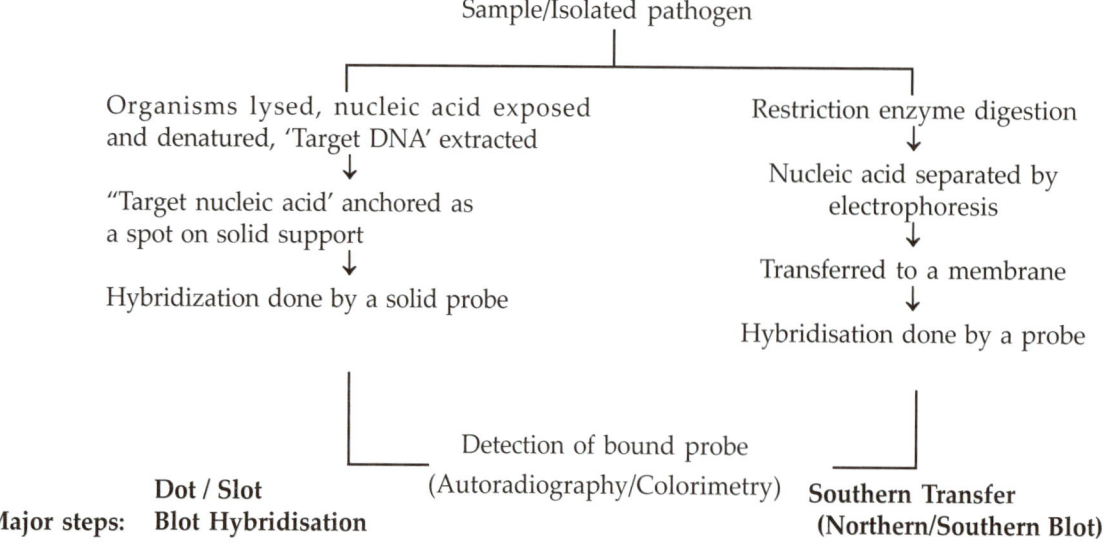

I. Electrophoretic separation:

Sample is placed in electrophoretic media –

1. Liquid
2. Solid (gel) – polyacrylamide gel (PAG) tubes or thin slabs (1 mm).

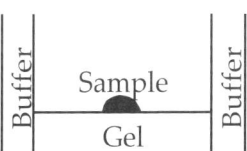

Electrolysis is done through gel – separation of protein/nucleic acid – proportional to molecular weight by length of molecule – detected by staining proteins with dye.

II. Transfer techniques:

A. Capillary blotting	B. Contact diffusion (protein migration bidirectional)	C. Electro-blotting (protein migration undirectional)

Advantages

1. Essential material readily available	1. No special apparatus required	1. Higher transfer efficacy
2. Several transfer experiments at one time	2. Two identical blots produced – for permanent record and for probe challenge	2. Shorter blotting time
		3. Transfer condition precise and reproducible

Disadvantage

Very slow	Very slow	Expensive apparatus

Transfer media

Nitrocellulose filters Most widely used Cheap, simple Longer shelf life (1 year at 4°C)	Diazo papers Preferred for electroblotting Preparation is difficult Reusable (4-5 times)	Cyanogen bromide activated papers

III. Hybridization:

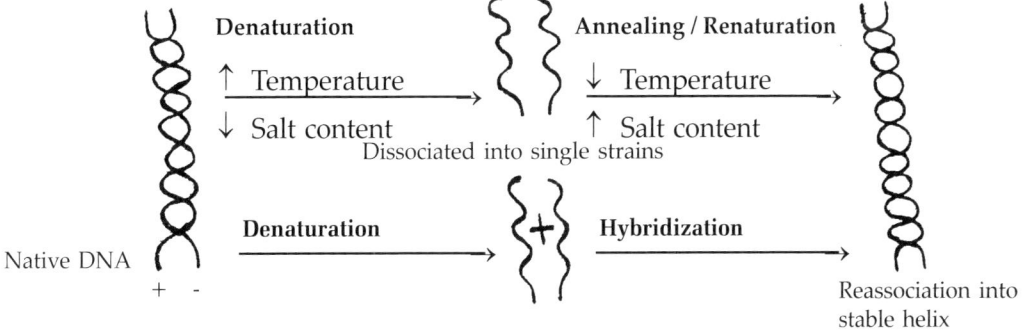

Both strains annealed are derived from different DNA molecules.

Stringency – Degree of tolerance of base pair mismatching in a duplex structure.

Low - ↓ Temperatutre, ↑ Salt, No formamide.

High - ↑ Temperature, ↓ Salt, ↑ Formamide

Hybdridisation assay – Process in which hybridisation reaction is used to analyse the nucleic acid content of an unknown sample.

Basic elements – Probe; Sample.

False positive ELISA	False negative ELISA
Autoantibodies	Acute HIV infection (window
Passive immunisation	period)
Immunosuppressive	Late stage AIDS
therapy	with HIV Ab positive Ig
Alcoholic liver disease	Presence of Rheumatoid
Influenza vaccination	Factor (RF)
Chronic renal disease	
Lymphoma	
Acute viral infection	
Stevens Johnson syndrome	

Requires controlled conditions permisive to complementary base pairing. Probe sample hybrids are detected by autoradiography/colorimetry.

Electrode
Protein support screen
Filter paper
Polyacrylamide gel

Protein Blotting Sandwich

IV. **Detection of hybrids by:**

1. Autoradiography – ^{32}P labeled probe is bound to nucleic acid – X-ray film is placed in tight contact with probe bound material – developed film shows radioactive probe sites.

2. Colorimetric enzymatic detection
   – Biotin labeled probe bound to nucleic acid
   – Avidin/Biotin enzyme complexes are formed
   – Complexes bound to biotinylated probes
   – Enzyme substrate
   – Colour reaction at site of probes

Southern Blot	Northern Blot	Western Blot
Southern, 1975	Alwine et al, 1979	Tawbin et al and Renart et al,1979 Burnette, 1981
Detects DNA fragment by binding them to nitrocellular paper	Detects RNA fragment	Protein blotting followed by detection of antibody

Applications of Southern Blot

1. Epidemiology :
   a) Detects carrier state, e.g. in *N. gonorrhoeae, S. typhi.*
   b) Detects source of infection in outbreaks, e.g. in *N. meningitidis.*

2. Microbiology :
   a) Culture confirmation of isolates which are difficult to identify phenotypically.
   b) Identifies slow growing microorganisms.
   c) Detects genetically coded factors in isolates/ samples.
   d) Identifies toxins, e.g. LT, ST, SLT, Clostridial toxins, *S. aureus* exotoxins.
   e) Detects genes coding for antibiotic resistance.
   f) Detect invasive vs non-invasive property of bacteria, e.g. *S. epidermidis.*
   g) Typing can be done to show genetic relatedness.
   h) Direct detection of microorganisms

Virus	Bacteria	Parasites
Rotavirus	*Mobiluncus*	*T. cruzi*
Herpes virus	*Legionella*	*T. gondii*
Adenovirus	*H. ducreyi*	
HBV	*C. trachomatis*	
EBV	*M. tuberculosis*	

3. Cancer biology
4. Metabolic disorders – detection of biochemical and cellular alterations.

Western Blot

Application of protein blotting and immunochemical identification of transferred molecules.

Step I. Preparation of strip.

Virus is grown in vitro – denatured – disruption of virions – placed in PAG – proteins separated – electroblot – proteins transferred to nitrocellulose paper – paper cut into strips with the proteins of the virus.

Step II.

 Ab

 Ag   Nitrocellular strip incubated with dilution of patient's serum

↓ **Conjugate 1**

Antibodies (Abs) get attached to separated viral antigen (Ags)

   ↓ **Conjugate 2**

Secondary detecting antibodies

(Anti human Ig antibody which enzyme is attached)

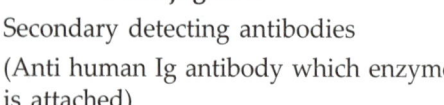   ↓ **Substrate**

Substrate detect enzyme antibody by permanently staining the strip.

Applications:

1. Virology :
   a) Expression of viral proteins during infection
   b) Serodiagnosis of viral diseases
   c) Diagnosis of HIV infection.

2. Immunology :
   a) Determination of specificity of monoclonal antibody production
   b) Detection of IgE against certain allergens
   c) Diagnosis of autoimmune diseases
   d) Identification of red blood cell antigen
   e) Diagnosis of infertility – antibodies against spermatozoa
   f) To monitor antibody production after immunisation.

Advantages:
- Minimal false positive
- Antibodies directed to specific major proteins, visualised directly.

Disadvantages:
- Expensive
- Technically demanding
- Time consuming.

Western Blot should be done in cases of false positive or false negative ELISA for detection of HIV antibodies.

Interpretation of Western Blot for HIV infection (WHO 1990)

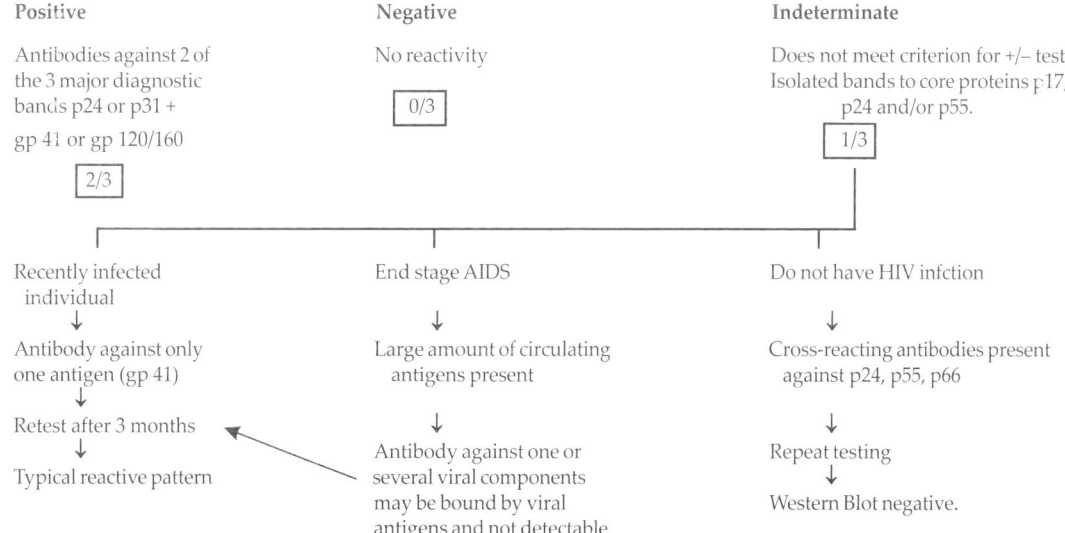

Positive	Negative	Indeterminate
Antibodies against 2 of the 3 major diagnostic bands p24 or p31 + gp 41 or gp 120/160	No reactivity	Does not meet criterion for +/– test. Isolated bands to core proteins p17, p24 and/or p55.
2/3	0/3	1/3

Recently infected individual → Antibody against only one antigen (gp 41) → Retest after 3 months → Typical reactive pattern

End stage AIDS → Large amount of circulating antigens present → Antibody against one or several viral components may be bound by viral antigens and not detectable

Do not have HIV infction → Cross-reacting antibodies present against p24, p55, p66 → Repeat testing → Western Blot negative.

C. **Immunoelectron microscopy** – Ag-Ab reactions are visualised by electron microscope with the help of appropriate markers e.g. ferritin (electron dense substance). Conjugated Ab is used to trace intracellular viral Ags. Some enzymes and colloid particles can be conjugated with Abs.

D. **Radioallergosorbent test (RAST)** – This test measures Ag-specific IgE in a radioimmunoassay where the ligand is a labeled anti-IgE Ab. It is identical to standard RIA except that the Ag (allergen) is covalently bound to a cellulose disc, rather than non-covalently to a radiolabeled plate.

II. MOLECULAR METHODS

A. **Electrophoretic protein typing:** Sodium dodecyl sulphate-polyacrylamide gel electrophoresis (SDS-PAGE) was first described by Laemmli (1970), as a method for the separation of polypeptides in complex mixtures and the determination of their molecular weights. Electrophoretic protein typing was performed by isolating whole-cell or cell-surface proteins, separating the SDS-PAGE and staining the gel to determine the resulting pattern. Alternatively, the proteins can be radiolabelled during isolation and the pattern detected by autoradiography.

B. **Restriction endonuclease analysis (REA) of whole-cell DNA:** It has the potential to give good strain differentiation. Like plasmids, chromosomal DNA can be digested with restriction enzymes. In restriction enzymes analysis (REA) endonucleases with relatively frequent restriction sites are used to digest the bacterial DNA thereby generating hundreds of fragments ranging from ~0.5 kb to 50 kb in length. Such fragments can be separated by size with the use of constant field agarose gel electrophoresis, and the patterns can be detected by staining the gel with ethidium bromide.

Some restriction endonucleases

ECoRI	E. coli
Hind III	H. influenzae
San 3A	S. aureus

The major limitation of this technique is the difficulty of interpreting the complex profiles, which consist of hundreds of bands that may be unresolved and overlapping. Furthermore, REA patterns may be confounded by the presence of plasmids, whose DNA can readily contaminate genomic DNA preparations. REA has been applied successfully to many species, in particular streptococci and *C. difficile, Listeria monocytogenes, Bordetella species*, and non typeable *H. influenzae.*

Haertl and Bandlow (1990), modified the conventional REA by performing the electrophoresis of the restriction digests of whole-cell DNA using polyacrylamide gels instead of agarose gels for resolving the DNA fragments and employing ultrasensitive silver staining for visualizing the fragments. This modified method was named as small fragment-restriction endonuclease analysis (SF-REA), since the typing was based on analysis of small molecular weight fragments (0.6-4.0kb) resolved in the polyacrylamide gels.

C. Nucleic acid probes:

A nucleic acid probe is a sequence of single-stranded nucleic acid that can hytridize specifically with its complementary strand via nucleic acid base pairing. Hybridisation involves denaturation of dsDNA into single strands, and detecting of ssDNA with a labeled complementary ssDNA probe.

When DNA probe technology is performed on tissues or biopsies that are prepared for histopathologic examination in surgical and anatomic pathology, this technique is known as **in-situ hybridization.**

Probe. A probe is a fragment of DNA/RNA typically labelled which forms strong covalently bonded hybrids with the specific complementary strand of nucleic acid. Production is by **Recombinant DNA technology.**

Types of probes

Requirements:

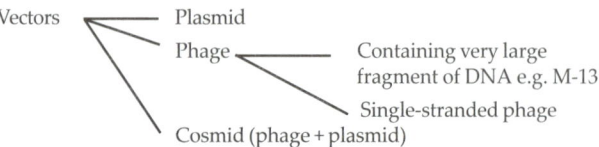

1. Vector PBR_{322} (universal acceptor plasmid)
2. Insert (desired gene) which is sliced and taken out by restriction enzyme ECoRI.

   The circular DNA becomes linear.

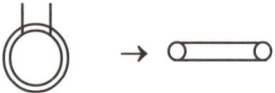

Then the insert is ligated in the PBR_{322} by ligase and again a circular DNA is formed.

The vector is then inserted into a host (*E. coli* K12) by transformation and grown on plates. The transferred bacteria is selected by antibiotic sensitivity testing.

Example. PBR_{322} is sliced in the gene for tetracycline resistance. So the transformed bacteria will show tetracycline sensitivity. The selected bacteria is grown and from them the desired gene is extracted.

E. coli F^- NA_2^+ Lac^- K12

Steps:

i.   Desired fragment is inserted into plasmid vector

ii. Plasmid multiplies in bacterial host

iii. Plasmids are isolated

iv. DNA probe fragment is purified

v.   Labeled with avidin/biotin

vi. Fragment is dissociated into single strands.

Natural	Preparation	Characteristics
1. dsDNA	Bacteriophage Bacterial plasmid Sequence cloned	High specificity and sensitivity Limited production Dedicated laboratory required. Denaturation required prior to hybridisation
2. ss Antisense RNA (riboprobes)	Transcription of RNA from a DNA template using RNA polymerase	Stable hybrids Non-specific binding Dedicated laboratories required Susceptible to degradation by RNA.
3. ssDNA	By PCR	Stable probes Do not require denaturation before use Dedicated molecular laboratories not required.
4. Oligoriboprobes	By transcribing RNA from DNA template using RNA polymerase	Ready access to tissue Stable Susceptible to RNA.
Synthetic		
5. Oligodeoxy- nucleotides	Synthetic	Wide availability Ready access to fixed tissue Less sensitivity, may give false negative result.

Uses of probes:

1. **For culture confirmation** – *Campylobacter spp., Enterococcus spp., H. influenzae, L. monocytogenes, M. tuberculosis, M. avium-intracellulare, M. kansasii, N. gonorrhoeae, S. agalactiae, B. dermatitidis, C. immitis, C. neoformans, H. capsulatum.*

2. **For direct detection in clinical specimens** – *C. trachomatis, G. vaginalis, L. pneumophila, N. gonorrhoeae, S. pyogenes, Candida spp., T. vaginalis,* Human papilloma virus.

3. For detection of enterotoxic genes – *E. coli, Cl. perfringens,* Shiga toxins I and II, *S. aureus* strains encoding enterotoxins A,B,C and TSS-T1.

4. For detection of genes encoding resitance to antimicrobial agents.

5. For epidemiologic studies to pinpoint bacterial strains involved in outbreaks.

D. Amplification techniques :

Polymerase chain reaction (PCR). PCR is a "target amplification" technique, i.e. a target sequence of DNA is identified and amplified to such an extent that it can be detected. It uses an automated, computerised "hot block" called a thermal cycler. Ingredients required are :

dsDNA, ss oligonucleotide "primers", deoxynucleotide triphosphates (usually labelled with ^{32}P) and "Taq DNA polymerase".

a) Initial denaturation step melts the ds DNA at 95°-100°C.

b) Single-stranded oligonucleotide primers that needs tobe amplified are allowed toannual to the denatrued DNA strands during a cooling step.

c) Extension of the primers by DNA synthesis is done by thermostable **Taq polymerase** (purified from *Thermus aquaticus,* a thermophilic bacterium living in hot springs at temperatures of 70-75°C).

Repetition of the heat denaturing primer annealing – primer extension reaction sequence for 30-50 treatment cycles, results in the amplification of the DNA sequence located between the primers.

The PCR end-products may be detected by electrophoretically separating the components of the final amplified sample in agarose or acrylamide gels, followed by staining of the gel for 15 minutes in running buffer containing 2 drops of 2mg/ml ethidium bromide. The gel thus separated can be visualised on a shortwave ultraviolet radiation transilluminator, comparing the separated bands with those of a standard run in parallel. DNA can be amplified exponentially by an average of a factor of 10^7. It is a highly sensitive technique by which minute quantities of specific DNA or RNA sequences can be enzymaticaly amplified.

Modifications of PCR technique:

1. **Multiplex PCR** – Multiple primer pairs for different target molecules are included in the same amplification mixture. One set of primers is used to amplify an internal "control" sequence while the other set is used to prime the amplification of the DNA sequence of interest.

   Uses– To detect structural genes for toxins A and B in *Clostridium difficile.*

   To detect structural gene mecA in oxacillin resistant strains of staphylococci.

2. **Nested PCR** – Several rounds of amplification are performed with one set of primers, and the product of this amplification is subsequently amplified using another set of primers that lie within the internal sequence amplified by the first primer set, i.e. second set of primers is "nested" within the sequence amplified with the first primer set.

3. **TaqMan PCR assay**

   Detects *Y. enterocolitica* in blood and can detect contamination in stored blood units. As few as 6 bacteria/200µl (30 bacteria/ml) can be detected within 2 hours.

   An oligonucleotide probe having a reporter fluorescent dye attached to its 5′ end and a quencher dye attached to its 3′ end is used. Because of the proximity of reportar and quencher dyes, the unbound probe is not able to omit a fluorescent signal. When the probe hybridizes to its target template, the reporter dye is cleaved by the 5′ nuclease activity of Taq polymerase and becomes capable of emitting a fluorescent signal without the suppressor activity of the quencher dye.

   Other uses – Used in reverse transcriptase PCR using 16S rRNA as a template for detection of live bacteria and also viability of bacteria in blood.

   Advantage – It eliminates the use of multiplex PCR, Southern blotting or agarose gel electrophoresis.

4. **Transcription based amplification or Nucleic acid sequence** – based amplification **(NASBA) or 3 SR (self-sustaining sequence replication)**. It amplifies ssRNA, rather than DNA. Used for detection of point mutations resulting in zidovudine resistance in strains of HIV-1.

   Reverse transcriptase makes a cDNA copy of the target, thus forming a cDNA/target RNA hybrid. The RNAse H enzyme (from *E. coli*) then degrades the strand of RNA in the latter, leaving the ss cDNA, The primaer is extended using the cDNA strand as a template, resulting via dsDNA copy of the target sequence. Both strands of DNA copy are flanked by bacteriophage T7 DNA-dependent RNA polymerase promoter regions and both strand can serve as templates for this enzyme, producing several antisense RNA copies of the target sequence.

5. **Strand displacement amplification (SDA) technique** – After "matching" of a single strand of dsDNA by a site-specific restriction endonuclease, DNA polymerase can bind and synthesise a complementary copy of the ssDNA; the nicked strand becomes displaced from the strand being copied during the process of DNA synthesis. Then subsequently polymerisation and strand displacement continue to occur because of continual regeneration of unaltered, ss "nick-able" sites in duplex molecules. *Mycobacterium tuberculosis* can be detected from sputum specimens by this method.

6. **Q beta replicase method.** This uses a "replicase" enzyme that is able to synthesise the genomic RNA of bacteriophage Q beta. A naturally occuring variant of the phage particle called MDV-1 behaves as a substrate for the enzyme and is manipulated so that an oligonucleotide "probe" sequence can be inserted into one of the "loops", and this probe region anneals to its recognition sequence, thus becoming resistant to hydrolysis by RNase. Then there is specific amplification of the probe. It is used to detect DNA or RNA targets.

7. **Ligase chain reaction (LCR).** ss target DNA is incubated with oligonucleotide probes that bind to the target in an end-to-end fashion. A thermostable DNA ligase then "ligates" or joins the two probes together. The duplex is heated causing denaturation and separation of the target ssDNA and the ligated probes. The latter and the target again bind probe sequences in an end-to-end fashion, followed by ligation to form another duplex and these steps are repeated several times.

8. **Branched-chain DNA method** is an example of **signal amplification.** The target DNA is denatured to single strands binds to small oligonucleotide "capture probes" that anneal to continuous runs of nucleotides. Extender probes also anneal forming the "tail" of a large "multimer", ultimately forming a dsDNA "tree". Addition of enzyme substrate (dioxetane) generates the signal –chemiluminescence. It is used for detection of HIV-1, HCV and CMV to assess viral load and the response of patients to antiviral agents.

9. **Random amplified polymorphic DNA (RADP)** or arbitrary primed PCR (AP-PCR) is similar to RFLP as it assays DNA sequence variations in short regions, and instead of analysing RE recognition sequence, it focuses on PCR priming regions. It is less time consuming.

10. **Repetitive sequence based PCR.** Simple repetitive DNA (micro and minisatellite) sequences in the genome when used as single primers with PCR, generate individual band pattern called PCR fingerprinting. This is more reliable than RAPD as primers are longer and there is strong homology between primers and the target sequences.

11. **Direct sequencing.** A highly conserved region of DNA can be studied for distantly related organism while a very variable part of DNA may be analysed for closely related ones. The internal transcribed spacer (ITs) region of rDNA is suitable for the latter.

Problems of PCR

- False positive reactions caused by introduction of nucleic acids into the reaction mixtures.
- Use of electrophoresis, filtermembrane hybridization and radioactive probes is a time-consuming and labor-intensive activity, requiring highly trained and dedicated personnel.

Applications of PCR

A. In Bacteriology:

a) Identification and detection – *M. tuberculosis* (PCR, SDA, Q beta) *T. pallidum* (Multiplex PCR), *Brucella spp.*, *B. pseudomallei*, *Campylobacter spp.* (by analysis of 16S rRNA), *C.trachomatis*, *H. ducreyi* (Multiplex PCR), *H. pylori*, *L. interrogans*, *N. gonorrhoeae* (LCR), *S. typhi*, *B. pertussis* (PCR, Nested PCR), *B. burgdorferi* (Nested PCR), *S. pneumoniae*, Typing of resistant plasmids, *Y. enterocolitica* (TaqMan PCR assay)

b) Strain typing – *S. pyogenes*, *Y. enterocolitica*, *L. pneumophila* (Molecular typing), CDC group IV-C-2.

c) Toxigenic strains detection – *C. diphtheriae*, *E. coli* 0157:H7 (Multiplex PCR).

d) Detection of methicillin resistance – *S. aureus* and CONS (both Multiplex PCR).

B. In Mycology:

a) Detection – *Aspergillus spp.*, *P. jirovecii*

b) Identification and characterization – *Candida spp.*

C. In Virology:

a) Detection – HSV1 and 2 (Multiplex PCR), CMV, Enteroviruses, HCV (PCR, bDNA), HEV, HHV6 and 7, HPV (in situ PCR), Parvovirus B19, Measles

b) Quantitation in serum – HIV-1 RNA and proviral DNA (NASBA, PCR, bDNA)

c) Subtyping of isodates – RSV

d) Diagnosis in utero and infancy – Rubella

e) Subgenus identification.

D. In Parasitology:

a) Detection – *Plasmoium spp.*, *T. cruzi*, *E. bieneusi*

b) Diagnosis – *Encephalitozoon hellem*, *T. gondii*.

E. Restriction fragment length polymorphism (RFLP) analysis using DNA probes:

One way to simplify RFLPs generated by chromosomal DNA is to transfer the DNA fragments generated by frequent cutting enzymes, onto a nitrocellulose or nylon membrane. The DNA on the membrane then is hybridized with specific chemically or radioactively labelled piece of DNA or RNA (a probe which binds to the relatively few fragments on the membrane that have complementary nucleic acid sequences). Variations in the number and size of the fragments detected by hybridization are referred to as RFLPs. All strains carrying loci homologous to the probe are typeable, and the results are highly reproducible.

RFLP analysis using the DNA insertion element IS 6110 currently is the method of choice for typing isolates of *Mycobacterium tuberculosis*.

Reduction in the number of bands to be analyzed can be achieved by using the labeled DNA probes for distinguishing individual strains of bacteria. However, these probes are species specific or even strain specific for some strains within a given species.

F. Ribotyping:

The ubiquitous and polymorphic ribosomal DNA loci are highly conserved in the eubacteria. This finding led to the use of *E. coli* rRNA as a universal probe for studying the restriction patterns for taxonomic purposes (Grimont and Grimont, 1986). Later Stull and co-workers (1988) independently developed this concept of universal probe, demonstrating its utility for epidemiological studies and coined the term ribotyping. The three species of rRNA (23S, 16S, and 5S) are present in equimolar amounts. in 70S ribosome particles. They are co-transcribed as a large 30S precursor RNA, which undergoes maturation process to yield the three different mature species of rRNA. The corresponding genes are thus organized into a polycistronic transcription unit; the rrn operon. Starting from the promoter, the individual genes are arranged in the following order. 5'-16S-23S-5S-3'. A typical *E.coli* rrn operon spans 6,000 to 7,000 bp of DNA. While most genes in bacteria are present in one copy, rrn operons are unique in the property that they are present in multiple copies. Depending on the species, the number of copies vary from 2 to 11 per bacterial cell.

Between the genes of 16S and 23S rRNAs is a spacer region that contains genes encoding several tRNAs and several direct repeat sequences. These regions are heterogeneous. In addition some rrn operons also have tRNA genes in their 3'-flanking regions. A high degree of homology for these sequences exists among various bacteria, despite the genetic diversity of the remainder of the chromosome.

The probe most often used is DNA. It can be cDNA, synthesized directly from rRNA by reverse transcriptase, a recombinant plasmid in which the rrn DNA has been cloned or a synthetic oligonucleotide constructed from the 16S or 16S and 23S rRNA gene sequence of *E.coli* rRNA is not the only heterologous probe that has been used for the study of other species. Some workers have used homologous probes derived from the specific rRNA sequences of the organism under study, as has been done with studies with *Mycoplasma species, Pseudomonas cepacia, Haemophilus influenzae, Legionella spp.* and *Campylobacter spp.*

Initially the probes used were labeled with ^{32}P. For RNA probes and oligonucleotides, end labeling is used, and nick translation or random oligopriming is used for cloned probes. But the effective non-isotopic cold-labeling systems that have now been developed offer many advantages over the radioactive systems including safety, easy disposal, and stability of the probes, which could be stable for several months. Moreover the results are as good as, or superior to those obtained with isotopic labeling. Ribotyping was particularly well suited for studying the epidemiology of nosocomial *S. marcescens, Enterobacter cloacae, Salmonella typhi, Aeromonas spp. and Helicobacter pylori* strains.

Ribotyping has also been applied to other bacterial species among which some might be involved in nosocomial infections. These include *E.coli, Proteus mirabilis, Campylobacter spp., Yersina enterocolitica, Vibrio cholerae, Corynebacterium spp. and Shigella spp.*

G. Pulsed-field gel electrophoresis (PFGE) of whole-cell DNA:

Pulsed-field gel electrophoresis was first described in 1984 as a tool for examining the chromosomal DNA of eukaryotic organisms. In this method, the bacterial genome is digested with restriction enzyme that has relatively few recognition sites and thus generates approximately 10 to 30 restriction fragments ranging from 10 to 800 kb. These fragments can be resolved using a specially designed chamber that positions the agarose gel between three sets of electrodes that form a hexagon around the gel. DNA molecules are separated by an electric field that alternates in direction. The isolation of chromosomal DNA is technically difficult. To avoid shearing of the high molecular weight DNA, it is digested in an agarose plug, which is then inserted into a slab agarose gel in a special electrophoresis tank. Certain organisms such as *Clostridium difficile* and *Aspergillus spp.* may not be typeable by PFGE because DNA cannot be extracted intact. Some bacterial species such as *H. influenzae*, have limited genetic variability and cannot be typed by PFGE. In general, PFGE is one of the most reproducible and highly discriminative typing technique. PFGE has been applied to a wide range of bacterial species.

The main disadvantages of the PFGE are the cost of the equipment and the time consumed. Moreover, the entire PFGE protocol takes about one week.

H. Fluorescent amplified fragment length polymorphisms (FAFLP):

FAFLP analysis belongs to the category of selective restriction fragment amplification techniques, which are based on the ligation of adapters (i.e. linkers and indexers) to genomic restriction fragments followed by a PCR-based amplification with adapter-specific primers. For FAFLP analysis only a small amount of purified genomic DNA is needed. This is digested with two restriction enzymes, one with an average cutting frequency (like EcoRI) and a second one with a higher cutting frequency (like MseI or TaqI). Double-stranded oligonucleotide adapters are designed in such a way that the initial restriction site is not restored after ligation, which allows simultaneous restriction and ligation, while religated fragments are cleaved again. An aliquot is then subjected to two subsequent PCR amplifications under highly stringent conditions with adapter-specific primers that have at their 3' ends an extension of one to three nucleotides running into the unknown chromosomal restriction fragment. An extension of one selective nucleotide amplifies 1 of 4 of the ligated fragments, whereas three selective nucleotides in both primers amplify 1 of 4,096 of the fragments. The PCR primer which spans the average frequency restriction site is labeled. After polyacrylamide gel electrophoresis, a highly informative pattern of 40 to 200 bands is obtained. The patterns obtained from different strains are polymorphic due to

i.   Mutations in the restriction sites
ii. Mutations in the sequences adjacent to the restriction sites and complementary to the selective primer extensions
iii. Insertions or deletions within the amplified fragments.

Fluorescent labels with different emission spectra (FAM, ROX, JOE, TAMRA) for analysis on a Perkin - Elmer ABI automated sequencer are used.

Applications of FAFLP technology in infectious disease management

1. Molecular epidemiology: Tracking clonal expansions in hospital outbreaks
2. Identification of outbreak strains and sporadic infections
3. Evolutionary studies
4. Toxonomic studies
5. Genome mapping and characterization of BAC libraries
6. Preventing duplicate strain accessions in bacterial culture repositories
7. Analysis of paired isolates and serial isolates for analysing selection pressure on total chromosome of the pathogen.

Most molecular methods are still used only in reasearch or reference laboratories and their use in routine clinical laboratory is rather expensive and requires expertise.

I. Multilocus enzyme electrophoresis:

MLEE detects differences in the electrophoretic mobility of individual soluble metabolic enzymes. A variation indicates difference in the gene encoding the enzyme. Multiple metabolic enzymes are used such as a-G ALA, LDH, GDH, MDH and SOD.

J. DNA hybridisation analysis:

Extension of restriction endonuclease analysis done by transferring the DNA to nitrocellulose or nylon membrane (Southern blotting) and then hybridised with specific probes. This decreases the number of bands to be compared and is easier for interpretation.

K. Karyotyping:

Complementary or more useful than REA and DNA hybridisation, Chromosomal size variation is assayed by separating chromosomal DNA using PFGE. It deals with expensive equipment and the time required is also long.

L. Single stranded conformational polymorphism (SSCP):

Minor sequence variation in small ssDNA in a highly conserved segment causes changes in tertiary structure in short stranded DNA fragments, which are then detected by separation on a gel under non-denaturing conditions.

REFERENCES

1. Koneman EW, Allen SD, Janda WM, Schreckenberger P, Winn Jr. WC. In Color Atlas & Textbook of Diagnostic Microbiology. 6th Ed. Lippincott, Philadelphia. 2006.

2. Topley & Wilson's, Microbiology and Microbial Infections. Immunology. 10th Ed. Eds. Kaufmann SHE, Steward MW. John Wiley & Sons Ltd., West Sussex, UK. 2009.

3. Roitt I, Brostoff J, Male D. Immunology. 5th Ed. Mosby - Year Book, Inc. USA. 1998.

34

Chapter

Investigation of presumed outbreaks of bacterial infections in hospitals often requires strain typing data to identify outbreaks and distinguish epidemic from endemic or sporadic.

Typing is done in the following situations:

- An outbreak in which patients are infected/colonized within a defined time period or due to some common or point source such as food, health personnel or patient-associated shared apparatus.

- Differentiation of non-outbreak sporadic cases from outbreak.

- Monitoring of patients for identifying the source of specific strains.

- Distinguishing between reinfection with new strains and recurrence of the original strains following therapy.

- Can also demonstrate an association of certain types with a disease or clinical presentation, mixed infecting agents, antibiotic resistant strains and the distribution of resistance mechanism in strains.

A useful and effective typing system should be :

- Standardized
- Reproducible
- Sensitive
- Stable
- Easily available
- Inexpensive
- Applicable to a wide range of microorganisms
- Field-tested

Typing system	Microorganisms
Biotype	*Salmonella, Shigella, Vibrio cholerae,* Other species of *Enterobacteriaceae, Pseudomonas, Staphylococcus,* Fungi
Antibiogram	*Enterobacteriaceae, Pseudomonas* and other nonfermenters, *Staphylococcus*
Serotype	Viruses, Enterobacteriaceae, *Pseudomonas,* Streptococci, *Legionella, Chlamydia*
Bacteriocin type	*Shigella, Pseudomonas, Serratia, E. coli, Klebsiella, Proteus, Clostridium*
Bacteriophage type	*Staphylococcus, Salmonella, Pseudomonas, Mycobacterium, Clostridium*
Plasmid profiles	All bacteria, Resistant strains
Restriction endonuclease analysis	Herpes simplex virus, Adenovirus, Bacterial plasmids, Bacterial DNA, HIV

High resolution PAGE of bacterial proteins has been used for identification at the species, subspecies and intrasubspecies levels, e.g. *Aeromonas species, H. influenzae, Pseudomonas aeruginosa, Campylobacter jejuni, Helicobacter pylori, Acinetobacter calcoaceticus, Branhamella catarrhalis.*

Plasmid profile analysis

Limitation – some bacteria do not carry plasmids, so are non-typeable by this method.

The same plasmid can migrate to one or more locations, depending on the conditions under which the gel is run. This is overcome by digesting the plasmid DNA with a restriction enzyme – known as **Restriction endonuclease analysis of plasmid DNA (REAP)**. *S. epidermidis* carry several plasmids, *Pseudomonas aeruginosa* and streptococci rarely or never carry plasmids. It is also used in *S. aureus* which harbour a single plasmid that migrates only a short distance into the gel.

PCR based typing techniques:

1. **PCR-RFLP.** It amplifies a defined fragment of DNA and subsequent digestion of amplification product with one or more restriction enzymes to generate restriction fragment length polymorphism (RFLP).

Examples – Urease gene of *Helicobacter pylori,* Flagellin gene of *Campylobacter jejuni,* Tox gene of *Corynebacterium diphtheriae*. It has excellent reproducibility, but discriminatory power varies depending on the locus amplified and the restriction enzymes used.

2. **AP-PCR/RAPD. Random amplification of polymorphic DNA (RAPD)**, also known as **Arbitrarily primed PCR (AP-PCR)** has increased the applicability of PCR for epidemiological purposes. RAPD typing is based on the principle, that short primers, whose sequences are not directed at any specific site in the genome, will hybridize at random sites of the genome. The proximity, number and location of those priming sites vary among strains, and the method produces DNA fingerprints that differ when electrophoresed and visualized. The template DNA can be crude and the same primer set can be used to type most organisms. AP-PCR uses primers of lengths comparable with those used in standard amplification reactions (18-24 bases). RAPD uses shorter primers (10 bases in length). **DNA Amplification Fingerprinting (DAF)** uses primers 5 to 8 bases in length. The term **Multiple Arbitrary Amplicon Profiling (MAAP)** was coined for all these techniques and differences in DNA fingerprints were defined as **Amplification Fragment Length Polymorphism (AFLP)**.

Disadvantage – Low inter-laboratory reproducibility, due to low stringency conditions used for primer annealing.

3. **PCR ribotyping.** Primers are designed to anneal to the 16S and 23S rRNA genes and amplifies the intervening spacer sequences, which may vary in length. Differences in the size of PCR products can be detected after gel electrophoresis, while restriction enzyme digestion or PCR sequencing can differentiate products of similar size. The amplified patterns are stable and reproducible.

Advantage – Oligonucleotide primers used are complementary to sequences which are highly conserved amongst eubacteria and therefore can be used to type different bacterial species, e.g. *Burkholderia cepacia, S. aureus, Cl. difficile*, etc. A new typing method based on long PCR ribotyping capable of amplifying approximately 6kb rDNA operons of nontypeable *H. influenzae* is also there.

4. **PCR of repetitive chromosomal elements. Enterobacterial Repetitive Intergenic Consensus Sequences (ERICS)** and **Repetitive Extragenic Palindromic (REP)**. ERIC elements are 126bp long and occur singly throughout the genome. REP sequences are 38bp in length. There can be multiple copies at each chomosomal location, with more than 700 in the *E. coli* genome. Both the methods have been successfully used for typing *Listeria monocytogenes* strains isolated from humans, animals and food.

5. **Nucleotide sequence analysis.** Multiple isolates can be compared by sequencing same locus from each one. PCR-based sequence analysis of bacterial ribosomal RNA has been used.

REFERENCES

1. Koneman EW, Allen SD, Janda WM, Schreckenberger P, Winn Jr. WC. In Color Atlas & Textbook of Diagnostic Microbiology. 6th Ed. Lippincott, Philadelphia. 2006.

2. Topley & Wilson's, Microbiology and Microbial Infections. Bacteriology, Vol. 1. 10th Ed. Eds. Borriello SP, Murray PR, Funkay G. John Wiley.

BIOTECHNOLOGY AND NANOTECHNOLOGY IN MEDICINE

Chapter

BIOTECHNOLOGY IN MEDICINE

Biotechnology is not a new subject, although it has received more attention in the recent years. It can be defined as use of living organisms in industry or industrial type processes. As it stands today, biotechnology has a signigicant impact on diagnosis, treatment and prevention of diseases. It has opened up new research avenues for discovering how healthy bodies work and what goes wrong when problems arise. Biotechnology - based diagnostic tools detect diseases more quickly and with greater accuracy.

Applications in Diagnostic Microbiology:

- Monochonal antibody conjugates are now used in many commercial systems in both immuno-fluorescence and EIA formats. In addition to direct detection of microbial structural antigens, they are also being used for identification of enteric toxins produced by *Vibrio, Campylobacter, Shigella,* etc; Uropathogenic and toxigenic strains of *Escherichia coli*; Diagnosis of *Chlamydia trachomatis* infection, *H. pylori* infection, Leptospirosis, Malaria, Systemic fungal infections; HIV by antibody detection by ELISA and Western Blot, etc.

- Recombinant DNA based technologies for detection of *Leptospira* antigens and antibodies, *Bacillus anthrax* antibodies, etc.

- RIBA – Recombinant Immunoblot Assay for diagnosis of HIV, HCV, etc.

- Hybrid capture assay for diagnosis of CMV, HPV.

- Commercially available nucleic acid probes using chemiluminescence for detection e.g. *Chlamydia trachomatis, N. gonorrhoeae.*

- In situ hybridization for organisms that are difficult to grow in culture e.g. HPV, E-B virus, HBV and HIV-1.

- Assay of viral load in HIV disease to assess the progress and/or prognosis of the disease.

Applications in therapeutics:

Therapies with fewer side effects.

- Therapeutic use of endogenous proteins and antibodies

- Discovery of new therapeutic compounds in plants and other organisms

- Production of Insulin, Colony stimulating factors, Erythropoietin, Interferons, etc. IFN α2b for treatment of HBV, HCV and leukemias. (e.g. Shanferon by Shanta Biotech).

- Production of Natural Pro-Vitamin A (β-carotene) and natural mixed carotenoids from aquatic organisms – Nutraceuticals.

- Gene therapy.

- Uses of monoclonal antibodies – To block cytokines, to kill cancer cells without affecting normal cells; For treatment of acute renal failure; Fertility enhancement (r-human LH), for fractures difficult to heal (r-osteogenic protein 1), etc.

Applications in prevention of diseases:

Development of India's first genetically engineered newer and safer vaccines. Hep B vaccine (Shanvac-B by Shanta Biotech) has revolutionised biotechnology in India.

We now are aware of the precise mechanisms that drive and direct biological processes, have detailed information about molecular basis of diseases, nucleotide and protein sequences of specific organisms, gene mapping of the human genome, etc. through biotechnology.

The National Centre for Biotechnology Information (NCBI) at National Library of Medicine(NLM), Bethesda, Maryland was established in November 1988. It is a multidisciplinary research group comprising of computer scientists, molecular biologists, mathematicians, biochemists, research physicians and structural biologists concentrating on basic and applied research.

In India, apart from Research and Development, Govt. of India, DRDO Gwalior has also come out with certain diagnostic kits at cheaper rates. So let us hope for the best in the near future that newer avenues open up in the field of biotechnology and a new era in medical research and health care provision begin.

RE (round eastern) – Genetic scissors (cut nucleotide base at predesigned site)

Applications of Genetic Engineering in Bacteria:

Impact on agricultural set up.

- *B. thuringensis* – Increased crop production, increased protection.

Protein toxic to larvae of insects, parasites, nematodes and mosquitoes.

- *P. fluorescens, Clavibacter xyli* – Insect control.

- *Serratia, Streptomyces and Vibrio spp.* – Extracellular chitinases for biological control of soil born fungal infection of plants.

- *P. putida* – Biodegradation of toxic waste, toluene, chlorobenzene, chlorophenols.

- Degradation and/or sequencing of environmental pollution.

Genetic Engineering in filamentous fungi – Medicine, industry, basic research, agriculture, food, food additives and beverages.

Genetic Engineering in humans is controversial and hotly debatable, but has shown a surprisingly rapid rate of progress from "fiction to fact".

DNA research in humans – 4 major areas.

- Carrier detection and prenatal diagnosis

- Molecular basis of genetic diseases – molecular oncology

- Production of human biologicals – blood components, vaccines, proteins, hormones

- **Gene therapy**. Gene splicing' – Paul Berg 1973.

'Human Genome Project' – Dr. Hargobind Khurana.

It is the transfer of new genetic material, i.e. recombinant DNA, transiently or permanently, into the cells of an individual, with resulting therapeutic benefit to the individual.

- Advent of recombinant DNA technology in mid 1970s led to the isolation and manipulation of genes.

- Development of retroviral based gene transfer systems in the early 1980s.

- Richard Morgan and W. F. Anderson of Maryland and California proposed an outline for the first human gene therapy protocol in 1987.

- Pilot gene transfer experiment in May 1989.

- First true gene therapy experiment in September 1990.

Vector DNA

↓ Transported to

Cell nucleus

↓ Transcription machinery of cell

Recombinant mRNA

↓

Nucleus to ribosomes

↓ Translation

Therapeutic Protein Product

- Corrects cellular defects in target cells

- Secreted to alter cellular metabolism at distant site.

Types of Gene Therapy

- **Germline gene therapy** – Permanent introduction of genetic material to germ cells, which allows generational passage of the genes to the offspring, allowing permanent introduction of new or altered traits in human population. It is essentially banned worldwide. "Scientists playing God?"

- **Somatic gene therapy** – Transfer of DNA to somatic cells, that cannot be transmitted to new generations. Target of therapy is individual affected by or predisposed to the development of significant life - threatening disease.

Gene augmentation is trying to augment a gene by supplying a good gene (not replacement or correction). Clinical trials are going on in malignancy and rheumatoid arthritis cases.

2 types – Targetted gene augmentation – at site of defect
          – Non-targetted gene augmentation– functional.

Bone marrow, skin and liver offer most realistic opportunities for gene therapy as these are accessible, manipulated in vitro.

Pre requisite for a viral vector – nontoxic, monogenic, replication defective, modification of genome of wild type viruses before use – **'crippled virus'** (can not replicate in vivo), cell lines allowing replication and encapsulation of defective recombinant virus is required in vitro.

Vectors (or gene transfer vehicles) for gene therapy

1. **Non-viral vectors**

   a) Microinjection; b) Electropolation; c) High velocity; d) Receptor mediated gene transfer (DNA, ligands – target organ is liver); e) Co-precipitation with calcium phosphates; f) Polycations, – DNA complex; g) Erythrocytic ghosts (used as carrier particles); h) Site – directed recombination;

   i) **Liposome mediated gene transfer** (plasmid DNA incorporated into liposomes).

   Lipid – DNA complex

   ↓ Electrostatic interaction

   Binds to plasma membrane of target cells

   ↓

   Internalization into endosomes and lysosomes of target cells

   ↓

   Some escapes degradation and is released to cytosol

   ↓

   Maintained as 'episome' (not integrated into genome)

   ↓

   Transient effects, which is beneficial in the gene therapy of cancer.

Advantages of liposome mediated gene transfer

– Not associated with extraviral sequences

– Capable of mediating direct in vivo gene transfer

– Have minimal toxicity

Disadvantages of all physical methods

– Unstability of unintegrated sequences

– Multiple, tandemly operated copies

– Inactivation by genetic and epigenetic mechanisms.

   a) to d) – physical; e and f – chemical.

2. **Viral (Biological) Vectors**

   - **DNA viruses** – SV 40, Polyoma virus, Adenovirus 2 and 5, AAV2, HSV1, CMV, Pseudorabies, Vaccinia, BPV (Bovine papilloma virus), EBV. Adenoviral vectors are stable, facilitates purification and production in laboratory, efficient binding, internalization and degradation of endosomes.

   - **RNA viruses** – Retroviruses (murine is most often used, avian, human), 'C' types of retroviruses are commonly used.

     Mostly all protocols are based on Moloney murine leukemic virus (MMLV) and gibbon – leukemic virus (GLV). Retroviral vectors – integration into human genome in random, insertion near a tumor suppressor gene could alter its expression, potential of combining with endogenous retroviruses leading to production of infective retroviral particles.

3. **Combined Vectors**

   Biological + Physical gene transfer systems

   Plasmid DNA + Polylysine – conjugated antibody + Adenovirus vector (trimolecular complex). Achieves high level of gene transfer.

   ### Organoids

   Genetically altered cells that secrete therapeutic gene products systemically or locally, when implanted.

   - Fator VIII cDNA transplanted in hemophiliacs

   - β-glucuronidase for correction of mucopoly-saccharidosis VII

   - Cytokines in cancer therapy.

Clinical Trials with Gene Therapy

Disease	Gene inserted
ADA deficiency	Adenosine deaminase
Advanced cancers	TNF α and IL2
Liver disease	α Dh receptor
Ovarian cancer	Thymidine kinase
Malignant melanoma	HLA - B7
AIDS	TK
Hemophilia	Factor VIII

Both humoral and cellular immune responses remain a major hurdle in the application of gene transfer for therapy of genetic and acquired diseases.

Advantages of gene therapy – Possible cure of genetic/nongenetic diseases

   – Modification of pathogenesis through genes/products (e.g. HIV suppression)

   – No need for HLA matched donor

   – No immunosuppression, so no graft versus host (GVH) reaction especially in bone marrow transplants.

   – Correction of germline defect is possible, but ethically prohibited.

Risks of gene therapy – Foreign DNA insert is detrimental. May stimulate oncogenes or inactivation of suppressor genes.

   – Affection of regulatory pathway of cells

   – Affection of cell differentiation

   – Affection of germ line.

Limitations of gene therapy – No ideal vector, low transferred efficiency of corrected gene

   – Low stability of cured genes

   – Technical uncertainty

   – Not effective in multigene disorders where the products or genes not known/cloned

   – High cost

NANOTECHNOLOGY

"The power of nanotechnology is rooted in its potential to transform and revolutionize multiple technology and industry sectors, including aerospace, agriculture, biotechnology, homeland security and national defense, energy, environmental improvement, information technology, medicine and transportation. Discovery in some of these areas has advanced to the point where it is now possible to identify applications that will impact the world we live in".

- National Nanotechnology Initiative

Nanoscience will change the nature of almost every human-made object in the next century. Economic Impact of Nanotechnology will be $1 Trillion per year by 2015. The highest growth rates will be in the convergence between bio- and nanotechnologies in the healthcare and pharmaceutical sectors.

- Nanotechnology is the art and science of manipulating atoms and molecules to create new systems, materials and devices. It is the manipulation of matter with at least one dimension sized from 1 to 100 nanometers. One nanometer (nm) is one billionth, or 10^{-9}, of a meter. One inch = 25,400,000 nanometers.

- The term "nano-technology" was first used by Norio Taniguchi in 1974. In 1980, nanotechnology emerged as a field.

- Small particles – more surface area, more atoms to contact a surface
- Extremely precise – materials can be made close to perfection to the point that exact number of atoms can be measured.
- Two major breakthroughs sparked the growth of nanotechnology in modern era.
  - The invention of the scanning tunneling microscope in 1981 which provided unprecedented visualization of individual atoms and bonds
  - Invention of the analogous atomic force microscope in 1989.
- Nanotechnology is the engineering of functional systems at the molecular scale. It refers to the projected ability to construct items from the bottom up, using techniques and tools being developed today to make complete, high performance products.
- Two main approaches used in nanotechnology:
  - In the **"bottom-up" approach**, smaller components are arranged into more complex assemblie, i.e. materials and devices are built from molecular components which assemble themselves chemically by principles of molecular recognition.
  - In the **"top-down" approach**, smaller devices are created by using larger ones to direct their assembly, i.e. Nano-objects are constructed from larger entities without atomic-level control.
- Recent application of nanomaterials include a range of biomedical applications, such as tissue engineering, drug delivery and biosensors.
- Bionanotechnology is the use of biomolecules for applications in nanotechnology, including use of viruses and lipid assemblies.
- A regulatory framework to assess and control risks associated with the release of nanoparticles and nanotubes is lacking and have drawn parallels with bovine spongiform encephalopathy ("mad cow" disease), thalidomide, genetically modified food, nuclear energy, reproductive technologies, biotechnology and asbestosis.
- The Royal Society report identified a risk of nanoparticles or nanotubes being released during disposal, destruction and recycling and recommended that "manufacturers of products that fall under extended producer responsibility regimes such as end-of-life regulations, publish procedures outlining how these materials will be managed to minimize possible human and environmental exposure".

Nanomedicine

Nanomedicine is the medical application of nanotechnology. Nanomedicine ranges from the medical applications of nanomaterials and biological devices, to nanoelectronic biosensors and even possible future applications of molecular nanotechnology, such as biological machines. Current problems for nanomedicine involve understanding the issues related to toxicity and environmental impact of nanoscale materials. The **National Nanotechnology Initiative** expects new commercial applications in the pharmaceutical industry that may include advanced drug delivery systems, new therapies and in vivo imaging. Nanomedicine research is receiving funding from the US **National Institutes of Health**, including the funding in 2005 of a five-year plan to set up four nanomedicine centers.

A) Drug delivery system

Nanotechnology has provided the possibility of delivering drugs to specific cells using nanoparticles. The overall drug consumption and side-effects may be lowered significantly by depositing the active agent in the morbid region only and in no higher dose than needed. Targeted drug delivery is intended to reduce the side effects of drugs with concomitant decreases in consumption and treatment expenses. The efficacy of drug delivery through nanomedicine is largely based upon:

- Efficient encapsulation of the drugs
- Successful delivery of drug to the targeted region of the body
- Successful release of the drug.
- Drug delivery systems, lipid- or polymer-based nanoparticles, can be designed to improve the pharmacokinetics and biodistribution of the drug. However, the pharmacokinetics and pharmacodynamics of nanomedicine is highly variable among different patients.
- When designed to avoid the body's defense mechanisms, nanoparticles have beneficial properties that can be used to improve drug delivery. Complex drug delivery mechanisms are being developed, including the ability to get drugs through cell membranes and into cell cytoplasm. Triggered response is one way for drug molecules to be used more efficiently.

B) Nanoparticles can be used in combination therapy for decreasing antibiotic resistance or for their antimicrobial properties. Nanoparticles might also used to circumvent multidrug resistance (MDR) mechanisms.

C) Three forms of nanomedicine that have already been tested in mice and are awaiting human trials that will be using gold **nanoshells** to help diagnose and treat cancer (cancer treatment with iron nanoparticles or gold shells) using liposomes as vaccine adjuvants and as vehicles for drug transport and Silver Nano platform for using the silver nanoparticles as an antibacterial agent.

D) Drug detoxification is also another application for nanomedicine which has shown promising results in rats. Advances in Lipid nanotechnology was also instrumental in engineering medical nanodevices and novel drug delivery systems as well as in developing sensing applications. Another example can be found in dendrimers and nanoporous materials or to use block co-polymers, which form micelles for drug encapsulation.

E) Protein and peptides exert multiple biological actions in the human body and they have been identified as showing great promise for treatment of various diseases and disorders. These macromolecules are called biopharmaceuticals. Targeted and/or controlled delivery of these biopharmaceuticals using nanomaterials like nanoparticles and dendrimers is an emerging field called **nanobiopharmaceutics** and these products are called nanobiopharmaceuticals.

F) Another vision is based on small electromechanical systems - **nanoelectromechanical** systems are being investigated for the active release of drugs. Nanotechnology is also opening up new opportunities in implantable delivery systems, which are often preferable to the use of injectable drugs, because the latter frequently display first-order kinetics (the blood concentration goes up rapidly, but drops exponentially over time).

Nanoparticles

- Quantum dots (nanoparticles of cadmium selenide), when used in conjunction with MRI, can produce exceptional images of tumor sites. Quantum dots glow when exposed to ultraviolet light. When injected, they seep into cancer tumors. The surgeon can see the glowing tumor, and use it as a guide for more accurate tumor removal. These nanoparticles are much brighter than organic dyes and only need one light source for excitation. This means that the use of fluorescent quantum dots could produce a higher contrast image and at a lower cost than today's organic dyes used as contrast media. The downside, however, is that quantum dots are usually made of quite toxic elements.

- Nanotechnology-on-a-chip is one more dimension of **lab-on-a-chip technology**. Magnetic nanoparticles, bound to a suitable antibody, are used to label specific molecules, structures or microorganisms.

- Gold nanoparticles tagged with short segments of DNA can be used for detection of genetic sequence in a sample.

- **Super chips** – Combination of Silicon and Galium arsenide create wireless chips

- Sensor test chips containing thousands of nanowires, able to detect proteins and other biomarkers left behind by cancer cells, could enable the detection and diagnosis of cancer in the early stages from a few drops of a patient's blood.

- Nanomaterials that can see inside vessels for plaque build up

- **Nanosensors** for disease detection – 10 times more faster, 100,000 times more accurate

- Nanofilters will help create impurity free drugs

- Nanotechnology is using arthroscopes, which are pencil-sized devices that are used in surgeries with lights and cameras, so surgeons can do the surgeries with smaller incisions. The smaller the incisions, the faster the healing time which is better for the patients. It is also helping to make an arthroscope smaller than a strand of hair.

- Magnetic micro particles are proven research instruments for the separation of cells and proteins from complex media. The technology is available under the name Magnetic-activated cell sorting or Dynabeads among others. More recently it was shown in animal models that magnetic nanoparticles can be used for the removal of various noxious compounds including toxins, pathogens, and proteins from whole blood in an extracorporeal circuit similar to dialysis.

- Nanoparticles such as graphene, carbon nanotubes, molybdenum disulfide and tungsten disulfide are being used as reinforcing agents to fabricate mechanically strong biodegradable polymeric nanocomposites for bone tissue engineering applications. The addition of these nanoparticles in the polymer matrix at low concentrations (~0.2 weight%) leads to significant improvements in the compressive and flexural mechanical properties of polymeric nanocomposites. Potentially, these nanocomposites may be used as a novel, mechanically strong, light weight composite as bone implants.

Applications of Nanotechnology

Materials Science	Powders, Coatings, Carbon Nano-materials, C-NanoFabrics
Energy	Solar Power & PhotoVoltaics, Hydrogen Fuel Cells, LED White Light
Medicine/Biotech	Genomics, Proteomics, Lab-on-a-Chip, Carbon-Nanotubes, BuckyBalls
Electronics	MRAM/Spintronics, NRAM, Q-Dots, Q-Bits
Devices	Lithography, Dip Pen Lithography, AFM, MEMS

Important Nanoparticles

Nanoparticles	Nature	Property	Thermal/Electrical conductivity
Carbon-Nanotubes	4 nm width (smaller diameter than DNA)	100 times stronger than steel	Thermal/Electrically conductive
BuckyBalls – C_{60}	Third major form of pure carbon	Compressed – becomes stronger than diamond	Heat resistance & Electrical conductivity
MEMS (Micro-Electro-Mechanical systems)	Common silicon substrate	Integration of mechanical elements, sensors, actuators and electronics	
Quantum Dots	2-10 nm in diameter	Nanocrystals - Quantum dots 3.2 nm in diameter have blue emission; Quantum dots 5 nm in diameter have red emission.	Semi-conductors

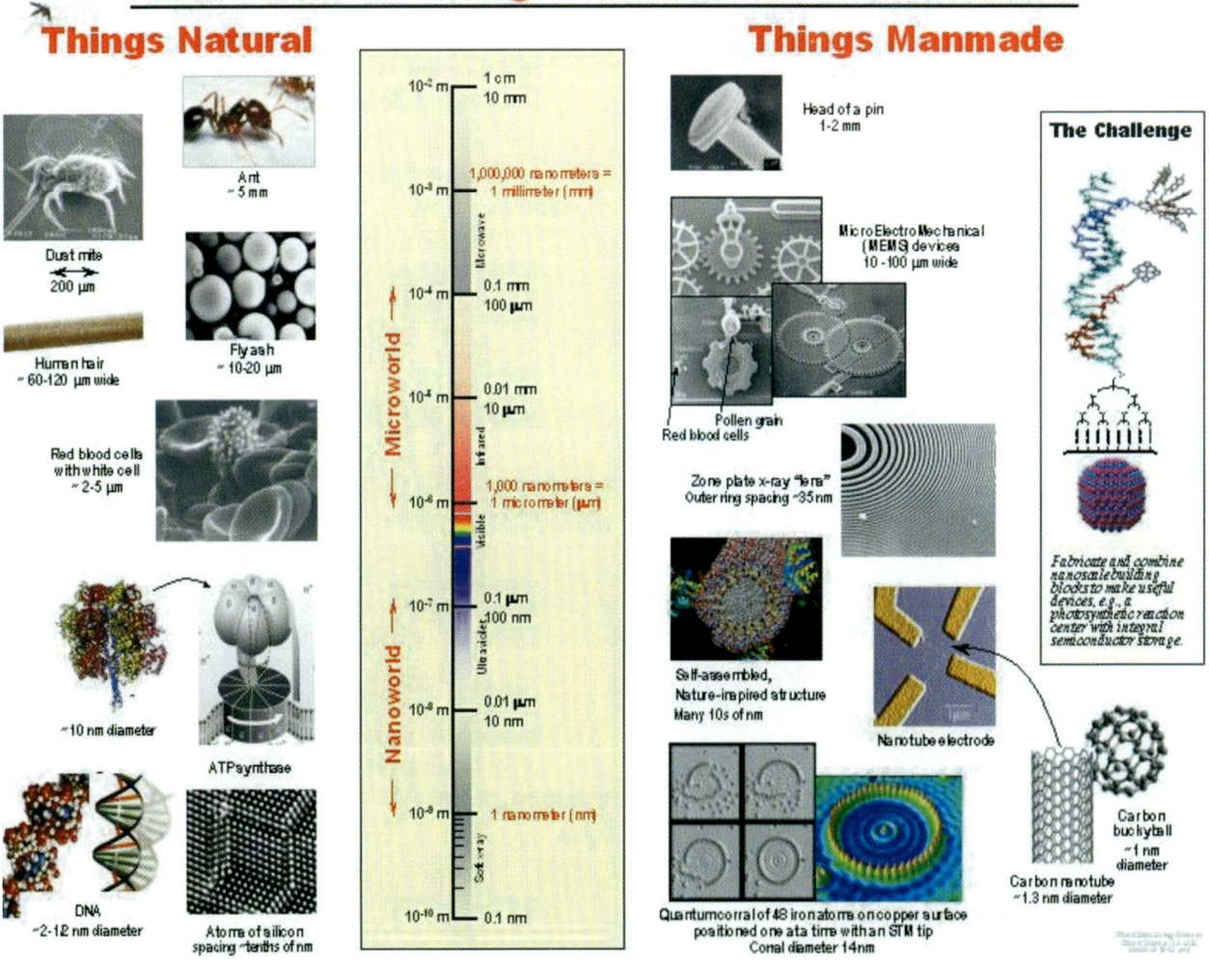

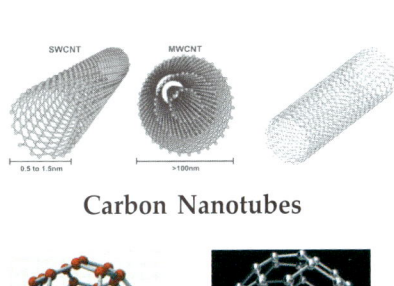

Carbon Nanotubes

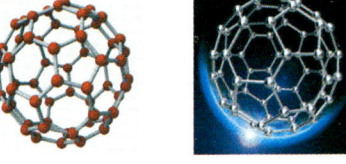

C$_{60}$-BuckyBalls

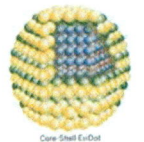

Left: Quantum Dots; Centre & Right: MEMS

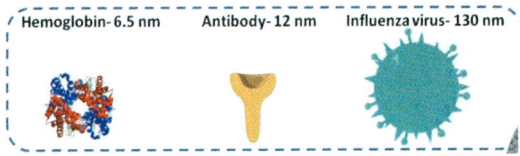

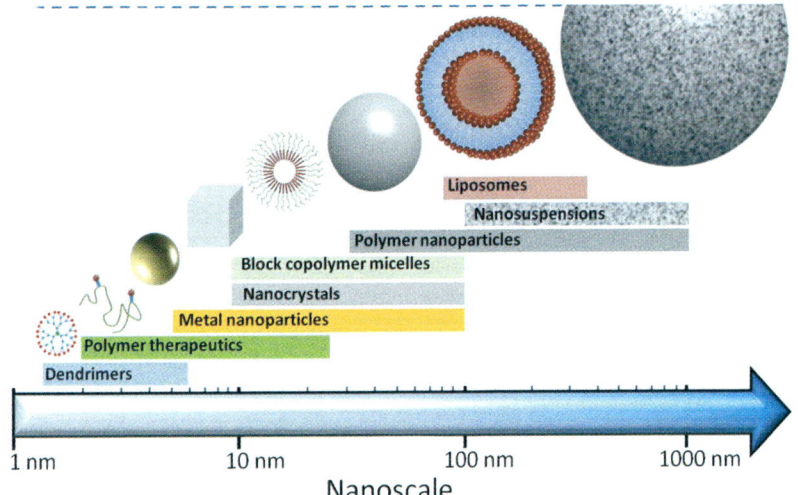

Applications of Nanomedicine

- Abraxane, approved by the U.S. Food and Drug Administration (FDA) to treat breast cancer, non-small-cell lung cancer (NSCLC) and pancreatic cancer, is the nanoparticle albumin bound paclitaxel.

- Doxil was originally approved by the FDA for use on HIV-related Kaposi's sarcoma. It is now being used to also treat ovarian cancer and multiple myeloma. The drug is encased in liposomes, which helps to extend the life of the drug that is being distributed. Liposomes are self-assembling, spherical, closed colloidal structures that are composed of lipid bilayers that surround an aqueous space. The liposomes also help to increase the functionality and it helps to decrease the damage that the drug does to the heart muscles specifically.

- C-dots (Cornell dots) are the smallest silica-based nanoparticles with the size <10 nm. The particles are infused with organic dye which will light up with fluorescence. Clinical trial is underway since 2011 to use the C-dots as diagnostic tool to assist surgeons to identify the location of tumor cells.

- Nanosilver wound dressings is used for burn victims.

- Burn dressings are coated with nanocapsules containing antibiotics. The harmful bacteria in the wound causes the nanocapsules to break open, releasing the antibiotics.

- Nanosponges are polymer nanoparticles coated with a red blood cell membrane, which absorb the toxins and remove them from the blood stream.

- Nanocrystalline silver is an antimicrobial agent for the treatment of wounds.

- A nanoparticle cream contains nitric oxide gas which is known to kill bacteria significantly reduces staphylococcal abscesses.

- Nanoparticles composed of polyethylene glycol-hydrophilic carbon clusters (PEG-HCC) can absorb free radicals at a much higher rate than the proteins.

- A lens coated with carbon nanotubes can convert light from a laser to powerful focused sound waves, which can be used for noninvasive surgery.

- Using nanoparticle contrast agents, images such as ultrasound and MRI (magnetic resonance imaging), have a favorable distribution and improved contrast. This might be accomplished by self-assembled biocompatible nanodevices that will detect, evaluate, treat and report to the clinical doctor automatically.

- Gelatin nanoparticles can be used to deliver drugs to damaged brain tissue.

- Nanoparticles deliver vaccine, protect the vaccine, allowing the vaccine time to trigger a stronger immune response.

Ongoing trials with nanoparticles

- An early phase clinical trial using the platform of 'Minicell' nanoparticle for drug delivery have been tested on patients with advanced and untreatable cancer. Built from the membranes of mutant bacteria, the minicells were loaded with paclitaxel and coated with cetuximab, antibodies that bind the **epidermal growth factor receptor (EGFR)** which is often overexpressed in a number of cancers, as a 'homing' device to the tumor cells. The tumor cells recognize the bacteria from which the minicells have been derived, regard it as invading microorganism and engulf it. Once inside, the payload of anti-cancer drug kills the tumor cells. Measured at 400 nanometers, the minicell is bigger than synthetic particles developed for drug delivery. The researchers indicated that this larger size gives the minicells a better profile in side-effects because the minicells will preferentially leak out of the porous blood vessels around the tumor cells and do not reach the liver, digestive system and skin. This Phase 1 clinical trial demonstrated that this treatment is well tolerated by the patients. As a platform technology, the minicell drug delivery system can be used to treat a number of different cancers with different anti-cancer drugs with the benefit of lower dose and less side-effects.

- In 2014, a Phase 3 clinical trial for treating inflammation and pain after cataract surgery, and a Phase 2 trial for treating dry eye disease were initiated using nanoparticle loteprednol etabonate

- In 2015, the product, KPI-121 was found to produce statistically significant positive results for the post-surgery treatment.

- Nanoparticles have high surface area to volume ratio, allows many functional groups to be attached to a nanoparticle, which can seek out and bind to certain tumor cells. Additionally, the small size of nanoparticles (10 to 100 nanometers), allows them to preferentially accumulate at tumor sites (because tumors lack an effective lymphatic drainage system). Limitations to conventional cancer chemotherapy include drug resistance, lack of selectivity and lack of solubility. Nanoparticles have the potential to overcome these problems.

- In photodynamic therapy, a particle is placed within the body and is illuminated with light from the outside. The light gets absorbed by the particle and if the particle is metal, energy from the light will heat the particle and surrounding tissue. Light may also be used to produce high energy oxygen molecules which will chemically react with and destroy most organic molecules that are next to

them (like tumors). This therapy is appealing for many reasons. It does not leave a "toxic trail" of reactive molecules throughout the body (chemotherapy), because it is directed where only the light is shined and the particles exist. **Photodynamic therapy** has potential for a noninvasive procedure for dealing with diseases, growth and tumors. Kanzius RF therapy is one example of such therapy. Also, gold nanoparticles have the potential to join numerous therapeutic functions into a single platform, by targeting specific tumor cells, tissues and organs.

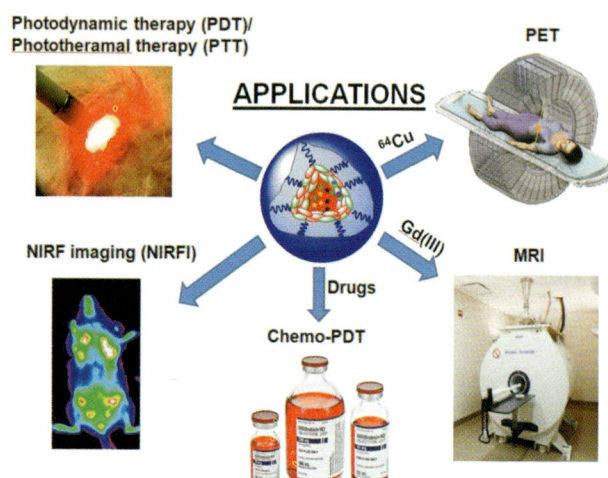

- Nanotechnology may be used as part of **tissue engineering** to help reproduce or repair damaged tissue using suitable nanomaterial-based scaffolds and growth factors. Tissue engineering if successful may replace conventional treatments like organ transplants or artificial implants.

- **Medical devices** - Neuro-electronic interfacing is a visionary goal dealing with the construction of Nano devices that will permit computers to be joined and linked to the nervous system. This requires the building of a molecular structure that will permit control and detect nerve impulses by an external computer.

- Nanoparticles that deliver chemotherapy drugs directly to cancer cells are under development. Tests are in progress for targeted delivery of chemotherapy drugs and their final approval for their use in cancer patients.

- Research is going on whether bismuth nanoparticles can concentrate radiation used in radiation therapy to treat cancer tumors.

- Targeted heat therapy is being developed to destroy breast cancer tumors. Nanotubes accumulate at the tumor. Infrared light from a laser is absorbed by the nanotubes and produces heat that incinerates the tumor.

- **Nanorobots** can actually be programed to repair specific diseased cells, functioning in a similar way to antibodies in our natural healing processes and can eliminate bacterial infection in a patient within minutes.

- The small size (< 100 nm) and large surface area of functionalized nanomagnets leads to advantageous properties compared to hemoperfusion, which is a clinically used technique for the purification of blood and is based on surface adsorption. These advantages are high loading and accessibility of the binding agents, high selectivity towards the target compound, fast diffusion, small hydrodynamic

resistance and low dosage. This approach offers new therapeutic possibilities for the treatment of systemic infections such as sepsis by directly removing the pathogen.

- It can also be used to selectively remove cytokines or endotoxins, or for the dialysis of compounds which are not accessible by traditional dialysis methods. However, the technology is still in a preclinical phase and first clinical trials are not expected before 2017.

REFERENCES

1. Topley & Wilson's, Microbiology and Microbial Infections. Bacteriology, Vol. 1. 10th Ed. Eds. Borriello SP, Murray PR, Funkay G. John Wiley & Sons Ltd., West Sussex, UK. 2009.

2. Koneman EW, Allen SD, Janda WM, Schreckenberger P, Winn Jr. WC. In Color Atlas & Textbook of Diagnostic Microbiology. 6th Ed. Lippincott, Philadelphia. 2006.

3. Biotechnology in Medicine. https://www.boundless.com/biology/textbooks/boundless-biology-textbook/biotechnology-and-genomics-17/biotechnology-119/biotechnology-in-medicine-480-11702/

4. Applications of Biotechnology in Medicine. http://www.biotechonweb.com/Application-of-biotech-in-Medical.html

5. Ehrhardt A, Haase R, Schepers A, Deutsch MJ, Lipps HJ, Baiker A. Episomal vectors for gene therapy. Curr Gene Therapy. 2008; 8: 147-61.doi:10.2174/156652308784746440. PMID 18537590.

6. Productive Nanosystems A Technology Roadmap, 2007, Battelle Memorial Institute and Foresight Nanotech Institute.

7. IWGN Workshop Report: Nanotechnology Research Directions: Vision for Nanotechnology in the Next Decade, 2000, Edited by M.C. Roco, R.S. Williams, P. Alivisatos, Springer.

8. Investor's guide to Nanotechnology and Micromachines - Glenn Fishbine.

9. https://en.wikipedia.org/wiki/Nanotechnology

10. www.nano.gov

11. www.science.doe.gov/nano

12. www.nanotechgroup.org

BIOFILMS AND QUORUM SENSING

36

Chapter

BIOFILMS

- Biofilms are communities of microorganisms in a matrix that joins them together and to living or inert substrates
- Biofilms are surface-attached communities of bacteria, encased in an extracellular matrix of secreted proteins, carbohydrates, and/or DNA, that assume phenotypes distinct from those of planktonic cells
- Vary in thickness from mono cell layer to 6-8 cm thick, average 100μm
- Biofilms are diverse from their formation on teeth as plaques and submerged rocks in a stream
- Biofilms offer their member cells several benefits
- Biofilms may form:
  - On solid substratums in contact with moisture
  - On soft tissue surfaces in living organisms
  - At liquid-air interfaces.
- Submerged biofilms seems to form columns and mushroom like projections that are separated by water-filled channels
- Floating biofilms form a skin or pellicle at the air-liquid interface – shows organization of cells with the matrix at the outside
- Films that form on the surface of solid media such as agar or other surfaces.
- Pioneering studies by Cholodny, Henrici and Zo Bell > 50-60 years ago.
- Imersion of glass slides into natural environments and observing the biofilms developed on them under microscope.
- Formation of multicellular communities depends on the production of extracellular substances (glycocalyx).

Bacterial cell characteristics

- A slower growth rate
- Increased antibiotic resistance
- Top to bottom gradient of increasing antibiotic resistance
- Proximity of cells lead to elevated frequency of horizontal transfer of genes for resistance.

Environmental and cultural characteristics which affect the selection of biofilms multispecies

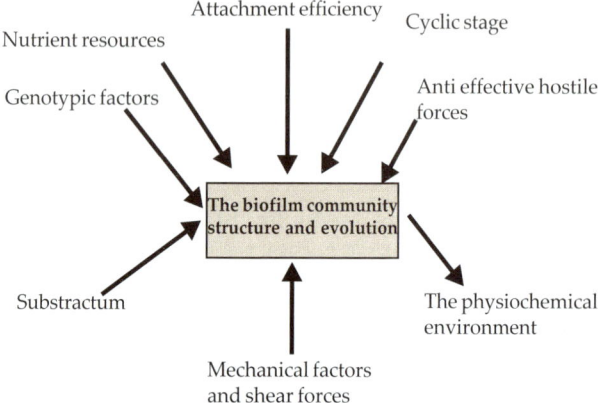

- Colonization of the surface by bacteria
- Factors affecting bacterial attachment to surfaces:
  - Nutrient concentration
  - pH
  - Temperature
  - Electrolyte concentration
  - Surface types
- High surface energy materials – negatively charged hydrophilic materials like glass, metal, minerals
- Low surface energy materials – low positively or low negatively charged hydrophobic materials like plastics.

Steps in biofilm formation

1. Initial attachment of cells with a surface or with each other
2. Establishment of microcolony
3. Formation of an extracellular matrix.
4. Convertion of homogenous environment into heterogenous one.
5. Attraction of other bacterial species to heterogenous environment

Many bacteria attach to the surface resulting in the formation of biofilm until series of complex communities result.

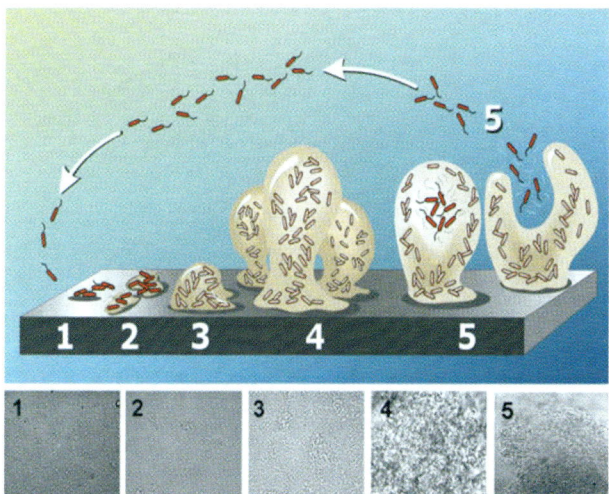

Quorum sensing

- Certain species of bacteria communicate with each other within the biofilm.

- As their density increases, the organisms secrete low molecular weight molecules that signal when the population has reached a critical threshold.

- This process, called quorum sensing, is responsible for the expression of virulence factors.

- Exopolysaccharides in the glycocalyx contribute to biofilm formation

- **Polysaccharides:**
  - Carbohydrates significantly impact bacterial virulence

  - Bacteria have capsular polysaccharides and exopolysaccharides

  - The polysaccharides are not soluble and do not disassociate with the bacterial cells

  - Many bacteria have been found to produce cellulose

  - This is a novel finding in the case of *Salmonella typhimurium* and *E. coli*

  - The bacterium *Gluconacetobacter xylinus* has been recognized as a cellulose producer

  - Many other bacteria have genes homologous to the bcs, bacterial cellulose synthesis genes

  - *Vibrio cholerae* does not appear to have a gene which encodes a cellulose – but the bacterium has two domains homologous to *Gluconeacetobacter.*

- **Matrix:**
  - Key components of the matrix are extracellular polysaccharides and proteins

  - Dead cells have also been identified in biofilms

  - Extracellular DNA is also important.

Pathogens that have been studied for the formation of biofilms

- *Staphylococcus aureus*
- *Staphylococcus epidermidis*
- *Pseudomonas aeruginosa*
- *Escherichia coli*
- *Enterococcus faecalis*
- *Vibrio cholerae*
- *Salmonella typhi*

Now scientists believe biofilm formation is a universal feature of all bacteria.

Staphylococcus aureus

- Proteins involved in biofilm formation – Accumulation-Associated protein (AAP),

- The Clumping factor A (Clf A),

- Staphylococcal Surface Protein (SSP1) and

- The Biofilm associated protein (Bap).

- Intercellular adhesions with in biofilms of *Staphylococcus epidermidis*, a major cause of medical device related infections, is mediated by the PIA (polysaccharide intercellular adhesin).

Pseudomonas aeruginosa

- Two new polymers have been found in *Pseudomonas*

- pelA-G gene produces a glucose rich polymer

- pslA-O genes produce a mannose rich polymer

- Quorum sensing hierarchy plays a central role in the late stages of biofilm maturation.

- Two quorum sensing systems, Las RI and RhlR1.

- Third Qsc R quorum sensing system has been identified that further modulates their expression.

- Mus20 protein similar to Bap protein of *S. aureus.*

Escherichia coli

- Extracellular proteinaceous fibers called curli.

- Curli are involved in the formation of biofilms and are potent inducers of the host inflammatory respons.

- Curli are structurally and biochemically defined as amyloids.

- PIA polymers

- Adhesins are molecules that are attached to bacterial fimbriae

- Conjugative pili greatly accelerates initial adhesion and biofilm development

Enterococcus fecalis

- *E. fecalis* biofilms on dental root canals, urethral catheters, uretheral stents and heart valves have been observed.

- While it is not clear that the ability of *E. fecalis* to form biofilms is essential for virulence, it appears that a majority of clinical isolates do possess the ability to form a biofilm in vitro

- Esp protein of *Enterococcus fecalis* (surface protein similar to Bap of *S. aureus*).

Other bacteria

- *Vibrio cholerae* - Major extracellular polysaccharide in the biofilm are VPS and the genes, vps and hapR
- *Salmonella typhi* - sty2875 protein similar to Bap protein of *S. aureus*.

Candida species

- Major genes that contribute to drug resistance in *C. albicans* and *C. dubliniensis* are CDR genes (CDR 1 and CDR 2) and MDR genes.
- These genes have been demonstrated to be upregulated during biofilm formation and development.
- Three distinct developmental phases:
  - Early (0-11 h)
  - Intermediate (12-30 h)
  - Mature (38-72 h)
- The detailed structure of mature *C. albicans* biofilms consists of a dense network of yeast, hyphae and pseudohyphae.

Dental plaque

- Found on the enamel of teeth
- Epithelial cells of the oral mucosa
- Participate in coaggregation which occurs between different species of bacteria.

Mechanisms

Two distinct mechanisms:

- The multiplication of bacteria already attached to the tooth surface
- The subsequent attachment and multiplication of new bacterial species to cells of bacteria already present in the plaque mass.

Chronic infections

- The study has been boosted on discovering their relation to chronic infections associated to medical implants
- These include those tissues involving infections of the endocardium, pneumonia in patients with cystic fibrosis, chronic kidney infections in patients with urinary catheters, osteomyelitis in patients with orthopaedic implants.

Endocarditis

- Biofilm of bacteria + host components on valve = vegetation
- Requires prior valve injury
- 200 X increase in antibiotic resistance

Infectious Kidney Stones

- 15-20% involve urinary tract infection
- Bacterium → biofilm → mineralization
- Causative organisms have urease
- Urea → NH_4 + H_2CO_3
- Biofilm concentrates urease → crystal formation

Cystic Fibrosis

- Mutation in chloride channel in epithelial cells of the lung
- 1st stage: intermittent infections
- 2nd stage: permanent infection with *Pseudomonas aeruginosa*
- Mucoid type - overproduce alginate
- Antibiotic resistance

Biofilms and contact lenses

Bacterial biofilm formation on contact lenses and contact lens storage cases may be a risk factor in contact lens-associated corneal infections. Studies have shown that contamination of lens cases by bacteria, fungi, and amoebae is common with 20% to 80% Antibiotic resistance in biofilmsof lens wearers having a contaminated lens case.

Biofilms and Antibiotics

- There is a growing concern for antibiotic resistance in bacteria growing in surface-adherent biofilms
- Many antibiotic assays for susceptibility and resistance are based upon planktonic or free cells rather than attached
- Chronic and device related infections go unresolved even when the organisms do indeed test for antibiotic sensitivity.
- The diffusion of antibiotics in biofilms has been studied
- Beta lactamase producing bacteria increase enzyme production in response to antibiotic treatment
- The enzyme accumulates in the matrix of the biofilm thereby inactivating the antibiotic.

Mar Operon

Multiple Antibiotic Resistance operon (*Mar*) chromosomal and encodes for permease proteins (AcrB) which actively export a wide range of xenobiotics from bacterial cells.

- *Mar* is widely distributed. Recent reports show that *Mar* can be regulated not only by exposure to sub-MIC levels of antibiotic, but also through slow growth rate, the stringent response and a number of other unrelated stimuli.
- It is regulated through part of a global regulatory system that also controls exopolymer biosynthesis.

- This operon is of major interest since it is likely that this would be switched on in biofilms and might be a major factor in the high level antibiotic resistance observed in biofilms.

Biofilm detection

- False negative cultures
- Visible but non cultivable organisms
- Underestimated or low colony count
- Inappropriate specimen and loss of or decreased antimicrobial susceptibility
- Biofilms are resilient, adherent and with EPS, quite resistant to culturing by swabs.
- Cooper et al developed a Gram staining technique of the catheter tips, the technique depends on optical properties of the different catheters but it is time consuming since it requires the microscopical examination of at least 200 oil immersion fields.
- By direct acridine orange staining of the catheter tips.
- By scanning electron microscope.
- A semiquantitative method for culturing vascular cannulas on solid media.

Tissue Culture Plate Method (TCPM)

- The microorganisms are grown in polystyrene tissue culture plates for 24 hours
- Then after washing fixed with sodium acetate (2%) and stained with crystal violet (0.1% w/v).
- Biofilm formation is detected by measuring optical density with ELISA reader.

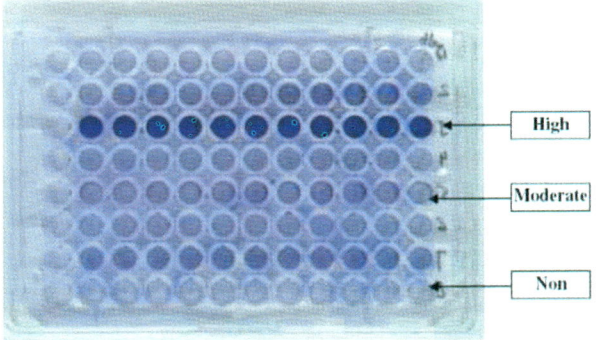

High, moderate and non slime producers differentiated with crystal violet staining

Tube Method

- The microorganisms are grown in trypticase soy broth with 1% glucose in tubes for 24 hours.
- The tubes are then decanted and washed with PBS (phosphate buffer saline) and stained with crystal violet (0.1%).
- The tubes are then washed and dried and biofilm formation is considered positive when a visible film lines the wall and bottom of the tube.

Congo Red Agar Method (CRA)

The microorganisms are grown on brain heart infusion agar with 5% sucrose and congo red. Positive results are indicated by black colonies with a dry crystalline consistency.

Bioluminescent Assay

- Attenuated Total Reflecting Spectroscopy (ATR)
- It has been used to monitor the conditioning films that are an early harbinger of biofilm formation.

Piezoelectric Sensors

- Such as quartz with crystal microbalances monitor frequency shifts as mass accumulates on the **sensor surface.**

Microarrays

Used to assess the genes present in different stages of biofilm formation

- In Staphylococcal biofilms the same genes are active the sar A staphylococcal accessory regulator and the ica ADBC regulator
- One of the best ways to evaluate gene expression
- DNA chips are used for a solid support
- These are made of silicon or glass
- They have DNA attached in orderly arrays
- The DNA is deliverd to specific position on the chip using tiny pins to apply a solution
- The spots are treated and dried in order to bind the DNA
- Usually cDNA
- cDNA is prepared from mRNA
- These pieces of DNA are usually 500-5000 nucleotides long.

Commercial chips (II)

- Oligonucleotides about 25 bases in length can be synthesized and placed directly on the chip
- The chip is 1.3 cm on a side and can have over 200,000 addressable positions
- The probes are often expressed sequence tags(ESTs)
- The nucleic acids to be analyzed are isolated and labeled with fluorescent reporter groups
- The DNA or target nucleotides are incubated with the fluorescent groups and then washed
- The chip is scanned with lasers.

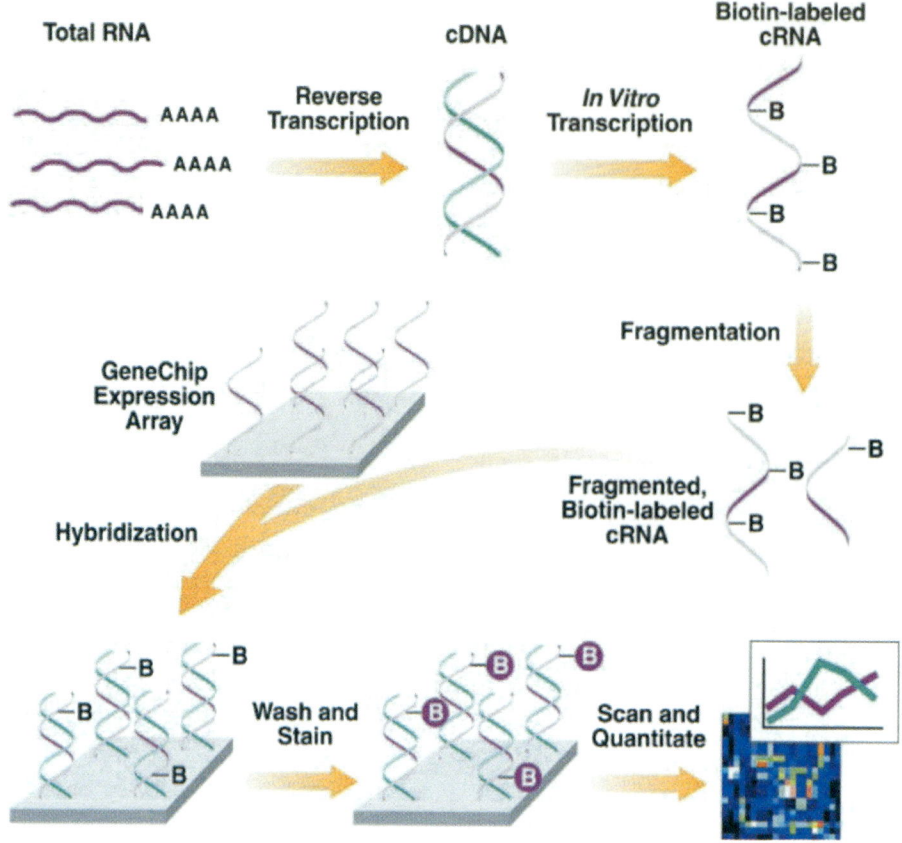

Preparation of Microarrays

Time of Flight Secondary Ion Mass Spectrometry (ToFSIMS)

- The combination of morphology and biochemical signature may put biofilms into the category of indisputable biomarkers

- In this way Bacterial biofilms could be traced and evident that they were present in rocks.

Agents for the destruction of biofilms

- Alexidine (chlorhexidine, polyhexamethylene biguanides)

- Monophenylethers (phenoxyethanol)

- Quaternary amonium compounds (cetrimide, benzalkoniums) have demonstrated biochemical bases for the activities and associated mammalian cell toxicities of thiol interactive agents (bronopol, isothiazolones).

- Mechanical disruption/removal (sonication)

- Immune modulation (Azithromycin and low dose doxycycline)

- Antimicrobial agents (silver and tobramycin)

- Amphotericin B lipid formulations and the Echinocandins against the *Candida* biofilms.

QUORUM SENSING

Quorum sensing is the phenomenon of microbial communication. It is a cell-to-cell communication:

- Way of exchange information through 'communication molecule'

- Refers to Small Molecules (SMs) or specifically autoinducers

- A phenomenon to regulate phenotypes

- This type of bacterial communication is achieved only at higher cell densities.

- Quorum sensing blockers interfere with signalling system. This signalling system is known as 'Quorum Sensing'. It is the process by which bacteria coordinate their behaviours in a density-dependent manner, using signalling molecules.

- Bacteria release various types of molecules called as autoinducers in the extracellular medium, these molecules are mediators of quorum sensing.

- When concentration of these signalling molecules exceed a particular threshold value, these molecules are internalized in the cell and activate particular set of genes in all bacterial population, such as genes responsible for virulence, competence, stationary phase, etc.

Bacteria are single celled organisms with no nervous system or brain. As environmental conditions change rapidly, bacteria need to respond quickly in order to survive. So how do individual bacterial cells "decide" between different physiological processes, living as part of a complex community called a 'biofilm'? Quorum sensing (QS) has been found to play a role in *Pseudomonas aeruginosa* biofilm formation. QS roles in biofilms – attachment, maturation, aggregation and dissolution, dispersal.

Density-dependent Quorum Sensing

Single cell Low (QS molecules), Target genes off → Growing aggregate Increasing (QS molecules), Target genes off → Microbial quorum inducing (QS molecules), Target genes on.

Quorum sensing Signal mechanism involves subsequent interaction of the signal with intracellular effectors. Signal molecule → Cell surface receptors → Secondary messengers.

Bacteria constantly produce and secrete certain signalling molecules, called as auto-inducers or pheromones. Density dependent phenomena → positive feedback loop.

Quorum sensing signal molecules consist of three categories:

- Fatty Acid derivative, called Homoserine lactones (HSLs)
- Amino Acids and Short Peptide derivatives
- Autoinducer-2 (AI-2) System

Acyl Homoserine Lactones (AHLs)

- Mediate quorum sensing in Gram negative bacteria
- Intraspecies communication
- Able to diffuse through membrane
- Induce synthesis of compounds interacting with host organism
- Several types depending on the length of the acyl side chain
- Many bacterial species can produce more than one type of Acyl HSL.

Autoinducer Peptides (AIPs)

- These are small peptides, regulate gene expression in Gram positive bacteria such as *Bacillus subtilis, Staphylococcus aureus, etc.*
- Regulates competence and sporulating gene expressions.

- Three part system – gene cluster, an enzyme and 2-component response system
- Interact with membrane bound histidine kinase as receptor.

AI-2 System

- In Gram positive and Gram negative bacteria
- Interspecies communication
- Furanosylborate diester

AI-2 controlled processes

- Induces mini cell formation
- Induces expression of stationary phase genes
- Inhibition of initiation of DNA replication

QS in bacterial pathogenesis

- QS is involved in expression of virulence genes in various bacteria, indicating the possible role of quorum sensing as a drug target.
- Several QS system mutant bacteria show the heavily reduced pathogenicity
- *Pseudomonas aeruginosa* mutant in synthesis of autoinducer molecules shows heavy reduction in pathogenesis. In *P. aeruginosa* QS molecules are synthesized by two autoinducer synthase – LasI and RhlI.
- In an *in-vivo* study, using two strains *P. aeruginosa*; PAO1 (virulent), and PAOR (*lasI* and *rhlI* double mutant, avirulent), it was seen that rats infected with PAOR are much immunologically active and number of *P. aeruginosa* also reduced.

Complications of QS

- Metabolic state – Influence signal production rate by positive autoregulation
- Point of induction of QS – *P. aeruginosa* gene regulation primarily occurs at substratum and *S. aureus*
- Important for predicting rates of signal synthesis through biofilm – point of immediate induction and how it induces remaining population.

Inhibition of quorum sensing by bacterial virulence inhibition

- QS inhibitors have been synthesized and have been isolated from several natural extracts such as garlic extract.
- QS inhibitors have shown to be potent virulence inhibitor both in *in-vitro* and *in-vivo*, using infection animal models.

Strategies for quorum sensing inhibition

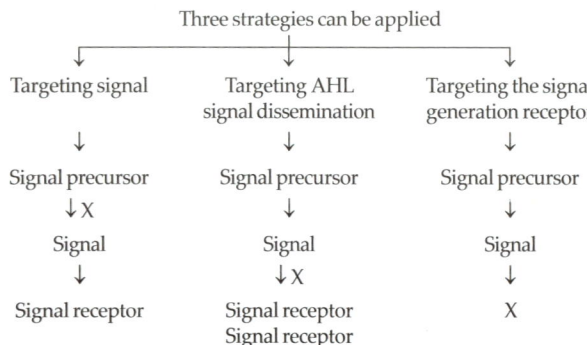

Three strategies can be applied

Targeting AHL signal dissemination

QS molecules can be degraded by:

- Increasing pH (> 7): as at higher pH, AHL molecules undergo lactonolysis in which its biological activity is lost.

- At higher temperature AHL undergoes lactonolysis

- Some plants infected by pathogenic bacteria *E. carotovora*, increase the pH at the site of infection, resulting in lactonolysis of AHL molecules

- Some bacteria produces lactonolysing enzymes, such as AiiA, e.g. *Bacillus cereus, B. thuringiensis*.

Quorum sensing inhibitors (QSIs)

- Garlic extract

- 4-nitro-pyridine-*N*-oxide (4-NPO)

These two QSIs also significantly reduced *P. aeruginosa* biofilm tolerance to tobramycin treatment as well as virulence in a *Caenorhabditis elegans* pathogenesis model.

Other QSIs

- Used bacterial mutant strain

- *Chromobacterium violaceum* (CV 26)

- Violacein synthesis by bacteria.

Bioassay

Media – 0.55 yeast extract, 1% tryptone, filter paper (0.22 mm), Sea water, QS blockers (FUR1, FUR2), Spectrophotometer.

Concentration of N-hexanoyl homoserine lactones (HHL) 3.7 to 10.8 M, Incubated at 27pC for 16 hours.

No QS blocker + HHL = Positive control; No QS blocker + No HHL = Negative control.

Turbidty measurance at 660 nm.

Result – Violacein inhibited in presence of HHL. FUR1 & FUR2 inhibited bacterial QS at concentrations of 10^{-3} – 10^{-5} M. Violacein formation reduced upto 90%.

Quorum sensing controlled processes

Bioluminescence	It occurs in various marine bacteria such as *Vibrio harveyi* and *Vibrio fischeri*. Takes place at high cell density
Biofilm formation	It is compact mass of differentiated microbial cells, enclosed in a matrix of polysaccharides. Biofilm resident bacteria are antibiotic resistant. Quorum sensing is responsible for development of thick layered biofilm
Virulence gene expression	QS upregulates virulence gene expression
Sporulation	QS upregulates spore-forming genes in *Bacillus subtilis*
Competence	It is ability to take up exogenous DNAQS Increase competence in *Bacillus subtilis*

Applications of Quorum Sensing

A promising application of QS is **Bioluminescence**. Most studied example is *Vibrio harveyi*.

Commonly forms symbiotic relationship with some fishes. These marine animals carry a specialized organ called light organ, in which bacteria like *V. harveyi* are housed.

$$FMN + NADH + H_+ \xrightarrow{\text{FMN Reductase}} FMNH_2 + NAD^+$$

$$FMNH_2 + RCHO + O_2 \xrightarrow{\text{Luciferase}} FMN + RCOOH + H_2O + Light$$

Isolation of Bioluminescence bacteria and its application in toxicity detection

A) Isolation of bacteria from environment

Media – Peptone 5g/l, Yeast extract 2g/l, Agar agar 15g/l, 75% sea water & 25% distilled water, pH 7, Temp. 22pC – 25pC for 24 hours. Gram nature and motility should be checked.

B) Enhancement in natural Bioluminescence

Media – peptone, yeast extract, glycerol, 22pC – 25pC for 18 hours, Temp. 25pC, 30pC, 37pC for 18 hours.

Effect of autoinducer

- 200 ml optimized sea water + media in two flasks and autoclaved 2% inoculum at 22pC – 25pC for 18 hours

- One flask centrifuged at 1000 rpm for 10 mins and one kept at roomtemperature.

- Supernatant added in another flask and luminescence is observed.

C) Study of effect of toxic chemicals on Bioluminescence

100 ml optimized sea water

↓

2% inoculum at 22pC – 25pC for 18 hours

↓

40 ml crude autoinducer (isolated)

↓

10 mins incubation

↓

0.002 gm/ml $CuSO_4.5H_2O$

↓

EC100 value is 0.1 gm/l; EC50 value is 0.05 gm/l

Other applications

- Gene therapy to check protein production
- Cancer cells, tissue damage deduction
- Inhibition of biofilm development by RNA III Inhibiting Peptide (RIP)
- Sewage Sludge therapy.

Future perspectives

- Isolation of proper QS inhibitors may replace the antibiotics
- Luminometer – An instrument for studying toxicity-bioluminescence relationship
- Development of a kit by using bacterial phenomena, which can give precise toxicity status of any water body.

REFERENCES

1. Donlan RM. Biofilms: Microbial Life on Surfaces. Emerging Infectious Diseases 2002; 8: 881-90.

2. Havarstein LS. Microbial Biofilm. Norwegian University of Life Sciences MDCCCIIX. http://www.umb.no/statisk/konferanser/lsh_nov25.pdf.

3. Kokare CR, Chakraborty S, Khopade AN, Mahadik KR. Biofilm: Importance and applications. Indian J Biotecnol 2009; 8: 159-68.

4. www.ncbi.nlm.nih.gov/pubmed/11544353.

5. Rutherford ST, Bassler BL. Bacterial Quorum Sensing: Its Role in Virulence and Possibilities for its Control. Cold Spring Harbor Perspectives in Medicine. perspectivesinmedicine.cshlp.org/content/2/11/a012427.

STORAGE AND PRESERVATION OF MICROORGANISMS

37

Chapter

- Prevention of multiplication of microorganism in formulated product thereby preventing spoilage or contamination.
- Stock culture – A culture of microorganism maintained solely to keep it viablefor s/c into frozen medium.

Aim:

- To preserve the typical biological characteristics of organism as they have capability to undergo variation.

Objectives:

- By teaching institute and type collecting agencies
- Use in testing new prepared batches of reagent.
- To allow work at length with culture isolation from clinical material that present difficulties and identification.

Preservation Program

- Stocks of preserved cultures are subjected for routine viability checks & survival assesment.
- Initially new cultures are checked one or two weekly then preserved for long term
- For unpredictable losses, advisable to preserve culture by more than one method
- Ampoules of stock culture-ampoules regularly taken from stock for use or distribution.

Protocol for Preservation of Cultures

- Culture purity check
- Preparation of the ampoules (labeling, sterilizing)
- Growth of the culture
- Suspension of the cells in preservation medium
- Dispensing of cell suspension into ampoules.
- Preservation (by method of choice)
- Ampoule stock storage
- Update ampoule stock records
- Ampoule recovery and testing (viability, purity, genetic stability)

National and International Culture Collection Centres

- ATCC (American Type Culture Collection Centre, Maryland, U.S.A.)
- NCIB (National Collection of Industrial Bacteria, Britain)
- DSM (Deutsche Sammlung von Mikroorganismen and Zelkulturen, Germany)
- NCTC (National Collection of Type Culture, London)
- MTCC (Microbial Type Culture Collection, Osaka Japan)
- MTCC (Microbial Type Culture Collection and Gene bank Institute of Microbial Technology, Chandigarh)
- ICIM (Indian Culture of Industrial Microorganisms, National Chemical Laboratory, Pune.

Repeated subcultures

- Periodic transfer on fresh, sterile media
- Cultures maintained by alternate cycles of active growth
- Aseptic technique without special equipment
- Frequency of transfer varies with organism
- Frequency of transfer can be lowered by using media with minimal nutrition which lowers metabolism of organism.
- Solid media should be chosen in preference to liquid media as there are higher chances of contamination in liquid media
- Slope cultures are used as oxygen sensitive bacteria may benefit from stab culture.
- Sealed tube hermetically, not with cotton wooled plug as media may dry out & culture may be lost.

Disadvantages

- Change of characteristics – may lost, reduced, intensified it occurs when intervals between transfer are short
- Contamination – if large amount of culture taken.
- Mislabeling – wrong names, label distorted
- Loss of culture – common with delicate organism, temperature fluctuation of incubator or refrigerator.

Paraffin Method

- Very simple and cost effective
- Infusion agar, blood agar slanted and covered with sterile, heavy paraffin oil or mineral oil
- Height 1 cm above the top of slanted surface
- Oil should be sterile and autoclaved at 15 psi for 1 hour
- Quality of oil is important as rancidity and toxic sub are harmful to organism
- Heating in dry oven at 110°C for 1 hour to drive out any entrapped moisture
- Layer of paraffin prevents dehydration and provides anaerobic condition
- Reduces metabolic activity as reduced growth and reduced oxygen tension
- Also with mineral oil stored at 0-5°C for 15-20 years
- Oil should be above the uppermost level of media so cant dry out and separate from wall of tube
- It is preferable to sterilise by hot air oven at 120°-150°C as during autoclaving moisture mix with oil, giving milky appearance.

Preservation of cultures under Paraffin oil

Advantages:

- Less chance of contamination
- Single colonies of mass culture can be used
- No seals such as rubber caps/waxes needed
- No special apparatus needed, e.g. centrifuge, vaccum pump, desiccator

Method:

- Required solid media e.g. infusion agar/blood agar slanted
- Covered with sterile heavy paraffin oil or mineral oil
- Height 1cm above the top of slanted surface
- Oil should be above the uppermost level of media
- Cannot dry out and separate from wall of tube or float to surface of oil
- Quality of oil is important, rancidity and toxic substances harmful to organism
- Sterilize oil by autoclave 15 psi for 1 hour
- Heating in a drying oven 110°C for 1 hour to drive out any entrapped

Storage in soil

- Inoculating 1 ml suspension of spore into soil
- Which is autoclaved twice and incubated at room temp for 5-10 days
- Initial growth period allows fungus to use moisture and then grow as dormant
- Bottles are stored in refrigerator

- Examples are *Fusarium, Penicillium, Rhizopus, Aspergillus, Alternaria* species.

Storage in Silica Gel

- Bacteria, yeast an be stored in silica gel powder at low temp for 1-2 years
- Finely powered, heat sterilised, cooled silica powder is mixed with thick suspention of cell
- Quick dessicate at lower temp so that cell can can remain viable for longer period.

Cold storage

- Live cultures are preserved at 4°C
- Low metabolic rate at lower temp reduce growth.
- Short term storage as toxins produced slowly which can kill microbes.
- For the preservation of blood, serum, culture media, cultures, vaccines, etc.
- Viruses require low temperature, so commercial type of deep freeze refrigerators working at -100°C to -400°C.

Storage by freezing

- Thick bacterial suspension stored at -30°C
- Storage in liquid nitrogen at -196°C
- Freezing and thawing cycle
- As water is removed from cell during freezing as ice, concentration of electrolytes in unfrozen water is more; this too may be harmful.
- Cultures frozen in presence of cryoprotectant have reduced damage from ice crystal, e.g. Glycerol or DMSO (dimethylsulphoxide)
- Two ml of glycerol solution added to agar slant cultures; shaking causes emulsification
- Emulsion added to ampoule containing 5 ml of culture
- Ampoules placed in methylated spirit and CO_2 and stored at -70°C rapidly; then placed in deep freezer at -40°C to use as stock culture
- Ampoules are kept in water bath at 45°C before plate culture
- Liquid nitrogen storage speed of recovery fast and better but disadvantage is cost of apparatus and supply of liquid nitrogen
- Risk of explosion when ampoules are brought to room temp
- Possible contamination of liquid nitrogen.

Storage by freeze drying

- It is multistage process
- Freezing temporary to stop metabolic activity
- Removal of water without thawing (sublimation)
- Ends as dried product

- Sealed under vacuum and stored at room temperature
- Rapid freezing required as slow freezing causes high salt concentration and denaturation of protein and death of organism.
- Liquid frozen in shallow layers with large surface for evaporation.

Two types

1) Centrifugal
2) Shelf

- Freezing by evaporation under vacuum
- Centrifugation prevents frothing by reducing surface area
- Advantages:
    - Lower chances of contamination
    - Viability and integrity of ampoules prevents changes in pressure and temperature
- Disadvantage: High cost of commercial equipments

Storage by drying

- Applicable for organisms which are sensitive to freezing.
- Drying from liquid state
- Paper disk – thick suspension of bacteria placed over sterile, thick absorbent paper dried over phosphorus pentoxide in a desiccator under vacuum
- Gelatin disc – thick suspension of bacteria added to nutrient gelatin placed over absorbant paper.
- Predried plugs – thick bacterial suspension placed on sterile cellophane or predried plug of peptone, starch, dextran
- L-drying – drying from liquid state by using vacuum, desiccator, water bath used for vesicular arbusculer mycorrhizal fungi.
- By spreading small suspension over large surface area
- Freezing prevented by restricting water vapor flow by inserting cotton wool plug or by controlling vacuum by means of valve
- Immersion of ampoules in water bath.

Lyophilization

Definition: A process used for preserving biological material, by removing the water from the sample, which involves first freezing the sample and then drying it, under a vacuum, at very low temperatures

Lyophilization, or freeze drying of bacterial cultures stabilizes them for long-term storage while minimizing the damage that may be caused by strictly drying the sample. Many microorganisms survive well when freeze-dried and can be easily rehydrated and grown in culture media, after prolonged periods of time in storage. Lyophilization is also used in the biotechnology and biomedical industries to preserve vaccines, blood samples, purified proteins and other biological material.

Preservation of fungi

- Storage of fungi – shows peculiarities like transition to pleomorphism/contamination by bacteria/mites or other fungi

Methods of preservation:

- Subculturing – Troublesome; Time consuming; Irregularity leads to loss of viability
- Covering with mineral oil
- Storage in distilled water
- Lyophilization
- Stock culture
    - Source of material of quality control
    - For comparative identification
    - For production of metabolites
    - Teaching collection.
- For longer period more than 2 methods are used as chances of loosing stain is minimized.
- Repeated sub culture on Saboraud's dextrose agar after a gap of few weeks.

References

1. Collee JG, Marion BP, Fraser AG, Simmons A. Mackie and McCartney's Practical Medical Microbiology. 14th Ed. Churchill Livigstone, U.K. 1996.

2. Koneman EW, Allen SD, Janda WM, Schreckenberger P, Winn Jr. WC. In Color Atlas & Textbook of Diagnostic Microbiology. 6th Ed. Lippincott, Philadelphia. 2006.

3. Gulati AK, Pal D. Diagnostic and Practical Microbiology. 1st Ed. Medical Allied Agency, Calcutta. 1985.

QUALITY ASSURANCE AND QUALITY CONTROL IN MICROBIOLOGY

38
Chapter

Quality is degree of congruence between expectation and realization.

WHO definition: Quality means meeting the predetermined requirements of users for a particular substance or service.

Total Laboratory Quality Program

- **Total Quality Management (TQM)** – It evolved as an activity to improve patient care by having the laboratory monitor its work to detect deficiencies and subsequently correct them.

- **Continuous Quality Improvement (CQI) or Performance Improvement (PI)** – CQI and PI went a step further by seeking to improve patient care by placing emphasis on not making mistake in the first place. CQI and PI advocate continuous training to guard against having to correct deficiencies.

- **Quality Control (QC)** – Control of errors in the performance of tests and verification of test results. It must cover all aspects of every procedure within the department. It must be practicable, achievable and affordable.

- **Quality Assurance (QA)** – Total process whereby the quality of laboratory reports can be guaranteed. "The right result, at the right time, on the right specimen from the right patient, with the result interpretation based on correct reference data and at the right price".

- **Internal Quality Control (IQC)** – Set of procedures undertaken by the staff of a laboratory for continuously and concurrently assessing laboratory work and the emergent results.

- **External Quality Assurance (EQA)** – System of objectively assessing the laboratory performance by an outside agency.

QA = IQC + EQA

	IQC	EQA
Nature	Concurrent & Continuous	Retrospective & Periodic
Performed by	Laboratory Staff	Independent Agency
Objective	Release Of Reliable Results	Ensure Inter-laboratory
	On Day To Day Basis – Precision	Comparability – Quality

Components of a QA Program

- Personnel with adequate training and experience
- Proper specimen collection
- Employment of techniques with precision and accuracy
- Proper performance of tests
- Efficient processing of results
- Reagents and equipments of good quality
- Methods of detecting errors
- Corrective steps when analyses go out of control
- Continuous training of staff
- Documentation
- Co-ordination
- Timely feedback

Facilities at Laboratory

- **Separate space for:**
  - Sample Collection
  - Sample Analysis
  - Storage of Samples, Reagents, Chemicals, Stationery, Records, etc.
  - Washing
  - Media Preparation & Autoclaving
  - Seminar Room & Library
  - Staff Room & Toilets
- **Adequate and qualified staff**

Standard Operating Procedure Manual (SOPM)

- Each section of the laboratory should have a copy of SOP which should be easily accessible to all
- SOP should be available on the work bench area
- It should be reviewed annually
- SOP should contain only those procedures which are currently in use
- Any change in the SOP must be documented by recording it and having it duly signed by the Chief of the Laboratory.

Essential components of SOPM

- Abbreviated administrative structure diagram
- Laboratory safety instructions
- Specimen collection
- Inoculation procedure and Details of procedures
- Differential tests and Antibiotic susceptibility testing (AST)
- Serological testing
- Reference to higher laboratory
- Quality control
- Reporting

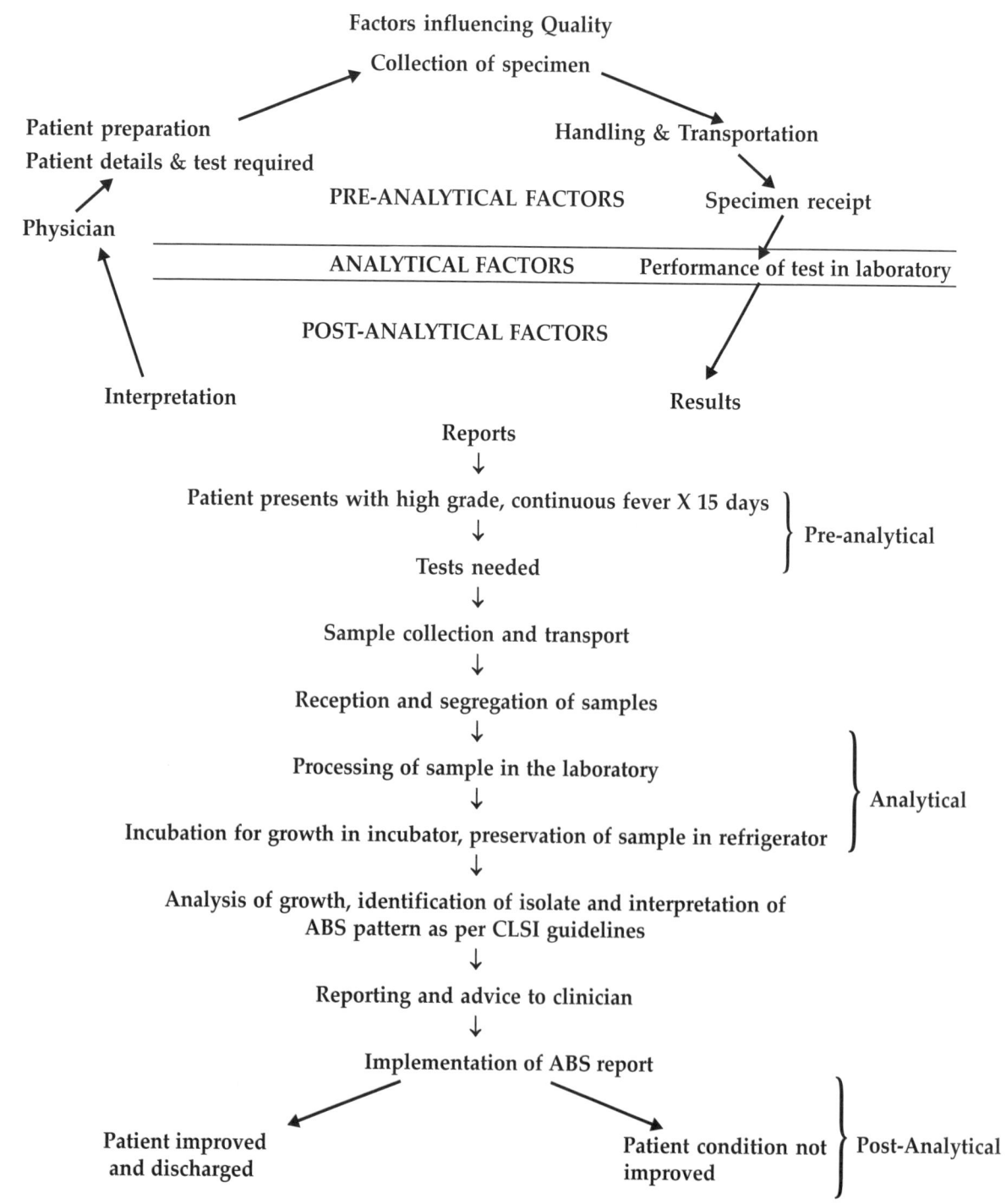

Pre-analytical factors

- Right investigations
- Right Sample
- Right Technique
- Right Laboratory
- Right Transportation
- Right Quantity
- Right labeling

Test needed: Blood Culture and Sensitivity

Requisition form:

- Name: Mr. XYZ
- Age: 50 yrs.
- Sex: Male
- Registration no.: 15/24351
- Investigation needed: Blood C/S
- Brief S/S: c/o high grade, continuous fever X 15 days
- Treatment details: Patient not on any antimicrobial therapy
- Signature of physician: Dr. ABC
- Date: 21/06/2015

Sample transport: Immediate transport in upright position to the laboratory

Sample receipt: Sample is received at the laboratory and all details of the patient are counterchecked, both on the requisition form and the sample itself.

Specimen Collection and Transport

- The laboratory is responsible for providing instructions for the proper collection and transport of specimens.
- These instructions should be available to the clinical staff for use whenever and wherever the specimens are collected.

Written Collection Instructions

- Test selection criteria
- Patient selection criteria
- Timing of specimen collection (e.g. before antimicrobials are administered)
- Optimal specimen collection site
- Approved specimen collection method
- Specimen transport medium
- Specimen transport time and temperature

- Specimen holding instructions if it cannot be transported immediately (e.g. hold at 4°C for 24 hours)
- Availability of test (on-site or sent to reference laboratory)
- Hours during which test is performed (daily or batch)
- Turn-around time
- Result reporting procedures

Information that should be filled

- Patient name
- Age, gender and occupation
- Hospital or laboratory number
- Ordering physician
- Whether the patient is receiving antimicrobial therapy
- Suspected agent or syndrome

Criteria for unacceptable specimens

- Unlabeled or mislabeled specimens
- Use of improper transport medium/container
- Excessive transport time
- Improper temperature during transport or storage
- Improper collection site for test
- Specimen leakage out of transport container
- Sera that are excessively hemolyzed, lipemic or contaminated with bacteria

Transportation of Specimens

- For short distance (by hand) – specimen placed upright in appropriate racks
- For long distance – placed in three containers
- Most specimens should be processed within 1-2 hours after collection

Triple Packaging System

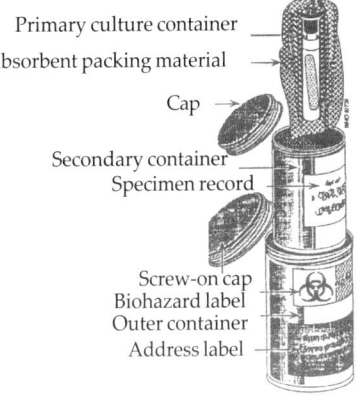

Primary culture container
Absorbent packing material
Cap
Secondary container
Specimen record
Screw-on cap
Biohazard label
Outer container
Address label

Analytical Factors

- **Reagents stability, integrity and efficiency:** Stable; Efficient; Desired quality; Continuously available; Validated

- **Equipment reliability:** Meet technical needs; Compatible; User and maintenance friendly; Cost effective; Validated

- **Specificity and sensitivity of selected test:** Adequate sensitivity; Sufficient specificity; Cost effective; Compatible with available infrastructure and expertise; Interpretable; Meets the needs/ objectives; Validated

- **Procedural reliability:** Using Standard Operating Procedures

- **Documentation:** All the written policies, plans, procedures, instructions and records, quality control procedures and recorded test results involved in providing a service or the manufacture of a product

- **Use of appropriate controls:**
  - Internal: Laboratories, Calibrated against national standards
  - External: Supplied by manufacturer, National, International

- **Proficiency of personnel:** Education, Training, Aptitude, Competence, Commitment, Adequate number, CME, Supervision, Motivation.

Sample processing:

- Gram stain: Pus cells, Gram negative bacilli seen
- Culture: Using in-house prepared media:
  - Blood agar, Chocolate agar and MacConkey agar
  - Biochemicals for organism identification
- Antibiotic sensitivity using Kirby Bauer Disc Diffusion Method
- Incubation: incubated in an incubator at 37°C for 18 hours.

Post-analytical factors

- Accurate recording
- Range of normal values
- Age and sex related variation
- Turn-around time
- Availability of guidance.

Levy-Jennings Chart

- A graphical method for displaying control results and evaluating whether a procedure is in-control or out-of-control
- Control values are plotted versus time
- Lines are drawn from point to point to accent any trends, shifts, or random excursions.

Findings over Time

- Ideally should have control values clustered about the mean (± 2 SD) with little variation in the upward or downward direction

- **Imprecision** = Large amount of scatter about the mean. Usually caused by errors in technique

- **Inaccuracy** = May be seen as a trend or a shift, usually caused by change in the testing process

- **Random error** = No pattern. Usually poor technique, malfunctioning equipment.

Documentation

"If you have not documented it, you have NOT done it!"

- Name and serial number of instrument/media/ reagent

- Elements to be checked and kind of data to be collected

- Frequency of checking

- Record of data

- Comments on data

- Changes made to restore accuracy and precision, if any

- Signature with date of the person performing these tasks

Maintenance of QC records

- All QC results should be recorded on an appropriate QC form.

- Corrective action should be noted on this form. If temperature is adjusted or a biochemical test repeated, the new reading within the tolerance limits should be listed.

- In many laboratories the supervisor reviews and initials all forms weekly and the director then reviews each one monthly.

- QC records should be maintained for at least 2 years except those on equipment, which must be saved for the life of instrument.

Personnel

- It is laboratory director's responsibility to employ sufficient qualified personnel for the volume and complexity of the work performed.

- Documentation of competency

- Training of staff twice a year

- Continuing education program should be provided

- All documentation should be maintained in personnel file.

Staff and qualification

Supervisory personnel

- Small and Medium laboratories – MBBS/MSc in concerned specialty with at least 5 years experience

- Large laboratories – MD/PhD in respective discipline

Technical personnel

- Graduate in MLT

- Science graduate with 1 year experience in medium sized laboratory.

- Diploma in MLT with 2 years experience in medium sized laboratory.

Type of laboratories

- Small lab: deals with up to 50 patients/day

- Medium lab: 51-500 patients/day

- Large lab: > 500 patients/day 3

Quality Control of Laboratory Materials

The reason for bad analysis is the poor quality of the following materials:

- A specimen that is not collected properly or preserved adequately

- Contaminated or deteriorated reagents

- Glassware that is not properly cleaned

- A pipette that is not calibrated before use

- An instrument that is not calibrated or maintained properly.

Proper use of reagents and standards

- Store all aqueous solutions in plastic bottles that can be tightly closed.

- Colored aqueous solutions are preferably stored in amber-colored plastic bottles.

- Never store organic liquids and solutions in plastic bottles.

- Keep all reagents and standards requiring refrigeration in the refrigerator.

- Never introduce a pipette, a glass rod, or any other substance into the reagent or standard bottle.

Quality control of stains

- Each lot of newly prepared stain tested with positive and negative controls

- Regular testing done:
  - Weekly: Gram stain
  - Each use: Fluorescent stain
  - Each day of use: Z-N stain and other stains

- Discard stains if outdated or deteriorated.

Quality control of stains

Stain	Control organism/material	ATCC No.	Expected result
Ziehl-Neelsen	*Mycobacterium sp.; E. coli*	25177; 25922	Red bacilli; Blue bacilli
Acridine orange	*E. coli; S. aureus*	25922; 25923	Fluorescent bacilli; Fluorescent cocci
PAS	Dermatophytes	-	Magenta: fungi; Pink: background
Gram	*E. coli; S. aureus*	25922; 25923	Gram –ve bacilli; Gram +ve cocci
Iodine solution	Stool with cysts	-	Visible cyst nuclei
Romanowsky stain	Thin film blood smear	-	Distinct staining of RBCs and WBCs

Quality control of reagents

- QC done with each new batch - daily for catalase, oxidase, coagulase tests; weekly for bacitracin, optochin, ONPG, X and V factors

- Mycobacteriology: each day of use

- Mycology: weekly with positive controls – each new batch of serum for germ tube test

- Antisera: each new batch tested in parallel with previously tested batch using positive and negative controls

- Antigen detection kits: each day of use.

Procedure	Control	Expected reaction
Catalase	*S. aureus; Streptococcus species*	Bubbling (+); No bubbling (-)
Coagulase	*S. aureus; S. epidermidis*	Clot in 4 hrs (+); No clot (-)
Oxidase	*P. aeruginosa; E. coli*	Purple colour in 30 secs (+); No change (-)
Bacitracin	*S. pyogenes; S. fecalis*	Zone of inhibition (+); No zone of inhibition (-)
Optochin	*S. pneumoniae; S. viridans*	Zone of inhibition (+); No zone of inhibition (-)
Indole	*E. coli; K. pneumoniae*	Red ring (+); Yellow ring (-)

Commercially Prepared Culture Media

The CLSI subcommittee has published a list of media that do not require re-testing in the user's laboratory if they have been purchased from manufacturers who follow CLSI guidelines.

User-prepared and nonexempt, commercially prepared media

- QC forms for user-prepared media should contain:
  - The amount of prepared
  - The source of each ingredient
  - The lot number
  - Sterilization methods
  - The preparation date
  - The expiration date (Usually 1 month for agar plate and 6 month for tube media)
  - The name of preparer
- All user prepared culture media also should be checked for:
  - Proper color
  - Depth
  - Smoothness
  - Hemolysis
  - Excessive bubbles
  - Contamination

Medium	Incubation	Control	Result
Blood Agar	24 hrs, CO_2	S. aureus S. pneumoniae	Beta haemolytic Alpha hemolytic
Chocolate Agar	24-48 hrs, CO_2	H. influenzae N. gonorrhoeae	Growth Growth
MacConkey Agar	24 hrs	E. coli P. mirabilis	Pink colony Colorless colony without swarming
Simmon's citrate agar	48 hrs	E. coli K. pneumoniae	No growth Growth, blue
Mannitol salt agar	24-48hrs	S. aureus S. epidermidis	Yellow Red
Christensen's urease agar	4-18 hrs	P. vulgaris E. coli	Pink No change
Xylose lysine deoxycholate agar	24 hrs	Salmonella Shigella E. coli	Red colony with black centre Red colony Yellow colony
Bile esculin agar	48 hrs	E. fecalis S. pyogenes	Growth, blackening No growth, no blackening

Sterility Check

- A representative sample of the lot should be test for sterility.
- 5% of any lot is tested when a batch of 100 or fewer units is prepared.
- A maximum of 10 units are tested for large batches.
- Sterility is routinely checked by incubating the medium for 48 hours at the temperature at which it will be used.

Contamination parameter

- Media autoclaved completely, incubate 2 plates at 37°C for 24 hours. If contaminated, incubate 2 more plates. Still contaminated, discard whole batch
- Media containing non-sterilized additives e.g. blood, serum etc., incubate entire batch for 24 hours. If contamination exceeds > 10% (WHO > 3-5%), discard the entire batch

Use of Stock Cultures

- To operate a quality control program, stock culture must be maintained by all laboratories. They are available from many sources
- Commercial sources
- Proficiency testing
- Patients isolates
- American Type Culture Collection (ATCC)
- When quality control testing appears have failed, it is usually the stock culture rather than the test itself that has failed
- Organisms may mutate with repeated sub culturing. For best results, a stock culture should be grown in a large volume of broth ,then divided among enough small freezer vials to last a year
- With this technique, a new vial can be removed from the freezer weekly so that organisms do not have to be continually subcultured

- Media selection for freezing should not contain sugar. If organism utilize sugar while being maintained, the acid products that result may kill organism with time.

Popular Media for Stock Cultures

- Schaeler broth with glycerol
- Chopped meat (anaerobes)
- Tryptic Soy agar with mineral oil overlay (at room temperature)
- Cystein-tryptic agar (CTA) without carbohydrates
- 10% skimmed milk
- Trypticase Soy Broth with 15% glycerol
- Nonfastidious (rapidly growing), aerobic bacterial organisms can be saved up to 1 year on TSA slant.
- Long –term storage of aerobes or anaerobes can be accomplished either by lyophilizartion (freeze drying) or freezing at -70°C.
- Frozen, non-fastidious organism should be thawed, re-isolated and re-frozen every 5 years.
- Fastidious organisms should be thawed, re-isolated and re-frozen every 3 years.

Functional checks for equipments

- **Calibration:** Process which is applied to quantitative measuring or metering of equipment to assure its accurate operation throughout its measuring limits.

- **Validation:** Steps taken to confirm and record the proper operation of equipment at a given point of time in the range in which tests are performed.

Performance Checks

- Instrument
- Equipment logs should contain the following information
- Instrument name, serial number, and date put use
- Procedure and periodicity (daily, weekly, monthly for routine function check)
- Acceptable performance ranges
- Instrument function failure ,including specific details of steps taken to correct the problems (corrective action)
- Date and time of services requests and response
- Date of routine preventive maintence (PM) which should follow manufactures recommendations
- Maintenance records should be retained in the laboratory for the life of instrument.
- Specific guidelines regarding periodicity of testing for autoclaves, biological safety cabinets, centrifuges, incubators, microscope, refrigerators, freezers, water baths and other microbiology laboratory should be checked in standard reference books and followed accordingly.

Equipment	Procedure	Schedule	Tolerance limit
Refrigerator	Temp. check	Daily	2-8°C
Freezer	Temp. check	Daily	± 5°C
Incubator	Temp. check	Daily	± 1°C
Water bath	Temp. check	Daily	± 0.5°C
Anaerobic jar	Methylene blue; *P. aeruginosa*	Each use	Colourless; No growth
Serology	Count of revolutions/min	Each use	180 ± 10 rpm
Autoclave	Bowie dick strips; Spore strips	Each run monthly	Colour change; No growth
Hot air oven	Spore Bio-safety hood strips	Weekly	No growth
Centrifuge	Revolution check by tachometer	Monthly	-
Microscope	Stage and lenses	Each use	-
Pipettes	Volume delivery check 10 times	Initially	-
Balances	Checked against known weights	Annual	-
Bio-safety hood	Air velocity check	6 monthly	Flow 50 ± 5 ft/min

Quality Control in AST

- According to CLSI, Kirby Bauer method is recommended for antimicrobial testing by disc diffusion.
- Standard strains:
  - *Staphylococcus aureus* ATCC 25923
  - *Escherichia coli* ATCC 25922
  - *Pseudomonas aeruginosa* ATCC 27853

- **Variables to be monitored**
  - Antibiotics potency (disc size 6 mm)
  - Storage of discs (-20°C)
  - Inoculum standardization (0.5 McFarland)
  - Standard methodology
  - Incubation temperature (35°C) and time (16-18 hours)
  - Measure zone size precisely and interpret by referring to CLSI guidelines.

Factors influencing zone size in AST by disc diffusion method

Composition of medium	Affects rate of growth, diffusion of antibiotics and activity of antibiotics
Acidic pH of medium	Aminoglycosides, erythromycin zones are decreased
Alkaline pH of medium	Tetracycline, novobiocin, methicillin zones are increased
Depth of the agar medium	Thin media yield excessively large inhibition zones and vice versa
Inoculum density	Larger zones with light inoculum and vice versa
Timing of disc application	If after application of disc, the plate is kept for longer time at room temperature, small zones may form
Incubation time	Ideal 16-18 hours; less time does not give reliable results
Incubation in the presence of CO_2	Increases zone size of tetracycline and methicillin
Addition of thymidine to medium	Decreases activity of trimethoprim
Chelating agents such as calcium, magnesium and iron	Decrease diffusion of tetracycline and gentamicin
Proper spacing of the discs	Avoids overlapping of zones
Potency of antibiotic discs	Deterioration in contents leads to reduced size
Temperature of incubation	Larger zones are seen with temperatures < 35°C

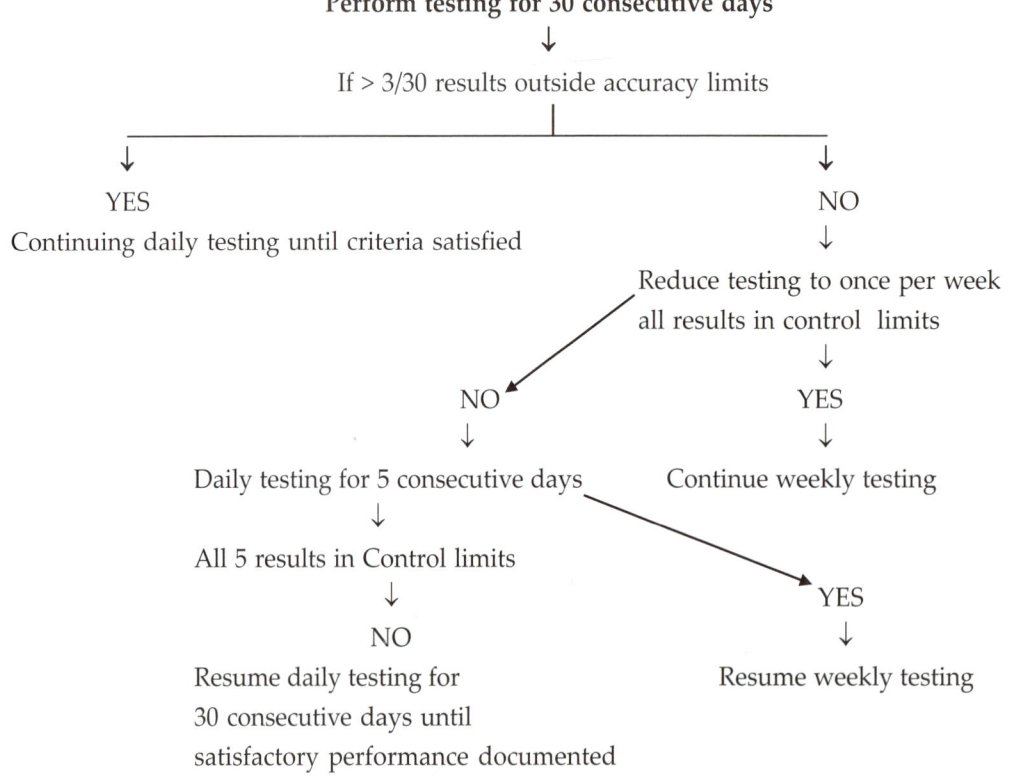

Perform testing for 30 consecutive days

↓

If > 3/30 results outside accuracy limits

YES — Continuing daily testing until criteria satisfied

NO ↓

Reduce testing to once per week all results in control limits

↓

NO ↓ — Daily testing for 5 consecutive days

↓

All 5 results in Control limits

↓

NO — Resume daily testing for 30 consecutive days until satisfactory performance documented

YES ↓ — Continue weekly testing

YES ↓ — Resume weekly testing

Antisera

- Lot number
- Date received
- Condition received
- Expiry date must be recorded for all shipments of antisera.
- In addition, the antisera should be dated when opened.
- New lots must be tested concurrently with previous lots.
- Testing must include positive and negative controls.

Quality control in Serology

- Most appropriate test must be chosen
- SOPM should be available all the time
- Hemolysed blood samples are not suitable for serological tests
- Sera to be used as controls should be kept sterile to avoid deterioration
- Each procedure should have positive and negative controls

External quality control

- Provide laboratory management with an insight into their performance

- Improve both local and national standards
- Reveals unsuspected area of difficulty
- Provides an educational stimulate for improvements
- Acts as a check on the efficacy of internal quality control procedures
- Demonstrates to colleagues and customers a commitment to quality.

Audit

- Planned and documented activity performed in accordance with written procedures and checklists to verify by investigation, examination and evaluation of objective evidence, that applicable elements of QA program have been developed, documented and implemented.
- Internal audits: carried out by laboratory's own staff.

Accreditation

- Procedure by which an authorative body gives formal recognition that a laboratory is competent to carry out specific tasks.
- Mandatory in developed countries, voluntary in India.

- Require accreditation board (ISO, NABL), set of standards, inspectors/assessors.

Procedure

Application by laboratory → Acknowledgement by NABL board → Assessors → Assessment report to NABL → Feedback to laboratory → Corrective measures by laboratory and reapplication to NABL → Approval and issue of accreditation certificate.

Preventive measures against laboratory-acquired infections

- Protect workers, patients and cultures
- Perform adequate sterilization before washing or disposing waste
- Provide safety hoods
- Ensure that tissues are handled and disposed of properly
- Promote regular hand washing and cleaning of bench tops
- Ensure use of gloves
- Provide mechanical pipetting devices
- Provide special disposal containers for needles and lancets.

Quality control of equipments

Equipment	Routine care	Monitoring	Technical maintenance and inspection
Anaerobic jar	Clean inside of jar each week Reactivate catalyst after each run (160°C, 2 h) Replace catalyst every 3 months	Use methylene blue indicator strip with each run Note and record decolorization time of indicator each week	Inspect gasket sealing in the lid weekly
Autoclave	Clean and change water monthly	Check and adjust water level before each run Record time and temperature or pressure for each run Record performance with spore strips once a week	Every 6 months
Centrifuge	Wipe inner walls with solution weekly or after breakage of glass tubes or spillage	—	Replace brushes annually
Hot-air oven for sterilization of glassware	Clean inside monthly	Record time and temperature for each run	Every 6 months
Incubator	Clean inside walls and shelves monthly	Record temperature at the start of each working day (allowance 35 ± 2°C)	Every 6 months
Microscope	Wipe lenses with tissue or lens paper after each day's work Clean and lubricate mechanical stage weekly Protect with dust cover when not in use	Check alignment of condenser monthly	Annually
Refrigerator	Clean and defrost every 2 months and after power failure	Record temperature on first day of each week (allowance 2-8°C)	Every 6 months
Water-bath	Wipe inside walls and change water monthly	Check water level daily Record temperature on first day of each week (allowance 54-57°C)	Every 6 months

REFERENCES

1. Forbes BA, Sahm DF, Weissfeld AS. In Bailey & Scott's Diagnostic Microbiology. 12th Ed. Mosby, Inc. An Affiliate of Elsevier Science. 2007.

2. Collee JG, Marion BP, Fraser AG, Simmons A. Mackie and McCartney's Practical Medical Microbiology. 14th Ed. Churchill Livigstone, U.K. 1996.

3. College of Physicians & Surgeons of Saskatchewan - Laboratory Quality Assurance Program, Procedures/ Guidelines for the Microbiology Laboratory.https://www.cps.sk.ca/CPSS/Programs_and_Services Laboratory_Quality_Assurance.aspx?

   LabQualityCCO=Laboratory%20 Quality%20Assurance%20Program%20Overview.

4. Arora DR. Quality assurance in Microbiology. Indian J Med Microbiol 2004; 22: 81-6.

5. Practice of Quality Assurance in laboratory medicine in developing countries. In Health laboratory services in support of primary health care in developing countries. World Health Organization, New Delhi. 1994; pp 77-137.

BIOSAFETY MEASURES

39
Chapter

The application of combination of laboratory practice and procedures, laboratory facilities and safety equipment used when working with potentially infectious micro-organisms.

Biosafety measures should be followed against:

- All infectious organisms (e.g. bacteria, viruses, fungi, parasites, prions, rickettsiae, etc.)
- Human blood, tissues, fluids, cell lines
- Potentially infectious animals or animal tissue
- Certain types of recombinant DNA.

Goal
Provide the highest practical protection and the lowest practical exposure.

Identifying Risk
- Understanding the biology of the agent
- Susceptibility and transmission within the host
- Hazards associated with equipment and procedures

Most common modes of exposure
Inhalation
- Infected aerosols or droplet nuclei or dust
- Aerosol is a cloud of small droplets of liquid (< 0.1mm in diameter) in air
- All laboratory procedures are capable of generating infectious aerosols:
    - Withdrawal of a loopful from a broth culture
    - Mixing of suspension with a loop
    - Removal of wet stopper or cotton plug
    - Expulsion of residual fluid from pipette
    - Centrifuging overfill tubes
    - Breakages in centrifuge
    - Pouring
    - Opening culture vials
    - Breaking culture plates or tubes

Ingestion
- Splashing on lips
- Touching mouth with contaminated fingers, pen or pencil

- Consumption of food at work place
- Mouth pipetting of fluids

Inoculation
- Splashing on eye
- Rubbing with contaminated fingers
- Incision with sharp instrument or broken glass
- Animal and insect bite; scratches

Classification of Organisms into Risk Groups is based on:
- Pathogenic potential
- Mode of transmission
- Epidemiological consequences of escape from the laboratory
- Relation to host susceptibility (preventive measures and effective treatment)

RISK GROUP 1
- Unlikely to cause disease in humans or animals
- Low individual or community risk
    - *E. coli*
    - *B. subtilis*
    - *Naegleria gruberi*

RISK GROUP 2
- May cause disease but not serious
- Moderate individual risk
- Low community risk
- Treatment and preventive measures available
    - **Bacteria**: *Staphylococcus, Streptococcus, Clostridium difficile*
    - **Viruses**: Hepatitis A, B virus, measles, mumps
    - **Parasite**: *Toxoplasma species*
    - **Fungi**: *Candida species*
      Human body fluids, blood, tissues

RISK GROUP 3
- May cause serious disease
- High individual risk
- Low community risk

- Treatable and preventive measures available
    - **Bacteria**: *B. anthracis, Brucella abortus, Brucella canis, Brucella suis, Mycobacterium sp.; M. tuberculosis, Coxiella burnetii, Chlamydia psitticae, Rickettsiae, Salmonella typhi and paratyphi, Shigella dysenteriae type*1, *Y. pestis, Burkholderia mallei* and *pseudomallei*
    - **Viruses**: Hepatitis C,D,E, Dengue type 1-4, Yellow fever, HIV, Rabies, Chikungunya, Japanese B encephalitis,West Nile fever virus, CJD, Kuru
    - **Parasites**: *P. falciparum, E. granulosis, L. brasiliensis and donovani, Naegleri fowleri, T. solium, Trypanosomes*
    - **Fungi**: *P.marneffi, Paracoccidides brasiliensis, Coccidioides immitis, Histoplasma capsulatum, Blastomyces dermatitidis.*

RISK GROUP 4

- Serious or fatal disease, often not treatable
- Easy transmission
- High individual and community risk
    - Filoviridae: Ebola virus, Marburg virus, Reston virus
    - Arenaviridae: Lassa fever virus
    - Poxviridae: Variola major and minor
    - Flaviviridae: Kyasanur forest disease

Biosafety practices provide protection to:

- Worker
- Product
- Lab support personnel
- Environment

Principles of Biosafety

These include:

- Standard microbiological practices
- Safety equipment (Primary barrier) i.e. Biological Safety Cabinet (BSC) and Personal Protective Equipment (PPE)
- Lab facilities (Secondary barrier) i.e. controlled access, airlocks, specialized ventilation systems to ensure directional air flow, air treatment systems to decontaminate exhaust air
- Building (Tertiary barrier)

Standard microbiological practices

- Restricted access
- Handwashing
- Prohibit eating, drinking, smoking, applying cosmetics, nail-biting
- Prohibit mouth pipetting

- Safe handling of sharps
- Use of Personal Protective Equipment
- Decontaminate work surface daily
- Maintain insect and rodent programme

Personal protective equipment

- A solid front, back-closing laboratory gown provides better protection
- Gloves should be pulled over the wrists of the gown rather than worn inside.
- Mask
- Safety glasses
- Shoe covers, boots
- Respirators, face shields

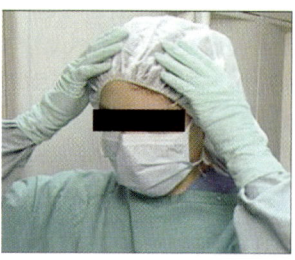

Biosafety Levels (BSLs)

- Different than the Risk Groups
- BSLs are used in risk management
- BSLs are ways to control the agent
- Once risk is assessed then the appropriate BSLs is determined

Access

- The international biohazard warning symbol and sign (Fig.1) must be displayed on the doors of the rooms where microorganisms of Risk Group 2 or higher risk groups are handled.
- Only authorized persons should be allowed to enter the laboratory working areas.
- Laboratory doors should be kept closed.
- Children should not be authorized or allowed to enter laboratory working areas.
- Access to animal houses should be specially authorized.
- No animals should be admitted other than those involved in the work of the laboratory.

Biosafety Level 1

- Non-pathogenic organisms
- Open bench - no containment
- Use good laboratory practices, waste disposal, and aseptic techniques

Biosafety Level 2

- Most pathogens encountered in clinical laboratory fall into risk group 2
- Basic lab practices, PPE
- Procedures likely to generate aerosols should be performed in BSC
- Lab facilities:
  - Restricted access to lab
  - Biohazard sign
  - Lockable doors
  - Bio Safety Cabinets (Class-2)
  - Eyewash readily available
  - Autoclave in proximity to lab
  - Disinfect benches after work

Biosafety Level 3

- Agents of high hazard to personnel and environment

- Lab facilities:
  - Controlled access
  - Bio Safety Cabinets
  - Respiratory protection needed
  - Double door entry
  - Directional inward airflow
  - Autoclave within
  - Separate building/isolated zone
  - Decontaminate spills promptly
- Lab personnel:
  - Strictly follow guidelines
  - Report incidents
  - Medical surveillance

Biosafety Level 4

- Total containment
- Lab facilities:
  - Airlock entry
  - Changing room
  - Entry and exit showers

Class III BSC, or positive pressure moon-suits in conjunction with class II BSCs
  - Double door autoclave
  - Supply and exhaust air filter
  - Bio-waste liquid sterilizer

Summary of biosafety level requirements

	1	2	3	4
Isolation of laboratory	No	No	Yes	Yes
Room sealable for decontamination	No	No	Yes	Yes
Ventilation :				
- inward ariflow	No	Desirable	Yes	Yes
- controlled ventilating system	No	Deirable	Yes	Yes
- HEPA-filtered air exhaust	No	No	Yes/No[a]	Yes
Double-door entry	No	No	Yes	Yes
Airlock	No	No	No	Yes
Airlock with shower	No	No	No	Yes
Anteroom	No	No	No	–
Anteroom with shower	No	No	Yes/No[b]	No
Effluent treatment	No	No	Yes/No[b]	Yes
Autoclove:				
- on site	No	Desirable	Yes	Yes
- in laboratory room	No	No	Desirable	Yes
- double-ended	No	Desirable	Yes	Yes
Biological safety cabinets	No	Desirable	Yes	Yes
Personal safety monitoring capability[c]	No	No	Desirable	Yes

a – Dependent on location of exhaust

b – Dependent on agents(s) used in the laboratory

c – Window, closed-circuit T.V., two-way communication.

Relation of risk groups to biosafety levels, practices and equipment

Risk group	Biosafety level	Lab type	Lab practices	Safety equipment
1	Basic biosafety level 1	Basic teaching, research	Good microbiological techniques (GMT)	None, open bench work
2	Basic biosafety level 2	Primary health services, diagnostic services, research	GMT plus protective clothing, biohazard sign	Open bench plus BSC for potential aerosols
3	Containment biosafety level 3	Special diagnostic services, research	As level 2 plus special clothing, controlled access, directional airflow	BSC
4	Maximum containment biosafety level 4	Dangerous pathogen units	As level 3 plus airlock entry, shower exit, special waste disposal	Class III BSC, or positive pressure suits in conjunction with classII BSCs, double ended autoclave, filtered air.

Biosafety cabinets

Containment and protection devices used in laboratories working with biological agents with a primary purpose of protecting laboratory personnel, product and environment.

Classification of BSCs

BSCs are classified according to air volumes which are re-circulated or exhausted:

- Class I
- Class II (A, B1, B2, B3)
- Class III

Class I Biosafety cabinet

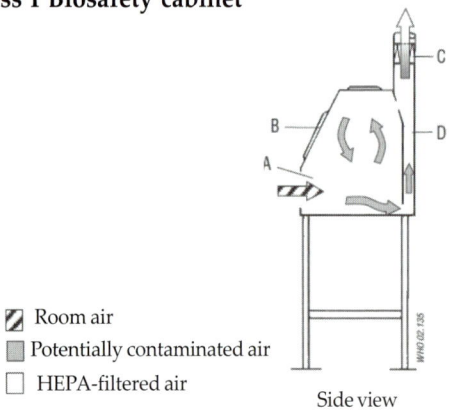

■ Room air
■ Potentially contaminated air
□ HEPA-filtered air

Side view

A, front opening; B, sash; C, exhaust HEPA filter; D, exhaust plenum.

- Room air is drawn in through the front opening at a minimum velocity of 75 lfpm
- Passes over the work surface
- Air from the cabinet is exhausted through a HEPA filter

- The HEPA filter may be located in the exhaust plenum of the BSC or in the building exhaust.
- Protection factor of 1.5×10^5, represents number of particles which if liberated into the air of the cabinet will not escape into the room.
- Unsterilized room air is drawn over the work surface through the front opening, therefore no product protection.

Class II Biosafety cabinet

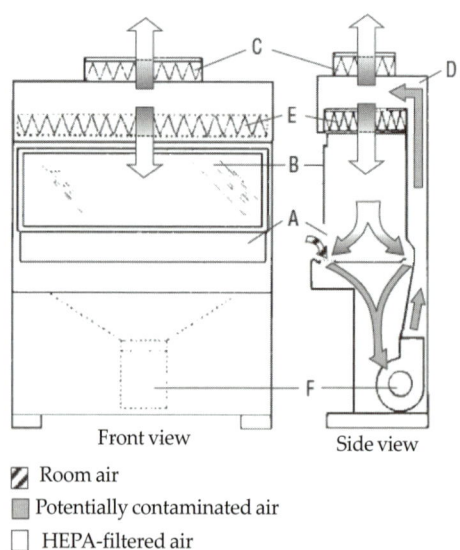

Front view   Side view

■ Room air
■ Potentially contaminated air
□ HEPA-filtered air

A, front opening; B, sash; C, exhaust HEPA filter; D, exhaust plenum; E, supply HEPA filter; F, blower.

- The Class II BSC differ from Class I BSCs by allowing only HEPA-filtered air (sterile) to flow over the work surface providing both product and personnel protection.

- They are classified into four types (A, B1, B2, B3), on the basis of construction type, airflow velocities and patterns, and exhaust systems.
- Used for working with infectious agents in Risk Groups 2 and 3 and with infectious agents in Risk Group 4 when positive-pressure suits are used.
- An internal fan draws room air into the cabinet through the front opening and into the front intake grill.
- Inflow velocity of air 75 lfpm.
- The supply air then passes through a supply HEPA filter before flowing downwards over the work surface.
- Air flows downwards and splits about 6–18 cm from the work surface, one half passing through the front exhaust grill, and the other half through the rear exhaust grill.
- Any aerosol particles generated at the work surface are captured and passed through the front or rear exhaust grills, providing product protection.
- The air is discharged through the rear plenum into the space between the supply and exhaust filters located at the top of the cabinet.
- Owing to the relative size of these filters, about 70% of the air recirculates through the supply HEPA filter back into the work zone; the remaining 30% passes through the exhaust filter outside.
- Are not suitable for work with low levels of volatile toxic chemicals and volatile radionuclides.

Class II type B1 BSC

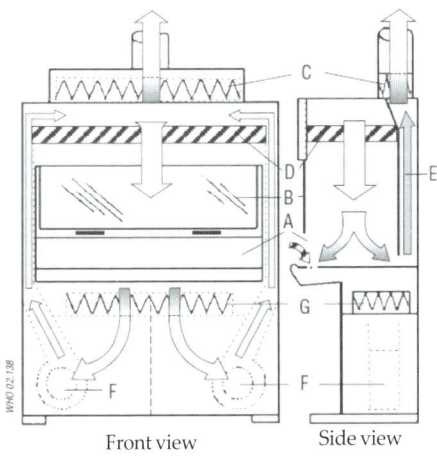

Front view Side view

- ▨ Room air
- ▨ Potentially contaminated air
- ☐ HEPA-filtered air

A, front opening; B, sash; C, exhaust HEPA filter; D, supply HEPA filter; E, negative-pressure exhaust plenum; F, blower; G, HEPA filter for supply air; Connection of the cabinet exhaust to the building exhaust system is required.

- Air enters the cabinet from the room through a front opening at a velocity of 100 lfpm

- Recirculate 30% of air through a sump and recirculating plenum passing through 2 HEPA filters before being exhausted through a HEPA filter into negative pressure ducts to the outside
- Because of minimal re-circulation, suitable for work with low levels of volatile toxic chemicals and trace amounts of radionuclides.

CLASS II TYPE B2 BSC

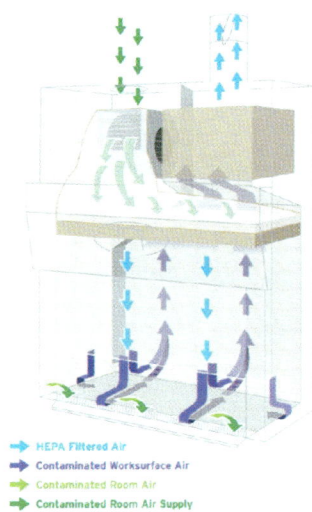

- ➡ HEPA Filtered Air
- ➡ Contaminated Worksurface Air
- ➡ Contaminated Room Air
- ➡ Contaminated Room Air Supply

Air inflow 0% recirculated vs 100% exhausted.

- Room air enters through front opening, drawn into cabinet plenum and then exhausted outside through HEPA filter and negative pressure duct system
- Simultaneously, room air enters through a second port and HEPA filter which passes down the work area and then joins the outflow stream
- It can be used for toxic chemicals.

Class II type B3 BSC

Similar to Class II A biosafety cabinet except:
- Face velocity of 100 lfpm
- Ducts and plenums under negative pressure
- Cabinet air exhausted to the outside of building
- Suitable for work with volatile toxic chemicals and radionuclides.

Class III Biosafety cabinet
- Highest level of personnel protection
- Used for Risk Group 4 agents
- Class III BSCs are suitable for work in Biosafety Level 3 and 4 laboratories.
- All penetrations are sealed "air tight"
- Samples are passed into the box through an interlock
- Disadvantage: Difficult to perform fine manipulations through thick rubber gloves

- Supply air is HEPA-filtered
- Exhaust air passes through two HEPA filters into negative pressure duct system to the outside
- Cabinet interior under negative pressure (about 124.5 Pa).

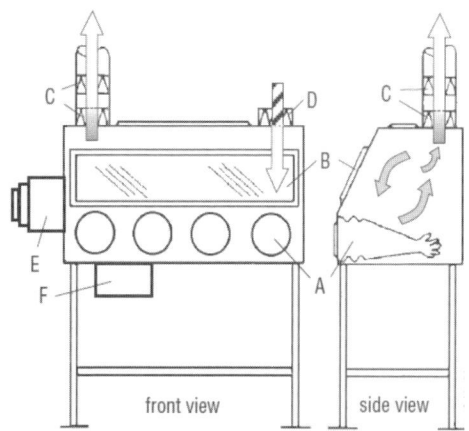

- ▨ Room air
- ▦ Potentially contaminated air
- ▢ HEPA-filtered air

A, glove ports for arm-length gloves; B, sash; C, double exhaust HEPA filters; D, supply HEPA filter; E, doble-ended autoclave or pass-through box; F, chemical dunk tank. Connection of the cabinet exhaust to an independent building exhaust air system is required.

Selection of a Biological safety cabinet

A BSC should be selected primarily in accordance with the type of protection needed:

- Product protection
- Personnel protection against Risk group 1-4 micro-organisms
- Personnel protection against exposure to radionuclides and volatile toxic chemicals
- Combination of these.

Material placement

- The front intake grill of BSCs must not be blocked.
- All materials should be placed as far back in the cabinet, without blocking the rear grill.
- Aerosol-generating equipment (e.g. centrifuges) should be placed towards the rear of the cabinet.
- Bulky items, such as biohazard bags, discard pipette trays should be placed to one side of the interior of the cabinet. If they are placed outside the cabinet, frequent in-and-out movement can compromise both personnel and product protection.
- Active work should flow from clean to contaminated areas.

BSC Operating Procedures

- **Pre-disinfect:** spray or swab all interior surfaces with appropriate disinfectant
- **Assemble material:** introduce material required to perform the procedure, place material such that clean and contaminated items do not meet, place contaminated material container at right rear
- **Pre-purge cabinet:** Cabinets should be turned on at least 5 min before beginning work and after completion of work to allow the cabinet to "purge", i.e. to allow time for contaminated air to be removed from the cabinet environment
- **Prepare self:** put protective clothing
- **Do the procedures:** do not remove hands from work space until procedures are complete and all critical material is secured, remove gloves into contaminated material container
- **Post-purge cabinet:** allow air purge period with no activity inside
- **Finish personally:** remove protective clothing, mask, and wash hands
- **Post-disinfect:** move materials to incubator, to biohazard bag, autoclave as appropriate, spray or swab all interior surfaces with appropriate disinfectant
- **Shutdown cabinet:** turn off blower and fluorescent lamp, turn on UV lamp

Alarms

- Sash alarms: found only on cabinets with sliding sashes. The alarm signifies that the operator has moved the sash to an improper position. Corrective action for this type of alarm is returning the sash to the proper position.
- Airflow alarms: indicate a disruption in the cabinet's normal airflow pattern. This represents an immediate danger to the operator or product. When an airflow alarm sounds, work should cease immediately and the laboratory supervisor should be notified

Safe Work Practices for BSC Use

- Do not use the top of the cabinet for storage. The HEPA filter could be damaged and the airflow is disrupted.
- Make sure the cabinet is at an even level. If the cabinet base is uneven, airflow can be affected.
- Never completely close the window sash with the motor running as this condition may cause motor burnout.
- Cabinets should be placed away from doors, windows, vents or high traffic areas to reduce air turbulence.

- For BSC without fixed exhaust, the cabinet exhaust should have a twelve inch (30-35cm) clearance from the ceiling for proper exhaust air flow. Also, allow a twelve inch clearance on both sides of the cabinet for maintenance purposes.
- Perform all work using a limited number of slow movements, as quick movements disrupt the air barrier. Try to minimize entering and exiting your arms from the cabinet, but if you need to, do it directly, straight out and slowly.
- If a bunsen burner has to be used, place it at the rear of the work area where the air turbulence from the flame will have the least possible effect on the air stream.

Cleaning and disinfection

- All items within BSCs, including equipment, should be surface-decontaminated and removed from the cabinet when work is completed.
- At the end of the work day, the final surface decontamination should include a wipe-down of the work surface, the sides, back and interior of the glass.
- A solution of bleach or 70% alcohol should be used.
- A second wiping with sterile water is needed when a corrosive disinfectant, such as bleach, is used.
- It is recommended that the cabinet is left running. If not, it should be run for 5 min in order to purge the atmosphere inside before it is switched off.

Spills

- When a spill of bio-hazardous material occurs within a BSC, clean-up should begin immediately, while the cabinet continues to operate.
- All materials that come into contact with the spilled agent should be disinfected and/or autoclaved.
- Group 1 & 2 – disinfect and clean
- Group 3:
    - Spillage outside Safety Cabinet: evacuate the room immediately for 1 hour and then fumigate the room
    - Spillage inside Safety Cabinet: fumigate
- Even in the absence of spillages, disinfect benches at the end of each day work by 70% isopropyl alcohol or 1% hypochlorite solution.

Decontamination

- BSCs must be decontaminated before filter changes and before being moved.
- The most common decontamination method is by heat autoclaving or fumigation with formaldehyde gas.
- BSC decontamination should be performed by a qualified professional.

Certification

- The functional operation and integrity of each BSC should be certified to national or international performance standards at the time of installation and regularly thereafter by qualified technicians, according to the manufacturer's instructions.
- Evaluation of the effectiveness of cabinet containment should include tests for cabinet integrity, HEPA filter leaks, downflow velocity profile, face velocity, negative pressure/ventilation rate, air-flow smoke pattern, and alarms and interlocks.
- Optional tests for electrical leaks, lighting intensity, ultraviolet light intensity may also be conducted.
- These tests should be undertaken by a qualified professional.

DOP/PAO test

- Tests integrity of HEPA filters using DOP/PAO solutions

    DOP (Dispersed Oil Particulate): dispersed aerosol of dioctylphthalate oil

    PAO (Poly Alpha Olefin): non-carcinogenic liquid which is the most common replacement for DOP
- These solutions generate smoke and then gas particles of the size greater than 0.3 μ.

When to do DOP test:

Any equipment that has HEPA filter and is moved, going to be used for new project every 4 weeks if on same project.

- This test certify that HEPA filter is functioning properly and there is no leakage or damage.
- Aerosol generator produces oil aerosol
- Introduced into cabinet
- Passes through HEPA filter
- Aerosol photometer measures amount of aerosol upstream and downstream the filter.

BSC	Face velocity l/fpm	Airflow % (Re-circulated)	Airflow % (exhausted)	BSL	Radionuclides/ Toxic chemicals
Class I	75	0	100, HEPA filter	2, 3	No
Class II A	75	70	30, HEPA filter	2, 3	No
Class II B1	100	30	70, HEPA filter	2, 3	Yes, but low levels
Class II B2	100	0	100, HEPA filter	2, 3	Yes
Class II B3	100	70	30		
			Plena under negative pressure to room and exhaust air ducted	2, 3	Yes
Class III	NA	0	100, 2 HEPA filters	3, 4	Yes

REFERENCES

1. Koneman EW, Allen SD, Janda WM, Schreckenberger P, Winn Jr. WC. Administrative aspects of Microbiology Laboratory. In Color Atlas and Textbook of Diagnostic Microbiology, 6th Ed. Lippincott, Philadelphia. 2006.

2. Collee JG, Marion BP, Fraser AG, Simmons A. Safety in Microbiology Laboratory. Mackie and McCartney's Practical Medical Microbiology. 14th Ed. Churchill Livigstone, U.K. 1996.

3. Laboratory Biosafety Manual. WHO. 3rd Ed. 1983. http://www.who.int/csr/resources/publications/biosafety/WHO_CDS_CSR_LYO_2004_11/en/

INDEX